EDITION 2

Social Problems

To B. D. S. and B. W. S.

Social Problems

Community, Policy, and Social Action

Anna Leon-Guerrero

Pacific Lutheran University

PINE FORGE PRESS
An Imprint of SAGE Publications, Inc.
Los Angeles • London • New Delhi • Singapore

For information:

Pine Forge Press
An Imprint of SAGE Publications, Inc.
2455 Teller Road
Thousand Oaks, California 91320
E-mail: order@sagepub.com

SAGE Publications Ltd.
1 Oliver's Yard
55 City Road
London EC1Y 1SP
United Kingdom

SAGE Publications India Pvt. Ltd.
B 1/I 1 Mohan Cooperative Industrial Area
Mathura Road, New Delhi 110 044
India

SAGE Publications Asia-Pacific Pte. Ltd.
33 Pekin Street #02-01
Far East Square
Singapore 048763

Printed in Canada

Library of Congress Cataloging-in-Publication Data

Leon-Guerrero, Anna.
Social problems: community, policy, and social action/Anna Leon-Guerrero.—2nd ed.
    p. cm.
Includes bibliographical references and index.
ISBN 978-1-4129-5966-7 (pbk. : acid-free paper)
    1. Social problems—United States. 2. Social problems. 3. Critical thinking. I. Title.

HN59.2.L46 2009
361.10973—dc22   2008004785

This book is printed on acid-free paper.

08 09 10 11 12   10 9 8 7 6 5 4 3 2 1

Acquisitions Editor:	Jerry Westby
Editorial Assistant:	Eve Oettinger
Production Editor:	Laureen A. Shea
Interior Designer:	Ravi Balasuriya
Copy Editor:	Robin Gold
Typesetter:	C&M Digitals (P) Ltd.
Proofreader:	Theresa Kay
Indexer:	Michael Ferreira
Cover Designer:	Candice Harman
Marketing Manager:	Jennifer Reed Banando

Brief Contents

Detailed Contents

Part II. The Bases of Inequality

Part III. Our Social Institutions

Part IV. Our Social and Physical Worlds

Part V. Individual Action and Social Change

Preface

During the 2008 presidential campaign, most Americans identified the economy, health care, and the U.S.-Iraq war as our most important social problems. (It is worth noting that the list is relatively unchanged from the 2004 presidential campaign.) When surveyed, individuals explained that they were supporting a particular candidate because of how well they thought he or she would handle these issues.

Although social problems are an important part of our lives, even determining the outcome of presidential elections, many still do not understand the problems we face. Perhaps you are like the many students in my classroom who have never met a homeless person, never been a victim of a violent crime, or never experienced discrimination. How much do you really know about homelessness, violent crime, or discrimination?

I wrote this text with two goals in mind: to offer a better understanding of social problems and to begin working toward real solutions. In the pages that follow, I present three connections to achieve these goals. The first connection is the one between sociology and the study of social problems. Using your sociological imagination (which you'll learn more about in Chapter 1), you will be able to identify the social and structural forces that determine our social problems. I think you'll discover that this course will be interesting, challenging, and sometimes frustrating. After you review these different social problems, you may ask, "What can be done about all this?" The second connection that will be made is between social problems and their solutions. In each chapter, we will review selected social policies along with innovative community programs that attempt to address or correct these problems. The final connection is the one that I ask you to make yourself: recognizing the social problems in your community and identifying how you can be part of the solution. To assist your learning, this text includes a variety of special features.

- *A focus on the basis of social inequalities.* Throughout the text, we will examine how race and ethnicity, gender, social class, sexual orientation, and age determine our life chances. Chapters 2 through 6 focus on the bases of social inequality and how each contributes to our experience of social problems.
- *A focus on social policy and social action.* Each chapter includes a discussion on relevant social policies or programs. In addition, each chapter highlights how individuals or groups have made a difference in their community. The text concludes with a chapter titled "Social Problems and Social Action" that identifies ways you can become more involved.

- *Voices in the Community.* The chapters include personal stories from people attempting to make a difference in their community. Some of these stories come from professionals in their field; others come from ordinary individuals who accomplish extraordinary things. For example, in Chapter 4, you'll be introduced to Bernice R. Sandler, the woman behind Title IX, and in Chapter 13, Max Kenner, student founder of the Bard Prison Initiative, an educational program for prisoners.
- *What Does It Mean to Me? and Internet and Community Exercises.* Each chapter includes questions or activities that can be completed by small student groups or on your own. Some of the exercises ask you to reflect on the material in the chapter. But many of the exercises require you to collect data and information on what is going on in your own state, city, or campus. These exercises take you out of the classroom, away from the textbook, and into your community!
- *Taking a World View.* In this boxed feature, social problems are examined from a global perspective. We will look at Japan's educational tracking system (Chapter 8), Mexico's maquiladoras (Chapter 9), and India's all female international news organization (Chapter 11).

I wanted to write a book that captured the experiences that I've shared with students in my own social problems course. I sensed the frustration and futility that many felt by the end of the semester—imagine all those weeks of discussing nothing else but "problems"! I decided that my message about the importance of *understanding social problems* should be complemented with a message on the importance of *taking social action.*

Social action doesn't just happen in Washington, D.C., or in your state's capital, and political leaders aren't the only ones engaged in such efforts. Social action takes place on your campus, in your neighborhood, in your town, in whatever you define as your "community." I knew that there were stories to be told by ordinary people—community, church, business, or student leaders—who recognized that they had the power to make a difference in the community. Each semester, I brought these individuals into the classroom to share their stories, but also to illustrate that despite the persistence of many social problems, members of our community have not given up. Their stories inspired me and my students to find our own paths to social action.

I hope that by the time you reach the end of this text, with your newfound sociological imagination, you will find your own path to social action. Wherever it leads you, good luck.

Acknowledgments

My heartfelt appreciation goes to Jerry Westby and his team at Pine Forge Press. In particular, I would like to thank Jerry and Denise Simon for always supporting my vision for this text, but also for their wisdom and patience as I explored new (a.k.a. global) ways to build upon this vision. I am grateful for the kind and consistent production support provided by Eve Oettinger. I am indebted to Laureen Shea for her guidance during the production process, to Robin Gold for her fine editing, and to Mark Guillette, for his work on the instructor's manual and ancillary materials.

The following sociologists served as the first audience and reviewers for this text. Thank you all for your encouragement and for your insightful comments and suggestions, many of which have been incorporated in this second edition.

For the second edition:

Donna Abrams, *Georgia Highlands College*

Brian C. Aldrich, *Winona State University*

Carl Backman, *Auburn University*

Janet Cosbey, *Eastern Illinois University*

Janine Dewitt-Heffner, *Marymount University*

Ronald Ferguson, *Ridgewater College*

Mark Guillette, *Valencia Community College*

Gaetano Guzzo, *Wright State University*

Jason Hendrickson, *University at Albany*

Judith Hennessy, *Central Washington University*

Ronald Huskin, *Del Mar College*

Richard Jenks, *Indiana University Southeast*

Rohald Meneses, *University of Florida*

Paul Mills, *University of Alabama*

Adam Moskowitz, *Columbus State Community College*

Wendy Ng, *San Jose State University*

Robert Parker, *University of Nevada Las Vegas*

James Roberts, *University of Scranton*

Katherine R. Rowell, *Sinclair Community College*

Rita Sakitt, *Suffolk County Community College*

Frank Salamone, *Iona College*

Jim Sikora, *Illinois Wesleyan College*

For the first edition:

Arfa Aflatooni, *Linn-Benton Community College*

Joanne Ardovini, *Sam Houston State University*

Bernadette Barton, *Morehead State University*

Allison Camelot, *California State University, Fullerton*

Janine Dewitt-Heffner, *Marymount University*

Dan Dexheimer, *University of Florida*

Woody Doane, *University of Hartford*

Joe Dupris, *California State University, Humboldt*

Rachel Einwohner, *Purdue University*

Heather Smith Feldhaus, *Bloomsburg University*

Jim Fenelon, *California State University, San Bernardino*

Bobbie Fields, *Central Piedmont Community College*

Debbie Franzman, *Allan Hancock College*

Marcie Goodman, *University of Utah*

George Gross, *Northern Michigan University*

Mark J. Guillette, *Valencia Community College*

Julia Hall, *Drexel University*

Dan W. Hayden, *University of Southern Indiana*

Chuck Holm, *San Diego State University*

Leslie Houts, *University of Florida*

James R. Hunter, *Indiana University-Purdue University at Indianapolis*

K. Land, *Duke University*

Nick Larson, *Chapman University*

Kari Lerum, *Seattle University*

Stephen Light, *SUNY Plattsburgh*

Dennis Loo, *Cal Poly Pomona*

Scott Lukas, *Lake Tahoe Community College*

Christina Myers, *Oklahoma State University*

Paul Roof, *San Juan College*

Kim Saliba, *Portland Community College*

Norma K. Simmons, *Washington State University*

Deborah Sullivan, *Arizona State University*

Mary Texeira, *California State University, San Bernardino*

Linda A. Treiber, *North Carolina State University*

Gailynn White, *Citrus College*

Anthony W. Zumpetta, *West Chester University*

I wish to express my appreciation to my family, friends, and colleagues, all of whom endured my never-ending stories about this project. Mahalo nui loa for supporting my work.

I dedicate this book to the two people who have been with me from the beginning of this journey: to my mentor, Byron D. Steiger, and to my husband, Brian W. Sullivan. From Byron, I learned the importance of loving one's work. Thank you for showing me what an excellent teacher can and should be. From Brian, I learned the value of caring for one's community and the environment. Thank you for all that you do—this book would not have been possible without you.

PART I

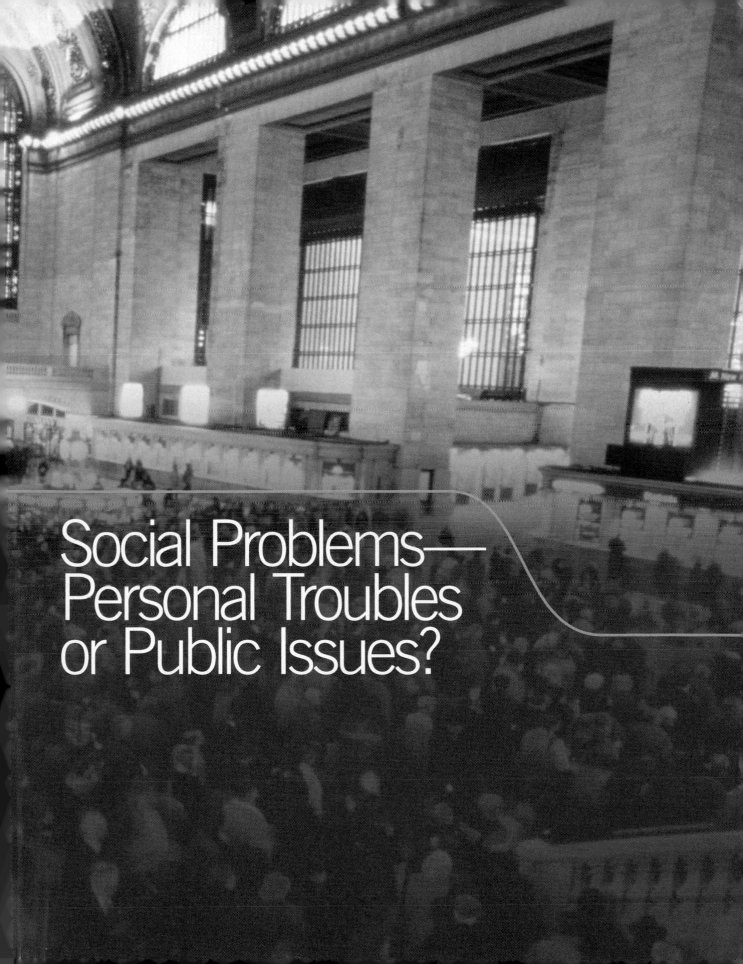

Social Problems—
Personal Troubles
or Public Issues?

Seeing Problems Sociologically

▶ We often speculate about the causes for the moods and behaviors we observe in others. If we saw this unhappy little boy, we might assume that he's spoiled or tired or sick or perhaps even a temperamental, bratty type.

◀ But if we think sociologically and expand our focus beyond this boy to include the social context in which he exists, we begin to notice a few things. The social context gives us additional information to explain the individual and his experiences. One thing we notice is that the boy is part of a family.

Another thing we might notice is that his family appears rather poor, at least judging from their clothing and their home and car, seen in the background. We could speculate about the causes of the family's poverty. We might conclude that their poverty is a result of laziness or a lack of ambition.

What we have done, however, is to identify personal shortcomings or failures as the source of problems and to define the family's poverty as a personal trouble, affecting just one boy and his family. The sociological imagination provides us with an awareness that personal troubles are often caused by institutional or structural forces. Take another look at the family, and this time note what is in the background.

The boy's father used to work in the lumber mill but lost his job when the factory closed. The sociological imagination reminds us that a social problem is not based simply on individual failures but rather is rooted in society. In this case, unemployment is not just experienced by one boy and his family but by all in the community.

Which makes more sense to you: Is it better to try to solve the problem of poverty by helping this boy and his family, and others like them, one family at a time? Or is it better to seek long-term solutions through structural changes?

CHAPTER 1

Sociology and the Study of Social Problems

I f I asked everyone in your class what they believe is the most important social problem facing the United States, there would be many different answers. Terrorism. Increasing gas prices. School violence. Poverty. Homelessness. Illegal immigration. Most would agree that some or all of these are social problems. But which is the most important, and how would we solve it?

Suppose I asked the same question in a South African college classroom. AIDS is likely to be one of the responses from South African college students. Effective risk reduction strategies, along with new treatments for HIV/AIDS, have saved countless lives in the United States. During the early 1980s, nearly 150,000 Americans were infected with the disease each year, but by the early 1990s, the number of infected cases had dropped to 40,000 per year where it remains today (Centers for Disease Control 2007a). However, Africa remains the epicenter of the pandemic with more than 25 million infected with the disease. The prevalence of HIV is predicted to triple during the next decade especially in Africa, but also in the former USSR, China, and India. The AIDS pandemic has been described as a threat to global stability, a global security concern (Lichtenstein 2004).

Globalization, defined as the process of increasing transborder connectedness (Hytrek and Zentgraf 2007)—whether economically, politically, environmentally, or socially—poses new challenges and opportunities to understanding and solving social problems. In this era of globalization, we recognize that what happens in our country may affect other countries

Photos 1.1a & 1.1b

Social problems have no boundaries, yet some capture our attention more than others. What makes a social problem real to you?

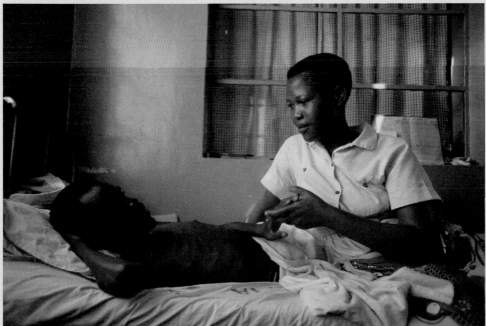

and vice versa. We are not the only country to experience social problems. Knowledge based on research to understand and policies to address social problems here could be applied in other countries and what other countries have learned based on their social problems could be applied in the United States. Finally, we all need a little help from our neighbors—we can increase our connectedness and goodwill with other countries through implementing solutions collaboratively rather than alone. So what do you think? Is HIV/AIDS in South Africa a problem only for South Africans or is it also a problem for those living in the United States?

This is how we spend much of our public conversation—on the Senate floor, on afternoon talk shows, at work, or in the classroom—arguing, analyzing, and just trying to figure out which problem is most serious and what needs to be done about it. In casual or sometimes heated conversations, we offer opinions about whether the United States should have invaded Iraq, the causes of school violence, the scope of the immigration problem, or appropriate policies for the African AIDS pandemic. Often, these explanations are not based on firsthand data collection or on an exhaustive review of the literature. For the most part, they are based on our opinions and life experiences, or they are just good guesses.

What this text and your course offer is a sociological perspective on social problems. Unlike any other discipline, sociology provides us with a form of self-consciousness, an awareness that our personal experiences are often caused by structural or social forces. **Sociology** is the systematic study of individuals, groups, and social structures. A sociologist examines the relationship between individuals and society, which includes such social institutions as the family, military, economy, and education. As a social science, sociology offers an objective and systematic approach to understanding the causes of social problems. From a sociological perspective, problems and their solutions don't just involve individuals, but also have a great deal to do with the social structures in our society. C. Wright Mills first promoted this perspective in his 1959 essay, "The Promise" (2000).

Using Our Sociological Imagination

According to Mills, the sociological imagination can help us distinguish between personal troubles and public issues. The **sociological imagination** links our personal lives and experiences with our social world. Mills describes how personal troubles occur within the "character of the individual and within the range of his immediate relationships with others" (1959/2000:8), whereas public issues are a "public matter: some value cherished by publics is felt to be threatened" (p. 8). As a result, the individual or those he is in contact with can resolve a trouble, but the resolution of an issue requires public debate about what values are being threatened and the source of such a threat.

Let's consider unemployment. One man unemployed is his own personal trouble. Resolving his unemployment involves reviewing his current situation, reassessing his skills, considering his job opportunities, and submitting his résumés or job applications to employers. Once he has a new job, his personal trouble is over. However, what happens when your city or state experiences high levels of unemployment? What happens when there is a nationwide problem of unemployment? This does not affect just one person but, rather, thousands or millions. A personal trouble has been transformed into a public issue. This is the case not just because of how many people it affects; something becomes an issue because of the public values it threatens. Unemployment threatens our sense of economic security. It challenges our belief that everyone can work hard to succeed. Unemployment raises questions about society's obligations to help those without a job.

We can make the personal trouble–public issue connection with regard to another issue, one that you might already be aware of—the increasing cost of college tuition. Salvador Henriquez works three jobs, and his wife, Colleen, works two. But even with five jobs between them, they are unable to support their daughter, Ana, a sophomore at New York

University. She graduated in the top 5 percent of her class and receives a $14,300 scholarship, but it does not cover all of her school expenses. Each year, the family takes out an additional $25,000 in loans for Ana's school expenses (Fresco 2004). Ana and her family may have found a way to support her education, but what will the Henriquez family do when Ana's three younger siblings are ready for college? Is this a personal trouble facing only the Henriquez family? Or is this a public issue?

The cost of tuition is rising at a faster rate than family income or student financial aid. During the 1980s, the cost of attending college rose three times as fast as median family income. Between 1981 and 2003, the cost of a public four-year education increased by 202 percent, and the consumer price index (the change in the cost of living) increased 80 percent (Boehner and McKeon 2003). During the 2007–2008 academic year, the average total fees (tuition, room and board) increased at double the rate of inflation, more rapidly than the consumer price index. At a four-year public institution, total fees were $13,589, but at four-year private institutions, the average cost was $23,712 (Glater 2007).

Although most Americans believe that all students have the opportunity to earn a college degree, a recent study concluded that the promise of a college education is a hollow one for low- and moderate-income students. Nearly one-half of all college-qualified, low- and moderate-income high school graduates are unable to afford college. During the first decade of the twenty-first century, 4.4 million high school graduates will not attend a four-year college, and about 2 million will attend no college at all (Advisory Committee on Student Financial Assistance 2002). In 2000, poor families spent an average of 25 percent of their annual income for their children to attend public four-year colleges. In comparison, middle-income families spent 7 percent of their income, and the wealthiest families spent 2 percent of their annual income (National Center for Public Policy and Higher Education 2002). College cost has become a serious social problem because the "barriers that make higher education unaffordable serve to erode our economic well being, our civic values, and our democratic ideals" (Callan and Finney 2002:10).

As Mills explains, "To be aware of the idea of social structure and to use it with sensibility is to be capable of tracing such linkages among a great variety of milieus. To be able to do that is to possess the sociological imagination" (1959/2000:10–11). The sociological imagination challenges the claim that the problem is "natural" or based on individual failures, instead reminding us how the problem is rooted in society, in our social structures themselves (Irwin 2001). We understand that we cannot resolve unemployment by changing one individual at a time. In the same way, we know that the Henriquez family should not be blamed for not being able to support Ana's education. The sociological imagination emphasizes the structural bases of social problems, making us aware of the economic, political, and social structures that govern employment and unemployment trends and the cost of higher education. Throughout this text, we will apply our sociological imagination to the study of social problems. Before we proceed, we need to understand what a social problem is.

What Does It Mean to Me?

Apply your sociological imagination to the problem of HIV/AIDS. Would AIDS have been on the top of your social problems list? Is HIV/AIDS in the United States a personal trouble or a public issue? Is HIV/AIDS in South Africa a personal trouble, a public issue, or even a global issue? Why or why not?

What Is a Social Problem?

The Negative Consequences of Social Problems

A **social problem** is a social condition that has negative consequences for individuals, our social world, or our physical world. If there were only positive consequences, there would be no problem. A social problem such as unemployment, alcoholism, drug abuse, or HIV/AIDS may negatively affect a person's life and health, along with the well-being of that person's family and friends. Problems can threaten our social institutions, for example, the family (spousal abuse), education (the rising cost of college tuition), or the economy (unemployment and underemployment). Our physical and social worlds can be threatened by problems related to urbanization (lack of affordable housing) and the environment (global warming).

Objective and Subjective Realities of Social Problems

A social problem has objective and subjective realities. A social condition does not have to be personally experienced by every individual to be considered a social problem. The **objective reality** of a social problem comes from acknowledging that a particular social condition does exist. Objective realities of a social problem can be confirmed by collection of data. For example, we know from the Centers for Disease Control that more than 1.2 million Americans were living with HIV/AIDS at the end of 2003 (Centers for Disease Control 2007b). You or I do not have to have to be infected with HIV to know that the disease is real, with real human and social consequences. We can confirm the realities of HIV/AIDS by observing infected individuals and their families in our own community, at AIDS programs, shelters, or hospitals.

The **subjective reality** of a social problem addresses how a problem becomes defined as a problem. This idea is based on the concept of the **social construction of reality**. Coined by Peter Berger and Thomas Luckmann (1966), the term refers to how our world is a social creation, originating and evolving through our everyday thoughts and actions. Most of the time, we assume and act as though the world is a given, objectively predetermined outside of our existence. However, according to Berger and Luckmann, we also apply subjective meanings to our existence and experience. In other words, our experiences don't just happen to us. Good, bad, positive, or negative—we also attach meanings to our reality.

From this perspective, social problems are not objectively predetermined. They become real only when they are subjectively defined or perceived as problematic. This perspective is known as **social**

Photo 1.2

The use of antiretroviral therapy has slowed the spread of HIV/AIDS in the United States. The Centers for Disease Control estimates that more than one million Americans are living with HIV/AIDS.

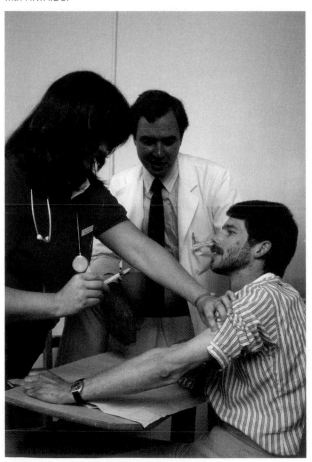

constructionism. Recognizing the subjective aspects of social problems allows us to understand how a social condition may be defined as a problem by one segment of society but be completely ignored by another. For example, do you believe AIDS is a social problem? Some may argue that it is a problem only if you are the one infected with the disease or AIDS is a problem if you are morally corrupt or sexually promiscuous. Actually, some would not consider AIDS a problem at all, considering the medical and public health advances that have successfully reduced the spread of the disease in the United States. Yet others would argue that AIDS still qualifies as a social problem.

Sociologist Denise Loseke explains, "Conditions might exist, people might be hurt by them, but conditions are not social problems until humans categorize them as troublesome and in need of repair" (2003:14). To frame their work, social constructionists ask the following set of questions:

> What do people say or do to convince others that a troublesome condition exists that must be changed? What are the consequences of the typical ways that social problems attract concern? How do our subjective understandings of social problems change the objective characteristics of our world? How do these understandings change how we think about our own lives and the lives of those around us? (Loseke and Best 2003:3–4)

The social constructionist perspective focuses on how a problem becomes defined. In particular, it examines how powerful groups, such as politicians, religious leaders, and the media, can influence our opinions and conceptions of what is a social problem. For example, in an effort to preserve their definition of the "traditional family," conservative political and religious groups encourage laws and practices that discriminate against parents with a gay or lesbian sexual orientation and the families they build. Such groups continue to offer support for the Defense of Marriage Act of 1996, which denies federal recognition of same-sex marriages and gives states the right to refuse to recognize same-sex marriages performed in other states. The act also created a federal definition of marriage: the legal union between one man and one woman as husband and wife. When the act passed, Senator Philip Gramm (R-Texas) explained, "The traditional family has stood for 5,000 years. Are we so wise today that we are ready to reject 5,000 years of recorded history? I don't think so" (CNN 1996). Although conservatives considered the act a victory, opponents expressed concern that the act created a social problem, specifically legislating discrimination against gay and lesbian couples and their families. According to Senator Carol Moseley Braun (D-Illinois), the act was really about "the politics of fear and division and about inciting people in an area which is admittedly controversial" (CNN 1996). From the social constructionist perspective, problems are in the "eye of the beholder" (Konradi and Schmidt 2001).

What Does It Mean to Me?

What are the objective realities of HIV/AIDS in your state? According to the Centers for Disease Control (2007b), approximately 40,000 new cases of HIV are reported each year. Use the Internet (or local resources) to determine the number of HIV/AIDS cases in your city or state. What subjective realities of HIV/AIDS can you identify?

Taking a World View

A Social Constructionist Approach to AIDS in Africa

Bronwen Lichtenstein (2004) examines AIDS as a socially constructed phenomenon. In the following discussion, she describes how the women of Africa were falsely blamed for the spread of the disease.

Much of the scientific and public speculation about the origins of the HIV/AIDS has centered on the African subcontinent. From the outset, Dr. Robert Gallo (codiscoverer of HIV) stated that he could not conceive of AIDS coming from anywhere else but Africa. The African theory is now widely accepted in the scientific literature, especially after it was linked to a single species of African ape.... Gilman (1988) and Fee (1988) have noted that understandings about STI (sexually transmitted infections) and Africa began during colonial times when syphilis was linked with African slavery. They also note that while the Europeans had infected colonized people with STI, blame was attributed to non-Europeans in the racist ideology of the times. Gilman (1988) has argued that such attributions about STI are still very much in line with Western notions that blacks are inherently different and have a fundamentally different relationship to disease. Blame has also been attributed to women as "natural reservoirs" of such disease, particularly in colonized countries (Kehoe 1992).

The AIDS epidemic in Africa has been defined as heterosexual or Type II in epidemiologic terms (Smallman-Raynor, Cliff, and Haggett 1992). For this reason, and in the tradition of the STI (sexually transmitted infections) epidemics, attributions of blame had again centered on women. Treichler has noted, "African women, whose exotic bodies, sexual practices or who knows what, are seen to be so radically different from those of

[other] women that anything could happen in them" (1988a:46). The African "Eve" icon has emerged from this conceptualization as sexually promiscuous and often a prostitute. This iconography is perpetuated in common lore, but also in HIV surveillance that focuses mainly on women as sentinel populations (i.e., pregnant women, female sex workers). This gendered approach has identified the problem of women's greater HIV risk, but the lack of surveillance data for men reinforces the notion of women being the locus of the disease and perpetuates largely unexamined assumptions about heterosexuality being the only mode of transmission in Africa. (Lichtenstein 2004:326)

Lichtenstein explains that as African women have been cast as the "female, heterosexual equivalent of the gay fast-laner; this focus on female promiscuity has proven a distraction from investigating the broader mechanisms of HIV/AIDS in Africa or men's role in the transmission and thus provided little predictive power for understanding the scope of the epidemic" (Lichtenstein 2004:326).

Lichtenstein warns,

Further outbreaks of HIV/AIDS in the twenty-first century will depend upon sexual cultures, war, famine, drug routes, and political unrest, as well as access to antiretroviral medicines and treated blood supplies. (Lichtenstein 2004:329)

Nations that have lacked the political will or resources for HIV prevention may face economic and social ruin, particularly in AIDS-ravaged countries in sub-Saharan Africa. The path of HIV transmission will depend on whether these countries perceive HIV/AIDS to be an intractable social problem, like poverty or crime, or whether AIDS as a "special case" will prevail in the interests of humanity. This question will be answered by how much AIDS is perceived as a threat in the twenty-first century and whether the disproportionate suffering of some citizens and countries will continued to be tolerated on a global scale. (Lichtenstein 2004:331).

The UNAIDS/WHO AIDS Epidemic Update estimated that 39.5 million people were living with HIV in 2006 (World Health Organization 2007). UNAIDS/WHO reported that there were 4.3 million new infections in 2006 with 2.8 million (or 65 percent) occurring in sub-Saharan Africa and dramatic infection rate increases in Eastern Europe and Central Asia (more than 50 percent since 2004).

In 2006, 2.9 million people died of AIDS-related illnesses worldwide.

In signing the Declaration of Commitment drafted by the 2001 United Nations General Assembly on HIV/AIDS, 189 nations unanimously acknowledged that AIDS is among the "greatest development crisis in human history" (UNAIDS 2006). Each nation committed resources to help the fight against AIDS nationally and internationally. In 2003, George W. Bush announced the creation of the President's Emergency Plan for AIDS Relief (PEPFAR/Emergency Plan). The United States, with a five-year $15 billion plan, is the world leader in its support for the fight against HIV/AIDS. With funding from the president's plan, the Centers for Disease Control Global AIDS Program has implemented HIV/AIDS prevention, treatment, and educational programming in 25 countries.

The History of Social Problems

Problems don't appear overnight; rather, as Malcolm Spector and John Kituse (1987) argue, the identification of a social problem is part of a subjective process. Spector and Kituse identify four stages to the process. Stage 1 is defined as a transformation process: taking a private trouble and transforming it into a public issue. In this stage, an influential group, activists, or advocates call attention to and define an issue as a social problem. The first HIV infection cases were documented in the United States in 1979. The disease was originally referred to as the "gay plague" because the first group to be identified with the disease was gay men from San Francisco, Los Angeles, and New York. The association of HIV/AIDS with this specific population led to its first being defined as a sexual epidemic rather than a public health threat. Gay activists and public health officials mobilized to increase awareness and began to change the public's perception of the disease in the early 1980s.

Stage 2 is the legitimization process: formalizing the manner in which the social problems or complaints generated by the problem are handled. For example, an organization or public policy could be created to respond to the condition. An existing organization, such as a federal or state agency, could also be charged with taking care of the situation. In either instance, these organizations begin to legitimize the problem by creating and implementing a formal response. In the early 1980s, HIV/AIDS task forces were created in the Centers for Disease Control and the World Health Organization. Similar groups were convened in the United Kingdom, France, and other countries. Although no single organization or country was in charge, all were intent on identifying the disease and finding a cure. Activists looked for public legitimization of the disease from then President Ronald Reagan. But Reagan did not acknowledge AIDS until 1985 when he was asked directly about the disease during a press conference. His first public statement about the disease came in 1987 at the Third International Conference on AIDS. By then nearly 36,000 Americans had been diagnosed with AIDS and more than 20,000 had died. AIDS advocates blamed Reagan's slow and ineffective response for these deaths and the increasing spread of the disease.

Individuals come together in public rallies such as this one at the United Nations HIV/AIDS meeting, to voice their concerns about HIV/AIDS policies and funding. These demonstrations galvanize the efforts of advocacy and activist groups, as well as educate the public about HIV/AIDS.

Stage 3 is a conflict stage, when Stage 2 routines are unable to address the problem. During Stage 3, activists, advocates, and victims of the problem experience feelings of distrust and cynicism toward the formal response organizations. Stage 3 activities include readjusting the formal response system: renegotiating procedures, reforming practices, and engaging in administrative or organizational restructuring. Many early public health protocols were revised in response to increased understanding about how HIV/AIDS is spread and treated. For example, patient isolation was common during the first stages of the disease. Teenager Ryan White had to petition for the right to attend public school with his classmates. Ryan and his mother's experiences also shed light on the difficulties faced by low-income, uninsured, or underinsured individuals and families living with HIV/AIDS. After his death in 1990, U.S. Congress passed the Ryan White CARE Act to provide for the unmet health needs of individuals with HIV/AIDS. The act continues to provide support for nearly 500,000 individuals annually, making it the largest federal government program for those living with the disease.

Finally, Stage 4 begins when groups believe that they can no longer work within the established system. Advocates or activists are faced with two options, to radically change the existing system or to work outside of the system. As an alternative to the government and public health agencies' response to HIV/AIDS, numerous independent advocacy and research groups were formed. Among them, the most critical and vocal voice has come from the AIDS Coalition to Unleash Power (ACT UP). Established in 1987, ACT UP called for direct action to end the epidemic, demanding better access to drugs, increasing AIDS public education, and prohibiting AIDS-related discrimination. Today, ACT UP has widened its focus to advocate on behalf of universal health care and prescription drug pricing controls.

Understanding the Sociological Perspective

The way sociologists conduct sociology and study social problems begins first with their view on how the world works. Based on a **theory**—a set of assumptions and propositions used for explanation, prediction, and understanding—sociologists begin to define the relationship between society and individuals. Theories vary in their level of analysis, focusing on a **macro** (societal) or a **micro** (individual) level. Theories help inform the direction of sociological research and data analysis. In the following section, we will review four theoretical perspectives: functionalist, conflict, feminist, and interactionist. Research methods used by sociologists are summarized in the next section.

Functionalist Perspective

Among the theorists most associated with the functionalist perspective is French sociologist Émile Durkheim. Borrowing from biology, Durkheim likened society to a human body. As the body has essential organs, each with a specific function in the body, he theorized that society has its own organs: the institutions of the family, economy, politics, education, and religion. These organs or social structures have essential and specialized functions. For example, the institution of the family maintains the health and socialization of our young and creates a basic economic unit. The institution of education provides knowledge and skills for women and men to work and live in society. No other institution can do what the family or education does.

Durkheim proposed that the function of society was to civilize or control individual actions. He wrote, "It is civilization that has made man what he is; it is what distinguishes him from the animal: man is man only because he is civilized" (Durkheim 1914/1973:149). The social order can be threatened during periods of rapid social change, such as industrialization or political upheaval, when social norms and values are likely to be in transition. During this state of normlessness or **anomie**, Durkheim believed society was particularly prone to social problems. As a result, social problems cannot be solved by changing the individual; rather, the problem has to be solved at the societal level. The entire social structure or the affected part of the social structure needs to be repaired.

The **functionalist perspective**, as its name suggests, examines the functions or consequences of the structure of society. Functionalists use a macro perspective, focusing on how society creates and maintains social order. Social problems are not analyzed in terms of how "bad" it is for society. Rather, a functionalist asks, How does the social problem emerge from society? Does the social problem serve a function?

The systematic study of social problems began with the sociologists at the University of Chicago. Part of what has been called the Chicago School of Sociology, scholars such as Ernest W. Burgess, Homer Hoyt, Robert E. Park, Edward Ullman, and Louis Wirth used their city as an urban laboratory, pursuing field studies of poverty, crime, and drug abuse during the 1920s and 1930s. Through their research, they captured the real experiences of individuals experiencing social problems, noting the positive and negative consequences of urbanization and industrialization (Ritzer 2000). Taking it one step further, sociologists Jane Addams and

Charlotte Gilman studied urban life in Chicago and developed programs to assist the poor and lobbied for legislative and political reform (Adams and Sydie 2001).

According to Robert Merton (1957), social structures can have positive benefits as well as negative consequences, which he called **dysfunctions**. A social problem such as homelessness has a clear set of dysfunctions but can also have positive consequences or functions. One could argue that homelessness is clearly dysfunctional and unpleasant for the women, men, and children who experience it, and for a city or community, homelessness can serve as a public embarrassment. Yet, a functionalist would say that homelessness is beneficial for at least one part of society, or else it would cease to exist. The population of the homeless supports an industry of social service agencies, religious organizations, and community groups and service workers. In addition, the homeless also highlight problems in other parts of our social structure, namely the problems of the lack of a livable wage or affordable housing.

What Does It Mean to Me?

Merton (1957) separated functions into two categories: **manifest** and **latent**. Manifest functions are the consequences that are intended and recognized, whereas latent functions are the consequences that are unintended and often hidden. What are the manifest and latent functions of HIV/AIDS?

Conflict Perspective

Like functionalism, conflict theories examine the macro level of our society, its structures and institutions. Whereas functionalists argue that society is held together by norms, values, and a common morality, **conflict** theorists consider how society is held together by power and coercion (Ritzer 2000) for the benefit of those in power. In this view, social problems emerge from the continuing conflict between groups in our society—based on social class, gender, race, or ethnicity—and in the conflict, the dominant groups usually win. There are multiple levels of domination; as Patricia Hill Collins describes, domination "operates not only by structuring power from the top down but by simultaneously annexing the power as energy of those on the bottom for its own ends" (1990:227–228).

As a result, this perspective offers no easy solutions to social problems. The system could be completely overhauled, but that is unlikely to happen. We could reform parts of the structure, but those in power would retain their control. The biggest social problem from this perspective is the system itself and the inequality it perpetuates.

The first to make this argument was German philosopher and activist Karl Marx. Conflict, according to Marx, emerged from the economic substructure of capitalism, which defined all other social structures and social relations. He focused on the conflict based on social class, created by the tension between the **proletariat** (workers) and the **bourgeoisie** (owners). Capitalism did more than separate the haves and have-nots. Unlike Durkheim, who believed that society created a civilized man, Marx argued that a capitalist society created a man alienated from his **species being**, from his true self. **Alienation** occurred on multiple levels: Man would become increasingly alienated from his work, the product of his work, other workers, and finally, his own human potential. For example, a salesperson could be so involved in the process of her work that she doesn't spend quality time with her coworkers, talk with her customers, or stop and appreciate the merchandise. Each sale transaction is the same; all customers and workers are treated alike. The salesperson could not achieve her human potential through this type of mindless unfulfilling labor. According to Marx, workers needed to achieve a **class consciousness**, an awareness of their social position and oppression so they can unite and overthrow capitalism, replacing it with a more egalitarian socialist and eventually communist structure.

Widening Marx's emphasis on the capitalist class structure, contemporary conflict theorists have argued that conflict emerges from other social bases, such as values, resources, and interests. Mills (1959/2000) argued the existence of a "power elite," a small group of political, business, and military leaders who control our society. Ralf Dahrendorf (1959) explained that conflict of interest is inherent in any relationship because those in powerful positions will always seek to maintain their dominance. Lewis Coser (1956) focused on the functional aspects of conflict, arguing that conflict creates and maintains group solidarity by clarifying the positions and boundaries between groups. Conflict theorists may also take a social constructionist approach, examining how powerful political, economic, and social interest groups subjectively define social problems.

Feminist Perspective

Rosemarie Tong explains, "Feminist theory is not one, but many, theories or perspectives and that each feminist theory or perspective attempts to describe women's oppression, to explain its causes and consequences, and to prescribe strategies for women's liberation" (1989:1). By analyzing the situations and lives of women in society, **feminist** theory defines gender (and sometimes race or social class) as a source of social inequality, group conflict, and social problems. For feminists, the patriarchal society is the basis of social problems. **Patriarchy** refers to a society in which men dominate women and justify their domination through devaluation; however, the definition of patriarchy has been broadened to include societies in which powerful groups dominate and devalue the powerless (Kaplan 1994).

Patricia Madoo Lengermann and Jill Niebrugge-Brantley (2004) explain that feminist theory was established as a new sociological perspective in the 1970s, largely because of the growing presence of women in the discipline and the strength of the women's movement. Feminist theory treats the experiences of women as the starting point in all sociological investigations, seeing the world from the vantage point of women in the social world and seeking to promote a better world for women and for humankind.

Although the study of social problems is not the center of feminist theory, throughout its history, feminist theory has been critical of existing social arrangements and has focused on such concepts as social change, power, and social inequality (Madoo Lengermann and Niebrugge-Brantley 2004). Major research in the field has included Jessie Bernard's (1972/1982) study of gender inequality in marriage, Collins's (1990) development of Black feminist thought, Dorothy Smith's (1987) sociology from the standpoint of women, and Nancy Chodorow's (1978) psychoanalytic feminism and the reproduction of mothering. Although sociologists in this perspective may adopt a conflict, functionalist, or interactionist perspective, their focus remains on how men and women are situated in society, not just differently but also unequally (Madoo Lengermann and Niebrugge-Brantley 2004).

Interactionist Perspective

An **interactionist** focuses on how we use language, words, and symbols to create and maintain our social reality. This micro-level perspective highlights what we take for granted: the expectations, rules, and norms that we learn and practice without even noticing. In our interaction with others, we become the products and creators of our social reality. Through our interaction, social problems are created and defined.

George Herbert Mead provided the foundation of this perspective. Also a member of the Chicago School of Sociology, Mead (1934/1962) argued that society was the organized and patterned interactions among individuals. As Mead defined it, the self is a mental and social process, the reflective ability to see others in relation to ourselves and to see ourselves in relation to others. The term **symbolic interactionism** was coined by Herbert Blumer in 1937. Building on Mead's work, Blumer emphasized how the existence of mind and self emerges from interaction and the use of symbols (Turner 1998).

How does the self emerge from interaction? Consider the roles that you and I play. As a university professor, I am aware of what is expected of me; as university students, you are aware of what it means to be a student. There are no posted guides in the classroom that instruct us where to stand, how to dress, or what to bring into class. Even before we enter the classroom, we know how we are supposed to behave and even our places in the classroom. We act based on our past experiences and based on what we have come to accept as definitions of each role. But we need each other to create this reality; our interaction in the classroom reaffirms each of our roles and the larger educational institution. Imagine what it takes to maintain this reality: consensus not just between a single professor and her students but between every professor and every student on campus, on every university campus, ultimately reaffirming the structure of a university classroom and higher education.

So, how do social problems emerge from interaction? First, for social problems such as alcoholism or juvenile delinquency, an interactionist would argue that the problem behavior is learned from others. According to this perspective, no one is born a juvenile delinquent. Like any other role we play, people learn how to become juvenile delinquents. Although the perspective does not answer the question of where or from whom the first delinquent child learned this behavior, it attempts to explain how deviant behavior is learned through interaction with others.

Second, social problems emerge from the definitions themselves. Objective social problems do not exist; they become real only in how they are defined or labeled. A sociologist using this perspective would examine who or what group is defining the problem and who or what is being defined as deviant or a social problem. As we have already seen with the HIV/AIDS epidemic in the United States, the problem became real only when activists and public health workers called attention to the disease.

Third, the solutions to social problems also emerge from our definitions. Schneider and Ingram (1993) argue that the social construction of target populations influences the distribution of policy benefits or policy burdens. Target populations are groups of individuals experiencing a specific social problem; these groups gain policy attention through their socially constructed identity and political power. The authors identify four categories: advantaged target populations are positively constructed and politically powerful (likely to receive policy benefits); contenders are politically powerful, yet negatively constructed (likely to receive policy benefits when public interest is high); dependent target populations have positive social construction, but low political power (few policy resources would be allocated to this group); and deviant target populations are both politically weak and negatively constructed (least likely to receive any benefits).

Jean Schroedel and Daniel Jordan (1998) applied the target population model to U.S. Senate voting patterns between 1982 and 1992, examining the allocation of federal funds to four distinct HIV/AIDS groups. As Schneider and Ingram's theory would predict, the group receiving the most funding were those in the advantaged group (war veterans and health care workers), followed by contenders (gay and bisexual men and the general population with AIDS), dependents (spouses and the public), and finally, deviants (IV drug users, criminals, and prisoners).

What Does It Mean to Me?

A summary of these sociological perspectives is presented in Table 1.1. These sociological perspectives will be reintroduced in each chapter as we examine a new social problem or set of problems. As you review each perspective, do not attempt to classify one as the definitive explanation. Consider how each perspective focuses on different aspects of society and its social problems. Which perspective(s) best fits with your understanding of society? your understanding of social problems?

Table 1.1

Summary of Sociological Perspectives: A General Approach to Examining Social Problems

	Functionalist	Conflict/Feminist	Interactionist
Level of analysis	**Macro**	**Macro**	**Micro**
Assumptions about society	Order.	Conflict.	Interaction.
	Society is held together by a set of social institutions, each of which has a specific function in society.	Society is held together by power and coercion. Conflict and inequality are inherent in the social structure.	Society is created through social interaction.
Questions asked about social problems	How does the problem originate from the social structure? How does the problem reflect changes among social institutions and structures? What are the functions and dysfunctions of the problem?	How does the problem originate from the competition between groups and from the social structure itself? What groups are in competition and why?	How is the problem socially constructed and defined? How is problem behavior learned through interaction? How is the problem labeled by those concerned about it?

The Science of Sociology

Sociology is not commonsense guessing about how the world works. The social sciences rely on "scientific methods to investigate individuals, societies, and social process; the knowledge produced by these investigations" (Shutt 2006:9).

Sociological research is divided into two areas: basic and applied. The knowledge we gain through **basic research** expands our understanding of the causes and consequences of a social problem, for example, identifying the predictors of HIV/AIDS or examining the rate of homelessness among AIDS patients. Conversely, **applied research** involves the pursuit of knowledge for program application or policy evaluation (Katzer, Cook, and Crouch 1998). Often, social programs are evaluated for how effective they are in reducing a problem or in creating some desired change. The information gained through applied research can be incorporated into social programs serving HIV/AIDS patients.

All research begins with a theory to help identify the phenomenon we're trying to explain and provide explanations for the social patterns or causal relationships between variables (Frankfort-Nachmias and Leon-Guerrero 2006). **Variables** are a property of people or objects that can take on two or more values. For example, as we try to explain HIV/AIDS, we may have a specific explanation about the relationship between two variables—social class and HIV infection. Social class could be measured according to household or individual income, whereas HIV infection could be measured as a positive test for the HIV antibodies. The relationship between these variables can be stated in a **hypothesis**, a tentative statement about how the variables are related to each other. We could predict that HIV infection would be higher among lower-income men and women. In this hypothesis statement, we've identified a **dependent variable** (the variable to be explained, HIV infection) along with an **independent variable** (the variable expected to account for the cause of the dependent variable, social class). Data, the information we collect, may confirm or refute this hypothesis.

Research methods can include quantitative or qualitative approaches or a combination. Quantitative methods rely on the collection of statistical data. They require the specification of variables and scales collected through surveys, interviews, or questionnaires. Qualitative methods are designed to capture social life as participants experience it. These methods involve field observation, depth interviews, or focus groups. Following are definitions of each specific method.

Survey research: Data collection based on responses to a series of questions. Surveys can be offered in several formats: a self-administered mailed survey, group surveys, in-person interviews, or telephone surveys. For example, information from HIV/AIDS patients may be collected by a survey sent directly in the mail or by a telephone or in-person interview (e.g., Simoni et al. 2006).

Qualitative methods: This category includes data collection conducted in the field, emphasizing the observations about natural behavior as experienced or witnessed by the researcher. Methods include participant observation (method for gathering data that involves developing a sustained relationship with people while they go about their normal activities), focus groups (unstructured group interviews in which a focus group leader actively encourages discussion among participants on the topics of interest), or intensive (depth) interviewing (open-ended, relatively unstructured questioning in which the interviewer seeks in-depth information on the interviewee's feelings, experiences, and perceptions). Sociologists have used various qualitative methods in HIV/AIDS research—collecting data through participant observation at clinics or support groups and focus group or depth interviews with patients or key informants (e.g., Chakrapani et al. 2007).

Historical and comparative methods: Research that focuses on one historical period (historical events research) or traces a sequence of events over time (historical process research). Comparative research involves multiple cases or data from more than one time period. For example, researchers have examined the effectiveness of HIV/AIDS treatments over time (e.g., Fumaz et al. 2007) and compared infection rates between men and women (e.g., Ballesteros et al. 2006).

Secondary data analysis: Secondary data analysis usually involves the analysis of previously collected data that are used in a new analysis. Large public survey data sets, such as the U.S. Census, the General Social Survey, the National Election Survey, or the International Social Survey Programme, can be used, as can data collected in experimental studies or with qualitative datasets. For HIV/AIDS research, a secondary data analysis could be based on existing medical records (e.g., Tabi and Vogel 2006). The key to secondary data analysis is that the data was not originally collected by the researcher, but was collected by another researcher and for a different purpose.

The Transformation From Problem to Solution

Although Mills identified the relationship between a personal trouble and a public issue more than 50 years ago, less has been said about the transformation of issue to solution. Mills leads us in the right direction by identifying the relationship between public issues and social institutions. By continuing to use our sociological imagination and recognizing the role of larger social, cultural, and structural forces, we can identify appropriate measures to address these social problems.

Let's consider homelessness. It does not arise from mysterious or special circumstances; rather, it emerges from familiar life experiences. The loss of a job, the illness of a family member, domestic violence, or divorce could make a family more susceptible to homelessness. And among those with HIV/AIDS, more than half will need housing assistance at some point in their lives. Without informal social support, a savings account, or suitable and adequate employment—and with the increasing cost of health care and the lack of affordable housing a family's economic and emotional resources can quickly be tapped out. What would it take to prevent homelessness in these situations? The answers are not based in each individual or each family; rather, the long-term solutions are structural solutions such as affordable health care, livable wages, and affordable low-income housing.

Modern history reveals that Americans do not like to stand by and do nothing about social problems. Actually, most Americans support efforts to reduce homelessness, improve the quality of education, or find a cure for HIV/AIDS. In some cases, there are no limits to our efforts. Helping our nation's poor has been an administrative priority of many U.S. presidents. President Franklin Roosevelt proposed sweeping social reforms during his New Deal in 1935, and President Lyndon Johnson declared the War on Poverty in 1964. President Bill Clinton offered to "change welfare as we know it" with broad reforms outlined in the Personal Responsibility and Work Opportunity Reconciliation Act of 1996. And in 2003, President George W. Bush supported the reauthorization of the 1996 welfare reform bill. No president or Congress has ever promised to eliminate poverty; instead, each promised only to improve the system serving the poor or to reduce the number of poor in our society.

Solutions require social action—in the form of social policy, advocacy, and innovation— to address problems at their structural or individual levels. **Social policy** is the enactment of a course of action through a formal law or program. Policy-making usually begins with identification of a problem that should be addressed; then, specific guidelines are developed regarding what should be done to address the problem. Policy directly changes the social structure, particularly how our government, an organization, or community responds to a social problem. In addition, policy governs the behavior and interaction of individuals, controlling who has access to benefits and aid (Ellis 2003). An example of homeless AIDS social policy was established in 1990. The Housing Opportunities for Persons with AIDS (HOPWA) program was established as part of the Cranston-Gonzales National Affordable Housing Act of 1990. HOPWA identifies long-term housing strategies for persons living with HIV/AIDS that prevent them from becoming homeless and ensures that persons have access to essential medical care and other services. Social policies are always being enacted. More recent homeless policy was proposed in 2007, when U.S. Senators Jack Reed (D-RI) and Richard Burr (R-NC) introduced

Photo 1.4

Lawmakers determine how we choose to respond to a social problem, setting social policy that directly changes our social structure. As a result, social policies govern the behavior and interaction of individuals, controlling who has access to programs and aid.

the Services for Ending Long-Term Homelessness Act (SELHA). If passed, the act would fund mental health, substance abuse, and employment services for the chronically homeless.

Social advocates use their resources to support, educate, and empower individuals and their communities. Advocates work to improve social services, change social policies, and mobilize individuals. National organizations such as the National Coalition for the Homeless and the National AIDS Housing Coalition or local organizations such as Project H.O.M.E. in Philadelphia provide service, outreach, education, and legal support for the homeless and for homeless AIDS patients.

Social innovation may take the form of a policy, a program, or advocacy that features an untested or unique approach. Innovation usually starts at the community level, but it can grow into national and international programming. Millard and Linda Fuller developed the concept of "partnership housing" in 1965, partnering those in need of adequate shelter with community volunteers to build simple interest-free houses. In 1976, the Fullers' concept became Habitat for Humanity International, a nonprofit, ecumenical Christian housing program responsible for building more than 125,000 houses worldwide. When Millard Fuller was awarded the Presidential Medal of Freedom, the nation's highest civilian honor, President Clinton described Habitat as "the most successful continuous community service project in the history of the United States" (Habitat for Humanity 2004).

Photo 1.5

As of September 2007, Habitat for Humanity volunteers in New Orleans completed 59 homes with 115 more under construction In this photo, volunteers from the PacifiCare Corporation based in Southern California are building a home for a family that lost theirs during Hurricane Katrina.

What Does It Mean to Me?

Investigate if there are community groups or organizations that serve the HIV/AIDS population in your city. What activities or programs do they sponsor? How do they provide assistance to those living with HIV/AIDS in your area?

Making Sociological Connections

In his book, *Social Things: An Introduction to the Sociological Life,* Charles Lemert (1997) tells us that sociology is often presented as a thing to be studied. Instead, he argues that sociology is something to be "lived," becoming a way of life. Lemert writes,

> To use one's sociological imagination, whether to practical or professional end, is to look at the events in one's life, to see them for what they truly are, then to figure out how the structures of the wider world make social things the way they are. No one is a sociologist until she does this the best she can. (1997:105)

We can use our sociological imagination as Lemert recommends, but we can also take it a step further. As Marx maintained, "The philosophers have only interpreted the world, in various ways; the point, however is to change it" (1972:107).

Throughout this text, we will explore three connections. The first connection is the one between personal troubles and public issues. Each sociological perspective—functionalism, conflict, feminist, and interactionist—highlights how social problems emerge from our social structure or social interaction. While maintaining its primary focus on problems within the United States, this text will also address the experience of social problems in other countries and nations. The comparative perspective will enhance your understanding of the social problems we experience here.

The sociological imagination will also help us make a second connection: the one between social problems and social solutions. Mills believed that the most important value of sociology was in its potential to enrich and encourage the lives of all individuals (Lemert 1997). In each chapter, we will review selected social policies, advocacy programs, and innovative approaches that attempt to address or solve these problems.

Textbooks on this subject present neat individual chapters on a social problem, reviewing the sociological issues and sometimes providing some suggestions about how it can and should be addressed. This book follows the same outline but takes a closer look at community-based approaches, ultimately identifying how YOU can be part of the solution in your community.

I should warn you that this text will not identify a perfect set of solutions to our social problems. Solutions, like the problems they address, are embedded within complex interconnected social systems (Fine 2006). Sometimes solutions create other problems. Some solutions fail. A program may have worked, but might no longer exist because of lack of funding or political and public support. Nevertheless, some policies and programs have been proven effective in helping individuals, families, or groups improve their lives and in reducing or completely eliminating the prevalence of particular problems.

In communities such as yours and mine, individuals and community groups are taking action against social problems. They are women, men, and children, common citizens and professionals, from different backgrounds and experiences. Whether they are working within the system or working to change the system, these individuals are part of their community's solution to a problem. Sociologist Gary Fine observes, "Those who care about social problems are obligated to use their best knowledge to increase the store of freedom, justice and equality" (2007:14). In the end, I hope you agree that it is important that we continue to do something about the social problems we experience.

In addition, I will ask you to make the final connection to social problems and solutions in your community. For this quarter or semester, instead of focusing only on problems reported in your local newspaper or the morning news program, start paying attention to the solutions offered by professionals, leaders, and advocates. Through the Internet or through local programs and agencies, take this opportunity to investigate what social action is taking place in your community. Regardless of whether you

Photo 1.6

Throughout the United States, college and university students such as these are volunteering in their community. Student volunteering may involve completing a service learning option linked to a specific course, participating in a university-sponsored volunteer event, or students finding service opportunities on their own.

define your "community" as your campus, your residential neighborhood, or the city where your college is located, consider what avenues of change can be taken and whether you can be part of that effort.

I often tell my students that the problem with being a sociologist is that my sociological imagination has no "off" switch. In almost everything I read, see, or do, there is some sociological application, a link between my personal experiences and the broader social experience that I share with everyone else, including you. As you progress through this text and your course, I hope that you will begin to use your own sociological imagination and see connections between problems and their solutions that you never saw before.

Main Points

- **Sociology** is the systematic study of individuals and social structures. A sociologist examines the relationship between individuals and our society, which includes institutions (the family), organizations (the military), and systems (our economy). As a social science, sociology offers an objective and systematic approach to understanding the causes of social problems.

- The **sociological imagination** is a way of recognizing the links between our personal lives and experiences and our social world.

- A **social problem** is a social condition that has negative consequences for individuals, our social world, or physical world.

- A social problem has objective and subjective realities. The **objective reality** of a social problem comes from acknowledging that a particular social condition negatively affects human lives. The **subjective reality** of a social problem addresses how a problem becomes defined as a problem. Social problems are not objectively predetermined. They become real only when they are subjectively defined or perceived as problematic. This perspective is known as **social constructionism**.

- Problems don't just appear overnight; rather, as Spector and Kituse (1987) argue, the identification of a social problem is a process.

- Sociologists use four theoretical perspectives: **functionalist**, **conflict**, **feminist**, and **interactionist**. The functionalist perspective examines the functions or consequences of the structure of society. Functionalists use a macro perspective, focusing on how society creates and maintains social order. A social problem is not analyzed in terms of how "bad" it is for parts of society. Rather, a functionalist asks how the social problem emerges from the society. What function does the social problem serve?

- Conflict, according to Marx, emerged from the economic substructure of capitalism, which defined all other social structures and social relations. He focused on the conflict based on social class, created by the tension between the **proletariat** (workers) and the **bourgeoisie** (owners). Marx argued that a capitalist society created a man alienated from his **species being**, from his true self. **Alienation** occurred on multiple levels: Man would become increasingly alienated from his work, from the product of his work, from other workers, and finally, from his own human potential.

- By analyzing the situations and lives of women in society, feminist theory defines gender (and sometimes race or social class) as a source of social inequality, group conflict, and social problems. For feminists, the patriarchal society is the basis of social problems.

- An interactionist focuses on how we use language, words, and symbols to create and maintain our social reality. This perspective highlights what we take for granted: the expectations, rules, and norms that we learn and practice without even noticing. In our interaction with others, we become the products and creators of our social reality. Through our interaction, social problems and their solutions are created and defined.

■ Solutions require social action—in the form of social policy, advocacy, and innovation—to address problems at their structural or individual levels. **Social policy** is the enactment of a course of action through a formal law or program. **Social advocates** use their resources to support, educate, and empower individuals and their communities. **Social innovation** may take the form of a policy, a program, or advocacy that features an untested or unique approach. Innovation usually starts at the community level, but it can grow into national and international programming.

On Your Own

Log on to the Web-based student study site at www.pineforge.com/leonguerrero2study for interactive quizzes, E-flashcards, journal articles, Community and Policy Guides, a Service Learning Guide, the end-of-chapter Web exercises, and additional Web resources.

Internet and Community Exercises

1. Review copies of your local newspaper from the past 90 days. Based on the front page or local section, what issues are important for your community? Crime? Job layoffs? Transportation? Pollution? Examine how the issue is defined and by whom. Is input from community leaders and neighborhood groups being included? Why or why not? Do these include the three elements of a social problem?

2. Social actions or responses are also linked to how we define the problem. If we believe the problem is structural, we'll find ways to change the structure. If the problem is defined at the individual level, a solution will attempt to change the person. Investigate the programs and resources that are available for the homeless in your community or state. Select three local programs and assess how each defines and responds to the homeless problem in your community.

3. What do you think is the most important social problem? Investigate what federal and state policies govern or regulate this problem and those it affects. What is the position of the main political parties—Democrats and Republicans—on this problem?

4. What do you think is the most important global social problem? Investigate how the governments of the United States and other nations have responded to this problem. How have citizens responded?

PART II

The Bases
of Inequality

"Like trees in a vast forest, humans come in a variety of sizes, shapes, and colors." Marilyn Loden and Judy Rosener (1991) wrote these words to describe human **diversity** or what the women call "**otherness** or those human qualities that are different from our own and outside the groups to which we belong" (p. 18). They created the Diversity Wheel (see Figure II.1) to represent our primary and secondary characteristics.

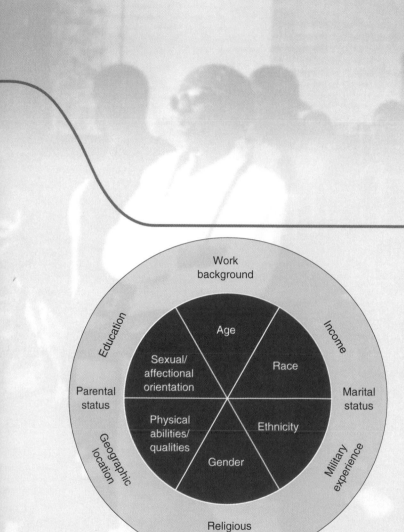

Figure II.1

Diversity Wheel

Source: From Loden, M., and Rosener, J. B., Workforce America! Managing Employee Diversity as a Vital Resource. *Copyright © 1991. Reprinted with permission of the McGraw-Hill Companies.*

Our primary characteristics are presented in the center of the wheel. These dimensions—age, ethnicity, gender, physical abilities, race, and sexual orientation—serve as our basic self-image and affect our lives from birth. As Loden and Rosener tell us, "There is no escaping the life-long impact of these six core elements" (1991:19). Secondary characteristics are less salient than primary characteristics. We can change these characteristics, such as our educational attainment, geographic location, income, marital status, military experience, parental status, religious beliefs, and work experience.

Each characteristic serves as a social boundary, letting us know who should be included in or excluded from our particular social group (Sernau 2001), defining the "otherness" each of us possesses. Sociologists use the term **social stratification** to refer to the ranking of individuals into social strata or groups. We are divided into groups such as women versus men or African Americans versus Asian Americans. Our lives are also transformed because of our group membership. In U.S. society, being different has also come to mean that we are unequal.

The differences between social strata become more apparent when we recognize how some individuals are more likely to experience social problems than others are. Attached to each social position are **life chances**, a term Max Weber used to describe the consequences of social stratification, how each social position provides particular access to goods and services such as wealth, food, clothing, shelter, education, and health care.

Sociologists refer to the unequal distribution of resources, services, and positions as **social inequality**. Certain social positions are subject to **prejudice** (a negative attitude based on the attributes of an individual) and **discrimination** (acts based on prejudiced beliefs against a specific individual or group) not experienced by others. And sadly, some social positions are also targets of violence.

In the next five chapters, we will explore two basic sociological questions: Why does social inequality exist, and how are we different from one another? We will review sociological theories that attempt to explain and examine the consequences of social inequality. Although the five bases of inequality are discussed in separate chapters, real life happens at the intersection of our social class, racial and ethnic identity, gender, sexual orientation, and age. These bases of inequality simultaneously define and affect us. We need to recognize how each social characteristic (class, race, ethnicity, gender, sexual orientation, or age) shapes the history, experiences, and opportunities of men, women, and children in the United States (Shapiro 2001). Your life experience may have less to do with your ability or your hard work and more to do with how (well) you are positioned in society. Ultimately, this includes one's experience of social problems.

If this is your first sociology course, these chapters will provide you with an overview of several basic sociological concepts. If you have already had a sociology course, welcome back; these chapters should provide a good review.

What Does It Mean to Me?

Based on some of the primary and secondary characteristics identified in the Diversity Wheel, I'd describe myself as a married, college educated, Asian Pacific Islander, heterosexual female. What social characteristics would you use to describe yourself? Which social characteristics are essential to your identity?

"Invisible" Dividing Lines

It is difficult to escape from our social differences and the system of stratification that separates us. However, society has attempted to eliminate some of the barriers between groups. In schools and workplaces in particular (where laws have been most stringent), we've had some success in overcoming stratification based on gender, race, and ethnicity.

Americans generally value diversity but remain committed to some dividing lines. For example, many forms of sex-based discrimination and segregation are rigidly enforced by mores, even among the most liberal Americans.

◀ Class distinctions also persist and can be compounded by differences based on ethnicity, race, and gender. About 45 percent of the U.S. poor are non-Hispanic Whites, with African Americans having the highest poverty rate among minority groups.

What "invisible" dividing lines can you think of that people may be taking for granted?

CHAPTER 2

Social Class and Poverty

Monday, August 29, 2005, 6:10 a.m.—Hurricane Katrina, classified as a Category 4 hurricane, hits Louisiana with 125 mile-per-hour winds. The following day two New Orleans' levees have broken and water begins to cover 80 percent of the city. Many who were not able to evacuate the city begin to climb to their roofs or higher ground for safety. The New Orleans' Superdome and convention center served as makeshift emergency housing for the city's displaced residents—mostly poor, minority, or physically/medically disabled residents who were unable to evacuate. More than 50,000 Louisianans were gathered at these sites with limited water and food supplies and medical care for several days. The last victims were evacuated from the Superdome on September 4, almost a week after the levees broke.

The agonizing plight of these men, women, and children received extensive global media coverage, with the overseas reaction comprising "sympathy mixed with shock and horror at what was seen by many as … a reminder of the extreme poverty in which many Americans live" (Haas 2005). Katrina was not an equal opportunity disaster (Dreier 2006). Before Katrina hit the city, New Orleans had a median household income of $18,477, with more than 31 percent of households having annual incomes less than $10,000 (Wright and Bullard 2007). The poorest neighborhoods were hit hardest by the hurricane and floodwaters revealing the vulnerability of America's poor to a worldwide audience.

The United States is perceived as one of the world's richest countries. As measured by its gross domestic product (GDP) and purchasing power parity (PPP), the United States is the third richest country. The GDP represents the total market value of all goods and services produced within a year and is used as a measure to assess a nation's standard of living. Purchasing power parity is a method of measuring the relative purchasing power of currencies from different countries for the same types of goods and services and is usually converted to U.S. dollar

Hurricane Katrina's victims were poor, minority, or physically or medically disabled residents. Unable to evacuate, those stranded were forced to wait for assistance at the New Orleans' Superdome and convention center.

measurements. Per capita (per person), our GDP/PPP was estimated at $41,950 for 2006. The top two countries were Luxembourg ($65,340) and Bermuda (based on an estimate, no dollar amount reported) (World Bank 2007).

Nonetheless, economic inequality is one of the most important and visible of America's social problems (McCall 2002). Martin Marger writes, "Measured in various ways, the gap between rich and poor in the United States is wider than [in] any other society with comparable economic institutions and standards of living" (2002:48). The overall distribution of wages and earnings has become more unequal and the distance between the wealthy and the poor has widened considerably in recent decades.

The U.S. Census examines income distribution by dividing the U.S. household population into fifths or quintiles. If all U.S. income were equally divided, each quintile would receive one fifth of the total income. However, based on U.S. Census 2005 data, 50 percent of the total U.S. income was earned by households in the highest quintile or among households making an average of $159,583. The lowest 20 percent of households (earning an average of $10,655 per year) had 3.4 percent of the total income (DeNavas-Walt, Proctor, and Lee 2006). Inequality grew between 1973 and 1999, when the top fifth of the distribution began to increase its share of aggregate income, while the bottom four-fifths began to lose their share (Jones and Weinberg 2000). (Refer to Table 2.1.)

However, wealth, rather than income, may be more important in determining one's economic inequality. **Wealth** is usually defined as the value of assets (checking and savings accounts,

Table 2.1

Share of Aggregate Income Received by Each Fifth, 2005

Fifth	Mean Income	Share
Top fifth	$159,583	50.4
Second fifth	$72,825	23.0
Third fifth	$46,301	14.6
Fourth fifth	$27,357	8.6
Lowest fifth	$10,655	3.4

Source: DeNavas-Walt, Proctor, and Lee 2006.

property, vehicles and stocks) owned by a household (Keister and Moller 2000) at a point in time. Wealth is measured in two ways: gross assets (the total value of the assets someone owns) and net worth (the value of assets owned minus the amount of debt owed) (Gilbert 2003). Wealth is more stable within families and across generations than are income, occupation, or education (Conley 1999) and can be used to secure or produce wealth, enhancing one's life chances.

As Melvin Oliver and Thomas Shapiro explain,

> Wealth is a particularly important indicator of individual and family access to life chances. Wealth is a special form of money not used to purchase milk and shoes and other life necessities. More often it is used to create opportunities, secure a desired stature and standard of living, or pass class status along to a one's children.... The command over resources that wealth entails is more encompassing than income or education, and closer in meaning and theoretical significance to our traditional notions of economic well-being and access to life chances. (1995:2)

Wealth preserves the division between the wealthy and the nonwealthy, providing an important mechanism for the intergenerational transmission of inequality (Gilbert 2003). Scott Sernau tells us,

> Wealth begets wealth.... It ensures that those near the bottom will be called on to spend almost all of their incomes and that what wealth they might acquire, such as an aging automobile or an aging house in a vulnerable neighborhood, will more likely depreciate than increase in value, and the poor will get nowhere. (2001:69)

Data reveal that wealth is more unequally distributed and more concentrated than income. Since the early 1920s, the top 1 percent of wealth holders has owned an average of 30 percent of household wealth. During the late 1980s and 1990s, the top 1 percent of wealth owners owned nearly 40 percent of all net worth and nearly 50 percent of all financial assets (Keister and Moller 2000).

What Does It Mean to Me?

Consider your own income and wealth status. How would you define your social class based on your own income and wealth? Your family's income and wealth? Which reveals more about your life chances?

What Does It Mean to Be Poor?

The often-cited definition of poverty offered by the World Bank is $1 per day. This represents "extreme poverty," the minimal amount necessary for a person to fulfill his or her basic needs. According to the organization (2008),

> Poverty is hunger. Poverty is lack of shelter. Poverty is being sick and not being able to see a doctor. Poverty is not being able to go to school and not knowing how to read. Poverty is not having a job, is fear for the future, living one day at a time. Poverty is losing a child to illness brought about by unclean water. Poverty is powerlessness, lack of representation and freedom.

The World Bank estimates that in 2001, 1.1 billion people had consumption levels below $1.00 a day, and more than 2.7 billion lived on less than $2.00 per day (World Bank 2008). By 2015, the number living in extreme poverty ($1.00 per day) should decrease to 753 million mostly because of significant improvements in education, gender equality, health care, environmental degradation, and hunger.

Sociologists offer two definitions of poverty, absolute and relative poverty. **Absolute poverty** refers to a lack of basic necessities, such as food, shelter, and income. **Relative poverty** refers to a situation in which some people fail to achieve the average income or lifestyle enjoyed by the rest of society. Our mainstream standard of living defines the "average" American lifestyle. Individuals living in relative poverty may be able to afford basic necessities, but they cannot maintain a standard of living comparable to that of other members of society. Relative poverty emphasizes the inequality of income and the growing gap between the richest and poorest Americans. A definition reflecting the relative nature of income inequality was adopted by the European (EU) Council of Ministers in 1984—"The poor shall be taken to mean persons, families and groups of persons whose resources (material, cultural and societal) are so limited as to exclude them from the minimum acceptable way of life in the member state in which they live."

Despite the economic growth of the 1980s and 1990s, the gap between the richest and poorest Americans has actually increased. The Center on Budget and Policy Priorities and the Economic Policy Institute reported that in most states, the gap between the incomes of the richest 20 percent of families and the incomes of the poorest 20 percent of families is wider than it was two decades ago. In 2004, the poorest 20 percent of families had an average income of $14,700 per year whereas the richest 20 percent had about 10 times as much, $155,200 (Sherman and Aron-Dine 2007). This increasing income gap has been attributed to the growth in wage inequality. Wages at the bottom or middle of the income scale have declined or remained constant during the past two decades. Additional attributing factors include increases in the number of families headed by a single person, specific federal tax policies, declines in investment income, and persistent unemployment.

The Federal Definitions of Poverty

There are two federal policy measures of poverty: the poverty threshold and the poverty guidelines. These measures are important for statistical purposes and for determining eligibility for social service programs.

Photo 2.2

Topping Forbes 2007 World's Billionaires List was Microsoft's Bill Gates, with a fortune estimated at $56 billion. With his wife Melinda, Gates created the Bill and Melinda Gates Foundation, which supports programs and research to reduce inequities in the United States and throughout the world.

Photo 2.3a & 2.3b

Not everyone in our society can attain the dream of owning a home. For more than 750,000 Americans, home is life on the streets, in shelters, and in transitional housing.

The **poverty threshold** is the original federal poverty measure developed by the Social Security Administration and updated each year by the U.S. Census Bureau. The threshold is used to estimate the number of people in poverty. Originally developed by Mollie Orshansky for the Social Security Administration in 1964, the original poverty threshold was based on the economy food plan, the least costly of four nutritionally adequate food plans designed by the U.S. Department of Agriculture (USDA). Based on the 1955 House Food Consumption Survey, the USDA determined that families of three or more people spent about one-third of their after-tax income on food. The poverty threshold was set at three times the cost of the economy food plan. The definition of the poverty threshold was revised in 1969 and 1981. Since 1969, annual adjustments in the levels have been based on the Consumer Price Index (CPI) instead of changes in the cost of foods in the economy food plan.

The poverty threshold considers money income before taxes and excludes capital gains and noncash benefits (public housing, Medicaid, and food stamps). The poverty threshold does not apply to people residing in military barracks or institutional group quarters or to unrelated individuals younger than age 15 (foster children). In addition, the definition of the poverty threshold does not vary geographically.

The **poverty guidelines**, issued each year by the U.S. Department of Health and Human Services, are used to determine family or individual eligibility for federal programs such as Head Start, the National School Lunch Program, or the Low Income Energy Assistance Program. The poverty guidelines are designated by the year in which they are issued. For example, the guidelines issued in January 2008 are designated as the 2008 poverty guidelines, but the guidelines reflect price changes through the calendar year 2007. There are separate poverty guidelines for Alaska and Hawaii. The current poverty threshold and guidelines are presented in Tables 2.2 and 2.3.

Photo 2.4

Urban youth are the most vulnerable poor population, barely surviving without shelter, food, and clothing. Cities in developing nations find themselves unable to provide basic services, such as education and health care, to these youth.

Who Are the Poor?

In 2006, the poverty rate was 12.3 percent or 36.5 million compared with the most recent low poverty rate of 11.3 percent or 31.6 million in 2000 (DeNavas et al. 2006; DeNavas-Walt, Proctor, and Smith 2007). (See Table 2.4 for a summary of poverty characteristics for 2006 and U.S. Data Map 2.1 for 2004–2006 poverty rate averages by state.)

Based on 2006 poverty figures and redefined racial and ethnic categories, Whites (who reported being White and no other race category, along with Whites who reported being White along with another race category) compose the largest group of poor individuals in the United States. About 43.9 percent of the U.S. poor were non-Hispanic Whites. However, the poverty rate for non-Hispanic Whites was the lowest, at 8.2 percent. Blacks continue to

Table 2.2

Poverty Threshold in 2007 by Size of Family and Number of Related Children Under 18 Years (dollars)

Size of Family Unit	Related Children Under 18 Years								
	None	1	2	3	4	5	6	7	8+
One person under 65	10,787								
65 years or older	9,944								
Two people									
Householder under 65	13,884	14,291							
Householder 65 or older	12,533	14,237							
Three	16,218	16,689	16,705						
Four	21,386	21,736	21,027	21,100					
Five	25,791	26,166	25,364	24,744	24,366				
Six	29,664	29,782	29,168	28,579	27,705	27,187			
Seven	34,132	34,345	33,610	33,098	32,144	31,031	29,810		
Eight	38,174	38,511	37,818	37,210	36,348	35,255	34,116	33,827	
Nine or more	45,921	46,143	45,529	45,104	44,168	43,004	41,952	41,691	40,085

Source: U.S. Census Bureau 2008.

Table 2.3

2008 Federal Poverty Guidelines

Size of Family Unit	48 Contiguous States and District of Columbia	Alaska	Hawaii
1	$10,400	$13,000	$11,960
2	14,000	17,500	16,100
3	17,600	22,000	20,240
4	21,200	26,500	24,380
5	24,800	31,000	28,520
6	28,400	35,500	32,660
7	32,000	40,000	36,800
8	35,600	44,500	40,940
For each additional person, add	3,600	4,500	4,140

Source: Federal Register 2008.

have the highest poverty rate, 24.3 percent, followed by Hispanics with a rate of 20.6 percent (DeNavas et al. 2007). Analysts predict that within a few years, Latinos will have a higher poverty rate than Blacks do. Racial segregation and discrimination have contributed to the high rate of minority poverty in the United States. Minority groups are disadvantaged by their

Table 2.4

Selected Poverty Characteristics, 2006 (numbers in thousands)

	Number	Poverty Rate
Total	36,460	12.3
Race/Ethnicity		
White alone or in combination	24,416	10.3
White alone, non-Hispanic	16,013	8.2
Black alone or in combination	9,048	24.3
Asian alone or in combination	1,353	10.3
Hispanic (of any race)	9,243	20.6
Age		
Under 18 years of age	12,827	17.4
18 to 64 years	20,239	10.8
65 years or older	3,394	9.4
Inside Metropolitan Areas		
Inside central cities	15,336	16.1
Outside central cities	13,947	9.1
Outside Metropolitan Areas	7,177	15.2

Source: DeNavas, Proctor, and Smith 2007.

Note: White alone, Black alone, and Asian alone include respondents who indicated only one race. White alone or in combination, Black alone or in combination, and Asian alone or in combination include respondents who indicated one race or a combination of races.

lower levels of education, lower levels of work experience, lower wages, and chronic health problems—all characteristics associated with higher poverty rates (Iceland 2003).

According to the National Center for Children in Poverty (2001), children are more likely to live in poverty than Americans in any other age group. The poverty rate among children is higher in the United States than in most other major Western industrialized nations. In 2005, the United Nations Children's Fund (UNICEF) released its report on the "Child Poverty League," identifying the percentage of children living in relative poverty (in households with income below 50 percent of the national median income). The United States ranks second highest at 16.6 percent. The country with the highest percentage of children living in relative poverty is Mexico (27.7 percent). (Refer to Figure 2.1 for additional data.)

Based on data collected by the U.S. Census, from a peak rate of 22.5 percent in 1993, the poverty rate for children declined to 16.7 percent in 2001 and 2002, but increased to 17.6 percent in 2003. In 2006, the poverty rate among children declined slightly to 17.4 percent (DeNavas-Walt et al. 2006; DeNavas-Walt et al. 2007). There remains a wide variation in children's poverty rates among states from 7 percent in New Hampshire to 27 percent in Mississippi. Rates remain high among specific groups—minority children and children of immigrants (Fass and Cauthen 2006).

Families with a female householder and no spouse present were more likely to be poor than were families with a male householder and no spouse present, 28.3 percent versus 13.2 percent. In contrast, the poverty rate for married-couple families was 4.9 percent (DeNavas-Walt et al. 2007). Single-parent families are more vulnerable to poverty because

U.S. DATA MAP 2.1

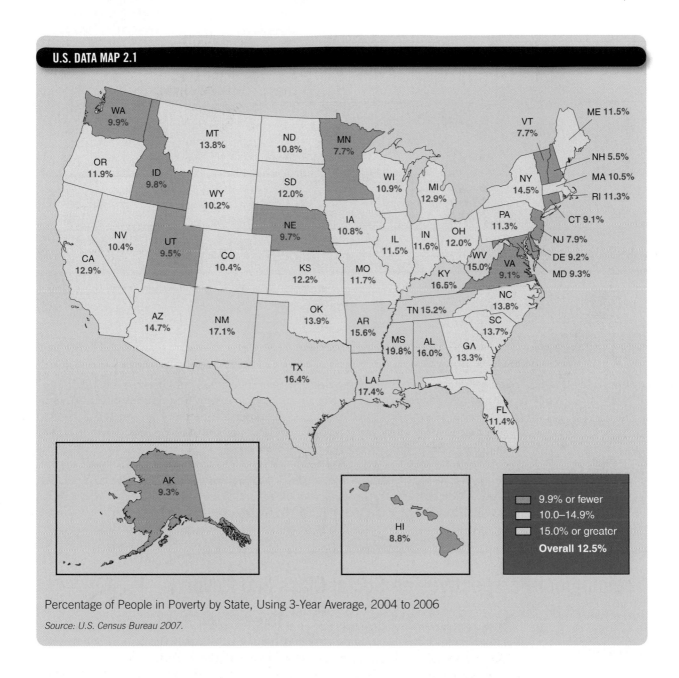

Percentage of People in Poverty by State, Using 3-Year Average, 2004 to 2006

Source: U.S. Census Bureau 2007.

there is only one adult income earner, and female heads of households are disadvantaged even further because women in general make less money than men do.

In their analysis of data from the Luxembourg Income Study, Lee Rainwater and Timothy Smeeding (2003) conclude that American single mothers' children fare worse than the majority of their global counterparts. The poverty rate among U.S. children living in single-mother families is close to 50 percent; the rate is slightly lower in Germany (48 percent) and Australia (46 percent). Countries with poverty rates below 20 percent include Sweden (7 percent), Finland (8 percent), Demark (11 percent), Belgium (13 percent), and Norway (14 percent). Generous social wage (e.g., unemployment) and social welfare programs reduce the poverty

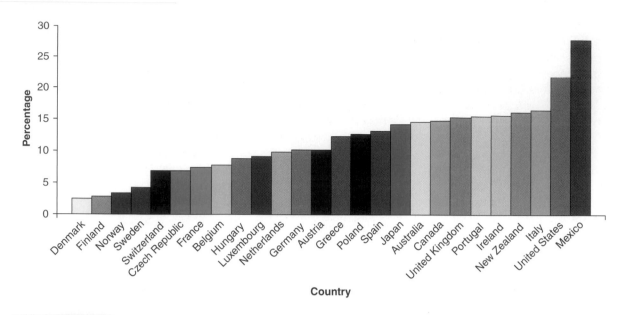

Country

Figure 2.1

The Child Poverty League, 2005. The Percentage of Children Living in Households With Income Below 50 Percent of the National Media Income

Source: United Nations Children's Fund 2005.

rate in these Nordic countries. Rainwater and Smeeding (2003) note that all combined, U.S. wage and welfare programs are much smaller than similar programs in other countries.

Although the majority of poor women and men live in cities and suburbs within metropolitan areas, poverty was still higher for residents outside metropolitan areas (15.2 percent) than for those living inside metropolitan areas (11.8 percent). Among those living in metropolitan areas, the poverty rate within central cities was 16.1 percent, almost twice as high as the rate in suburbs, 9.1 percent. The South had the highest poverty rate (13.8 percent), followed by the West (11.6 percent), the Northeast (11.5 percent), and the Midwest (11.2 percent) (DeNavas-Walt et al. 2007). The states with the highest poverty rates in 2006 were Arkansas, Mississippi, New Mexico, and West Virginia (Webster and Bishaw 2007).

Sociological Perspectives on Social Class and Poverty

Why do some prosper while others remain poor? Why does poverty persist in some families, but other families are able to improve their economic situation? In the next section, we will review the four sociological perspectives to understand the bases of class inequality.

Functionalist Perspective

Functionalists assume that not everyone in society can and should be equal. From this perspective, inequality is necessary for the social order. And what is equally important is how each of us recognizes and accepts our status in the social structure. Erving Goffman (1951), an interactionist, offers a functional explanation of social stratification, defining it as a universal characteristic of social life. Goffman argues that as we interact with one another, accepting our status in society and acknowledging the status of others, we provide "harmony" to the social order. But "this kind of harmony requires that the occupant of each status act towards others in a manner which conveys the impression that his conception of himself and of them is the same as their conception of themselves and him" (p. 294).

Functionalists contend that some individuals are more important to society because of their function to society. For example, society values the lifesaving work of a medical surgeon more than the retail function of a grocery store cashier. Based on the value of one's work or talent, society rewards individuals at the top of the social structure (surgeons) with more wealth, income, or power than those lower down in the social structure (grocery cashiers). According to this perspective, individuals are sorted according to their abilities or characteristics—their age, strength, intelligence, physical ability, or even sex—to play their particular role for society. Certain individuals are better suited for their positions in society than others. Our social institutions, especially education, sort everyone into their proper places and reward them accordingly. Because not all of us can (or should) become surgeons, the system ensures that only the most talented and qualified become surgeons. In many ways, the functionalist argument reinforces the belief that we are naturally different.

What Does It Mean to Me?

The functionalist perspective is often criticized for its value argument. Is society able to accurately assess and reward a position for its value, for its function? For example, the average salary for public teacher is $47,602 (for 2004–2005), whereas the average salary of an NFL player is $1.1 million (2006–2007). Before he became a Super Bowl champion, Indianapolis Colts' quarterback Peyton Manning signed a $98 million seven-year deal with his team. How does society determine the value of a professional football player? Of a public school teacher?

Functionalists observe that poverty is a product of our social structure. Specifically, rapid economic and technological changes eliminated the need for low skilled labor, creating a population of workers who were unskilled and untrained for this new economy. In many ways, theorists from this perspective expect this disparity among workers, arguing that only the most qualified should fill the important jobs in society and be rewarded for their talent.

Herbert Gans (1971) argued that poverty exists because it is functional for society. Gans explained that the poor uphold the legitimacy of dominant norms. The poor help reinforce cultural ideals of hard work and the notion that anyone can succeed if only they would try (so if you fail, it is your fault). Poverty helps preserve social boundaries. It separates the haves from the have-nots by their economics and according to their educational attainment, marriage, and residence. The poor also provide a low-wage labor pool to do the "dirty work" that no one else wants to do.

Our social welfare system, designed to address the problem of poverty, has been accused of being dysfunctional itself; critics suggest that the welfare bureaucracy is primarily concerned with its own survival. Poverty helps create jobs for the nonpoor, particularly the social welfare system designed to assist the poor. As a result, the social welfare bureaucracy will develop programs and structures that will only ensure its survival and legitimacy. Based on her personal experience working with and for the system, Theresa Funiciello observed, "Countless of middle class people were making money, building careers, becoming powerful and otherwise benefiting from poverty.... The poverty industry once again substituted its own interests for that of poor people" (1993:xix). We will discuss this further in the next perspective.

Conflict Perspective

Like the functionalist perspective, the conflict perspective argues that inequality is inevitable, but for different reasons. For a functionalist, inequality is necessary because of the different positions and roles needed in society. From a conflict perspective, inequality is systematically created and maintained by those trying to preserve their advantage over the system.

For Karl Marx, one's social class is solely determined by one's position in the economic system: You are either a worker or an owner of the means of production. But social class, according to Max Weber, is multidimensional. Economic factors include **income**, the money earned for one's work, and wealth, the value of one's personal assets such as savings and property. A person's social class is also influenced by **prestige**, the amount of social respect or standing given to an individual based on occupation. We assign higher prestige to occupations that require specialized education or training, that provide some social good to society, or that make more money. A final component of class is **power**. Weber defined power as the ability to achieve one's goals despite the opposition of others. Power is the ability to do whatever you want because no one can stop you.

Power is not limited to individuals. People with similar interests (or with similar income, wealth, and prestige backgrounds) often collaborate to increase their advantage in society. C. Wright Mills (1959/2000) argued that the United States is ruled by what he called a **power elite**. According to Mills, this elite group is composed of business, political, and military leaders. This elite group has absolute power because of its ability to withhold resources and prevent others from realizing their interests. Mills identified how the power elite effectively make decisions regarding economic policy and national security—controlling the difference between a boom or bust economy or peace or war abroad (Gilbert 2003). Refer to this chapter's In Focus feature for an in-depth look at the modern power elite.

Michael Harrington argues, "The real explanation of why the poor are where they are is that they made the mistake of being born to the wrong parents, in the wrong section of the country, in the wrong industry, or in the wrong racial or ethnic group" (1963:21). Inequalities built into our social structure create and perpetuate poverty.

Conflict theorists assert that poverty exists because those in power want to maintain and expand their base of power and interests, with little left to share with others. Welfare bureaucracies—local, state, and national—represent important interest groups that influence the creation and implementation of welfare policies. A welfare policy reflects the political economy of the community in which it is implemented (Handler and Hasenfeld 1991).

Francis Fox Piven and Richard A. Cloward (1993) state that the principal function of welfare is to allow the capitalist class to maintain control over labor. Welfare policy has been used by the state to stifle protest and to enforce submissive work norms. During periods of economic crisis, the state expands welfare rolls to pacify the poor and reduce the likelihood of serious uprising. However, during economic growth or stability, the state attempts to reduce the number of people on welfare, forcing the poor or dislocated workers back into the expanding labor force. Those who remain on welfare are condemned and stigmatized for their dependence on the system. In particular, the welfare state serves capitalist interests by promoting women's roles as reserve labor and as caretakers to reproduce the labor force (Kuhn and Wolpe 1978).

Feminist Perspective

Feminist scholars define the welfare state as an arena of political struggle. The drive to maintain male dominance and the patriarchal family is assumed to be the principal force shaping the formation, implementation, and outcomes of U.S. welfare policy (Neubeck and Cazenave 2001).

In Focus

Who Has the Real Power?

G. William Domhoff (2002) argues that real power is **distributive power**, the power individuals or groups have over other individuals or groups. Power matters when a group has the ability to control strategic resources and opportunities to obtain such resources. Money, land, information, and skills are strategic resources when they are needed by individuals to do what they want to do (Hachen 2001). Domhoff argues that distributive power is limited to an elite group of individuals whose economic, political, and social relationships are closely interrelated.

He writes about an elite power group composed of "members of the upper class who have taken on leadership roles in the corporate community and the policy network, along with high-level employees in corporations and policy-network organizations" (2002:9). Despite affirmative action policies and society's affirmation of the value of social equity, and although women and minorities are increasing their representation on corporate boards, presidential cabinets, and leadership groups, the power elite is still composed of White male Christians.

Power elites are members of various policy planning groups, corporate boards, advisory councils, and leadership forums. These individuals frequently exchange positions, moving back and forth between private and public spheres as advisers, lobbyists, or cabinet officials (Marger 2002). U.S. presidents rely on a small network of influential corporate and policy leaders. A study of presidential cabinet appointees from 1934 to 1980 (ending with Jimmy Carter's administration) found that 64 percent of appointees were top wealth holders or were on the corporate boards of the largest companies in their area of expertise.

The pattern of influence among the wealthy and the influent is evident in the selection of President George W. Bush's first-term cabinet appointees. Domhoff (2002) reports on the following cabinet members:

Before his term as President George W. Bush's vice president, Dick Cheney spent eight years as president of Halliburton, an oil drilling company. During the second U.S.–Iraq war, Halliburton received more than $8.2 billion in contracts, more than any other firm (Waxman 2004). Cheney was also a member of the board of directors of Electronic Data Systems, Procter & Gamble, and Union Pacific. Cheney also served on the board of the American Enterprise Institute, a pro-business think tank.

Secretary of State, and retired army general, Colin Powell served as a director of Gulfstream Aerospace until it merged with General Dynamics in 1999. Powell earned $1.49 million in stock options in exchange for helping the company sell corporate jets to Kuwait and Saudi Arabia. He was director of America Online at the time he was appointed. His net worth in 2001 was estimated at more than $28 million.

Donald Rumsfeld, Secretary of Defense, previously served as chief executive officer of G. D. Searle & Co. and at General Instruments. In 1998, he was a member of four corporate boards: Kellogg, Sears Roebuck, the Tribune Publishing Co., and Gulfstream Aerospace (like Powell, Rumsfeld made more than $1 million in stock options in exchange for helping sell corporate jets). He served as a trustee of two think tanks, the American Enterprise Institute and the Rand Corporation.

Former Secretary of Treasury Paul H. O'Neill was retired chair of Alcoa, the world's largest aluminum manufacturer (holding more than $50 million in stock) and director of Lucent Technologies. He was chair of the board of trustees at the Rand Corporation and was a trustee at the American Enterprise Institute, along with Cheney and Rumsfeld. O'Neill was forced to resign from office in 2002 after he publicly voiced his disapproval of tax cuts and increasing budget deficits.

Log on to *Study Site Chapter 2* for more information.

Social welfare scholar Mimi Abramovitz (1996) notes that welfare has historically distinguished between the deserving poor (widows with children) and the undeserving poor (single and divorced mothers). In the 1970s and 1980s, media and politicians created the image of the "Cadillac driving, champagne sipping, penthouse living welfare queens" (Zucchino 1999), suggesting that women—specifically, single mothers—were abusing welfare assistance. Women were accused of having more children to avoid work and to increase their welfare benefits. Marriage, hard work, honesty, and abstinence were offered as solutions to their poverty. The negative stereotypes of poor women stigmatized these women but also fueled support for punitive social policies (Abramovitz 1996).

As a group, teenage mothers were targeted during welfare reform in the 1990s. They were blamed for everything, as one journalist wrote: "The fact remains: every threat to the fabric of this country—from poverty to crime to homelessness—is connected to out-of-wedlock teen pregnancy" (Alder 1994). Even though little empirical evidence supports the link between birthrates and welfare benefits, the government seeks to punish these women, making it difficult for them to obtain and retain public assistance and to gain self-sufficiency.

Fraser (1989) argues that there are two types of welfare programs: masculine programs related to the labor market (social security, unemployment compensation) and feminine programs related to the family or household (Aid to Families with Dependent Children [AFDC], food stamps, and Medicaid). The welfare system is separate and unequal. Fraser believes that masculine programs are rational, generous, and nonintrusive whereas feminine programs are inadequate, intrusive, and humiliating. The quintessential program for women, AFDC, institutionalized the feminization of poverty by failing to provide adequate support, training, and income to ensure self-sufficiency for women (Gordon 1994). As described by Johnnie Tillmon (1972), a welfare recipient and welfare rights organizer,

> The truth is that AFDC is like a super-sexist marriage. You trade in a man for the man. But you can't divorce him if he treats you bad. He can divorce you, of course, cut you off anytime he wants. The man runs everything. In ordinary marriage sex is supposed to be for your husband. On AFDC you're not supposed to have any sex at all. You give [up] control of your own body.... You may even have to agree to get your tubes tied so you can never have more children just to avoid being cut off welfare.

Our current welfare system, Personal Responsibility and Work Opportunity Reconciliation Act (PRWORA), has been criticized for its treatment of women and their families. PRWORA created a pool of disciplined low-wage laborers: women who must take any job that is available or find themselves and their families penalized by the government (Piven 2002). With its emphasis on work as the path to self sufficiency, PRWORA forces women back to the same low-pay, low-skill jobs that may have led them to their poverty in the first place (Lafer 2002). The new policies fail to address the real barriers facing women: low job skills and educational attainment, racism and discrimination in the labor market, and the competing demands of work and caring for their children.

Interactionist Perspective

An interactionist would draw attention to how class differences are communicated through symbols, how the meaning of these symbols is constructed or constrained by social forces, and how these symbols reproduce social inequality. Our language reflects the quality of life that is associated with different amounts of economic resources. We distinguish the "very rich" from the "stinking rich" and someone who is "poor" versus "dirt poor" (Rainwater and Smeeding 2003).

Some sociologists have suggested that poverty is based on a **culture of poverty**, a set of norms, values, and beliefs that encourage and perpetuate poverty. In this view, moral deficiencies of individuals or their families lead to a life of poverty. Oscar Lewis (1969) and Edward Banfield (1974) argue that the poor are socialized differently (e.g., living from moment to moment) and are likely to pass these values on to their children. Patterns of generational poverty—poor parents have poor children, who in turn become poor adults, and so on—seem to support this theory.

Yet the culture of poverty explanation has been widely criticized. Opponents argue that there is no evidence that the poor have a different set of values and beliefs. This perspective defines poverty as a persistent state, that once you are poor, your values prohibit you from ever getting out of poverty. In fact, poverty data reveal that for most individuals and families, continuous spells of poverty are likely to last less than two years (Harris 1993).

Interactionists also focus on the public's perception of welfare and of welfare recipients. Most Americans do not know any welfare recipients personally or have any direct contact with the welfare system. Their views on welfare are likely to be shaped by what they see on television and what they read in newspapers and magazines (Weaver 2000). As a society, we have developed a sense of the "undeserving poor"; dependent mothers and fathers and nonworking recipients have become powerful negative symbols in society (Norris and Thompson 1995).

Martin Gilens (1999) explains that *welfare* has become a code word for *race*. Race and racism are important in understanding public and political support for antipoverty programs (Lieberman 1998; Neubeck and Cazenave 2001; Quadagno 1994). Gilens states that Americans perceive welfare as a Black phenomenon, believing that Blacks make up 50 percent of the poor population (compared with an actual 25 percent). This belief is exacerbated by the notion that Blacks are on welfare not because of blocked opportunities, but largely because of their lack of effort.

Gilens (1999) asserts that the news media are primarily responsible for building this image of Black poverty, for the "racialization of poverty." During the War on Poverty in the early 1960s, the media focused on White rural America, but as the civil rights movement began to build in the mid 1960s, the media turned attention to urban poverty, and the racial character of poverty coverage changed. Between 1965 and 1967, sensationalized portrayals of Black poverty were used to depict the waste, inefficiency, or abuse of the welfare system, whereas positive coverage of poverty was more likely to include pictures and portrayals of Whites. After 1967 and for most of the following three decades, larger proportions of Blacks appeared in news coverage of most poverty topics. "Black faces are unlikely to be found in media stories on the most sympathetic subgroups of the poor, just as they are comparatively absent from media coverage of poverty during times of heightened sympathy for the poor" (Gilens 1999:132). According to Gilens, this exaggerated link between Blacks and poverty is a serious obstacle to public support for antipoverty programs.

What Does It Mean to Me?

Note that none of these perspectives (see Table 2.5 for a summary) focus on the role of individual factors in determining poverty. Using our sociological imagination, we begin to recognize the social and institutional conditions that create poverty. Which of these perspectives best explains why poverty exists?

Table 2.5

Summary of Sociological Perspectives: Inequalities Based on Social Class

	Functionalist	Conflict/Feminist	Interactionist
Explanations of social class and poverty	Inequality is inevitable and emerges from the social structure. Poverty serves a social function.	Inequality is systematically maintained by those trying to preserve their class advantage. Class is based on multiple dimensions—income, wealth, prestige, and power. Welfare bureaucracies represent important interest groups that influence the creation and implementation of welfare policies.	Each social class has a specific set of norms, values, and beliefs. Poverty is a learned phenomenon based on a "culture of poverty" that encourages and perpetuates poverty. The public's perception of the welfare system and of welfare recipients is shaped by the media, political groups, and stereotypes.
Questions asked about social class and poverty	What are the functions and dysfunctions of inequality? What portions of society benefit from poverty?	What powerful interest groups determine class inequalities? How do our welfare policies reflect specific political, economic, and social interest groups?	Is poverty learned behavior? How are our perceptions of the poor determined by the media, news reports, and politicians? Has society created two images—the deserving versus the undeserving poor? Are these images accurate?

The Consequences of Poverty

The following section is not an exhaustive list of the consequences of poverty. Remaining chapters will also highlight the relationship between social class and the experience of a specific social problem (such as educational attainment, drug abuse, or access to health care). In addition, given the intersectionality of all the bases of inequality covered in this section of the book, there is a persistent overlap in the experience of social problems as a result of one's class, race, gender, sexual orientation, and age.

Health and Health Care

About 70 percent of children who live below the poverty line report being in excellent or good health in comparison with 85 percent of children who live at or above the poverty line. Children below the poverty line have lower rates of immunization and higher rates of activity limitation because of chronic illness than do children who live above the poverty line (Federal Interagency Forum on Child and Family Statistics 2002).

The increasing cost of quality health care and prescription drugs is a concern for many elderly. Though Medicare provides health coverage for men and women older than 65 years of age, most elderly still spend their own money for out-of-pocket (OOP) health care expenses, particularly for prescription drugs. In 2003, nearly one-third of all elderly spent $100 or more of their own money per month for prescription drugs (Safran et al. 2005). For more on the nation's Medicare program, refer to Chapter 6.

Most of the working poor are not eligible for public assistance medical programs, even though their employers do not provide health insurance (Seccombe and Amey 1995). Most individuals without health insurance coverage make less than $50,000 per year. According to the U.S. Census Bureau, in 2006, 25 percent of households with incomes less than $25,000 and 20 percent of households with incomes between $25,000 and $49,000 did not have health insurance coverage. As household income increases, so does the likelihood of having health insurance. Only 13 percent and 8 percent of households with incomes between $50,000–$74,999 and $75,000 or more, respectively, did not have health insurance (DeNavas-Walt et al. 2007). Employment does not guarantee health insurance coverage.

Although the poor are eligible for medical care under Medicaid, Medicaid coverage is not comprehensive, forcing most to seek treatment only in cases of medical emergencies. Having a routine physical or a regular family doctor is rare among poor and low-income families. Families were eligible for Medicaid only if they were also receiving government assistance.

In 1997, the State Children's Health Insurance program was established to help states expand Medicaid coverage, establish a separate child health insurance program, or combine approaches to cover children with family incomes up to 200 percent of the poverty line or higher. Beginning in 1998, the federal government also provides Transitional Medical Assistance for families leaving welfare for work. The program provides as much as 12 months of coverage when families lose regular Medicaid coverage because of an increase in earnings (leaving welfare for work). Despite these medical provisions, a review of data from 25 states where welfare recipients have left welfare for work indicates that there has been a decline in health coverage among eligible low-income families with children.

Jocelyn Guyer (2000) says that in most states, roughly half of parents in families that left welfare and more than one-third of children in those families lose Medicaid coverage. In some states, the problem is severe. In New York, 65 percent of children lost Medicaid after they left welfare; in Mississippi, 56 percent of children also lost coverage. Even among families who left welfare for work, fewer than half receive health coverage through their employers. As a result, families are more likely to have unmet medical needs after leaving welfare. Guyer (2000) notes that although little is known about why eligible families lose health coverage, some studies reveal that families were unaware of their eligibility for programs like Medicaid or Transitional Medical Assistance.

Food Insecurity and Hunger

Eleven percent of households, or 12.6 million American families, were food insecure at least some time during the year (USDA 2006). **Food insecure** means that these families did not always have access to enough food for all members of the household to enjoy active and healthy lives. Seven percent (or 8.2 million) avoided hunger by getting emergency food from community food pantries, eating less varied diets, and participating in the federal food assistance programs. The remaining 4 percent of households (or 4.4 million) had very low food security, meaning that normal eating patterns of one or more household members were disrupted because of insufficient resources for food. The prevalence of food insecurity is higher for some groups: single female–headed households with children (31 percent), Black households (22 percent), Hispanic households (18 percent), and households with income below the poverty line (36 percent) (USDA 2006).

The USDA provided food assistance to one in six Americans in 2001 through one of 14 public food assistance programs. Most of us are familiar with the U.S. food stamp program, the nation's largest nutrition program for low-income individuals and families. During 2006, the program served an average of 27 million low-income Americans each month. The average monthly benefit was $94.06 per person. Food stamps cannot be used to buy nonfood items (personal hygiene supplies, paper products), alcoholic beverages, vitamins and medicines, hot food products, or any food that will be eaten in the store. Although the food stamp program and other USDA programs have been shown to be effective in improving the purchasing power and nutritional status of specific populations, a large segment of low-income Americans are not being adequately served or served at all by these programs (Nicholas-Casebolt and McGrath Morris 2001). These families need to rely on private programs, such as food pantries or soup kitchens.

For one week, Governor Theodore R. Kulongoski (D-OR) challenged fellow Oregonians to join him and his wife Mary to live on an average Oregon food stamp budget of $21 per week per person or $3.00 a day. His efforts drew state, national, and global attention to food insecurity in his home state of Oregon, as well as the need for the federal government to preserve the current level of food stamps benefits. A sample of the governor and his wife's dinner items is listed in Table 2.6. Before his April 2007 challenge week, the governor and his wife spent an average of $51 per week on food, not including his meals while at work or during official functions; during his challenge week, their final food bill was $20.97 per person. Governor Kulongoski reported several challenges he and his wife experienced throughout the week—the demoralizing experience of not having enough to pay for all the food in his cart, having to make tough decisions on the quality and amount of food they could purchase, and experiencing hunger throughout the week as their food supply ran out (Kulongoski 2007).

As reported by Briefel et al. (2003), food pantries and emergency kitchens play an important role in the nutritional safety net for America's low-income and needy populations. These organizations are part of the Emergency Food Assistance System, a network of private organizations operating with some federal support. Almost one-third of pantry client households and two-fifths of kitchen client households are at or below 50 percent of the poverty line. The mean monthly income is $781 for pantry client households and $708 for kitchen client households. Food pantries were likely to serve families with children (45 percent of households included children), whereas emergency kitchens were likely to serve men living alone (38 percent) or single adults living with other adults (18 percent).

The U.S. Conference of Mayors (2006) reported that emergency food assistance increased by an average of 7 percent in 25 U.S. cities. Almost 50 percent of individuals requesting emergency food assistance were either children or their parents. Thirty-seven percent were employed adults. The U.S. Conference of Mayors attributed the increase primarily to unemployment and employment-related problems, but also identified the relationship between high housing costs, poverty, medical costs, substance abuse, transportation costs, and lack of education with emergency food assistance. In 2005, 23 percent of the demand for emergency food assistance is estimated to have gone unmet.

Poor families are disadvantaged because of their low income and because of their location. They are likely to experience higher food prices than the national average (Kaufman et al. 1997). Suburban supermarkets are likely to have lower food prices and a wider selection of food. Grocery stores in central cities and rural areas, where the poor reside, have higher operating costs and tend to be smaller. In rural areas, food prices are also higher because these markets are likely to be smaller and have lower sales volumes (Kaufman et al. 1997).

Table 2.6

The Kulongoskis' Dinner Items for One Week (food items for lunch and breakfast not included)

Meat

1 small chicken	$3.06

Produce

1 can of kidney beans	.49
1 can of corn	.50
1 can of tomatoes	.59
1 can of garbanzo beans	.79
1 carrot	.34
2 onions	.50
2 zucchini	1.10
1 10-lb bag of potatoes	.99
Lettuce	.99
Cabbage	.85

Pasta

2 Cup-o-Noodles	.76
3 boxes of macaroni/cheese	1.00
Macaroni	.99

Other

Pie crust	.89
Pesto	1.90
Salad dressing	1.19

Dairy

1 gallon of milk	1.00
1 package of imitation cheese	.99
Eggs	1.09

Source: Yardley 2007.

Lack of Affordable Housing

In the same way that access to child care, medical care, and transportation has been linked to economic self-sufficiency, research has also established the relationship between having affordable housing and leaving welfare. Welfare reform successes are greater among families with assisted housing than among other low-income families (Sard and Waller 2002). Unfortunately, most families who leave welfare for work do not earn enough to afford decent quality housing, and because of their employment status, they are not eligible for housing assistance.

According to U.S. Census data, 61 percent of renter households below the poverty line pay more than half of their income to cover housing costs. These families are on severely limited and strained budgets, forcing them to juggle expenses for food, transportation, child care, or other needs. They are more susceptible to falling behind on housing payments and risking eviction and homelessness (Rice and Sard 2007). Families that left welfare between

1997 and 1999 were more likely to report difficulty in paying a mortgage, rent, or utility bill than were families leaving welfare between 1995 and 1997. Among the more recent group of people leaving welfare, nearly 1 in 10 reported that they were forced to double up with other family members or friends because they could not afford the cost of housing (Loprest 2001).

The combination of low earnings and scarce housing assistance result in serious housing problems for the working poor. Nearly three-fifths of working-poor renters with children who do not have housing assistance pay more than 50 percent of their income on housing or live in seriously substandard housing, or both. Among unsubsidized poor renter families with at least full-time year-round minimum-wage earnings, 36 percent spent more than half their income on housing (Sard and Waller 2002).

According to the National Low Income Housing Coalition (Nelson, Treskon, and Pelletiere 2004), there is no state where a low-income worker can reasonably afford a modest one- or two-bedroom rental unit. More than 50 percent of low-income renter households paid more than half their income for gross rent expenses (rent plus utilities). In 40 states, workers needed to earn more than two times the minimum wage to afford basic housing. In some states—California, Connecticut, Maryland, Massachusetts, New Jersey, and New York—workers needed to earn three times the minimum wage to afford housing (Clemetson 2003). Federal housing programs only serve about one-quarter of the eligible low-income households, and few states invest in low-income housing programs (Sard and Waller 2002).

Although Hurricane Katrina struck advantaged and disadvantaged neighborhoods of New Orleans, the odds of living in a damaged area were greater for Blacks, renters, and the unemployed. In the Lower Ninth district, where many of the homes were destroyed by the breach in the Industrial Canal levee, the neighborhoods were more than 85 percent Black, more than a third of residents were below the poverty line, and 14 percent were unemployed before Katrina. Recovery of this neighborhood and others is hampered as the victims have fewer resources to return and rebuild their lives (Logan 2006). By 2007, only 7 percent of the Lower Ninth area had been reoccupied, whereas nearly half of the city's pre-Katrina population had returned (Bradberry 2007).

What Does It Mean to Me?

Of the consequences of poverty discussed in this section—health care, food insecurity, and housing—which do you think is most serious? What solutions can you offer to solve the problem? Explain whether your solutions are directed at the individual (micro) or social structure (macro) level.

Voices in the Community

Stephen Bradberry

In 2005, Stephen Bradberry received the Robert F. Kennedy Human Rights Award for his work for economic and social justice in the South. Bradberry was the first American bestowed the honor, which typically recognized international activists.

Originally from Chicago, Bradberry had been working in New Orleans for 19 years when Hurricane Katrina hit his city. At the time, Bradberry served as the organizer of the Louisiana chapter of Association of Community Organizations for Reform Now (ACORN), the largest community organization of low- and moderate-income families working for social justice and stronger organizations. As he describes their work,

> We're an issues-based organization. For instance, we've been fighting for the living wage for the past few years. We want to engage people in the electoral process now. As we get them engaged … we'll have a number of registered voters interested in passing the living-wage bill in Louisiana. (Robert F. Kennedy Memorial 2005)

In 2002, Bradberry, along with other New Orleans community leaders, supported the passage of an ordinance increasing the city's minimum wage to $6.15. Sixty-three percent of New Orleans voters supported the ordinance; however, the ordinance was later overturned by the Louisiana legislature.

Immediately after Katrina, Bradberry mobilized ACORN's network to provide assistance and shelter to her victims. Through cell phone text-messages, he directed local ACORN members to contact the chapter office for assistance. In addition, ACORN members and leaders visited displaced victims in other cities—Dallas, San Antonio, Little Rock, and Baton Rogue—to provide medical care, food, and clothing and to determine their relocation needs. In 2007, he reported that of 9,000 ACORN families, 7,500 still have not returned to their city.

Bradberry and his organization continue to play a vital role in the area's recovery. The organization has created or joined forces with other organizations for several initiatives: Home Clean-Out Demonstration Program (cleaning and removing damaged plaster and furniture for more than 2,000 families for the modest cost of $2,500 per home), Road Home Contract (establishing a program to assist eligible residents to apply for state funds), and ACORN Housing Corporation (developing affordable housing on 150 abandoned properties in New Orleans; the first homes built in the Lower Ninth since Katrina were ACORN Housing Corporation homes).

At his RFK award ceremony, Bradberry explained, "I certainly don't consider the things I do to be anything extraordinary. It's just a matter of putting on my pants and going to work everyday" (Miga 2005).

Responding to Class Inequalities

U.S. Welfare Policy

Throughout the twentieth century, U.S. welfare policy has been caught between two values: the desire to help those who could not help themselves and the concern that assistance could create dependency (Weil and Feingold 2002). The centerpiece of the social welfare system was established by the passage of the Social Security Act of 1935. The act endorsed a system of assistance programs that would provide for Americans who could not care for themselves: widows, the elderly, the unemployed, and the poor.

Under President Franklin D. Roosevelt's New Deal, assistance was provided in four categories: general relief, work relief, social insurance, and categorical assistance. General relief was given to those who were not able to work; most of the people receiving general relief were single men. Work relief programs gave government jobs to those who were unemployed through programs like the Civilian Conservation Corps and the Works Progress Administration. Social insurance programs included social security and unemployment compensation. Categorical assistance was given to poor families with dependent children, to the blind, and to the elderly. To serve this group, the original welfare assistance program, Aid to Dependent Children (later renamed Aid to Families with Dependent Children or AFDC), was created (Cammisa 1998).

Categorical programs became the most controversial, and the social insurance programs were the most popular. It was widely believed that social insurance paid people for working whereas categorical programs paid people for not working. Shortly after these programs were implemented, officials became concerned that individuals might become dependent on government relief (Cammisa 1998). Even President Roosevelt (quoted in Patterson 1981) expressed his doubts about the system he helped create: "Continued dependence upon relief induces a spiritual and moral disintegration fundamentally destructive to the national fibre. To dole out relief in this way is to administer a narcotic, a subtle destroyer of the human spirit" (p. 60).

The next great expansion of the welfare system occurred in the mid 1960s, when President Lyndon Johnson (1965) declared a War on Poverty and implemented his plan to create a Great Society. Rehabilitation of the poor was the cornerstone of Johnson's policies, and what followed was an explosion of social programs: Head Start, Upward Bound, Neighborhood Youth Corp, Job Corps, public housing, and affirmative action. Although poverty was not completely eliminated, defenders of the Great Society say that these programs alleviated poverty, reduced racial discrimination, reduced the stigma attached to being poor, and helped standardize government assistance to the poor. Conversely, opponents claim that these programs coddled the poor and created a generation that expected entitlements from the government (Cammisa 1998).

During the more than 50 years when the AFDC program operated, welfare rolls were increasing, and even worse, recipients were staying on government assistance for longer periods. In a strange irony, welfare, the solution for the problem of poverty, became a problem itself (Norris and Thompson 1995). Between 1986 and 1996, many states began to experiment with welfare reforms. Wisconsin was the first state to implement such a reform with a program that included work requirements, benefit limits, and employment goals.

In 1996, PRWORA was passed with a new focus on helping clients achieve self-sufficiency through employment. PRWORA was a bipartisan welfare reform plan to reduce recipients' dependence on government assistance through strict work requirements and welfare time limits. Replacing AFDC, the new welfare program is called Temporary Assistance for Needy Families (TANF). Instead of treating assistance as an entitlement, as it was under AFDC, TANF declares that government help is temporary and has to be earned. Under TANF, there is a federal lifetime limit of 60 months of assistance, although states may put shorter limits on benefits. PRWORA also gave states primary responsibility for designing their assistance programs and for determining eligibility and benefits. States are penalized for not meeting caseload targets, either reducing the number of welfare cases to below 1995 levels or increasing the percentage of cases involved in work activities.

The act had an immediate affect on the number of poor. When PRWORA became law, the poverty rate was 13.7 percent; 36.5 million individuals were poor, by the government's definition. A year later, the rate declined to 13.3 percent, and 35.6 million were poor. Rates declined to their lowest point in 2000, 11.3 percent or 31.6 million. According to the U.S. Census Bureau, the 2000 poverty rate was the lowest since 1979.

Although employment has increased among welfare recipients, many recipients have little education or work experience (Loprest 2002). As a group, welfare recipients are limited to jobs at the lower end of the labor market. In 1999, the median hourly wage of employed former welfare recipients was $7.15. About a quarter of those employed worked night shifts or irregular schedules. In addition, most former recipients have limited employment benefits: About one-third of employers offer health insurance, and only one-third or one-half provide sick leave. Studies indicate that one-quarter to one-third of those leaving welfare in a given three-month period were receiving benefits again within the next year (Acs and Loprest 2001). Though evaluation studies reveal overall increases in employment, income, and earnings of families formerly on welfare, many families remain poor or near poor and struggle to maintain employment (Hennessy 2005).

PRWORA was reauthorized under the Deficit Reduction Plan of 2005. The reauthorization requires states to engage more TANF clients in productive work activities leading to self-sufficiency. The five-year cumulative lifetime limit for TANF recipients remains unchanged. Funding of $150 million per year was also provided for healthy marriage and responsible fatherhood initiatives (U.S. Department of Health and Human Services 2006).

Advocates have asked for a reduction in poverty, not just caseloads, as the primary goal of welfare reform. They argue that by removing current limits on education and training, enhancing support services for families, restoring benefits to legal immigrants, and stopping the time clock penalty, TANF could be transformed into an effective poverty-reducing program (Coalition on Human Needs 2003). In addition, advocacy groups have encouraged federal lawmakers to increase funding for the TANF block grants to states, to keep pace with inflation at a minimum. Critics point to the economic fragility of low-income families and identify the need for sufficient health insurance, child care, and transportation assistance to help low-income families during times of crisis.

Others have argued for the importance of restoring support for postsecondary education as a means to promote self-sufficiency. According to the Center for Women Policy Studies (2002), after PRWORA, college enrollment among low-income women declined. Yet studies indicate that former TANF recipients with a college education are more likely to stay employed and less likely to return to welfare. For example, a study among former welfare recipients in Oregon found that only 52 percent of those with less than a high school diploma were employed after two years. In contrast, 90 percent of former TANF recipients with a bachelor's degree were still employed. Since 1996, 49 states—Oklahoma and the District of Columbia are exceptions—passed legislation to allow secondary education to count as activity under PRWORA. Georgia is the only state that allows recipients to enroll in graduate programs.

When Time Limits Run Out

Overwhelming evidence indicates that welfare caseloads have declined since the enactment of PRWORA. Two other factors—a strong economy and increased aid to low-income working families—may have also contributed to the early decline (Besharov 2002). Welfare officials often point to how the first to leave welfare were those with the most employable skills. However, research indicates that the early employment success of welfare reform has diminished as the economy faltered. According to the Urban Institute, 32 percent of welfare recipients were in paid jobs in 1999, but the number had fallen to 28 percent by 2002. Employment also declined for those who left welfare, from 50 percent in 1999 to 42 percent in 2002 (Zedlewski and Loprest 2003).

Under PRWORA, adults are limited to five years of federal welfare benefits, although 20 states impose shorter time limits. States are allowed to exempt as much as 20 percent of their caseloads from time limits and can provide state assistance to families that have used up their federal assistance. The first set of time limits expired on September 30, 2002. Are those who have left welfare able to sustain themselves? Angie O'Gorman (2002) describes the typical "leaver" as a single woman with two children. She is working full-time on minimum wage and makes about $10,300 annually or $178 per week. She continues to receive government assistance—food stamps, Medicaid, health insurance for her children, and other benefits. Is this self-sufficiency?

Taking a World View

Child Care in Europe

Child care is not just about fostering a child's development; it also supports parents' employment. Child care is essential to promote employment and reduce welfare use among low-income families. Among the working poor, child care arrangements that are difficult to pay for, difficult to access, do not cover parents' work hours, or are unreliable can interfere with parents' employment (Gennetian et al. 2002). In addition, child care policies and funding levels have implications for the development and safety of millions of low-income children and youth (Adams and Rohacek 2002). By 1999, child care represented the largest category of state expenditure of TANF funds after cash assistance. While welfare caseloads fell by half, child care caseloads doubled. In 2005, federal funding for child care was budgeted at $5 billion (U.S. Department of Health and Human Services 2006).

European countries offer comprehensive models of public child care. For example, in Scandinavian countries, there is an integrated system of child care centers and organized family day care serving children from birth till school age and managed by social welfare or educational authorities. Public child care is available to children ages 1 through 12. Nearly all employed parents have access to child care with little or no waiting time. In Sweden and Denmark, one-third to one-half of children under three years of age are in some form of full-day, publicly supported day care, along with 72 to 82 percent of children between the ages of three and five (Waldfogel 2001). Children of unemployed parents are not eligible for day care in about 40 percent of locations (Skolverket 2003). Ken Jaffee, director of the International Child Resource Institute in Sweden, says, "Child care is considered a necessity for the economic and social survival of a country, and there is more universal availability of child care. Every neighborhood has a center" (Polk 1997).

In France and Belgium, there is a two-phase system of child care. For younger children, full-day child care centers are provided under the authority of the social welfare system. Beginning at age two and a half to three, children are enrolled in full-day pre-primary programs within the educational system. Enrollment of younger children in child care is about 30 percent in Belgium and 24 percent in France. Nearly 100 percent of all eligible children are enrolled in pre-primary programs (Gornick and Meyers 2002).

Not only are child care services more available to European families, but child care staff are better trained and paid than they are in the United States. Child care workers and educators in European countries are required to have three to five years of vocational or university training. Funding for these extensive child care programs is provided by local, state, or national governments. Care for very young children is partially funded through parental co-payments that are scaled according to family income. Lower-income families generally pay nothing whereas more affluent families pay no more than 10 percent to 15 percent of their income toward child care (Gornick and Meyers 2002).

The United States has never embraced a national system for providing, funding, and regulating early childhood education and care. It is estimated that the U.S. government spends between 25 and 30 percent of the cost of child care for children under the age of three and for children age three to six. Denmark, Finland, France, Norway, and Sweden pay most of the costs of care, about 68 to 100 percent. One study found that the United States spends about $600 per year per preschool-age child, whereas France spends about $3,000 and Sweden spends more than $4,500 (Waldfogel 2001).

Studies that have traced people who leave welfare over longer periods indicate no significant earnings growth after leaving welfare (O'Gorman 2002). An analysis by the National Campaign for Jobs and Income Support (2001) found that leavers have trouble finding employment; lack child care, transportation, nutrition, and health care assistance; and remain concentrated in the low-wage labor market. It appears that individuals and families are cycling in and out of poverty, working when jobs are available but never achieving complete economic independence. Women who have multiple barriers to obtaining and retaining employment will be least likely to achieve economic self-sufficiency. Barriers may include low-level work experiences, less than a high school education, substance dependence, physical health problems, and other barriers such as domestic violence, transportation issues, or child care problems. Critics argue that welfare reform made little impact on the real barriers to self sufficiency: the need for a livable wage, employment opportunities, education or technical training, child care, health insurance, transportation, and affordable housing. Sanctioned families, families that will lose their welfare assistance, are seriously at risk. The parent in a sanctioned family is likely unable to comply with TANF requirements because of mental illness, developmental disability, or substance abuse (Blum and Farnsworth Francis 2002).

Problems are compounded for the rural poor, particularly those with little education, no phone, no car, and no jobs. Consider Tulare County, California's top agricultural county, home to Californians with the lowest education level and the highest rate of poverty. By January 2003, when lifetime welfare limits began to run out, Tulare's welfare families were sliding into deeper poverty. On a given day, Tulare County's job center will post 10 new jobs, whereas more than 6,000 new jobs may be posted in Los Angeles County. Realizing that the county cannot provide enough jobs for the poor, the More Opportunity for Viable Employment (MOVE) program paid to move more than 1,000 welfare recipients out of the county. Most of them moved to the Midwest, Las Vegas, or anywhere else jobs were available. Those who stay survive day by day, selling their automobiles for cash, working for minimum wage at fast food places, or attending job training classes. Kelly Young, a mother of two young girls, works two jobs: one for $7.25 an hour afternoons and weekends; the other a graveyard shift for $25 a night caring for an elderly client. Because she has no car, she needs to catch a ride to her workplaces. There is no local bus service in her rural community (Romney 2003).

Changing the Definition—Redefining the Poverty Line

In 1995, a panel of the National Academy of Sciences (NAS) called for a new poverty measure to include the three basic categories of food, clothing, and shelter (and utilities) and a small amount to cover other needs such as household supplies, child care, personal care, and non-work-related transportation. Because the census measure does not show how taxes, noncash benefits, and work-related child care and medical expenses affect people's well-being, the NAS panel cautioned that the current poverty measure cannot reflect how policy changes in these areas affect the poor. In addition, the measure does not consider how the cost of basic goods (food and shelter) has changed since the 1960s. As we have already discussed, the federal poverty measurement assumes that costs are the same across most of the states, except Hawaii and Alaska. It does not make sense that a family of four in Manhattan, New York, is expected to spend the same amount of money for food, clothing, and shelter as a family in Manhattan, Kansas (Bhargava and Kuriansky 2002).

The U.S. Census Bureau has been calculating experimental measures of poverty since 1999. For 2001, in measuring the overall poverty rate, the experimental measures report higher levels of poverty, especially when accounting for geographic differences in housing costs and for medical out-of-pocket expenses. Although the official rate is 11.7 percent, experimental measures vary from a low of 12.3 percent to 12.9 percent. When looking at the poverty rate for specific groups, the experimental measures tend to present a poverty population that looks more like the total population in terms of its mix of people: the elderly, White non-Hispanic individuals, and Hispanics (Short 2001).

Other alternative measures have been introduced. The Self-Sufficiency Standard developed by Diana Pearce defines the amount of income necessary to meet basic needs without public assistance (public housing, food stamps, etc.) or private/informal subsidies (free baby-sitting by a friend, food provided by churches or local food banks) (Family Economic Self-Sufficiency Project 2001). The Self-Sufficiency Standard determines the level of income necessary to raise a family out of welfare into total independence. The standard considers regional variations in cost and takes into account family size and composition (much like the U.S. federal measure) and the age of children in the household. The Economic Policy Institute offers a basic family budget calculator, a more realistic measure of the income required to support a basic standard of living. The budgets are individualized for 400 U.S. communities and for different family types.

Earned Income Tax Credit

Enacted in 1975, the Earned Income Tax Credit (EITC) program provides federal tax relief for low-income working families, especially those with children. The credit reduces the amount of federal tax owed and usually results in a tax refund for those who qualify. Similar programs are offered in the United Kingdom, Canada, France, and New Zealand. To qualify for the U.S. program, adults must be employed. A single parent with one child who had family income of less than $32,001 (or $34,001 for a married couple with one child) in 2006 could get a credit of as much as $2,662 (Center on Budget and Policy Priorities 2007). The EITC can be claimed for children under age 19 or under age 24 if they are still in college.

Expansions of the program in the late 1980s and early 1990s made the credit more generous for families with two or more children. In 1994, a small credit was made available to low-income families without children (Freidman 2000). Receipt of EITC credit does not affect receipt of other programs such as food stamp benefits, Medicaid, or housing subsidies.

Supporters of EITC argue that the program strengthens family self-sufficiency, provides families with more disposable income, and encourages work among welfare recipients. Families can use their credit to reduce debt, purchase a car, or pay for education (Freidman 2000). Almost half of EITC recipients planned to save all or part of their refund (Smeeding, Ross, and O'Conner 1999). The program is credited with lifting more children out of poverty than any other government program (Llobrera and Zahradnik 2004).

Nineteen states offer a state-level earned income credit for residents, usually a percentage of the federal credit. For more information, log on to *Study Site Chapter 2*.

Main Points

- **Absolute poverty** refers to a lack of basic necessities, such as food, shelter, and income. **Relative poverty** refers to a situation where people fail to achieve the average income or lifestyle enjoyed by the rest of society. Relative poverty emphasizes the inequality of income and the growing gap between the richest and poorest Americans.

- The **poverty threshold** is the original federal poverty measure developed by the Social Security Administration and updated each year by the U.S. Census Bureau. The threshold is used for estimating the number of people in poverty. **Poverty guidelines** are issued each year by the U.S. Department of Health and Human Services and are used for determining family or individual eligibility for federal programs. In 2006, the poverty rate was 12.3 percent or 36.5 million compared with the most recent low poverty rate of 11.3 percent or 31.6 million in 2000.

- Whites currently compose the largest group of poor individuals in the United States. However, the poverty rate for non-Hispanic Whites is the lowest, and Blacks continue to have the highest poverty rate, followed by Hispanics.

- Functionalists observe that class inequality is a product of our social structure. Rapid economic and technological changes eliminated the need for low-skilled labor, creating a population of workers who are unskilled and untrained for this new economy. In many ways, theorists from this perspective expect this disparity among workers. Lower wages and poverty are natural consequences of this system of stratification.

- Conflict theorists assert that poverty exists because those in power want to maintain and expand their base of power and interests, with little left to share with others. For Karl Marx, one's social class is solely determined by one's position in the economic system: You are either a worker or an owner of the means of production. According to Max Weber, however, social class is multidimensional. Economic factors include **income**, the money earned for one's work, and **wealth**, the value of one's personal assets such as savings and property. A person's social class is also influenced by **prestige**, the amount of social respect or standing given to an individual based on occupation. A final component of class is power. Weber defined **power** as the ability to achieve one's goals despite the opposition of others. Welfare bureaucracies—local, state, and national—represent important interest groups that influence the creation and implementation of welfare policies. Welfare policies reflect the political economy of the community in which they are implemented.

- Feminist scholars argue that the welfare state is an arena of political struggle. The drive to maintain male dominance and the patriarchal family is assumed to be the principal force shaping the formation, implementation, and outcomes of U.S. welfare policy.

- Interactionists attempt to explain how poverty is a learned phenomenon. Some sociologists have suggested that poverty is based on a "culture of poverty" and that moral deficiencies lead to a life of poverty. Opponents argue that there is no evidence that the poor have a different set of values and beliefs. Interactionists also focus on the public's perceptions: Most Americans do not have any direct contact with the welfare system or welfare recipients; thus, their views are likely to be shaped by the media.

- A new era of social welfare began in 1996 with the Personal Responsibility and Work Opportunity Reconciliation Act (PRWORA). PRWORA was a bipartisan welfare reform plan to reduce recipients' dependence on government assistance through strict work requirements and welfare time limits. The act had an immediate effect on the number of people on welfare. However, although employment has increased among welfare recipients, many recipients have little education or work experience, limited employment benefits (including health care), and continue to struggle with child care.

- Eleven percent of households were food insecure (insufficient food for all family members to enjoy active and healthy lives) at least for some time during the year. The U.S. Department of Agriculture provides food assistance programs, but families often need to rely on private programs (food pantries and soup kitchens) as well. For a variety of reasons, poor families encounter higher food prices and a smaller selection of food than other families. Housing is another problem; the combination of low earnings and scarce housing assistance results in serious housing problems for the working poor.

- Overwhelming evidence indicates that welfare caseloads have declined since the enactment of PRWORA, perhaps helped by a strong economy and increased aid to low-income working families. However, the early employment success of welfare reform diminished as the economy faltered, and time limits to welfare receipt seem to compound problems. It appears that individuals and families, particularly women and the rural poor, are cycling in and out of poverty, working when jobs are available but never achieving complete economic independence.

On Your Own

Log on to the Web-based student study site at www.pineforge.com/leonguerrero2study for interactive quizzes, e-flashcards, journal articles, Community and Policy Guides, a Service Learning Guide, the end-of-chapter Web exercises, and additional Web resources.

Internet and Community Exercises

1. For more information about New Orleans post Hurricane Katrina, visit the Web site for the Greater New Orleans Community Data Center (www.gnocdc.org). The Brookings Institution, in collaboration with the Center, tracks recovery in the area. The Web site features impact and needs assessment reports, as well as updates, regarding social and human services for New Orleans' residents.

2. Investigate the welfare assistance or TANF program in your state. First, determine the name and administrative agency for the program in your state. Most TANF programs are administered by the Department of Health and Social Services or Department of Human Services. Determine what time limits and work requirement provisions have been legislated in your state. What educational activities can count toward work requirements? Are family support services (parent skill training, housing assistance) provided? What is your state's record on welfare reform? Has welfare reform made a difference for the poor in your state?

3. What is the median income for your state? The median income is the exact point where 50 percent of all incomes are above and 50 percent of all incomes are below. Log on to *Study Site Chapter 2* to access the U.S. Census Bureau Web site.

4. The United Nation's *Human Development Report* collects available data on poverty rates, defined in several ways, living below $1, $2, $4, and $11 a day. In addition, the United Nations monitors the percent of the nation's population that lives below the poverty line. Data are available primarily for lesser developed nations. Log on to *Study Site Chapter 2* to identify and compare poverty rates for five selected nations.

CHAPTER 3

Race and Ethnicity

On a soccer field in Clarkston, Georgia, you may find the Fugees, a boys' soccer program at play. Though most may think they selected their team name to honor the hip-hop band, Fugees is actually short for "refugees." These 9- to 17-year-old players are refugees from 18 different countries—Shahir Anwar, an Afghan teen whose parents fled the Taliban; Santino Jerke, a Sudanese boy who arrived after three years as a refugee in Cairo; and Mohammed Mohammed, an Iraqi Kurd whose family fled Iraq five years ago (St. John 2007). They are among the more than 1.1 million refugees admitted to the United States since 1990.

The United States is a diverse racial and ethnic society. Joe Feagin and Pinar Batur (2004) report that by the 2050s the majority of the U.S. population will comprise African, Latino, Asian, Middle Eastern, and Native Americans. Currently, White Americans are a minority in half of the largest 100 U.S. cities and in Hawaii, New Mexico, Texas, and California. Between 2015 and 2040, White Americans are expected to become a statistical minority in many other states. The U.S. Census estimates that in 2003 about 12 percent of the U.S. population or about 33.5 million individuals were foreign born (Larsen 2004).

Adding to the diversity of our population are increasing numbers of immigrants, their migration to the United States and throughout the world spurred on by the global economy. Population mobility since the middle of the twentieth century has been characterized by unprecedented volume, speed, and geographical range (Collin and Lee 2003). At the end of 2005, nearly 200 million people or about 3 percent of the world's population lived in a country other than their birth country (DeParle 2007). The International Organization for Migration (2003) predicts that by 2030, this figure should increase to 230 million. As Zygmunt Bauman (2000) describes, "The world is on the move."

In this chapter, we will explore how one's racial and ethnic status serves as a basis of inequality. Like social class, depending on one's race or ethnicity, a person's life chances are altered and the likelihood of experiencing particular social problems increases. We begin first with understanding how race and ethnicity are defined.

Defining Race and Ethnicity

From a biological perspective, a **race** can be defined as a group or population that shares a set of genetic characteristics and physical features. The term has been applied broadly to groups with similar physical features (the White race), religion (the Jewish race), or the entire human species (the human race) (Marger 2002). However, generations of migration, intermarriage, and adaptations to different physical environments have produced a mixture of races. There is no such thing as a "pure" race.

Social scientists reject the biological notions of race, instead favoring an approach that treats race as a social construct. In *Racial Formations in the United States: From the 1960s to the 1980s,* Michael Omi and Howard Winant explain how race is a "concept which signifies and symbolizes social conflicts and interests by referring to different types of human bodies" (1994: :54). Instead of thinking of race as something "objective," the authors argue that we can imagine race as an "illusion," a subjective social, political, and cultural construct. According to the authors, "The meaning of race is defined and contested throughout society, in both collective action and personal practice. In the process, racial categories themselves are formed, transformed, destroyed, and reformed" (p. 21). Robert Redfield says it simply, "Race is, so to speak, a human invention" (1958:67).

Race may be a social construction, but that does not make race any less powerful and controlling (Myers 2005). Omi and Winant argue that although particular stereotypes and meanings can change, "The presence of a system of racial meaning and stereotypes, of racial ideology, seems to be a permanent feature of U.S. culture" (1994:63).

Ethnic groups are groups that are set off to some degree from other groups by displaying a unique set of cultural traits, such as their language, religion, or diet. Members of an ethnic group perceive themselves as members of an ethnic community, sharing common historical roots and experiences. All of us, to one extent or another, have an ethnic identity. Increasingly the terms race and ethnicity are presented as a single construct pointing to how both terms are being conflated (Budrys 2003).

Martin Marger (2002) explains how **ethnicity** serves as a basis of social ranking, ranking a person according to the status of his or her ethnic group. He states that although class and ethnicity are separate dimensions of stratification, they are closely related: "In virtually all multiethnic societies, people's ethnic classification becomes an important factor in the distribution of societal rewards and hence, their economic and political class positions.... The ethnic and class hierarchies are largely parallel and interwoven" (2002:286).

Photo 3.1

In an interview with Oprah Winfrey, golf champion Tiger Woods said that he doesn't like being called African American. He explained that as he was growing up, he created the term *Cablinasian* to describe his ethnicity. The word combines his Caucasian-Black-Indian-Asian ethnic backgrounds.

Table 3.1

Race and Hispanic Origin, United States, 2004 (numbers in thousands)

	Total	Percent
White	239,880	81.7
Black or African American	39,232	13.4
American Indian and Alaska Native	4,409	1.5
Asian	13,956	4.8
Native Hawaiian and Other Pacific Islander	976	0.3
Hispanic or Latino (any race)	41,322	14.1
White alone, not Hispanic or Latino	197,840	67.4

Source: U.S. Census Bureau 2005.

As of 2002, Hispanic Americans were the nation's largest ethnic minority group. (The U.S. Census treats Hispanic origin and race as separate and distinct concepts; as a result, Hispanics may be of any race.) The U.S. Census Bureau includes in this category women and men who are Mexican, Central and South American, Puerto Rican, Cuban, and other Hispanic. The growth in the number of Hispanic Americans has been attributed to increased international immigration and higher birthrates. The states with the highest proportion of Hispanics include New Mexico, California, Texas, and Arizona. The 2002 ethnic and racial composition of the United States is presented in Table 3.1.[1]

Most U.S. families have an immigration history whether it is based upon stories of relatives as long as four generations ago or as recent as the current generation. **Immigration** involves leaving one's country of origin to move to another. Though immigration has always been a part of U.S. history, the recent wave of immigration, particularly in the end of the twentieth century and the beginning of the twenty-first, has led to the observation that we are in the "age of migration" (Castles and Miller 1998).

Most immigrants are motivated by the global economics of immigration—men and women will move from low-wage to high-wage countries in search of better incomes and standards of living. **Labor migration**, the movement from one country to another for employment, has been a part of U.S. history beginning with Chinese male workers brought to build railroads in the 1800s. These men never brought their families or had any intention of staying after their work was completed. In their analysis of current migration trends, Gary Hytrek and Kristine Zentgraf (2007) note how an increasing number of highly skilled laborers are moving from less developed areas around the world to the United States and Europe. These migrants are more likely to return to their place of birth or move on to a third country. Migration tends to occur between geographically proximate countries—Turkey and North African migration to Western Europe and Mexican and Central American migration to the United States.

The U.S. Census distinguishes between native and foreign-born residents. A native refers to anyone born in the United States or a U.S. Island Area such as Puerto Rico or the Northern Marianas Islands or born abroad of a U.S. citizen parent; foreign born refers to anyone who is not a U.S. citizen at birth. In 2003, among the 33.5 million foreign born in the United States, most were from Latin America (53.3 percent), then Asia (25 percent), Europe (13.7 percent), and other regions in the world (8 percent) (Larsen 2004). Refer to Figure 3.1.

Refugees, like the players on the Fugees soccer team, are defined by the Immigration and Nationality Act of 1980 as "aliens outside the United States who are unable or unwilling to return to his/her country of origin for persecution or fear of persecution on account of race,

Figure 3.1

U.S. Foreign Born by World Region of Birth, 2003

Source: Larsen 2004.

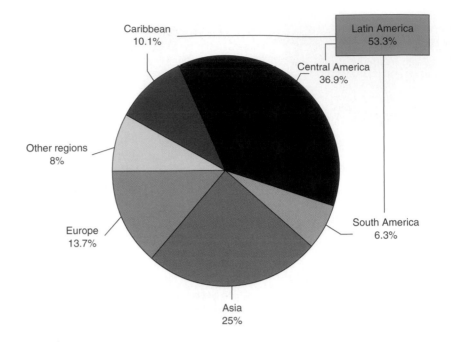

Caribbean 10.1%

Latin America 53.3%

Central America 36.9%

Other regions 8%

South America 6.3%

Europe 13.7%

Asia 25%

religion, nationality, membership in a particular social group, or political opinion." Data on the number of admitted refugees are collected annually by the U.S. Department of State. In 2005, 53,183 persons were admitted as refugees. Almost 60 percent of all admitted refugees were from just four countries: Somalia (19 percent), Laos (16 percent), Cuba (12 percent), and Russia (11 percent). The majority of admitted refugees were less than 25 years of age (55 percent), male (51 percent), and single (58 percent) (Jeffreys 2006).

What Does It Mean to Me?

You may not be able to tell from my last name (Leon-Guerrero), but I consider my ethnic identity to be Japanese. My middle name is Yuri, a Japanese name that means "Lily." I am Japanese not only because of my middle name or because of my Japanese mother, but also because of the Japanese traditions that I practice, the Japanese words that I use, and even the Japanese foods that I like to eat. Do you have an ethnic identity? If you do, how do you maintain it?

Patterns of Racial and Ethnic Integration

Sociologists explain that **ethnocentrism** is the belief that one's own group values and behaviors are right and even better than all others. Feeling positive about one's group is important for group solidarity and loyalty. However, it can lead groups and individuals to believe that certain racial or ethnic groups are inferior and that discriminatory practices against them are justified. This is called **racism**.

Though not all inequality can be attributed to racism, our nation's history reveals how particular groups have been singled out and subject to unfair treatment. Certain groups have been subject to **individual discrimination** and **institutional discrimination**. Individual discrimination

includes actions against minority members by individuals. Actions may range from avoiding contact with minority group members to physical or verbal attacks against minority group members. Institutional discrimination is practiced by the government, social institutions, and organizations. Institutional discrimination may include segregation, exclusion, or expulsion.

Segregation refers to the physical and social separation of ethnic or racial groups. Although we consider explicit segregation to be illegal and a thing of the past, ethnic and racial segregation still occurs in neighborhoods, schools, and personal relationships. According to Debra Van Ausdale and Feagin,

> Racial discrimination and segregation are still central organizing factors in contemporary U.S. society.... For the most part, Whites and Blacks do not live in the same neighborhoods, attend the same schools at all educational levels, enter into close friendships or other intimate relationships with one another, or share comparable opinions on a wide variety of political matters. The same is true, though sometimes to a lesser extent, for Whites and other Americans of color, such as most Latino, Native and Asian American groups. Despite progress since the 1960s, U.S. society remains intensely segregated across color lines. Generally speaking, Whites and people of color do not occupy the same social space or social status. (2001:29)

Exclusion refers to the practice of prohibiting or restricting the entry or participation of groups in society. In March 1882, U.S. Congressman Edward K. Valentine declared, "The [immigration] gate must be closed." That year, Valentine, along with other congressional leaders, approved the Chinese Exclusion Act. From 1882 to 1943, the United States prohibited Chinese immigration because of concerns that Chinese laborers would compete with American workers. Through the 1940s, immigration was defined as a hindrance rather than a benefit to the United States.

In 2005, the largest group of refugees in the United States was from Somalia. A male Somalian social worker is leading a Somalian family (not pictured) on a tour of a supermarket in Phoenix, Arizona.

Finally, **expulsion** is the removal of a group by direct force or intimidation. In 2006, journalist Eliot Jaspin documented the extent of racial expulsion that occurred in towns from Central Texas through Georgia. After the Civil War through the 1920s, White residents expelled nearly all Black persons from their communities, usually using direct physical force. Thirteen countywide expulsions were documented in eight states between 1864 and 1923 in which 4,000 Blacks were driven out of their communities.

What Does It Mean to Me?

After the 2007 NCAA Women's Basketball finals, radio personality Don Imus made a racial and sexist comment regarding the members of the Rutgers University Women's basketball team on his morning radio program. The statement ignited a national debate over racist and misogynistic language and lyrics. Imus was fired from his long-time position with NBC. Poll data from the Pew Research Center for the People and the Press revealed most White and Black Americans supported his termination, though others believed that his dismissal did not fit his "crime." Should Imus have lost his job for his remark?

Taking a World View

Caste Discrimination in India

It is estimated that more than 250 million people worldwide suffer from caste segregation. A caste system segregates or excludes individuals on the basis of their descent. Caste communities exist in Asia and parts of Africa and include groups such as the Dalits (or untouchables) of Nepal, Bangladesh, India, and Sri Lanka; the Buraku people of Japan; and the Osu of Nigeria. Despite formal legislation to abolish or combat abuses based on caste systems, the Human Rights Watch (2001) reports, "Discriminatory treatment remains endemic and discriminatory societal norms continue to be reinforced by government and private structures and practices." The inequalities suffered by these groups affect every level of their lives—social, physical, economic, and political.

The world's longest surviving social hierarchy is India's caste system. The caste is justified by the religious doctrine of "karma," a belief that an individual's place in life is determined by deeds in a previous life. The Dalits are the lowest members of the caste system, representing one-sixth of India's population or approximately 160 to 240 million people. The status of the Dalits is reinforced by state allocation of resources and facilities—separate facilities are provided for different castes. For example, electricity,

running water, or sanitation facilities may be installed for the upper caste section of town, but not for the Dalits and other lower castes. Dalits are prohibited from crossing from their side of the village. They cannot use the same water wells or visit the same temples as members of upper castes.

Children face discriminatory and abusive treatment at school by their teachers and their fellow students. Dalit children are often forced to sit in the back of the classroom. Most of the Indian schools with Dalit students are deficient in classrooms, teachers, and teaching aids. Dalit children also have the highest drop-out rates among Indian children—49.3 percent at the primary level, 67.8 percent for middle school, and 77.6 percent for secondary school.

Dalits are prohibited from performing marriage or funeral rites in public areas or from speaking directly to members of upper castes. Intermarriage among castes is still condemned in India, punishable by social ostracism or violence. The Dalits are usually employed as the removers of human waste and dead animals, leather workers, street cleaners, and cobblers. Dalit children are often sold into slavery to pay off debts to creditors. About 15 million children are working as slaves in order to pay off family debts.

In Focus

Japanese American Internment Camps

Asian Americans are often characterized as the "model minority," which focuses on their socio-economic achievement. Native-born Asian Americans as a group have achieved the same or better educations, occupations, and income levels as White Americans. Yet, social scientists observe how this image of success ignores Asian Americans' history and experience of discrimination, in the United States. Part of their history is the internment of Japanese Americans between 1942 and 1945.

In August 1941, U.S. Representative John Dingell (Michigan) wrote to President Franklin D. Roosevelt, suggesting that 10,000 Hawaiian Japanese be incarcerated to ensure "good behavior" by Japan. Roosevelt did not act on Dingell's suggestion. But two months later when Japan attacked Pearl Harbor in December 1941, Roosevelt signed Executive Order 9066, which allowed the Secretary of War to "prescribe military areas … from which any of all persons may be excluded." Though no single ethnic group was identified, the order targeted Japanese Americans, most residing in the Western states. War department officials argued that Japanese Americans could not be trusted to serve in the Pacific theater because it was difficult to separate "the sheep from the goats" (Asahina 2006).

Photo 3.3

From 1941 to 1945, about 120,000 Japanese Americans were relocated to internment camps. Like the Mochida family, most were instructed to pack their belongings in a single suitcase. The camps were closed at the end of World War II.

Scholars and historians have identified racism as the root explanation for Roosevelt's Executive Order and for the lack of public outcry against the action. Cheryl Greenberg (1995) explains how anti-Asian prejudice pre-dated Pearl Harbor, evidenced by immigration and citizenship restrictions, prohibition in several western states against property ownership or practicing certain trades, employment discrimination, and residential segregation. That there were no similar war time incarceration programs for German or Italian Americans suggests that Japanese Americans were singled out based on racist decisions.

For the next four years, 120,000 Japanese Americans were relocated to several sites—undeveloped federal reclamation projects (Tule Lake, Minidoka, and Heart Mountain), on land meant for subsistence homesteads (Rowher and Jerome), on Indian reservations (Poston and Gila River), or unused public city or county land (Manzanar and Topaz). Men, women, and children were given short notice to pack all their personal items in a single suitcase. It is estimated that as a group they left behind about $200,000,000 worth of real, personal, and commercial property (about $6.4 billion in current dollars) (Asahina 2006).

On February 1, 1942, Roosevelt authorized the formation of a regimental combat team made up Japanese American volunteers. The 442nd Regimental Combat Team would become the most decorated unit in World War II history for its size and length of service. In less than two years, the 442nd participated in seven major campaigns in Italy and France, received seven Presidential Distinguished Unit Citations, and was awarded 18,142 individual decorations. Approximately 22,500 Japanese Americans served in the Army during World War II in the 442nd, the 100th Battalion, and the 1399th Engineers Construction Battalion (Asahina 2006).

At the end of World War II, the internment camps began to shut down. Japanese Americans were instructed to return home, but the reality was that they had no homes to return to. Most were poor and homeless.

In 1948, President Truman signed the Japanese American Evacuation Claims Act, to compensate Japanese Americans for certain economic losses attributable to their forced evacuation. President Gerald Ford finally rescinded Executive Order 9066 (the original order that led to their internment) in 1976. In 1988, President Ronald Reagan signed a presidential apology to those interned. Reagan also signed HR 442 into law, providing for individual payments of $20,000 to each surviving internee and a $1.25 billion education fund among other provisions. Bruce Ackerman (2004:113) observes, "It took almost half a century before the Japanese-American victims of war time concentration camps gained financial compensation, and then only by a Special Act of Congress that awarded incredibly tiny sums."

In 2006, President George W. Bush signed a law providing federal funding to restore and preserve 10 internment camps for education and to serve as a memorial. U.S. Representative Doris Matsui (D-CA), who was born in the Poston, Arizona, camp, said at the law's signing, "Preserving these internment sites is a solemn task we all bear. Those who come after us will have a physical reminder of what they will never allow to happen again."

Sociological Perspectives on Inequalities Based on Race and Ethnicity

Functionalist Perspective

Theorists from this perspective believe that the differences between racial and ethnic groups are largely cultural. The solution is **assimilation**, a process where minority group members become part of the dominant group, losing their original group identity. This process is consistent with America's image as the "melting pot." Milton Gordon (1964) presents a seven-stage assimilation model that begins first with cultural assimilation (change of cultural patterns, e.g., learning the English language), followed by structural assimilation (interaction with members of the dominant group), marital assimilation (intermarriage), identification assimilation (developing a sense of national identity, e.g., identifying as an American, rather than as an Asian American), attitude receptional assimilation (absence of prejudiced thoughts among dominant and minority group members), behavioral receptional assimilation (absence of discrimination, e.g., lower wages for minorities would not exist), and finally civic assimilation (absence of value and power conflicts).

Assimilation is said to allow a society to maintain its equilibrium (a goal of the functionalist perspective) if all members of society, regardless of their racial or ethnic identity, adopt one dominant culture. This is often characterized as a voluntary process. Critics argue that this perspective assumes that social integration is a shared goal and that members of the minority group are willing to assume the dominant group's identity and culture, assuming that the dominant culture is the one and only preferred culture (Myers 2005). The perspective also assumes that assimilation is the same experience for all ethnic groups, ignoring the historical legacy of slavery and racial discrimination in our society.

Assimilation is not the only means to achieve racial-ethnic stability. Other countries maintain **pluralism**, where each ethnic or racial group maintains its own culture (cultural pluralism) or a separate set of social structures and institutions (structural pluralism). Switzerland, which has a number of different nationalities and religions, is an example of a pluralistic society. The country, also referred to as the Swiss Confederation, has four official languages—German, French, Italian, and Romansh. Relationships between each ethnic group are described for the most part as harmonious because each of the ethnically diverse parts joined the confederation voluntarily seeking protection (Farley 2005). In his examination of pluralism in the United States, Min Zhou notes, "As America becomes increasingly multiethnic, and as ethnic Americans become integral in our society, it becomes more and more evident that there is no contradiction between an ethnic identity and an American identity" (2004:153).

Conflict Perspective

According to sociologist W. E. B. Du Bois (1996), perhaps it is wrong "to speak of race at all as a concept, rather than as a group of contradictory forces, facts and tendencies." The problem of the twentieth century, wrote Du Bois, is "the color line."

Conflict theorists focus on how the dynamics of racial and ethnic relations divides groups while maintaining a dominant group. The dominant group may be defined according to racial or ethnic categories, but can also be defined according to social class. Instead of relationships based on consensus (or assimilation), relationships are based on power, force, and coercion. Ethnocentrism and racism maintain the status quo by dividing individuals along racial and ethnic lines (Myers 2005).

Drawing upon Marx's class analysis, Du Bois was one of the first theorists to observe the connection between racism and capitalist-class oppression in the United States and throughout the world. He noted the link between racist ideas and actions to maintain a Eurocentric system of domination (Feagin and Batur 2004). Du Bois wrote,

> Throughout the world today organized groups of men by monopoly of economic and physical power, legal enactment and intellectual training are limiting with great determination and unflagging zeal the development of other groups; and that the concentration particularly on economic power today puts the majority of mankind into a slavery to the rest. (1996:532)

Though most theorists from this perspective see conflict as emanating from one dominant group, conflict may also be mutual. Edna Bonacich (1972) offers a theory of ethnic antagonism, encompassing all levels of mutual intergroup conflict. She argues that this ethnic antagonism emerges from a labor market, split along ethnic and class lines. To be split, the labor market must include at least two groups of workers whose price of labor differs for the same work. Conflict develops between three classes: businesses or employers, higher paid labor, and cheaper labor. Bonacich explains that as businesses attempt to maintain a cheap workforce (not caring about who does the work as long as it gets done), higher paid workers attempt to maintain their prime labor position (resisting the threat of lower wage laborers), and cheaper laborers attempt to advance their position (threatening higher paid workers). Higher paid workers may use exclusionary practices (attempting to prevent the importation of cheaper non-native labor) or caste arrangements (excluding some groups from certain types of work) to maintain their advantage in the labor market. According to Bonacich, the presence of a cheaper labor group threatens the jobs of higher paid workers and the standard for wages in all jobs. Under these conditions, laborers remain in conflict with each other, and the interests of capitalist business owners are maintained.

Feminist Perspective

Feminist theory has attempted to account for and focus on the experiences of women and other marginalized groups in society. Feminist theory intersects with multiculturalism through the analysis of multiple systems of oppression, not just gender, but including categories of race, class, sexual orientation, nation of origin, language, culture, and ethnicity. Emerging from this is Patricia Hill Collins' Black feminist theory. Black feminists identify the value of a theoretical perspective that addresses the simultaneity of race, class, and gender oppression.

Black feminist scholars note that the misguided application of traditional feminist perspectives of "the family," "patriarchy," and "reproduction" to understand the experience of Black women's lives. Black women do not lead parallel lives, but rather different lives. British scholar Hazel Carby (1985:390) argues that because Black women are subject to simultaneous oppression based on class, race, and patriarchy, the application of traditional (White) feminist perspectives are not appropriate and are actually misleading to comprehend their true experience. She argues that White feminist theory has to recognize "White women stand in a power relation as oppressors of Black women" (p. 390).

As an example, Carby analyzes an article on women in Third World manufacturing. Carby highlights how the photographs accompanying the article are of "anonymous Black women." She observes, "This anonymity and the tendency to generalize into meaningless, the oppression of an amorphous category called 'Third World Women,' are symptomatic of the ways in which the specificity of our experiences and oppression are subsumed under inapplicable concepts and theories" (p. 394).

Interactionist Perspective

Sociologists believe that race is a social construct. We learn about racial and ethnic categories of White, Black, Latino, Asian, Native American, and immigrant through our social interaction. The meaning and values for these and other categories are provided by our social institutions, families, and friends (Ore 2003). As much as I and other social scientists inform our students about the unsubstantiated use of the term race, for most students, race is real. The term is loaded with social, cultural, and political baggage, making deconstructing it difficult to accomplish.

Social scientists have noted how people are raced, how race itself is not a category but a practice. Howard McGary (1999:83) defines practice as "a commonly accepted course of action that may be over time habitual in nature; a course of action that specifies certain forms of behavior as permissible and others impermissible, with rewards and penalties assigned accordingly." In this way, racial categories and identities serve as intersections of social beliefs, perceptions, and activities that are reinforced by enduring systems of rewards and penalties (Shuford 2001).

Table 3.2

Summary of Sociological Perspectives: Inequalities Based on Race and Ethnicity

	Functionalist	Conflict/Feminist	Interactionist
Explanations of racial and ethnic inequality	Assimilation into a dominant culture preserves the stability of society. Ethnic pluralism may also achieve stability.	Inequality is systematically maintained by those trying to preserve their advantaged positions. Class divisions overlap with racial and ethnic divisions. Feminist scholars advocate for a theoretical perspective that simultaneously considers the intersection of race, class, and gender.	Race is a social construct. Racial and ethnic categories are also linked with positions of privilege or marginalization.
Questions asked about racial and ethnic inequality	How can minority groups be assimilated into mainstream society? Can cultural and structural pluralism be maintained?	What powerful interest groups determine racial and ethnic inequalities? How are these structures maintained?	How do we learn about race and ethnicity? How are some groups more privileged than others? How do our perceptions and beliefs perpetuate racial and ethnic inequalities?

The practice of being raced includes with it the bestowing of power and privilege and what is granted to one group may be denied to another. For example, Madonna and Angelina Jolie were honored (in some circles) for their adoption of children from Cambodia, Ethiopia, and Malawi, spotlighting their adoptions as examples of international goodwill and charity. Yet, as Matthew Jacobsen (1998) asks, why can White women have Black children but Black women cannot adopt White children? The interactionist perspective reminds us that racial designations may be fictitious, but their consequences are real.

A summary of all theoretical perspectives is provided in Table 3.2.

The Consequences of Racial and Ethnic Inequalities

Income and Wealth

"Race is so associated with class in the United States that it might not be direct discrimination, but it still matters indirectly," says sociologist Dalton Conley (Ohlemacher 2006:6). Data reported by the U.S. Census reveal that Black households had the lowest median income in the 2005, $30,858, which was 61 percent of the median income for non-Hispanic White households, $50,784. The median income for Hispanic households was $35,967, 71 percent of the median for non-Hispanic White households. American Indians and Alaska Natives had a median income of $33,627 (based on a three-year estimate). Asian households had the highest median income, $61,094, 120 percent of the median for non-Hispanic White households (DeNavas-Walt, Proctor, and Lee 2006).

The U.S. Census presents income data in quintiles, dividing the population in fifths to determine the distribution of income. Households in the lowest quintile earned $19,178 or less in 2005, whereas households in the highest quintile earned $91,705 or more. Researchers calculate the Gini Index, an inequality measure that determines what percent of the total aggregate income earned by households that year is earned by each quintile. Their analysis revealed that non-Hispanic White households represented a larger proportion of households in the highest income quintile (81.2 percent) than in the lowest income quintile (61.4 percent). Despite their high median income, Asians represented only 5.8 percent of high-income quintile households, but only 3 percent of the lowest income quintile. Black and Hispanic households each represented a larger proportion of households in the lowest income quintile (20.6 and 13.4 percent, respectively) than in the highest quintile group (5.8 and 5.9 percent, respectively) (DeNavas-Walt et al. 2006).

Because of years of discrimination, low educational attainment, high unemployment, or underemployment, African Americans have not been able to achieve the same earnings or level of wealth as White Americans have. Studies indicate that for every dollar earned by White households, Black households earned 62 cents (Oliver and Shapiro 1995). Blacks have between $8 and $19 of wealth for every $100 possessed by Whites. Whites have nearly 12 times as much median net worth as Blacks, $43,000 compared with $3,700 (Oliver and Shapiro 1995).

One measure of wealth is home ownership. Home ownership is one of the primary means to accumulate wealth (Williams, Nesiba, and McConnell 2005). It enables families

Photo 3.4

Interactionists examine the privilege bestowed on one group above another. For example, why do we accept White women adopting children of color, but not women of color adopting White children? Angelina Jolie has adopted three children of color. Pictured here, with Jolie and Brad Pitt, are daughter Zahara from Ethiopia and son Maddox from Cambodia. A second son, named Pax, was adopted from Vietnam in 2007.

to finance college and invest in one's future. Historically, home ownership grew among White middle-class families after World War II, when veterans had access to government and credit programs making home ownership more affordable. However, Blacks and other minority groups have been denied similar access because of structural barriers such as discrimination, low income, and lack of credit access. Feagin (1999) identifies how inequality in homeownership has contributed to inequality in other aspects of American life. Specifically, Blacks have been disadvantaged because of their lack of homeownership, particularly in their inability to provide their children with "the kind of education or other cultural advantages necessary for their children to compete equally or fairly with Whites" (p. 86).

In 2004, U.S. homeownership reached a record high of 69.2 percent with nearly 73.4 million Americans owning their own homes. However, racial gaps in homeownership have increased in the past 25 years. In 2005, 75 percent of White households owned their own homes, compared with 46 percent of Black households and 48 percent of Hispanics.

Education

The U.S. Supreme Court's 1954 decision in *Brown v. Board of Education* ruled that racial segregation in public schools was illegal. Reaction to the ruling was mixed, with a strong response from the South. A major confrontation occurred in Arkansas, when Governor Orval

Faubus used to state's National Guard to block the admission of nine Black students into Little Rock Central High School. The students persisted and successfully gained entry into the school the next day with 1,000 U.S. Army paratroopers at their side. The Little Rock incident has been identified as a catalyst for school integration throughout the South. Despite resistance to the court's ruling, legally segregated education disappeared by the mid-1970s.

However, a different type of segregation persists, called **de facto segregation**. *De facto segregation* refers to a subtler process of segregation that is the result of other processes, such as housing segregation rather than because of an official policy (Farley 2005). Here, we clearly see the intersection of race and class. Schools have become economically segregated, with children of middle- or upper-class families attending predominantly White suburban schools and the children of poorer parents attending racially mixed urban schools (Gagné and Tewksbury 2003). Researchers, teachers, and policymakers have all observed a great disparity in the quality of education students receive in the United States (for more on social problems related to education, turn to Chapter 8). Educational systems reinforce patterns of social class inequality and, along with it, racial inequality (Farley 2005).

Latinos have the lowest educational achievement rates—for high school and college degrees—compared with all other major racial and ethnic groups in the United States. (Refer to Table 3.3.) Richard Fry (2004) explains that though more than 300,000 young Hispanics will graduate from high school each year, fewer than 60,000 will complete a bachelor's degree. Data indicate that about half of young Latinos who enroll in college are at least minimally prepared academically to succeed in a four-year college. Even among the best prepared Latino students, only 57 percent finish a bachelor's degree compared with 81 percent of their White counterparts. Hispanic undergraduates disproportionately enroll in "open-door" institutions that have lower degree completion rates (Fry 2004).

Much of the research on the achievement gap between Latinos and White students has focused on the characteristics of the students (family income, parents' level of education). However, according to the Pew Hispanic Center (Fry 2005), we need to also consider the social context of Hispanic students' learning, noting how educators and policymakers have more influence over the characteristics of their schools than over the characteristics of students.

Based on their state and national assessment of the basic characteristics of public high schools for Hispanic and other students, the Pew Hispanic Center found that Latinos were

Table 3.3

Educational Attainment by Race and Ethnicity (percentages reported), 2005

	Less Than 9th Grade	9th to 12th Grade	High School Graduate	Some College	College Graduate
Hispanic	24.0	16.5	27.0	20.2	12.3
Native born	9.8	15.0	30.6	28.8	15.8
Foreign born	34.4	17.6	24.4	13.9	9.8
White alone	3.2	7.8	30.3	28.7	30.0
Black alone	5.6	14.3	33.2	29.6	17.3
Asian alone	8.1	6.1	17.0	19.6	49.3
Other, not Hispanic	5.9	11.4	29.0	32.6	21.1

Source: Pew Hispanic Center 2006.

Note: Columns may not add to 100 percent because of rounding.

more likely than Whites or Blacks to attend the largest public high schools (enrollment of at least 1,838 students). More than 56 percent of Hispanics attend large schools, compared with 32 percent of Blacks and 26 percent of Whites. Schools with larger enrollments are associated with lower student achievement and higher drop-out rates. In addition, the center reported that Hispanics are more likely to be in high schools with lower instructional resources, which includes higher student-to-teacher ratios associated with lower academic performance. Nearly 37 percent of Hispanics are educated in public high schools with a student-teacher ratio greater than 22 to 1, compared with 14 percent of Blacks and 13 percent of White students (Fry 2005).

In June 2007, the U.S. Supreme Court, voting 5 to 4, invalidated the use of race to assign students to public schools, even if the goal was to achieve racial integration of its schools. The ruling addressed public school practices in Seattle, Washington (where 41 percent of all public school students are White), and Louisville, Kentucky (where two-thirds of all public school students are White). Legal experts and educators were divided about whether the ruling affirmed or betrayed *Brown v. Board of Education*. Though he voted with the majority, Justice Anthony M. Kennedy said in a separate statement that achieving racial diversity and addressing the problem of de facto segregation were issues that school districts could constitutionally pursue as long as the programs were "sufficiently 'narrowly tailored'" (Greenhouse 2007:A1). It is unclear how the ruling will affect integration strategies adopted in school systems across the country.

Health

Racial disparities in access to health care and outcomes are pervasive according to Sara Rosenbaum and Joel Teitelbaum (2004). The issue is twofold—access to health care and the quality of care received once in the system. First, the researchers point to this nation's approach to health insurance as a system that "significantly discriminates against racial and ethnic minorities" (p. 138). Data reveal how in a voluntary, employment-based health care system, racial and ethnic minority group members are more likely to be uninsured or publicly insured. In 2006, White non-Hispanics had the lowest uninsured rate (11 percent), compared with Blacks (20 percent), Asians (15 percent), and Hispanics (34 percent) (DeNavas-Walt, Proctor, and Smith 2007). These disparities continue into old age—among those 65 years or older, non-Latino White seniors are more likely to have a private or employer-based supplemental health policy in addition to their Medicare coverage, whereas minority seniors are six to seven times more likely to have Medicaid (public assistance) in addition to Medicare.

Second, the researchers observe that even after minority patients enter a particular facility, they are less likely to receive the level of care provided to nonminority patients for the same condition regardless of their insurance status. For example, Latino and African American patients with public insurance do not receive coronary artery bypass surgery at rates comparable to those of White, publicly insured patients. African American patients are also less likely to receive treatment for early stage lung cancer and as a result have a lower five-year survival rate. Medicaid-insured African American and Latino children use less primary care (depending usually on emergency treatment), experience higher rates of hospitalization, and die at significantly higher rates than do White children. Though the U.S. government has invested in community-based primary health centers and programs to address these health care gaps, Rosenbaum and Teitelbaum (2004) conclude that these programs can hardly overcome the immense and inaccessible system of specialized and extended health services.

W. Michael Byrd and Linda Clayton (2002) assert that the health crisis among African Americans and poor populations is fueled by a medical-social culture laden with ideological, intellectual and scientific, and discriminatory race and class problems. They believe that America's health system is predicated on the belief that the poor and "unworthy" of our society do not deserve decent health. Consequently, health professionals, research, and educational systems engage in what they describe as "self serving and elite behavior" that marginalizes and ignores the problems of health care for minority and disadvantaged groups. They caution that our

> failure to address, and eventually resolve, these race- and class-based health policy, structural, medical-social and cultural problems plaguing the American health care system could potentially undermine any possibility of a level playing field in health and health care for African American and other poor populations—eroding at the front end … the very foundations of American democracy. (2002:572–573)

U.S. Immigration

The U.S. Census 2005 American Community Survey revealed that there were 35.2 foreign-born individuals in the United States, composing 12.1 percent of the total population. This is the highest number of foreign born ever recorded in U.S. history. Between January 2000 and March 2005, a record number of 7.9 million immigrants arrived to the United States; nearly half were illegal aliens. Currently, one in eight U.S. residents is foreign born. If the current trend continues, within 10 years, the foreign-born share of the U.S. population will match the high of 14.7 percent reached in 1910 (Camarota 2003).

Our immigrant population is concentrated in five states. California, New York, Texas, Florida, and New Jersey account for 63 percent of the immigrant population and only 35 percent of the native-born population. Camarota (2005), reporting for the Center for Immigration Studies, notes that one of the striking patterns of recent immigration is the lack of diversity among immigrants themselves. Mexico accounts for the majority of immigrants, almost six times the combined total of immigrants from China, Hong Kong, and Taiwan.

The post–2000 wave of immigrants included men and women with lower educational attainment—34 percent have less than a high school education. Camarota observes that their lower educational attainment is related to their economic success. A larger proportion of immigrants than native-borns have low incomes, lack health insurance, and rely on social assistance programs. One in four immigrants and their families live in poverty.

Immigrant labor is concentrated in five occupations—farming (44 percent of all those employed), construction (26 percent), building cleaning and maintenance (34 percent), food preparation (24 percent), and general production (23 percent). Camarota (2005) states that millions of native-born Americans are employed in these same occupations and that there is no truth to the statement that immigrants only do jobs natives don't want; no single occupation comprises entirely immigrant labor.

In 2006, the debate about foreign-born individuals in the United States heated up amid Bush administration proposals for comprehensive immigration reforms. While acknowledging the country's immigration heritage, the administration proposed strengthening security at our southern borders with Mexico and establishing a temporary worker program without the benefit of amnesty. The plan was criticized for creating a class of workers who would never become fully integrated in U.S. society and for a plan that focused specifically on Mexican workers, ignoring all other immigrant groups.

In 2007, President George W. Bush pushed sweeping changes to the nation's immigration laws. The president stressed the importance of a four-pronged plan that included the importance of border security, better enforcement of immigration laws, establishment of a temporary worker program, and resolving the status of the millions of illegal immigrants presently in the United States. In that same year, the U.S. Senate was unable to pass a bipartisan

Photo 3.5

Working under a special harvest visa, these Mexican farm workers weed a tobacco field in North Carolina. It is estimated that 44 percent of all employed immigrants are working in agriculture, followed by building cleaning and maintenance (34 percent).

immigration bill that included increasing border security and enforcement, expanding the temporary worker program, and granting illegal immigrants who arrived before January 1, 2007, immediate work authorization under a "Z" visa that would put them on the path to U.S. citizenship. In response to the failure of the federal government, every state took matters into its own hands, debating similar immigration issues within its own state legislature. In 2007, 41 states adopted immigration laws ranging from extending health care and education to the children of illegal immigrants to curbing illegal immigrants' access to jobs (Preston 2007).

What Does It Mean to Me?

Is immigration a problem in the United States? Who is affected and how? What are the subjective and objective realities of immigration? What are the structural sources of the problem?

Global Immigration as a Social Problem

The United States is not the only country grappling with the issue of immigration. Great Britain, France, Germany, Australia, and other countries have seen an increase in pro- and anti-immigration protests, as well as increased hate crimes acts against immigrants in recent years.

Migration has been elevated to a top international policy concern (Düvell 2005) largely because of the threat of terrorism and the challenge of global politics. Migrants now depart from and arrive in almost every country in the world. During the past 30 years, the proportion of foreign-born residents living in developed countries has generally increased, whereas the proportion has remained stable or decreased in developing countries. Migrant labor has been used at both ends of the labor market—low wage/manual labor to high wage/knowledge-based labor. Though globalization has created wealth, lifting many out of poverty, it still has not narrowed the gap between the rich and the poor (Global Commission on Immigration 2005).

Consider France's immigration situation. One in every four French citizens has a non-French parent or grandparent (Silverman 1992). About 25 percent of France's immigrant population comprises men and women of color from North Africa or sub-Saharan Africa. During fall 2005, France experienced its worst civil unrest in decades. North African rioters targeted schools, hospitals, and cars, prompting authorities to declare a state of emergency and impose curfews. The unrest broke out following the deaths of two young North African men who were electrocuted when they hid from police in an electricity substation in a Paris suburb.

The riots were characterized as France's Katrina, exposing poverty and discrimination experienced by African French. Godoy (2002) reports that an invisible ceiling exists in France's social and economic life that "stops the rise of qualified individuals to middle and high-level executive positions." Discrimination in housing, employment, and education are commonplace for France's immigrants. Many bars and clubs remain closed to African French. As the rioting subsided, Prime Minister Dominique de Villepin acknowledged in his remarks to the French National Assembly that the violence was the result of France's failure to provide hope to thousands of young immigrants. Sweeping social and economic reforms were implemented after the riots.

The slogan "Russia is for Russians" was supported by more than 50 percent of Russians in a recent 2006 poll. In April 2007, a government decree banned immigrants from working as vendors or traders in Russia's public markets. The decree was in response to anti-immigrant rioting in the town of Kondopoga, when rioters set fire to immigrant-owned market stalls. As a result of the ban, only 68 percent of all Russian stalls are operating, leading to a shortage of goods and price increases. The targeted immigrant vendors are primarily Azeris, Uzbeks, Tajiks, and Chinese immigrants. In 2006, 539 attacks were recorded on members of ethnic minorities in Russia, including 54 racially motivated murders (Kramer 2007).

Globalization has intensified the need to coordinate and harmonize government policies. Migration flows are regarded as a threat to national and global stability with some calling for an international migration policy (Düvell 2005). The United States, Canada, Great Britain, France, Germany, Belgium, Italy, Spain, and Japan have increased policy coordination regarding immigration, refugee admissions, and programs to integrate foreigners and their family members already present in each country (Lee 2006).

Photo 3.6

This sign near San Ysidro, California, warns motorists to look out for immigrants crossing the Interstate 5 highway. According to the U.S. Border Patrol, 1,954 people died crossing the U.S.-Mexico border illegally between 1998 and 2004. The leading causes of death were heat stroke, dehydration, and hypothermia; others also die in car accidents or by other accidental causes.

Responding to Racial and Ethnic Inequalities

Encouraging Diversity and Multiculturalism

Accelerated global migration and a resurgence of racial/ethnic conflicts characterized the close of the twentieth century (Witting and Grant-Thompson 1998) and certainly the

In October 2005, violence broke out between North African immigrant youths and French security forces in a Paris suburb. Hundreds rioted for days, responding to the deaths of two teens accidentally electrocuted while fleeing police. In 2007, the deaths of two teens, sons of African immigrants, renewed violent protests throughout France.

beginning of the twenty-first. In an effort to reduce racial/ethnic conflict and to encourage multiculturalism, researchers, educators, political and community leaders, and community members have implemented programs targeting racism and prejudice. Acknowledging that both are complex phenomena with individual, cultural, and structural components, these strategies attempt to address some or most of the components.

Kathleen Korgen, J. Mahon, and Gabe Wang (2003) believe that colleges and universities have the potential to counter the effects of segregated neighborhoods and socialization in primary and secondary schools. Interaction among races thrust together on a college campus provides a unique opportunity for individuals to experience and discuss the aspects of racial/ethnic diversity in their lives, some for the first time (Odell, Korgen, and Wang 2005). Gordon Allport (1954) argues that intergroup contact can have a positive effect in reducing interracial prejudice and increasing tolerance if four conditions are met: (1) cooperative interdependence among the groups, (2) a common goal, (3) equal group status during contact, and (4) support by authority figures.

Increasing numbers of colleges and universities are instituting course requirements that encourage students to examine diversity in the United States and globally. The Association of American Colleges and Universities (2000) reported that 62 percent of schools have a diversity course requirement or were in the process of developing one. This is quite an increase from 1990, when only 15 percent of colleges and universities had such a requirement. Research is emerging on the effectiveness of diversity programming on college and university campuses. In one such study, D. A. Grinde (2001) found that more than 85 percent of University of

Vermont students believed that diversity courses strengthened their understanding of and appreciation of cultural diversity.

What Does It Mean to Me?

Does your college have a diversity requirement? If you have completed the course, do you believe the learning experience changed your diversity beliefs and values? Why or why not?

Educational programs are used most often to promote diversity in public and private workplaces. These programs attempt to eliminate incorrect stereotypes and unfounded prejudices by providing new information to participants (Farley 2005). Diversity training is thought to make managers aware of how their biases affect their actions in the workplace (Kalev, Dobbin, and Kelly 2006). Research indicates that such programs are effective when people are not made to feel defensive over past behavior, but are participating in a learning process of new (versus old) ideas. This has also been found to be effective in diversity simulation and experiential exercises (i.e., role playing) (Farley 2005). These programs are designed to familiarize employees with antidiscrimination laws, to suggest behavioral changes that could address bias, and to increase cultural awareness and cross-cultural communication among employees (Bendick, Egan, and Lothjelm 1998).

Business leaders are motivated to address diversity on principle and because they recognize how their company's productivity and success depend on it (Galagan 1993). General diversity and management programs have been established in companies such as Aetna, Ernst and Young, General Mills, and Hewlett Packard. All programs note the importance of creating an "inclusive" workforce and work environment. In addition to diversity training or sensitivity programs, businesses have successfully implemented diversity management programs, targeting the development and advancement of women and people of color in their organization.

AT&T is an example of a corporation that has attempted to address diversity in its organization and the communities it serves. AT&T supports the Hispanic Association of Communications Employees (HACE) of AT&T, a volunteer employee group that develops educational and community programs. Since 1990, the San Diego chapter of the association has awarded more than $600,000 in scholarships to high school and college students. AT&T San Diego has been a partner in community events and programming, including San Diego's Latino Film Festival, Fiesta Patrias, and World Soccer parties. The San Diego HACE chapter, along with other city chapters (Los Angeles and Dallas), offers Internet services and training for low-income and non-English-speaking communities.

Voices in the Community

Rosa Parks

Most would mark the beginning of the U.S. civil rights movement as December 1, 1955. On that day, a Black seamstress in Montgomery, Alabama, refused to give up her seat on the bus to a White person. Rosa Parks explains,

People always say that I didn't give up my seat because I was tired, but that isn't true. I was not tired physically, or no more tired than I usually was at the end of a working day. I was not old, although some people have an image of me as being old then. I was forty-two. No, the only tired I was, was tired of giving in. (Academy of Achievement 2005)

For her actions, Parks was arrested and fined for violating a city ordinance, thrusting her in the middle of America's civil rights movement.

The bus incident led to the formation of the Montgomery Improvement Association, with a young Martin Luther King Jr. elected as its leader. The association promoted its first non-violent protest, boycotting the city-owned bus company. The boycott lasted 381 days ending in November 1956, when the U.S. Supreme Court struck down the Montgomery ordinance under which Parks had been fined, outlawing racial segregation on public transportation.

When asked about the historical bus boycott, Parks remembers,

As I look back on those days, it's just like a dream. The only thing that bothered me was that we waited so long to make this protest and to let it be known wherever we go that all of us should be free and equal and have all opportunities that others should have. (Academy of Achievement 2005)

Parks remained active in the civil rights movement until her death in 2005. In 1957, she and her husband, Raymond, moved to Detroit, Michigan. Parks worked for U.S. Congressman John Conyers. President Bill Clinton honored Parks with the Presidential Medal of Freedom in 1996. Upon her death in 2005, her body was laid in state in the Capitol rotunda, the first woman to be so honored.

Parks was asked what advice she would give a young person who wants to make a difference. She replied,

The advice I would give any young person is, first of all, to rid themselves of prejudice against other people and to be concerned about what they can do to help others. And of course, to get a good education, and take advantage of the opportunities that they have. In fact, there are more opportunities today than when I was young. And whatever they do, to think positively and be concerned about other people, to think in terms of them being able to not succumb to many of the temptations, especially the use of drugs and substances that will destroy the physical health, as well as mental health. (Academy of Achievement 2005)

Affirmative Action

Since its inception nearly 40 years ago, affirmative action has been a "contentious issue on national, state, and local levels" (Yee 2001:135). Affirmative action is a policy that has attempted to improve minority access to occupational and educational opportunities (Woodhouse 2002). No federal initiatives enforced affirmative action until 1961, when President John Kennedy signed Executive Order 10925. The order created the Committee on Equal Employment Opportunity and forbade employers with federal contracts from discriminating on the basis of race, color, national origin, or religion in their hiring practices. In 1964, President Lyndon Johnson signed into law the Civil Rights Act, which prohibits discrimination based on race, color, religion, or national origin by private employers, agencies, and educational institutions receiving federal funds (Swink 2003).

In June 1965, during a graduation speech at Howard University, President Johnson spoke for the first time about the importance of providing opportunities to minority groups, an important objective of affirmative action. According to Johnson (1965),

You do not take a person who, for years, has been hobbled by chains and liberate him, bring him to the starting line of a race and then say, "You are free to compete with all others" and still justly believe you have been completely fair. Thus it is not enough just to open the gates of opportunity. All our citizens must have the ability to walk through those gates. This is the next and the more profound stage of the battle for civil rights. We seek not just freedom but opportunity.

Employment

In September 1965, Johnson signed Executive Order 11246, which required government contractors to "take affirmative action" toward prospective minority employees in all aspects of hiring and employment. Contractors are required to take specific proactive measures to ensure equality in hiring without regard to race, religion, and national origin. The order also established the Equal Employment Opportunities Commission (EEOC), charged with enforcing and monitoring compliance among federal contractors. In 1967, Johnson amended the order to include discrimination based on gender (Swink 2003). In 1969, President Richard Nixon initiated the Philadelphia Plan, which required federal contractors to develop affirmative action plans by setting minimum levels of minority participation on federal construction projects in Philadelphia and three other cities (Idelson 1995). This was the first order that endorsed the use of specific goals for desegregating the workplace (Kotlowski 1998), but it did not include fixed quotas (Woodhouse 2002). In 1970, the order was extended to all federal contractors (Idelson 1995).

According to Dawn Swink (2003), "While the initial efforts of affirmative action were directed primarily at federal government employment and private industry, affirmative action gradually extended into other areas, including admissions programs in higher education" (2003:214–215). State and local governments followed the lead of the federal government and took formal steps to encourage employers to diversify their workforces.

Opponents of affirmative action believe that such policies encourage preferential treatment for minorities (Woodhouse 2002), giving women and ethnic minorities an unfair advantage over White males (Yee 2001). Affirmative action, say its critics, promotes "reverse discrimination," the hiring of unqualified minorities and women at the expense of qualified White males. Some believe affirmative action has not worked and ultimately results in the stigmatization of those who benefit from the policies (Heilman, Block, and Stahatos 1997; Herring and Collins 1995).

Proponents argue that only through affirmative action policies can we address the historical societal discrimination that minorities experienced in the past (Kaplan and Lee 1995). Although these policies have not created true equality, there have been important accomplishments (Tsang and Dietz 2001). As a result of affirmative action, women and people of color have gained increased access to forms of public employment and education that were once closed to them (Yee 2001). Yet, research indicates that ethnic minorities and women do not have an unfair advantage over White men. Women and ethnic minorities are not receiving equal compensation compared with White males with similar education and background (Tsang and Dietz 2001). Although it may not be perfect, affirmative action has been the "only comprehensive set of policies that has given women and people of color opportunities for better paying jobs and access to higher education that did not exist before" (Yee 2001:137).

Shawn Woodhouse (1999, 2002) argues that the differences in individual perceptions of affirmative action policy may be related to the differences of racial group histories and socialization experiences. She writes,

Based upon these rationalizations, it is implicit that individuals interpret affirmative action through an ethnic specific lens. In other words, most individuals will assess their group condition when considering contentious legislation such as affirmative action because after all, a group's history impacts its view of American society. (2002:158)

Education

Based on Title VI of the 1964 Civil Rights Act, affirmative action policies have been applied to student recruitment, admissions, and financial aid programs. Title VI permits the consideration of race, national origin, sex, or disability to provide opportunities to a class of disqualified people, such as minorities and women, who have been denied educational opportunities. Affirmative action policies have been supported as remedies for past discrimination and as means to encourage diversity in higher education. Affirmative action practices were affirmed in the 1978 Supreme Court decision in the *Regents of the University of California v. Bakke,* suggesting that race-sensitive policies were necessary to create diverse campus environments (American Council on Education and American Association of University Professors 2000; Springer 2005).

Although affirmative action has been practiced since the *Bakke* decision, affirmative action has recently become vulnerable, particularly to challenges of the diversity argument in the Supreme Court's decision. The first challenge occurred in one of our most diverse states, California. In 1995, the California Board of Regents banned the use of affirmative action guidelines in admissions. In 1996, California voters followed and passed Proposition 209, the California Civil Rights Initiative, which effectively dismantled the state's affirmative action programs in education and employment. Also in 1996, a federal appeals court ruling struck down affirmative action in Texas. In the *Hopwood v. Texas* decision, the ruling referred to affirmative action policies as a form of discrimination against White students. State of Washington voters passed an initiative in 1998 that banned the use of race-conscious affirmative action in schools. In 1999, Florida Governor Jeb Bush banned the use of affirmative action in admission to state schools.

The *Hopwood* ruling led to a decline in the number of minority students enrolling in Texas A&M and the University of Texas (Yardley 2002). California's state universities experienced a similar drop in minority student applications and enrollment after the *Bakke* decision and the California Civil Rights Initiative. In response, states have instituted other practices with the goal of increasing minority student recruitment. For example, California and Texas have initiated percentage solutions. In Texas, the top 10 percent of all graduating seniors are automatically admitted into the University of Texas system. California initiated a similar plan, covering only the top 4 percent of students, and Florida recently announced the One Florida Initiative, allowing the top 20 percent of graduating high school seniors into the state's public colleges and universities. The University of Georgia increased its recruitment efforts among minority students, hoping to enlarge the pool of applications from minorities (Schemo 2001).

In 2000, a federal judge upheld the University of Michigan's affirmative action program, ruling that "a racially and ethnically diverse student body produces significant educational benefits such that diversity, in the context of higher education, constitutes a compelling governmental interest" (Wilgoren 2000). In 2003, the case was considered by the U.S. Supreme Court, and in a 5 to 4 vote, the Court upheld the University of Michigan's consideration of race for admission into its law school. Writing for the majority, Justice Sandra Day O'Conner stated, "In order to cultivate a set of leaders with legitimacy in the eyes of the citizenry, it is necessary that the path to leadership be visibly open to talent and qualified individuals of every race and ethnicity" (Greenhouse 2003:A1). In a separate decision, the U.S. Supreme Court voted 6 to 1, invalidating the university's affirmative action program for admission into its undergraduate program. Unlike the law school program, the undergraduate program used a point system based on race. Twenty points on a scale of 150 were awarded for membership in an underrepresented minority group; 100 points were necessary to gain admission into the university (Greenhouse 2003).

In November 2006, Michigan voters approved Proposal 2, a state law banning consideration of race or gender in public university admissions or government hiring or contracting. After asking the courts if it could delay complying with the new law until its current admission process had been completed, the University of Michigan announced in January 2007 that it would comply with the law and stop considering race or gender in its admissions decisions.

Main Points

- From a biological perspective, a **race** can be defined as a group or population that shares a set of genetic characteristics and physical features. Social scientists reject the biological notion of race, instead treating race as a social construct.

- **Ethnic groups** are groups that are set off to some degree from other groups by displaying a unique set of cultural traits, such as their language, religion, or diet. Members of an ethnic group perceive themselves as members of an ethnic community, sharing common historical roots and experiences.

- Sociologists explain that **ethnocentrism** is the belief that the values and behaviors of one's own group are right and actually better than all others. Although feeling positive about one's group is important for group solidarity and loyalty, it can lead groups to believe that certain racial or ethnic groups are inferior and that discriminatory practices against them are justified. This is called **racism**.

- Certain ethnic/racial groups have been subject to **institutional discrimination**, discrimination practiced by the government, social institutions, and organizations. Institutional discrimination may include segregation, exclusion, or expulsion. **Segregation** refers to the physical and social separation of ethnic or racial groups. **Exclusion** refers to the practice of prohibiting or restricting the entry or participation of groups in society. **Expulsion** is the removal of a group by using direct force or intimidation.

- The impact of race and social class has been documented in studies regarding income attainment and mobility, educational attainment, and health and medical care. In all three areas, racial and ethnic minorities are disadvantaged.

- The U.S. Census 2005 American Community Survey revealed that there were 35.2 foreign-born individuals in the United States, constituting 12.1 percent of the total population. This is the highest number of foreign born ever recorded in U.S. history. Between January 2000 and March 2005, a record number of 7.9 million immigrants arrived to the United States; nearly half were illegal aliens.

- Migration has been identified as a global issue, partly because of its impact to the global economy and terrorism. Globalization has intensified the need to coordinate and harmonize government policies worldwide.

- Multiculturalism is promoted in schools and businesses through educational programming. Most U.S. colleges and universities require a diversity course as a general requirement.

- Affirmative action is a policy that has attempted to improve minority access to occupational and educational opportunities. Since its inception nearly 40 years ago, it has been a controversial issue on federal, state, and local levels.

On Your Own

Log on to the Web-based student study site at www.pineforge.com/leonguerrero2study for interactive quizzes, e-flashcards, journal articles, Community and Policy Guides, a Service Learning Guide, the end-of-chapter Web exercises, and additional Web resources.

Internet and Community Exercises

1. In what ways does your college encourage or celebrate racial and ethnic diversity (among its students, faculty, and staff)? Consider specific college sponsored clubs, activities, or events that highlight diversity on your campus.

2. To learn more about the internment of Japanese Americans during World War II, visit the Web sites for the Manzanar War Relocation Center and internment camps located in Tule Lake, California, and Topaz, Utah. These Web sites feature virtual tours, photographs, and testimony from those interned. Log on to *Study Site Chapter 3* for links.

3. Identify the largest private employer in your city or state. Investigate through the Internet or direct contact whether a diversity program or development office is in place. What are the diversity goals of this business, and how does it implement these goals (what specific program practices are in place)?

4. The International Organization for Migration, established in 1951, is an intergovernmental organization working with 120 nations to promote and support human and effective migration management. The organization also conducts immigration research worldwide. Its Web site includes a global map, noting migration problems affecting a selected nation. Log on to *Study Site Chapter 3* for links.

Note

1. Tracy Ore (2003) acknowledges that externally created labels for some groups are not always accepted by those viewed as belonging to a particular group. For example, those of Latin American descent may not consider themselves to be "Hispanic." In this text, I've adopted Ore's practice regarding which racial and ethnic terms are used. In my own material, I will use Latina/o to refer to those of Latin American descent and will use Black and African American interchangeably. However, original terms used by authors or researchers (e.g., use of the term Hispanic by the U.S. Census Bureau) will not be altered.

CHAPTER 4

Gender

On January 4, 2007, U.S. Representative Nancy Pelosi (D-CA) was elected Speaker of the House, the first woman in its history. Pelosi is second in line for the succession of the presidency, after the vice president. On the day of her induction, Pelosi noted how her election was an important victory for her party, but also acknowledged the importance of her election for women. She said, "For our daughters and granddaughters, today we have broken the marble ceiling. For our daughters and our granddaughters now, the sky is the limit." Yet, a few months later as Pelosi celebrated Women's History Month, she is quoted as saying,

> Women want what men want: an equal opportunity to succeed, a safe and prosperous America, good paying jobs, better access to health care, and the best possible education for our children. . . . Yet in terms of policies to assist women, we are lagging behind. Paychecks for women have dropped three years in row by almost $1,000. On average, working women earn only 77 cents for every dollar working men earn, and over the last five years, the number of women living in poverty has grown by almost 3 million. (Pelosi 2007)

There is no society where men and women perform identical functions, nor are they ranked or treated equally. Regardless of their level of technological development or complexity of their social structure, all societies have some form of gender inequality (Marger 2008). Some may argue that there are fundamental differences between males and females, based on fixed physiological differences or our **sex**. Yes, there are biological differences—our sexual organs, our hormones, and other physiological aspects—that

Photo 4.1

Representative Nancy Pelosi (D-CA) was elected Speaker of the House in 2007, the first woman in the history of the U.S. Congress.

are relatively fixed at birth (Marger 2008), but more than that makes us unequal.

Sociologists focus on the differences determined by our society and our culture, our **gender**. Although we are born male and female, we must understand and learn masculine or feminine behaviors. Gender legitimates certain activities and ways of thinking over others; it grants privilege to one group over another (Tickner 2002). Social scientists believe that gender differences are not caused by biological differences; rather, they are a product of socialization, prejudice, discrimination, and other forms of social control (Bem 1993). **Sexism** refers to prejudice or discrimination based solely on someone's sex or gender. Although sexism has come to refer to negative beliefs and actions directed toward women, men can also be subject to sexism.

Consider the history of women in the U.S. Senate. There is nothing automatic at birth that makes men more suited to become senators than women. Yet in the 218-year history of the U.S. Senate, only 35 women have been elected or appointed as members (out of 1,895 senators). The first woman senator was Rebecca Latimer Felton, sworn into office on November 21, 1922. The Georgia senator was appointed to fill a vacancy and served for only one day. In the early 1990s, there were only two women senators. In 1992, Patty Murray, from my home state of Washington, was the first elected woman senator to have young children at home during her term in office (Stolberg 2003). But within 10 years, in 2002, there were 14 women senators; four were working moms with young or school-age children. In 2007, 16 women were serving as U.S. senators.

Something that we take for granted can be considered as evidence of subordination or exclusion. For many years, there was no women's restroom in the U.S. Senate chambers. The nearest available restroom was on the first floor, along with the public restrooms. Women senators had to "schlep downstairs and stand in line with the tourists" (Collins 1993:12). Women senators did not have their own bathroom located outside the Senate Chamber, next to the restroom for men senators, until 1993. To make room for the facility, the existing men's restroom was remodeled into two separate restrooms. When the restroom was built, Eleanor Smeal, president of the Fund of the Feminist Majority, declared, "[This] signifies the end of one of the last all-male bastions in the country that has real power" (Picker 1993).

But there remains another bastion—the U.S. House of Representatives. The only restroom a few steps away from the House floor is designated for men, with a shoe-shine stand and fireplace. The trip to the nearest restroom for women representatives is described as "traversing a hall where tourist gather, or entering the minority leader's office, navigating a corridor that winds past secretarial desks and punching in a keypad code to ensure restricted access" (Talev 2006:AA8). Beginning January 2007, a record 71 women representatives were in the 435-member House. (It is worth noting that despite the record number of women serving in Congress, the United States is behind other countries in female representation in national parliament or congress. Refer to Table 4.1 for more information.)

Table 4.1

Top Ten Countries With the Highest Percentage of Seats in National Parliaments Held by Women, 2007

Country	Percentage	Country	Percentage
Rwanda	48.8	Netherlands	36.7
Sweden	47.3	Cuba	36.0
Finland	42.0	Spain	36.0
Costa Rica	38.6	Argentina	35.0
Norway	37.9	Mozambique	34.8
Denmark	36.9		

Source: Inter-Parliamentary Union 2007.

Note: In the list compiled by the Inter-Parliamentary Union (2007), the United States ranks 68th with 16.3 percent.

What Does It Mean to Me?

In 2006, Senator Patty Murray, Senator Maria Cantwell, and Governor Christine Gregoire made U.S. history. Washington became the first state to have women in both U.S. Senate seats and in the governor's mansion. Identify the percentage of women who have been elected to your state legislature and to the U.S. Congress for the past five years. (The Center for American Women and Politics at Rutgers University Web site provides current and historical data for each state.) Has the number of women elected to public office increased or decreased in your state?

Sociological Perspectives on Gender Inequality

Functionalist Perspective

Functionalists argue that gender inequality is inevitable because of the biological division of labor in the household. According to Émile Durkheim, social evolution led to the exaggeration of sex differences in personalities and abilities. In the most basic of social institutions, the family, it became necessary for men and women to establish role differentiation as well as functional interdependence. In other words, men and women would have complimentary, but different, roles in the household.

Durkheim wrote of the biological differences between men and women, noting that women had smaller brain capacity than males: "Woman retired from warfare and public affairs and consecrated her entire life to her family" (2007:43). As a result, woman leads a completely different life from man. This division of labor applied both in and out of the home—women were charged with familial roles, taking care of their children and their home, while men were charged with public work roles, assuming their primary role as family breadwinner. Although this division of labor may have been best suited in preindustrial society, these roles remain gender specific in modern society (Marger 2008).

This gendered division of labor and gender roles is held as the standard for society. Gender inequality is not defined as a product of differential power but, rather, as a functional necessity (Marger 2008). Women who assert their rights for social and economic equality are seen as attacking the structure of American society (Bonvillian 2006). As a result, changing gender roles are blamed for a range of social problems. Theorists from this perspective note that as increasing numbers of women have entered the workforce, the number of divorces and nonmarital child bearing has also increased. They suggest that children are more likely to suffer from divorce, more likely to become delinquent without adequate parental supervision, and more likely to be disadvantaged economically and socially if born to a single mother (Bonvillian 2006; Farley 2005). From this perspective, a change in gender roles (among women in particular) undermines the stability of the family (Farley 2005) and, ultimately, society.

Conflict and Feminist Perspectives

Gender inequality exists because it benefits a group in power and with power to shape society—men. Theorists from both perspectives argue that women will remain in their subordinate position as long as men maintain their social, economic, and cultural advantage in society. A system where men are dominant over women is referred to as a **patriarchy**.

Women's subordinate position in society is linked to their relationship to the means of production. As the next section of this chapter describes, compared with women, men are rewarded in our capitalist economy with higher wages, more prestige, and greater authority in the workplace (Bonvillian 2006). At home, men are treated with deference by their wives and children.

Bonvillian (2006) explains the interrelationship between patriarchal social relations and capitalist economies. As capitalistic economies developed, they incorporated preexisting patriarchal relations. Capitalism benefits from women's subordinate position at home. Women are willing to work in different types of jobs and for different wages than men because they define themselves according to their familial relationships as supporters rather than breadwinners. Capitalism also takes advantage of men's adherence to patriarchal values, subordinating men to their employers just as patriarchal relations subordinate women to men. Men are duty-bound to their jobs because of a sense of self-worth and obligation tied to their ability to provide for their families (Bonvillian 2006).

Early feminist scholars treated gender as an individual attribute, as a property of individuals or as part of the role that was acquired through socialization. However, contemporary feminist theorists define gender as a system of social practices that creates and maintains gender distinctions and inequalities (Ridgeway and Smith-Lovin 1999). Gender is referred to as a process, where gender is continually produced and reproduced. This is not just an individual characteristic, but also exists within patterns of social interaction and social institutions (Wharton 2004).

Gender inequality is a product of a complex set of social forces—"these may include the actions of individuals, but they are also found in expectations that guide social interaction, the composition of social groups, and the structures and practices of the institutions" (Wharton 2004:157). Sexism may be an individual act, but it can also become institutionalized in our organizations or through laws and common practices.

Sociologist Rosabeth Moss Kanter (1977) identified how business corporations have a hidden gender structure: functioning, but unwritten rules about what positions can be occupied by women and how many women should be employed in the corporation. Organizations implicitly or explicitly withhold support from their women employees in the form of training,

Photo 4.2

In 2005, though Japan's labor force comprised more than 27 million women, only 10 percent were employed in management positions. For the same year, women held 42.5 percent of managerial positions in the United States (Fackler 2007).

promotion, or wages. Even if men and women have similar jobs within a company, men usually have more income and authority than women do (Beeghley 2005).

True gender equality is possible only if women are able to assume positions of power in the economy and political system (Marger 2008) and redefine the structures and practices that oppress them.

Interactionist Perspective

As interactionists explain, many social values and meanings are expressed in our language. Language, write Stephanie Wildman and Adrienne Davis, "contributes to the invisibility and regeneration of privilege" (2000:50). These scholars argue that we need to sort individuals into categories such as race and gender. Upon hearing that someone has a new baby, why is it important to ask if it's a girl or a boy? This type of social categorization is important because it sets into motion the production of gender difference and inequality. Norms, values, and beliefs about the differences between boys and girls and men and women are reinforced through the gender socialization process. We won't know how to relate to this child without knowing its gender and children won't understand what it means to be male or female in our society unless they are socialized accordingly. People respond to others based on what they believe is expected of them and assume that others will do the same (Wharton 2004).

Wildman and Davis note that characteristics of those who are privileged become societal norms—the standard of what is good, correct, and normal versus bad, incorrect, and aberrant. In terms of gender, men are privileged and serve as the standard. Wildman and Davis refer to Catharine MacKinnon's observation how among many things: "Men's physiology defines most sports, their health needs largely define insurance coverage . . .

Photo 4.3

Norms, values, and beliefs about the differences between boys and girls are reinforced through the gender socialization process. The process includes teaching children what is appropriate for each regarding their behavior, language, and dress.

their perspectives and concerns define quality in scholarship, their experiences and obsessions define merit, . . . their image defines god, and their genitals define sex" (2000:54).

Male privilege defines many aspects of American culture from a distinctly male point of view. For example, the use of "he" is accepted as an all inclusive pronoun, but a generic "she" is not permitted; some actually get upset if you try to use it (if you have any doubt, try referring to God as "she"). The response, according to Wildman and Davis, is not about incorrect grammar; rather, it is about challenging the system of male privilege.

Our language is subject to change. For example, as more women are elected to public office, the state legislatures in New York, Rhode Island, and Utah have revised their constitutions to gender-neutral language, eliminating the exclusive use of *he* or *him*. Such sexist language only accentuates the dominance of men in political affairs; however, the usage of *she* or *they* may reinforce and encourage the increasing participation of women in politics. Sandy Galef, an assemblywoman who led the drive to revise New York's state constitution, explained, "These constitutions were written about men because that was the history of our country. But that's not the history anymore" ("Some States" 2003:A28).

A summary of the sociological perspectives is presented in Table 4.2.

What Does It Mean to Me?

In 2007, Harvard University named Drew Gilpin Faust its first woman university president in its 371-year history. Examine the academic leadership in your university. How many top positions are held by men? By women? Is leadership gendered at your university?

The Consequences of Gender Inequality

Gender inequality is a persistent feature of all modern societies. In this section, we will review the consequences of inequality in the following areas: employment, income, and education. Additional discussions on gender inequality are presented in Chapter 8, "Education," and Chapter 9, "Work and the Economy."

Table 4.2

Summary of Sociological Perspectives: Inequalities Based on Gender

	Functionalist	Conflict/Feminist	Interactionist
Explanations of gender inequality	Gender inequality is a functional necessity. A gendered division of labor and gender roles is needed to ensure the stability of society.	Women will remain in their subordinate position as long as men maintain their social, economic, and cultural advantage. Conflict theorists identify how women's subordinate position is linked to their relationship to the means of production. Feminist theorists refer to gender as a process, a system of social practices that creates and maintains gender distinctions and inequalities.	Social values and meanings are expressed in our language. Our language reflects the privileged position granted to men.
Questions asked about gender inequality	What societal values are supported by our (traditional) gender roles? Can gender roles be changed without jeopardizing the stability of society?	How is gender inequality created and maintained? How can the gender inequality structures and practices be altered? Can they be eliminated?	How does our use of language identify the privileged position of men in society? Can language be changed to reflect gender equality?

Occupational Sex Segregation

Sex segregation in the workplace remains a historical and contemporary fact. Despite educational and occupational gains made by women, women continue to dominate traditionally female occupations, which is referred to as **occupational sex segregation**. These occupations include secretaries (96.6 percent are women), receptionists (92.7 percent), registered nurses (91.3 percent), and preschool teachers (97.7 percent) (U.S. Department of Labor, Women's Bureau 2006). (Sixteen women senators of 100 U.S. senators constitute only 16 percent. In contrast, Iraq's temporary constitution adopted in 2004 required 25 percent of national assembly seats to be filled by women. Yet, equal rights for Iraqi women regarding marriage and inheritance were not guaranteed by the constitution.) For more information on U.S. occupational sex segregation, refer to Table 4.3.

Social scientists examine two types of sex segregation in the workplace—horizontal and vertical. Horizontal segregation represents the separation of women into nonmanual labor and men into manual labor sectors. Vertical segregation identifies the elevation of men into the best-paid and most desirable occupations in nonmanual and manual labor sectors, whereas women remain in lower-paid positions with no job mobility.

Maria Charles and David Grusky (2004) identify several social factors that promote and reproduce horizontal segregation. Employer and institutional discrimination help maintain the separation of women and men in the workplace, for example, excluding women intentionally or unintentionally from physically strenuous jobs. The process of child socialization encourages girls and boys to internalize sex-typed expectations of others, which in turn shapes their occupational aspirations and preferences. Sociologists have examined how girls

Table 4.3

Ten Leading Occupations of Women, 2006

Occupation	Percent Women	Median Weekly Earnings
Secretaries and administrative assistants	96.9	$584
Registered nurses	91.3	971
Nursing, psychiatric, and home health aides	88.9	395
Bookkeeping, accounting, and auditing clerks	90.3	582
Child care workers	94.2	345
Receptionists and information clerks	92.7	467
Maids and housekeeping cleaners	90.3	348
Teacher assistants	92.3	405
Hairdressers, hairstylists, and cosmetologists	93.4	391
Preschool and kindergarten teachers	97.7	554

Source: U.S. Department of Labor, Women's Bureau 2006.

and boys are subject to differential gender socialization from birth. Traditional gender role stereotypes are reinforced through the family, school, peers, and the media with images of what is appropriate behavior for girls and boys. This includes defining appropriate occupations for women versus men.

Internalization of sex-typed expectations also leads workers to believe that if they transgress norms about gender appropriate labor, they are subject to sanctions (from disapproval from their parents to harassment from fellow workers). Years of horizontal segregation have given the advantage to men who have a disproportionate number of peers and network ties into the manual sector.

Vertical segregation is based on deeply rooted and widely shared cultural beliefs that men are more competent than women and are better suited than women for positions of power. According to Charles and Grusky (2004), vertical segregation is reproduced because it is consistent with the value of "male primacy."

Occupational sex segregation is a worldwide phenomenon. Many studies have examined segregation cross-nationally and have found that though it is a feature of all industrial societies, the degree to which it exists varies. In her analysis of vertical and horizontal segregation in the 10 countries including the United States, Maria Charles (2003) found that women were underrepresented in the manual sector and that within the manual and nonmanual sectors, women's occupations were of lower average status. She reports the highest levels of horizontal segregation in Sweden and France, where women are about 30 times more likely to work in the white-collar than in blue-collar sectors. The likelihood is lower in the United States—women are 14 times more likely to work in white-collar sectors. The highest level of vertical segregation was found in France and in the United Kingdom. Vertical segregation for the United States was third lowest among the 10 countries Charles examined. The countries with the lowest levels of horizontal and vertical segregation were Portugal and Italy, which Charles attributed to the countries' development of two main occupational groups—professionals and craft-operations workers.

Jane Elliot's (2005) research revealed that there was greater occupational segregation between men and women and between full-time and part-time working women in the United Kingdom than in the United States. She characterizes employed women in the

United Kingdom as having a "returner" pattern of labor participation (periodic employment versus continual employment as in the United States) and notes that U.K. women are primarily concentrated in occupational groups that rely on part-time labor. U.K. labor laws encourage the hiring of part-time employees versus full-time employees, which encourages women's employment patterns. Elliot suggests that U.K. women may be less attached to their employment than U.S. women are because national health services are available regardless of employment. In the United States, one's employer usually provides health insurance.

Income Inequality

According to the National Committee on Pay Equity (2007), in 2005, for every dollar earned by a man, a woman made 77 cents (refer to Table 4.4 for wage gap data from 1960 to 2005). Another way to measure the earning difference is to examine wage ratios, comparing the annual earnings of women and women who work full-time all year—what is the difference in men's and women's lifetime earnings? Stephen Rose and Heidi Hartmann (2004) examined data for 1983 to 1998 and concluded that women workers in their prime earning years make 38 percent of what men make. During the 15-year period, an average prime age working woman earned only $273,592 compared with $722,693 earned by the average working man (in 1999 dollars). (For information regarding the gap between college men and women graduates turn to this chapter's In Focus feature.)

Why do men earn more? Social scientists have attempted to answer the question, offering different explanations for the earning gap. Some have emphasized the role of **human capital**, the skills workers acquire through education and work experience.

Photo 4.4

The U.S. labor force continues to be segregated by gender lines. Although a small percentage of women, 25 percent or less, is employed in traditionally male blue-collar occupations (such as construction, truck driving, and manufacturing), women continue to dominate administrative (clerical) and service occupations, constituting more than 80 percent of employees in these occupations (U.S. Department of Labor, Women's Bureau 2006).

Human capital theory suggests that women earn less than men do because of differences in the kind and amount of human capital they acquire (Wharton 2004). In the United States, women do have less continuous work experience than men do; their labor is interrupted with child birth and child rearing. Yet, research indicates that even among women with continuous work experience, their earnings are less than men (England 2001).

Another explanation offered by social scientists focuses on the **devaluation of women's work**. A higher societal value is placed on men than on women, and this is reproduced within the workplace. The relative worth of men's and women's economic activities are assessed within this value system, with men and masculine activities valued more highly than women and feminine activities (Wharton 2004). Still, as Table 4.5 reveals, even in identical occupations, women earn less than men.

Table 4.4

Gender Wage Gap, 1960–2005

Year	Women's Earnings	Men's Earnings	Percent Difference
1960	$16,144	$26,608	60.7
1970	20,567	34,642	59.4
1980	22,279	37,033	60.2
1990	25,451	35,538	71.6
2000	27,355	37,339	73.3
2005	31,858	41,386	77.0

Source: National Committee on Pay Equity 2007.

Table 4.5

Twenty Occupations With the Highest Median Earnings by Sex, 1999

Men's Occupations	Median (dollars)	Women's Occupations	Median (dollars)
Physicians and surgeons	140,000	Physicians and surgeons	88,000
Dentists	110,000	Engineering managers	75,000
Chief executives	95,000	Dentists	68,000
Lawyers	90,000	Lawyers	66,000
Judges, magistrates, and other judicial workers	88,000	Optometrists	65,000
Natural science managers	84,000	Pharmacists	63,000
Optometrists	84,000	Chief executives	60,000
Actuaries	80,000	Economists	58,000
Engineering managers	80,000	Computer and information systems managers	58,000
Economists	73,000	Sales engineers	57,000
Astronomers and physicists	71,000	Actuaries	56,000
Chemical engineers	70,000	Air traffic controllers and airfield operations specialists	56,000
Computer and information systems managers	70,000	Chemical engineers	56,000
Financial analysts	70,000	Computer software engineers	55,000
Marketing and sales managers	70,000	Natural science managers	55,000
Pharmacists	70,000	Aerospace engineers	54,000
Veterinarians	70,000	Electrical and electronics engineers	54,000
Personal financial advisors	69,000	Astronomers and physicists	51,000
Air traffic controllers and airfield operations specialists	67,000	Engineers, all other	51,000
Management analysts	67,000	Computer programmers	50,000
		Environmental engineers	50,000
		Judges, magistrates, and other judicial workers	50,000
		Materials engineers	50,000
		Mechanical engineers	50,000

Source: Weinberg 2004.

Note: Fifteen occupations appear on both lists, and in all cases, women make less than men in the same occupation.

Educational Attainment

In fall 2003, nearly 10 million women and 7 million men enrolled in higher education at all levels. According to the American Council on Education (ACE), women now represent a 57 percent majority in higher education (college or university) enrollment (King 2006). In all measures—percentage of high school graduates completing college preparatory curriculum, percentage of high school graduates immediately enrolling in college, and total higher education enrollment—women rank higher than men. Refer to Figure 4.1 for the percentage distribution of conferred bachelor's degrees by gender and race.

As Figure 4.1 reveals, the share of conferred degrees has also increased for minority women and men. Does this mean that fewer White men are earning bachelor's degrees than women or minority students? ACE reports that this is not the case:

> The number of white men earning bachelor's degrees has been essentially flat during this period [1996–2004], while the number of white women, and men and women of color earning bachelor's degrees has grown. So, the story has been one of increasing attainment for these groups, but no less attainment by white men. (King 2006:17)

The report notes that there is still some gender segregation by major. Men constitute most majors in theology (77 percent), MBA (59 percent), noneducation doctorate programs (55 percent), law (54 percent), and master's of science (52 percent) programs. Women are overrepresented in education programs (80 percent at the master's level and 64 percent at the doctoral level), but have increased their enrollment in medicine (51 percent) and other health science professional programs (53 percent).

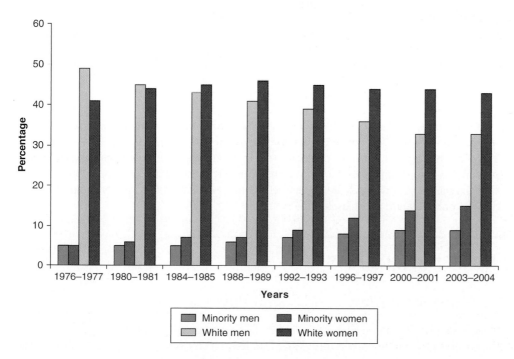

Figure 4.1

Percentage Distribution of Bachelor's Degrees Conferred by Gender and Race and Ethnicity, Selected Years, 1976–1977 to 2003–2004

Source: U.S. Department of Education 2004.

In Focus

The Pay Gap Between College-Educated Men and Women

During 2006 and 2007, two studies revealed surprising earning disparities between men and women college graduates. In separate analyses of data from the U.S. Department of Labor and the U.S. Department of Education, the Economic Policy Institute (a nonprofit nonpartisan think tank) and the American Association of University Women (AAUW) Educational Foundation reported the following:

- Though the wages of college-educated women grew rapidly since 1979, a woman college graduate still earned $6.76 less (24 percent) than a man college graduate in 2005 (Mishel, Bernstein, and Allegretto 2007).

- After one year out of college, women working full time earn only 80 percent as much as their male peers. Ten years after graduation, women's earnings drop to 69 percent as much as men earn. Despite controlling for hours of work, occupational type, and family status, college-educated women still earn less than college-educated men (Dey and Hill 2007).

- Students who graduate in women-dominated majors tend to get jobs that pay less than do jobs for students with men-dominated majors. For example, a woman education major earns 60 percent as much as a woman engineering major ($520 versus $872 per week). The earning ratio is the same, 60 percent, for men education majors and men engineering majors ($547 versus $915 per week) (Dey and Hill 2007).

- Even within the same major, after one year, women graduates make less than men graduates do. In education (a woman-dominated major and occupation), full-time working women earn 95 percent of what working men earn in the same field. In biological sciences and humanities (both characterized as a mixed gender major), women make 75 percent and 73 percent (respectively) of what men make (Dey and Hill 2007). (Refer to Figure 4.2.)

ACE attributes the increasing enrollment and degree attainment figures for women to the rising share of young women taking college preparatory courses during the 1990s and 2000s. Yet, ACE concludes that there is "no consensus on the causes of the gender gap and little comprehensive empirical research upon which to base firm conclusions" (King 2006:20). That other industrialized countries are experiencing similar educational gains for women suggests that this phenomenon is not just an American one.

Responding to Gender Inequalities

Feminist Movements and Social Policies

Historians mark the beginning of the feminist movement in United States and throughout the world in the nineteenth century. The U.S. feminist movement began in 1848 with the

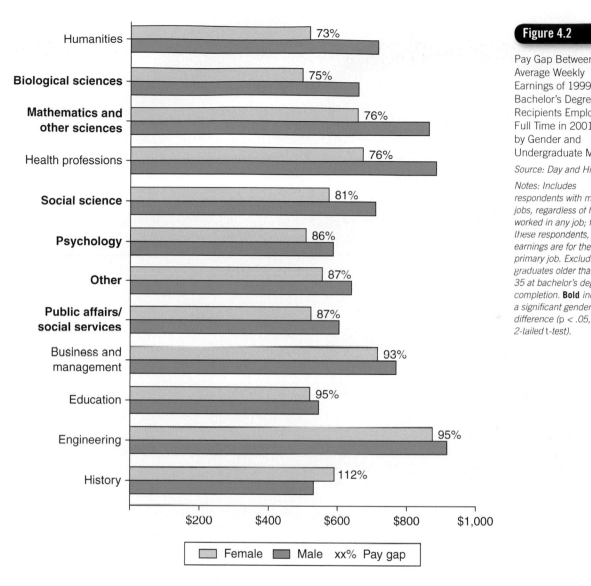

Figure 4.2

Pay Gap Between
Average Weekly
Earnings of 1999–2000
Bachelor's Degree
Recipients Employed
Full Time in 2001,
by Gender and
Undergraduate Major

Source: Day and Hill 2007.

*Notes: Includes
respondents with multiple
jobs, regardless of hours
worked in any job; for
these respondents,
earnings are for the
primary job. Excludes
graduates older than
35 at bachelor's degree
completion. **Bold** indicates
a significant gender
difference (p < .05,
2-tailed t-test).*

first Women's Rights Convention. A group of women, lead by Elizabeth Cady Stanton and Lucretia Mott, adopted a declaration of sentiments demanding, among other things, women's right to vote. During the same time, the women's suffrage movement began in Great Britain, with increasing demands for women's political and economic equality. The Nineteenth Amendment in 1920 affirmed for U.S. women the right to vote. In Great Britain, women were given the right to vote in 1918.

The feminist movement has been defined in "waves," the first beginning in the nineteenth century, followed by the second one during the twentieth century. Politically, the second wave movement focused on the expanding legal rights for women. During this period, Title VII of the Civil Rights Act of 1964 was passed prohibiting sexual harassment in the workplace and providing equal workplace opportunities for women and minorities (refer to Chapter 3, "Race and Ethnicity," for a discussion on Affirmative Action in Employment and Education)

Taking a World View

Violence Against Women

Femicide, sexual violence, sexual harassment, and sex trafficking and other forms of violence against women were recently documented in the United Nations (UN) Secretary-General's 2006 report on violence against women. "Punished for being female" is how Herbert (2006) described the findings of the UN report.

Based on surveys on violence against women conducted in 71 countries, the reported revealed that a significant proportion of women suffer physical, sexual, or psychological abuse, with the most common form of abuse being physical violence afflicted by an intimate partner. The UN stated that at least one of three women is subjected to intimate partner violence in their lifetime.

The systematic killing of women, **femicide**, is different from the murders of men and often involves sexual violence. The UN report indicates that between 40 and 70 percent of women murder victims are killed by husbands or boyfriends in Australia, Canada, Israel, South Africa, and the United States. One woman is reportedly killed by an intimate partner every six days in Columbia.

Female infanticide, prenatal sex selection, and the neglect of girls are widespread in South and East Asia, North Africa, and the Middle East. The UN estimates that more than 130 million girls have experienced or are dealing with the consequences of female genital mutilation.

The report attributes violence against women to historically unequal power relations between men and women and pervasive discrimination against women. Violence, said the UN, "is one of the key means through which male control over women's agency and sexuality are maintained" (United Nations 2006a:1). The report makes clear that violence is not confined to one nation, culture, or region; however, a woman's personal experience of violence is likely to be shaped by her ethnicity, class, age, sexual orientation, disability, nationality, or religion.

The study identifies the short- and long-term effects of violence against women. Women are more likely to suffer from physical, mental, chronic health, and reproductive health problems. Depression is a common response to sexual and physical violence. Violence against women may stop them from fulfilling their potential, restrict their economic activity and growth, and undermine their personal development. The costs of violence against women are both direct (services to treat and support women and their children and to bring their perpetrators to justice) and indirect (loss of productivity, as well as human suffering). The UN concluded by saying that violence against women "can only be eliminated, therefore, by addressing discrimination, promoting women's inequality and empowerment, and ensuring that women's human rights are fulfilled" (United Nations 2006b:1).

and Title IX of the Educational Amendments was passed in 1972 (refer to the next section). Still the movement was unsuccessful in passing the Twenty-Seventh Amendment to the U.S. Constitution, the Equal Rights Amendment (ERA), which proposed, "Equality of rights under law shall not be denied or abridged by the United States, or by any state, on account of sex." Though passed by the U.S. Congress in 1972, it was not ratified by the required 38 states to become a constitutional amendment. In 1977, Indiana became the 35th state, the last state, to ratify the amendment. ERA ratification bills have been introduced in remaining states without any success.

The third wave of feminism began in the 1990s. Though second-wave feminists are credited with achieving greater gender equality, they are criticized for assuming the universalization of the White women's experience and for focusing exclusively on oppression based solely on sex. The third wave of feminism attempts to address multiple sources of oppression—acknowledging oppression based on race and ethnicity, social class, and sexual orientation in addition to sex. Instead of focusing on gender equality within one country or nation, the focus has also expanded to a goal of global equality. Some even distance themselves from the use of the term feminist, believing the term is too confining or negative.

What Does It Mean to Me?

We take special note of female firsts—the first woman secretary of state (Madeleine Albright), the first woman speaker of the house (Nancy Pelosi), and the first women presidents of Princeton University (Shirley M. Caldwell Tilghman) and Harvard University (Drew Gilpin Faust). All of these events occurred during the twenty-first century, two centuries after the beginning of the feminist movement. In your opinion, has the feminist movement been successful? What remains to be achieved? Would you describe yourself as a feminist? Why or why not?

Title IX

Among the achievements of the second wave of feminism was the passage of Title IX of the Educational Amendments of 1972. Title IX of the Educational prohibits the exclusion of any person from participation in an educational program or the denial of benefits based on one's sex (Woodhouse 2002). The preamble to Title IX states, "No person in the United States shall, on the basis of sex, be excluded from participation in, be denied the benefits of, or be subject to discrimination under any educational programs or activity receiving federal financial assistance." In particular, the law requires that members of both sexes have equal opportunities to participate in sports and enjoy the benefits of competitive athletics (National Women's Law Center 2002b). In this chapter's Voices in the Community, Bernice Sandler explains how she became part of the Title IX legislation.

According to Title IX, schools are required to offer women and men equal opportunities to participate in athletics. This can be done in one of three ways: Schools demonstrate that the percentage of men and women athletes is about the same as the percentage of men and women students enrolled (also referred to as the "proportionality rule"), or the school has a history and a continuing practice of expanding opportunities for women students, or the school is fully and effectively meeting its women students' interests and abilities to participate in sports. In addition, schools must equitably allocate athletic scholarships. The overall share of financial aid going to women athletes should be the same as the percentage of women athletes participating in their athletic program. Finally, schools must treat men and women equally in all aspects of sports programming. This requirement applies to supplies and equipment, the scheduling of games and practices, financial support for travel, and the assignment and compensation of coaches (National Women's Law Center 2002a).

The law has been widely credited with increasing women's participation in high school and collegiate sports and for women's achievement in education. In 1997, data released by the U.S. Department of Education revealed the successes of Title IX after 25 years. In 1995, 37 percent of collegiate athletes were women, compared with 15 percent in 1972. In 1996, girls represented 39 percent of all high school athletes compared with only 7.5 percent in 1971 (U.S. Department of Education 1997). Today, more than 150,000 women participate in

Photo 4.5

The first time the Olympic Games included men's basketball as a medal sport was in 1936. Women's basketball became a medal sport 40 years later in 1976. A women's senior national team (like members of the team pictured here) regularly compete in national and international competitions in addition to the Olympic Games.

college sports, and about 3 million girls participate in high school sports (Garber 2002). In 1994, 63 percent of women high school graduates were enrolled in college, an increase from 43 percent in 1973. About 18 percent of women and 26 percent of men had completed four or more years of college in 1971 (U.S. Department of Education 1997). In 2002, 24.1 percent of men and 21.9 percent of women had completed a bachelor's degree or higher (U.S. Census Bureau 2002).

After more than 35 years, the controversy regarding Title IX continues. Many blame Title IX for the demise of some 400 collegiate men's programs. To achieve proportionality between the number of men and women athletes, schools have reduced the number of men athletes in minor sport programs such as wrestling, gymnastics, golf, and tennis (Garber 2002). In 2002, the National Wrestling Coaches Association filed a lawsuit against the U.S. Department of Education, claiming that by enforcing Title IX, the department was practicing sexual discrimination against men. In June 2003, a federal judge threw out the lawsuit, ruling that the coaches did not have standing to bring the lawsuit.

Evidence also indicates that not all colleges and universities are complying with the law. Although women in Division I colleges represent more than half the student body, women's sports receive only 43 percent of athletic scholarships, 32 percent of recruiting funds, and 36 percent of operating budgets (National Women's Law Center 2002a). Most surveyed Americans, about 70 percent, think Title IX should be strengthened or left alone (Brady 2003).

In June 2002, Secretary of Education Rod Paige announced the formation of the Commission on Opportunity in Athletics. Paige directed the 14-member commission to identify improvements in how Title IX is implemented. In January 2003, the Commission on

Opportunity in Athletics reviewed Title IX and recommended several changes. In general, the commission endorsed recommendations that would give schools more latitude in identifying athletic opportunities and scholarships. The commission deadlocked on a proposal to allow schools to allocate 43 percent of slots on varsity sports teams for women, although women make up 55.5 percent of college enrollments (Fletcher and Sandoval 2003). Department of Education officials confirm that they will retain Title IX enforcement, keeping the proportionality rule, but they will also begin emphasizing the other ways schools can meet the law through demonstrating a pattern of expanding opportunities for women or by proving the sports interests of women have been met (Associated Press 2003).

Voices in the Community

Bernice R. Sandler

In this feature, Bernice Sandler (1997), the woman behind Title IX, explains how she was transformed into a voice of change.

The year was 1969. I had been teaching part-time at the University of Maryland for several years during the time that I worked on my doctorate and shortly after I finished it. There were seven openings in the department, and I had just asked a fellow faculty member and friend why I was not considered for any of the openings. My qualifications were excellent, "But let's face it," he said, "You come on too strong for a woman." . . . I had no idea that this rejection would not only change my life but would ultimately change the lives of millions of women and girls because it triggered a series of events that would lead to the passage of Title IX. . . .

Although sex discrimination was illegal in certain circumstances, I quickly discovered that none of the laws prohibiting discrimination covered sex discrimination in education. I turned to the civil rights movement to see what African Americans had done to break down segregated school systems and employment discrimination, with the hope of learning what might be applicable to women's issues. I discovered a presidential Executive Order prohibiting federal contractors from discrimination in employment on the basis of race, color, religion and national origin that had been amended by President Johnson, effective October 13, 1968, to include discrimination based on sex. This discovery meant that there was a legal route to combat sex discrimination on campuses that held federal contracts.

The Director of the Office of Federal Contract Compliance at the Department of Labor, Vincent Macaluso, had been waiting for someone to use the Executive Order in regard to sex discrimination. Together we planned the first complaint against universities and colleges, and the strategies to bring about enforcement of the Executive Order.

Two months later under the auspices of the Women's Equity Action League (WEAL), I began what quickly became a national campaign to end discrimination in education and eventually culminated in the passage of Title IX. One January 31, 1970, WEAL filed a historic class action complaint against all universities and colleges in the country with specific charges against the University of Maryland. . . . During the next two years, I filed charges against approximately 250 institutions. Another 100 or so were filed by other individuals and organizations such as the National Organization for Women (NOW); in tandem with these administrative charges, we began a massive letter-writing campaign to members of Congress. . . .

Rep. Edith Green (OR), . . . chair of the subcommittee that dealt with higher education, agreed to hold congressional hearings in June and July of 1970 on education and employment of women. It was a time when there were virtually no books and only a few articles that addressed the issue of discrimination against women in education. There was little research or data, and barely a handful of unnoticed women's studies courses. There were no campus commissions on the status of women and only a few institutions had even begun to examine the status of women on their campus. . . .

In the spring of 1972, two years after the hearings, a portion of Rep. Green's original bill became law when Title VII of the Civil Rights Act was amended by Congress to cover all employees in educational institutions. Initially, Rep. Green had also sought to amend Title VI of the Civil Rights Act to include sex discrimination. However, at the urging of African American leaders and others, who were worried that opening Title VI for amendment could weaken its coverage, she proposed a separate and new title, which became Title IX.

. . . On June 23, 1972, Title IX of the Education Amendments of 1972, was passed by Congress and on July 1, was signed into law by President Richard Nixon. . . .

The words "too strong for a woman" turned me into a feminist. At that time, I had no legal, political or organizing experience. I was also extraordinarily naïve; I believed that if we passed Title IX, it would only take a year or two for the all the inequalities based on sex to be eliminated. Eventually, I realized that the women's movement was trying not simply to pass a piece of legislation, but to alter strongly embedded gender patterns of behavior and belief. To change all that would take masses of strong women and more than my lifetime to accomplish.

What Does It Mean to Me?

Investigate how Title IX is administered in your college or university. How many men and women athletes are at your university? How many programs for each? Have any programs been cut as a result of Title IX requirements? Interview coaches, athletes, and administrators on their view of Title IX: Has it made a difference for students and athletes at your school?

Main Points

- All societies have some form of gender inequality. Whereas **sex** refers to our biological differences (those set at birth), **gender** refers to our masculine and feminine behaviors (those set by our society or culture). Gender legitimates certain activities and ways of thinking over others; it grants privilege to one group over another.

- **Sexism** refers to prejudice or discrimination based solely on someone's sex. Sexism may be an individual act, but it can also become institutionalized in our organizations or through laws and common practices.

- According to functionalists, our gendered division of labor and gender roles is the standard for society. Gender inequality is not defined as a product of differential power but, rather, as a functional necessity. Women who assert their rights for social and economic equality are seen as attacking the structure of American society.

- Gender inequality exists because it benefits a group in power and with power to shape society—men. Theorists from both conflict and feminist perspectives argue that women will remain in their subordinate position as long as men maintain their social, economic, and cultural advantage in society. A system where men are dominant over women is referred to as a **patriarchy**. From a conflict perspective, women's subordinate position in society is linked to their relationship to the means of production. Contemporary feminist theorists refer to gender as a process, a system of social practices that creates and maintains gender distinctions and inequalities.

- From an interactionist's perspective, social values and meanings are expressed in our language. Language defines and maintains privilege in society; regarding gender, men are privileged and set many standards.

- Despite educational and occupational gains made by women, women continue to dominate traditionally female occupations, which is referred to as **occupational sex segregation**. Social scientists identify two types of sex segregation in the workplace—horizontal and vertical. Horizontal segregation represents the separation of women into nonmanual labor and men into manual labor sectors. Vertical segregation identifies the elevation of men into the best-paid and most desirable occupations in nonmanual and manual labor sectors whereas women remain in lower-paid positions with no job mobility.

- According to the National Committee on Pay Equity (2007), in 2005, for every dollar earned by a man, a woman made 77 cents. Social scientists have attempted to explain the earning gap between men and women. **Human capital theory** suggests that women earn less than men do because of differences in the kind and amount of human capital they acquire. Another explanation offered by social scientists focuses on the **devaluation of women's work**.

- For many educational measures, percentage of high school graduates completing college preparatory curriculum, percentage of high school graduates immediately enrolling in college, and total higher education enrollment, women exceed men. For example, women now represent a 57 percent majority in higher education enrollment.

- The feminist movement has been defined in "waves." The first began in the nineteenth century, with women's suffrage as its primary goal. The second wave began in the twentieth century with increased political focus on ensuring legal rights for women. The third wave of feminism began in the 1990s, attempting to address multiple sources of oppression—acknowledging oppression based on race and ethnicity, social class, sexual orientation, and sex.

- Title IX of the Educational Amendments of 1972 prohibits the exclusion of any person from participation in an educational program or the denial of benefits based on one's sex. As one of the provisions of Title IX, schools are required to offer women and men equal opportunities to participate in athletics. The law has been widely credited with increasing women's participation in high school and collegiate sports and for supporting women's achievement in education. However, some feel Title IX has been harmful to collegiate men's programs. Because of concern that not all colleges and universities were complying with Title IX, the U.S. Department of Education formed a commission in 2002 to make recommendations. In January 2003, the commission recommended several changes.

On Your Own

Log on to the Web-based student study site at **www.pineforge.com/leonguerrero2study** for interactive quizzes, e-flashcards, journal articles, Community and Policy Guides, a Service Learning Guide, the end-of-chapter Web exercises, and additional Web resources.

Internet and Community Exercises

1. Several policy and advocacy organizations focus on the earnings gap between men and women. The National Committee on Pay Equity (established in 1979) provides a state ranking of pay equity on its Web site. The committee also provides information on what you can do to promote pay equity in your state. The Institute for Women's Policy Research supports research on women's economic status. Its 2006 Status of Women in States is posted on its Web site. To learn more about your state's pay equity ranking or on the status of women in your state, log on to *Study Site Chapter 4.*

2. The Center for American Women and Politics (CAWP) at Rutgers University was founded in 1971. CAWP provides informational materials on women in federal and state political offices, convenes national forums for women public officials, and organizes educational programs to prepare young women for public leadership. To find out the history of women public officials in your state, go to the center's Web site (log on to *Study Site Chapter 4*). The site also provides current fact sheets about women in the U.S. Senate and House of Representatives. The site also provides links to other political women's groups such as Emily's List ("Early Money Is Like Yeast"), a political organization for pro-choice Democratic women, and the National Federation of Republican Women, a political organization for women in the Republican Party.

3. The United Nations Development Program created a gender empowerment measure, an index of women's participation in a nation's economic and political structures. Review the scores on the UN Web site (log on to *Study Site Chapter 4*) and identify the top five and bottom five nations on the list.

CHAPTER 5

Sexual Orientation

U.S. Marine Corps officer Antonio Agnone was deployed to Iraq in October 2005. Before his deployment, Agnone served as a logistics officer in Camp Lejeune, helping equip and deploy more than 800 Marines. In Iraq, he led a platoon in the Marine Corps' elite 22nd Marine Expeditionary Unit. During its deployment, the platoon was responsible for locating and disposing of Iraqi weapons, including improvised explosive devices (IEDs). Agnone's platoon was credited with saving many U.S. service personnel and Iraqi civilian lives and for his leadership; Agnone was honored with the Navy and Marine Corps Commendation Medal (Human Rights Campaign 2007a).

But during his Iraq deployment, Agnone, a gay man, knew that his partner would never be notified if Agnone were injured or killed. Because of the "don't ask, don't tell" (DADT) policy for homosexuals in the U.S. military, gay and lesbian couples must keep their relationships secret. Secrecy, paranoia, and frustration are common among homosexual couples and their families (Marquis 2003). Like Agnone and his partner, couples are forced to lie about their relationships and are not able to access supportive services provided to heterosexual married partners. Although their partners were deployed in the Middle East, stateside gay or lesbian partners are not eligible to use the base store or have access to support groups or status reports on the troops' whereabouts (Marquis 2003). Ultimately, Agnone decided not to reenlist because of DADT restrictions and returned home to be reunited with his partner and family.

One's sexual orientation serves as a basis of inequality. **Sexual orientation** is defined as the classification of individuals according to their preference for emotional-sexual relationships and lifestyle with persons of the same sex (**homosexuality**) or persons of the opposite sex (**heterosexuality**). **Bisexuality** refers to emotional and sexual attractions to persons of either sex. The term transgendered does not refer to a specific sexual orientation; rather, it refers to individuals whose gender identity is different from the one assigned to them at birth. The term **LGBT** is

Gary Owen (right) and Alan Spencer at their partnership ceremony in Manchester, England. Under Great Britain's 2004 Civil Partnerships Act, same-sex couples may register their partnership at civil ceremony, granting them similar legal rights as married couples have.

often used to refer to lesbians, gays, bisexuals, and **transgendered** individuals as a group.

There is no definitive study on the number of individuals who identify themselves as homosexual or bisexual. The study that is most often cited is one conducted in 1994 by Robert Michael and his colleagues. Based on a random survey of 3,432 U.S. adults age 18 to 59 years, Michael et al. found that 2.8 percent of males and 1.4 percent of females thought of themselves as homosexual or bisexual. About 5 percent of surveyed males and 4 percent of females said they had had sex with someone of the same gender after they turned 18. About 6 percent of males and 4 percent of females reported that they were sexually attracted to someone of the same gender. Based on the U.S. Census 2000 and election voter polls, David Smith and Gary Gates (2001) estimated the gay and lesbian population at 5 percent or more than 10 million men and women.

Gay rights have progressed in the United States and globally. Homosexuality was explicitly defined as a mental illness by the American Psychiatric Association until 1989. The World Health Organization removed a similar classification from its *International Classification of Diseases and Related Health Problems* in 1992. In 1996, South Africa became the first country to establish a constitutional ban against discrimination based on sexual orientation. In 2000, Vermont was the first U.S. state to recognize civil unions between same-sex partners; in 2004, Massachusetts was the first state to legalize same-sex marriage. And though national polls indicate increased support of gay and lesbian individuals, as a group they are still not immune to the experience of social problems. Based on their sexual orientation, gay and lesbian individuals continue to experience prejudice and discrimination regarding equal protection under family law and equal opportunities in the workplace.

Sociological Perspectives on Sexual Orientation and Inequality

Functionalist Perspective

Theorists in this perspective examine how society maintains our social order. Émile Durkheim argued that our social order depended on how well society could control individual behavior. Our most basic human behavior—our sexuality—is controlled by society's norms and values. Functionalists identify how society upholds heterosexuality and a marital union between a man and a woman as ideal normative behavior. This is also referred to as institutionalized heterosexuality, the set of ideas, institutions, and relationships that define the heterosexual family as the societal norm (Lind 2004).

Our legal, political, and social structures work in harmony to support these ideals (the conflict perspective of this is presented in the next section). The 1996 Defense of Marriage Act (DOMA) denies federal recognition of same-sex unions, defining marriage as a legal union only between a man and a woman. This legislation serves as a declaration about how the heterosexual family is valued and how all other family forms are not. Society grants legitimate kinship and familial obligations only through the heterosexual family. Consequently, society defines all other forms of sexuality and families that do not fit this ideal image as problematic. These forms are considered deviant or unnatural because they do not fit society's ideal.

Nonetheless, during the past decades, the gay rights movement has effectively influenced family rights, employment, and discrimination policies throughout the world. The movement has been successful largely because of its ability to affect institutional (macro) level changes—the focus of the functionalist perspective.

Conflict and Feminist Perspectives

Gore Vidal (1988) observes,

> In order for a ruling class to rule, there must be arbitrary prohibitions. Of all prohibitions, sexual taboo is the most useful because sex involves everyone … we have allowed our governors to divide the population into two teams. One team is good, godly, straight; the other is evil, sick and vicious.

Vidal's statement addresses the focus of both these perspectives, how conflict in our society is based on sexual orientation, with heterosexuals given the advantage.

Sociologists recognize that heterosexuals are granted a privileged place in our society. **Heterosexism** assumes that heterosexuality is the norm, encouraging discrimination in favor of heterosexuals and against homosexuals. Heterosexual privilege is defined as the set of privileges or advantages granted to some people because of their heterosexuality. For example, married couples receive more than 1,000 government benefits, ranging from the right to sue based on wrongful death of a partner, access to employment-based health benefits, and the ability to make medical decisions on behalf of a partner (Feigenbaum 2007).

From a conflict perspective, Amy Lind (2004) identifies how the DOMA helped institutionalize heterosexism because it blocks future proactive and protective legislation for gays and lesbians. She focuses specifically on heterosexual biases in social welfare policy, identifying its impact in three ways: through policies that explicitly target LGBT individuals as abnormal or deviant, through federal definitions that assume that all families are heterosexual, and through policies that overlook LGBT poverty and social needs because of stereotypes about affluence among LGBT families.

Evidence of the first type of heterosexual bias can be found in federal legislation such as DOMA and policy initiatives such as the healthy marriage promotion and fatherhood programs promoted by President George W. Bush. Current legislation funds abstinence only until marriage education programs in schools. Lind explains that gay, lesbian, and bisexual adolescents have no access to sexual education that pertains to their sexual experience. In an effort to preserve the traditional heterosexual family, these programs deny LGBT people their rights and needs.

The second type of heterosexual bias concerns how the U.S. Census defines the family and household. Lind refers to the 2003 definitions used by the U.S. Census. Family is defined as "a group of two or more (one of whom is the householder) related by birth, marriage, or adoption and residing together." Household "consists of all people who occupy a housing unit" and is distinguished by family versus nonfamily households. Family households are defined as "a household maintained by a householder who is in a family (as defined above) and includes any unrelated people who may be residing there," whereas a nonfamily household is "a householder living alone or where the householder shares a home exclusively with people to whom he/she is not related." Lind argues that these definitions privilege marital unions over domestic partnerships and the status of heterosexual families over other types of families.

Finally, the third type of heterosexual bias is based on stereotypes of lesbian, gay, and bisexual (LGB) individuals and families as affluent, despite evidence that LGB families are as economically diverse and stratified as heterosexual families are. Lesbians, gays, and bisexuals remain invisible in poverty studies or policies because they are assumed to be childless, have fewer family responsibilities, and thus higher overall incomes than heterosexual households have. With the exception of HIV/AIDS, LGB individuals are considered as not needing any economic, social, or health-related services.

From a feminist perspective, the question about gay marriage rights is bound to the ongoing critique of marriage as an institution (Bevacqua 2004). Scholars have argued that lesbian and gay marriages will positively disrupt the gendered definitions of marriage and the assumption that marriage is a prescribed hierarchy (Hunter 1995). However, just as feminists have criticized traditional marriage as an oppressive and dominating institution against women, feminists have also supported sexual freedom. Supporting gay marriages would mean that feminists would be supporting the very institution that perpetuates women's inequality.

Ann Ferguson (2007) explains that there are two main sides of the feminist argument: radical feminists who reject marriage outright on the basis of marriage as an oppressive institution versus liberal reform feminists who support the choice to marry on the understanding that men and women (or same-sex couples) can conduct their marriages in nontraditional ways. She supports the liberal reform side, arguing, "We should not simply reject marriage and hope it withers away, but instead should attempt to reform it as a better way to achieve these feminist goals [equality, freedom, and care]" (p. 52). On the topic of gay marriage, however, she concludes that some gay persons should not marry, not because it is a risky institution for women, but because the right to form one's family should not be tied to a one's marital status. "We should defend gay marriage as the formal right to access a basic citizen right" (p. 54).

What Does It Mean to Me?

Though same-sex couples would not have the same legal protection under civil unions or domestic partnerships as they would under marital law, these unions continue to be promoted by politicians and social leaders as viable alternatives to marriage. What do you think? Should civil unions or domestic partnerships be advocated as acceptable alternatives to marriage for same-sex couples? Why or why not?

Interactionist Perspective

In our society, no one gets "outed" for being straight. There is little controversy in identifying someone as heterosexual. Socially, culturally, and legally, the heterosexual lifestyle is promoted and praised. Although homosexuality has existed in most societies, it has usually been attached to a negative label—abnormal, sinful, or inappropriate. A socially determined prejudice, **homophobia**, is an irrational fear or intolerance of homosexuals (Lehne 1995). Homophobia is particularly directed at gay men.

Interactionists examine how sexual orientation is constructed within a social context. We tend to think of heterosexuality as unchanging and universal; however, Jonathan Katz explains how the term is a social invention that "designates a word and concept, a norm and role, an individual and group identity, a behavior and a feeling, and a peculiar sexual-political institution particular to the late nineteenth and twentieth centuries" (2003:145). Though heterosexuality existed before it was actually named in the early nineteenth century, "The titling and envisioning of heterosexuality did play an important role in consolidating the construction of the heterosexual's social existence" (p. 145). He argues that acknowledging heterosexuality as social invention—time bound and culturally specific—challenges the power of the heterosexual ideal.

Interactionists also examine the process of how individuals identify themselves as homosexual, what scholars describe as part of the development of a gay identity. Coming out (being gay and disclosing it to others) has come to symbolize the pursuit of individual rights and self-identification (Chou 2001). Coming out implies not just the disclosure of a gay identity, but also the individual's positive attitude toward and commitment to that identity (Dubé 2000). The disclosure of a gay identity merges a private sexual identity with a public social identity (Cass 1979). To come out successfully, a gay individual needs social and institutional support, in the form of support from family and friends, legal protection from discrimination and violence, cultural acceptance, financial equality, and access to health services (D'Augelli 1998).

The process of coming out to family members is particularly stressful for LGB youth. Fear of parental reactions has been identified as a major reason that LGB youth do not come out to their families (D'Augelli, Hershberger, and Pilkington 1998). Following disclosure, youth report verbal abuse and even physical attacks by family members. Youth who lived with their families and disclosed their sexual orientation were victimized by their families more often than were youth who had not disclosed (D'Augelli et al. 1998).

To avoid negative response from others, young lesbians and gay men hide their sexual orientation from family and friends (Rivers and Carragher 2003). Gay and lesbian youth may use one or more of the following concealment strategies: inhibiting behaviors and interests associated with homosexuality, limiting exposure to the opposite sex, avoiding exposure to information about homosexuality, assuming anti-gay positions, establishing heterosexual relationships, and avoiding homoerotic feelings through substance abuse (Radowsky and Siegel 1997). Research is inconclusive about how effective such concealment strategies are in reducing anxiety among lesbian and gay youth.

A summary of sociological theories regarding sexual orientation and inequality is presented in Table 5.1.

What Does It Mean to Me?

After the release of her last Harry Potter novel, author J. K. Rowling announced that Hogwarts' Headmaster Albus Dumbledore was gay. At a 2007 public event, Rowling described how as a young wizard Dumbledore was smitten with Grindewald, a childhood friend who turned into an evil wizard and was the predecessor to the infamous evil Lord Voldemort. This storyline was never included in any of the books in the Harry Potter series. Her statement was greeted with gasps, then cheers from the audience. Rowling responded, "If I'd known it would make you so happy, I would have announced it years ago." Why is her disclosure about Dumbledore's sexual orientation newsworthy? If she had made the announcement earlier, including it in one of her novels, would this have changed fans' response to the series or to the character?

Table 5.1

Summary of Sociological Perspectives: Inequalities Based on Sexual Orientation

	Functionalist	Conflict/Feminist	Interactionist
Explanations of sexual orientation and inequality	The heterosexual family is the way in which society identifies legitimate kinship and familial obligations. Society defines all other forms of sexuality and families that do not fit this ideal image as problematic in society.	Inequality is based on one's sexual orientation. Heterosexuality is privileged in our society. Debate about same-sex marriage is bound to the feminist critique about the oppressive nature of marriage.	Sexual orientation is a social construct—time bound and culturally specific. The development of a gay identity also depends on the support and reaction from others.
Questions asked about sexual orientation and inequality	How do our social structures endorse the heterosexual family? Is it possible to change the structures, to allow inclusiveness of other family forms and other sexual orientations?	In what ways do our social structures maintain heterosexism? What other forms of oppression exist for the LGBT community? Will same-sex marriage challenge the gendered definitions of marriage?	Who or what shapes our definitions of sexual orientation? How are LGBT individuals affected by inaccurate stereotypes or labels? How can negative stereotypes or labels be changed? What social support is necessary for someone to develop a positive gay identity?

Sexual Orientation and Inequality

U.S. Legislation on Homosexuality

Bisexual or homosexual men, women, and their families are subject to social inequalities through practices of discrimination and prejudice, many of them surprisingly institutionalized in formal law.

Sodomy laws criminalize oral and anal sex between two adults. Although the laws may apply to homosexuals and heterosexuals, sodomy laws are more vigorously applied against same-sex partners. Thirteen U.S. states still had state sodomy laws in 2003 (in 1960, sodomy was outlawed in every state).

In 1998, John Lawrence and Tyron Garner were fined $200 and spent a night in jail for violating a Texas statute that prohibits "deviate sexual intercourse" between two people of the same sex. The Texas statute does not apply to heterosexual couples. Their case was heard before the U. S. Supreme Court in March 2003. Attorneys for Lawrence and Garner argued that the Texas law was an invasion of their privacy and violated the equal protection clause of the Fourteenth Amendment because the law unfairly targets same-sex couples. Attorneys for the state argued that Texas has the right to set moral standards for its residents. In June 2003, the Court voted 6 to 3 to overrule the Texas law and all other remaining sodomy laws. Writing for the decision, Justice Anthony Kennedy said, "The state cannot demean their [homosexuals'] existence or control their destiny by making their private sexual conduct a crime" (Greenhouse 2003: A17). According to Kevin Cathcart, executive director of Lambda Legal, "This ruling starts an entirely new chapter in our fight for equality for lesbian, gay, bisexual, and transgendered people" (Lambda Legal 2003b).

In 2007, the U.S. House of Representatives approved legislation to grant gay individuals protection under hate-crime laws. The legislation was in response to the brutal deaths of James Byrd Jr., an African American who was dragged to death in Texas, and Matthew Shepard, a gay man who was beaten and left to die in Wyoming. If passed, the legislation would increase penalties against attackers. The federal hate-crime law, enacted in 1968, currently addresses crimes based on race, color, religion, and national original. Representative John Conyers (D-MI), who introduced the bill, said that the vote for the bill would not be a vote in favor of any sexual belief; rather, it would "provide basic rights and protections for individuals so they are protected from assaults based on their sexual orientation" (Hall 2007). Though supported by civil rights and law enforcement groups, some conservative and religious groups opposed the legislation, saying that the bill would create special classes of federally protected crime victims and would endanger First Amendment rights. President George W. Bush said he would veto the bill because it was unnecessary and constitutionally questionable.

The Rights and Recognition of Same-Sex Couples

The DOMA permits states to ban all recognition of same-sex marriages, and now more than 41 states have such bans in place. According to the law, the federal government will not accept marriage licenses granted to same-sex couples, regardless of whether a state provides equal license privileges to all types of partnerships. DOMA denies federal benefits available to or required for married opposite-sex couples. Gay and lesbian families are denied common legal protections that non-gay families take for granted such as adoption, custody, guardianship, social security, and inheritance (Lambda Legal 2003a).

In June 2002, however, President George W. Bush signed into law the Mychal Judge Act, which allows federal death benefits to be paid to the same-sex partners of firefighters and police officers who die in the line of duty (Bumiller 2002). The bill was named after the Reverend Mychal Judge, the New York City Fire Department chaplain who died in the collapse of the World Trade Center. Although White House officials say that President Bush did not consider the bill a gay issue, gay rights organizations considered its signing as a milestone for gays (Bumiller 2002). In 2006, the Federal Pension Protection Act became law, containing two key provisions that extend financial protections to same-sex couples and Americans who leave their retirement savings to non-spouse beneficiaries. Under the new law, an individual's retirement plan benefit can be transferred to a domestic partner or other non-spouse beneficiary. The second provision allows gay couples and others with non-spouse beneficiaries to draw on their retirement funds in the case of a medical or financial emergency.

Reacting to what he called the "divisiveness" of President George W. Bush's 2004 State of the Union Address on the issue of gay marriage, San Francisco Mayor Gavin Newsom secretly began planning to marry gay and lesbian couples and convinced longtime lesbian rights activists Del Martin, 83, and Phyllis Lyon, 79, to be the first to get married (Quittner 2004). Said Newsom, "We wanted to put a human face on this, and Phyllis and Del were critical.... To deny them the same protections as married couples would be to deny them as human beings, not as theory" (Quittner 2004). Between February 12 and March 10, 2004, more than 4,000 same-sex marriage licenses were issued in San Francisco. Although the majority of the couples were from California, couples from 45 other states and 8 countries also applied for marriage licenses in the city (Leff 2004).

After being together for 51 years, Phyllis Lyon (left) and Del Martin were married at San Francisco's City Hall in February 2004. They were the first legally married same-sex couple in the city. Their marriage license, along with 4,000 marriage licenses for same-sex couples issued in San Francisco, were ultimately voided by California's Supreme Court.

The California Supreme Court ordered San Francisco to stop issuing same-sex marriage licenses on March 11, 2004. In August 2004, the California Supreme Court ruled that San Francisco Mayor Newsom did not have the authority to issue marriage licenses to same-sex couples and declared the marriage licenses invalid. At the time of the decision, Phyllis Lyon said, "It is a terrible blow to have the rights and protections of marriage taken away from us. At our age, we do not have the luxury of time" (Lambda Legal 2004).

In response to the same-sex marriages in California, President Bush endorsed a constitutional amendment that would restrict marriage to two people of opposite sexes but left open the possibility that states could allow civil unions in same-sex relationships. However, in July 2004, the U.S. Senate failed to pass a constitutional amendment that would have declared marriage as a union only between a man and a woman.

For more on same-sex marriage legislation, refer to the discussion on family legislation in the last section of this chapter.

What Does It Mean to Me?

A public opinion survey conducted by the Gallup Organization in 2007 (Saad 2007) revealed the highest level of support for gay rights during the past three decades. The Gallup Organization reported that the majority of surveyed Americans believe that homosexuals should have equal rights for job opportunities (89 percent), homosexual relations should be legal (59 percent), and homosexuality should be considered an acceptable alternative life style (57 percent). Opposition regarding homosexual marriage remains high according to the data (53 percent saying that gay marriage should not be valid). With such strong support for gay rights in other life aspects, why does opposition regarding gay marriage remain?

Employment

The need to "manage a disreputable sexual identity at the workplace" has been called the most persistent problem facing lesbians and gay men (Schnieder 1986). Between 16 and 46 percent of gays, lesbians, or bisexuals have experienced discrimination based on sexual orientation (Katz and LaVan 2004). Title VII of the Civil Rights Act prohibits discrimination because of sex. Sex has been interpreted to mean gender, which means that protection for homosexuals based on sexual orientation is not yet covered.

M. V. Badgett and M. King (1997) note that, unlike discrimination based on easily observable characteristics such as skin color or gender, discrimination against gays and lesbians must be based on knowledge or suspicion of someone's sexual orientation. Lesbians and gay males who reveal their sexual orientation risk loss of income and lower chances at career advancement. A review of existing studies on workplace discrimination reveals that somewhere between one-quarter and two-thirds of LGB people report losing their jobs or missing promotions because of their sexual orientation. In addition, studies that compared gay and heterosexual workers with similar backgrounds and qualifications found that gay workers earn less than heterosexual workers do (Badgett 1997).

Protections against public and private workplace discrimination because of one's sexual orientation exist in 19 states, the District of Columbia, and several hundred U.S. cities and counties. More than 80 percent of all Fortune 500 companies have antidiscrimination policies protecting gay employees (Gunther 2006). In 2007, the U.S. House of Representatives approved a federal ban on job discrimination against gays, lesbians, and bisexuals. The law would make it illegal for employers to make decisions about hiring, firing, promoting, or paying an employee based on sexual orientation. A similar version of the bill was also introduced in the U.S. Senate (Miga 2007).

Photo 5.3

According to the Gay, Lesbian and Straight Education Network (GLSEN), there are an estimated 3,000 gay-straight alliance groups based in U.S. middle schools and high schools. Like this group of students from Arlington High School, the alliance provides support and education for gay and straight students.

In Focus

Gay Friendly Campuses

"What campuses do you consider to be LGBT-friendly?" This question was posed by the editors of the *Advocate College Guide for LGBT Students*. During 2006, the editors collected nominations from current LGBT college students and collected additional information based on interviews from students and faculty or staff members from the nominated school.

The editors based their 100 final selections on 10 criteria, as reported in the guide. According to editor Bruce Steele, the editors intended to assess "the effort that's being put forth by the colleges themselves to make their LGBT students comfortable" (Rosenbloom 2006, S2). Their ten criteria included the following:

1. Active LGBT student organization(s) on campus. Prospective LGBT students are looking for a sense of community with their peers and organizations that can offer social, educational, and leadership opportunities on campus.
2. Out LGBT students. Prospective students look for other LGBT students to be visible and active in academic and campus life settings.
3. Out LGBT faculty and staff. LGBT faculty and staff can serve as advisors and visible role models for LGBT students.
4. LGBT inclusive policies. Supportive campuses should have policies that include "sexual orientation" in their discrimination policy or have policies supporting same-sex domestic partner benefits.
5. Visible signs of pride. The prominent presence of rainbow flags and pink triangles can create a sense of openness, safety, and inclusion.
6. Out LGBT allies from the top down. Support from college administrators and alumni are essential to LGBT students.
7. LGBT-inclusive housing and gender-neutral bathrooms. Campuses may have options for LGBT-themed housing to foster a living and learning atmosphere for students.
8. Established LGBT campus center. What committed campus resources are available for LGBT students and organizations?
9. LGBT/Queer Studies academic major or minor. Students are looking for classes where they can learn about LGBT identity, politics, and history.
10. Liberal attitude and vibrant LGBT social scene. LGBT students want to be accepted fully. Students may want to live on a campus or in a city that offers queer entertainment.

The guide includes a list of top-20 U.S. campuses, described as "pioneering LGBT leaders in higher education":

American University	University of California–Berkeley
Duke University	University of California–Los Angeles
Indiana University	University of California–Santa Cruz
New York University	University of Massachusetts–Amherst
Oberlin College	University of Michigan
Ohio State University	University of Minnesota–Twin Cities
Pennsylvania State University	University of Oregon
Princeton University	University of Pennsylvania
Stanford University	University of Puget Sound
Tufts University	University of Southern California

What Does It Mean to Me?

How would you rate your school on these ten criteria? Is your campus LGBT-friendly? Why or why not?

What Does It Mean to Me?

Lambda Legal is a national organization committed to achieving full recognition of the civil rights of lesbians, gay men, the transgendered, and people with HIV or AIDS. The organization identifies each state that prohibits sexual orientation discrimination in employment. Investigate your state's discrimination laws by logging on to Lambda Legal's Web site (log on to *Study Site Chapter 5*). If your state does include such laws, a brief summary of the legislation is included.

Responding to Sexual Orientation Inequalities

Family Legislation

In 2003, the Massachusetts Supreme Judicial Court ruled that same-sex and opposite-sex couples must be given equal civil rights under the state's constitution. The ruling would allow same-sex couples in Massachusetts to obtain a civil marriage license, to make health and financial decisions for each other, to file joint state tax returns, and to receive other protections under state law. In early 2004, the court reaffirmed its decision, adding that only full marriage rights for gay couples, not just civil unions, would conform to the state's constitution. In May 2004, Massachusetts became the first state in the nation to let same-sex couples marry.

In May 2008, the California Supreme Court overturned the state's ban on same-sex marriage. The justices, in a 4 to 3 vote, declared that the fundamental "right to marry" extends equally to same-sex couples under the state constitution. Hawaii, Maine, Oregon, and Washington offer limited forms of domestic partnership to same-sex couples. The state of Vermont recognizes civil union partnerships, extending full and equal protection accorded to married couples under state law to lesbian and gay couples. On September 1, 2002, the *New York Times* printed its first same-sex marriage announcement for two men, Daniel Gross and Steven Goldstein. The couple was married in a civil ceremony in Vermont and in a commitment ceremony with Jewish vows. Connecticut, New Hampshire, and New Jersey also recognize civil union partnerships for same-sex couples.

Voices in the Community

Gay and Lesbian Families at the White House Egg Roll

The White House Easter egg roll has been held annually on the South Lawn since the mid-nineteenth century. The public event is open to all families, with tickets for the event distributed on a first-come, first-served basis. The egg roll took national stage, some say a political one, in 2006 when Family Pride, a national gay family organization, encouraged same-sex parents to bring their kids to the roll. About 50 gay and lesbian families signed up on the organization's Web site and more than 100 showed up for the ticket distribution.

Family Pride organizers explained that they were using the event to remind President George W. Bush that "gay families exist in this country and deserve the rights and protections that all families need" (Bumiller 2006). Religious conservatives accused the group of politicizing the family event. In a written statement, the White House reiterated that the egg roll was a public event and open to all families.

The idea to organize LGBT families began with Colleen Gillespie and her partner Alisa Surkis. A year earlier, Gillespie and Surkis stood in line with their daughter Ella, but were unable to attend when the event was cancelled due to rain. They were determined to return the following year and to organize other LGBT families to attend the event. Gillespie said, "It was really about wanting at some core level stake our right to participate in this great, historic tradition" (quoted in Bumiller 2006). Their own motivation to attend the egg roll began when Secretary of Education Margaret Spellings cancelled an episode of the children's show *Postcards from Buster* because one of the children profiled on the program had two mothers. According to the couple, Spellings' decision was a message to children of gay and lesbian parents that they should be ashamed of their families. They explain,

> We were being asked to make ourselves invisible in order to allow people to pretend—to themselves and to their children—that we weren't families at all. Given a choice between accommodating adults who were uncomfortable with our families and creating an atmosphere where our children could be open about and proud of their families, what kind of a parent would I be if I didn't choose the latter? (Surkis and Gillespie 2006)

More than 100 gay and lesbian families participated in the 2007 egg roll event.

Denmark was the first European country to recognize same-sex unions in 1989. Civil unions or gay marriages or partnerships are legal in Belgium (since 2002), France (2000), Germany (2001), Netherlands (2001), and Spain and Canada (2005). In 2006, South Africa passed its Civil Union Act, extending the legal rights of marriage to same-sex unions. The act is based on South Africa's constitution, which was the first in the world to prohibit discrimination based on sexual orientation. Marriages legal in these countries may not be recognized by other countries.

The number of gay and lesbian families in the United States is not certain. The 2000 U.S. Census estimated more than 600,000 gay and lesbian families; however, critics claim that the census undercounts the number of gay families by as much as 60 percent. The laws governing the rights and responsibilities of gay parents vary from state to state. Every state, except Florida, allows an individual to petition to adopt a child. Florida is the only state that explicitly forbids adoption by unmarried gay, lesbian, or bisexual individuals. Nine states, including the District of Columbia, approved joint parent adoption for gay and lesbian couples. In cases where one partner is the biological parent, the other parent may apply for second-parent or stepparent adoption. Second-parent adoption is legal in several states (Human Rights Campaign Foundation 2006).

Academic research has consistently indicated that gay parents and their children do not differ significantly from heterosexual parents and their children (McLeod and Crawford 1998). There is little or no evidence that the children of gay or lesbian parents are disadvantaged in any important way in comparison with children of heterosexual parents (Patterson and

Redding 1996). For their 2001 research, Judith Stacey and Timothy Biblarz examined the findings of 21 studies that explored how parental sexual orientation affected children. Based on the evidence, Stacey and Biblarz concluded that there are no significant differences between children of lesbian mothers and children of heterosexual mothers in measures of social and psychological adjustment, such as self-esteem, anxiety, and depression. Across studies, there was no relationship between parental sexual orientation and measures of children's cognitive ability. Also, levels of closeness and the quality of parent-child relationships did not vary significantly by parental sexual orientation. Stacey and Biblarz concluded,

> We propose that homophobia and discrimination are the chief reasons why parental sexual orientation matters at all. Because lesbigay parents do not enjoy the same rights, respect and recognition as heterosexual parents, their children contend with the burdens of vicarious sexual stigma. (2001:177)

Military Service

During his first presidential campaign, Bill Clinton promised to extend full civil rights to gays and lesbians, including military service (Belkin 2003). The military policy at that time banned gay and lesbian individuals from military service, stating that homosexuality was incompatible with military service. (For information on gay military service in other countries, refer to this chapter's Taking a World View feature.) In 1993, President Clinton suspended the policy and the National Defense Authorization Act became law.

Part of the law is the infamous DADT policy, a political compromise criticized since its inception. According to the policy, known homosexuals are not allowed to serve in the U.S. military, but the military is banned from asking enlistees questions about their sexual orientation. In addition, significant restrictions are placed on commanders wanting to investigate whether a soldier is gay (the complete policy is actually "Don't Ask, Don't Tell, Don't Pursue, Don't Harass"). Service members who disclose that they are homosexual are still subject to military discharge. Between its inception and the end of 2006, more than 11,700 service members have been discharged (Servicemembers Legal Defense Network 2007a). It is not known how many gay and lesbian service members have not reenlisted because of the policy without revealing their homosexuality.

Increasingly there has been a call to reexamine the DADT policy, from gay advocates, active and former military personnel, and administrators. In 2007, John Shalikashvili, retired Army general and former chairman of the Joint Chiefs of Staff, declared his support for the repeal of the DADT policy. Shalikashvili describes how his conversations with gay soldiers and marines, some with Iraq combat experience, "showed me how much the military has changed, and that gays and lesbians can be accepted by their peers" (2007:A19). He also cites evidence from a recent poll of 500 service members returning from service in Afghanistan and Iraq, revealing that 75 percent of those surveyed were comfortable interacting with gay people. Shalikashvili argues that given how our military forces have been stretched by increased and extended deployments in the Middle East, we cannot afford to lose the service of any American who is willing and able to do so.

In 2007, the Pentagon revealed that 58 Arabic language experts were discharged from military service since the inception of the policy because they were gay. U.S. House Representatives were critical of the Pentagon, believing that the policy and its actions were homophobic rather than focusing on the country's national security needs. Forty House members signed a letter asking the House Armed Services Committee chairman to investigate whether the DADT policy is serving the country well.

Taking a World View

Gay Military Service Policies

Among the 19 NATO countries, only 6—Greece, Hungary, Poland, Portugal, Turkey, and the United States—prohibit gays, lesbians, and bisexuals from open military service. The 24 NATO and non-NATO countries whose militaries have lifted their bans on gay military service are Australia, Austria, Belgium, Canada, the Czech Republic, Denmark, Estonia, Finland, France, Germany, Ireland, Israel, Italy, Lithuania, Luxembourg, the Netherlands, New Zealand, Norway, Slovenia, South Africa, Spain, Sweden, Switzerland, and the United Kingdom (Belkin 2003; Human Rights Campaign 2007b).

Aaron Belkin (2003) examined the experiences of four countries that lifted their bans on homosexual military personnel. In each country—Australia, Canada, Israel, and Great Britain—the bans were lifted in opposition from the military services.

Belkin explains that each country lifted its ban for different reasons. In Canada, the ban was lifted in 1992, after federal courts ruled that the military policy violated Canada's Charter of Rights and Freedoms. The ban was also lifted in 1992 in Australia, when Prime Minister

Paul Keating argued that the ban was not consistent with his country's integration of several international human rights conventions into its domestic laws and codes. Israel's military ban was lifted in 1993 after public response to Knesset hearings on the matter. In 1999, the European Court of Human Rights ruled that Great Britain's gay ban violated the right to privacy guaranteed in the European Convention on Human Rights.

All military personnel, academics, veterans, politicians, and nongovernmental observers interviewed for his research did not believe that lifting the gay bans undermined "military performance, readiness, or cohesion, lead to increased difficulties in recruiting or retention or increased the rate of HIV infection among the troops" (Belkin 2008:110). Grim predictions and anxiety about how military personnel would refuse to work with or share showers, undress, or sleep in the same room with gay soldiers were not substantiated in these countries. Many interviewed described the policy change as a "non-event" or "not that big a deal for us" and that the change was "accepted in 'true military tradition'" (pp. 110–111).

Interviewed military leaders stressed how all soldiers were held to the same standard of professional conduct regardless of sexual orientation or personal beliefs about homosexuality. Belkin states that none of the four militaries attempted to force military personnel to accept homosexuality. Data from the four countries confirm that soldiers refrain from the abuse and harassment of homosexual military personnel, though gay bashing and sexual harassment cases were documented in two of the countries (Australia and Israel).

In the war with Iraq, U.S. forces have served side by side with allied forces from nine countries that allow gays and lesbians to serve openly. In some cases, these forces work together in integrated units (Servicemembers Legal Defense Network 2007b).

Photo 5.4

As part of a coordinated campaign by gay rights lobbying groups, hundreds of American flags were displayed on the National Mall on the 2007 anniversary of the signing of the "Don't Ask, Don't Tell" law. Each flag represents a soldier discharged from military service because of his or her sexuality.

Main Points

- One's sexual orientation serves as a basis of inequality. **Sexual orientation** is defined as the classification of individuals according to their preference for emotional-sexual relationships and lifestyle with persons of the same sex (**homosexuality**) or persons of the opposite sex (**heterosexuality**). Bisexuality refers to emotional and sexual attractions to persons of either sex. The term transgendered does not refer to a specific sexual orientation; rather, it refers to individuals whose gender identity is different from the one assigned to them at birth. The term *LGBT* is often used to refer to lesbians, gays, bisexuals, and transgendered individuals as a group.

- There is no definitive study on the number of individuals who identify themselves as homosexual or bisexual.

- Functionalists identify how society upholds heterosexuality and a marital union between a man and a woman as ideal normative behavior. This is also referred to as institutionalized heterosexuality, the set of ideas, institutions, and relationships that define the heterosexual family as the societal norm.

- Sociologists recognize that heterosexuals are granted a privileged place in our society. **Heterosexism** assumes that heterosexuality is the norm, encouraging discrimination in favor of heterosexuals and against homosexuals. Heterosexual privilege is defined as the set of privileges or advantages granted to some people because of their heterosexuality.

- From a feminist perspective, scholars have argued that lesbian and gay marriages will positively disrupt the gendered definitions of marriage. However, just as feminists have criticized traditional marriage as an oppressive and dominating institution against women, feminists have also supported sexual freedom. Supporting gay marriages would mean that feminists would be supporting the very institution that perpetuates women's inequality.

- Although homosexuality has existed in most societies, it has usually been attached to a negative label—abnormal, sinful, or inappropriate. A socially determined prejudice, **homophobia**, is an irrational fear or intolerance of homosexuals. Interactionists examine how sexual orientation is constructed within a social context. The development of a gay identity has familial, social, legal, financial, religious, and health implications.

- Bisexual or homosexual men, women, and their families are subject to social inequalities through practices of discrimination and prejudice, many of them surprisingly institutionalized in formal law.

- Gay and lesbian couples are denied the same legal and social support given heterosexual couples. Their families are denied common legal protections that non-gay families take for granted such as adoption, custody, guardianship, social security, and inheritance.

- Between 16 and 46 percent of gays, lesbians, or bisexuals have experienced discrimination based on sexual orientation. Under current federal law, discrimination based on sexual orientation is not prohibited.

On Your Own

Log on to the Web-based student study site at www.pineforge.com/leonguerrero2study for interactive quizzes, e-flashcards, journal articles, Community and Policy Guides, a Service Learning Guide, the end-of-chapter Web exercises, and additional Web resources.

Internet and Community Exercises

1. The Human Rights Campaign, founded in 1980, is the largest national gay, lesbian, bisexual, and transgender advocacy organization in the United States. On its Web site, the organization posts information on specific LGBT laws and legislation. Go to www.hrc.org to learn about LGBT legislation in your state.

2. Does your campus have an LGBT organization? If it does, interview an administrator or student member to learn more about the organization. In what ways do the organization and your school's administration support LGBT students? If not, why does your campus not have an LGBT organization?

3. GLSEN, or the Gay, Lesbian & Straight Education Network, is the leading national education organization focused on ensuring safe schools for all students. The network focuses on educating teachers, students, and parents about how to reduce anti-LGBT name-calling, bullying, and harassment. GLSEN published the *State of the State Report 2004*, an analysis of safe school policies. Go to this chapter's *Study Site—Community and Policy Guide* to access information on your state's safe school policies.

CHAPTER 6

Age and Aging

When Frieda Birnbaum decided that her youngest child, a 6-year-old son, should have siblings closer to his age, she and her husband traveled to South Africa for a special in vitro fertilization treatment for older women. At 60 years of age, she became the oldest U.S. woman to give birth to twins, boys named Jake and Jared. Her older children, a 30-year-old son and 29-year-old daughter, did not support her decision to have the twins. Yet Birnbaum explained, "I hope I'm a role model for my daughter, that when she gets older she can make her own decision based on who she is, rather than what society dictates" (Springer 2007).

Age is both a biological and social classification (McConatha et al. 2003). Birnbaum is correct, there are social dictates regarding age—socially and culturally defined expectations about the meaning of age, our understanding of it, and our responses to it (Calasanti and Slevin 2001). We make a fuss over a 60-year-old woman giving birth to twins, the 77-year-old Ironman triathlete, or the 13-year-old college student because they are unexpected or deemed unusual for people of their age. Age distinguishes acceptable behavior for different social groups. Voting rights, the legal age to consume alcohol, military enlistment age, or the ability to hold certain elected offices (you can't be president of the United States until you are at least 35 years old) are examples of formal age norms. Informal age norms also demonstrate how a society defines what is considered appropriate by age (Calasanti and Slevin 2001).

Sociologists examine age and the process of aging through a **life course perspective**. This perspective examines the entire course of human life from childhood, adolescence, and adulthood to old age. The life course perspective tends to view "stages of life" as social constructions that reflect the broader structural conditions of society (Moody 2006). Aging occurs within a social context: one's social class, education, occupation, gender, and race will determine how one experiences adolescence or old age. However, there is also room for individuals to make

Photo 6.1

Age is both a biological and social classification. Active seniors, like Carol Johnston pictured here, expand our beliefs and expectations of elderly behavior and roles. As of December 2007, Johnston held the world record for pole vaulting, 2.24 meters for his age group (85 years).

their own choices in interpreting or embracing age-related roles (Moody 2006). **Gerontology** is the specific study of aging and the elderly, the primary focus of this chapter.

What Does It Mean to Me?

In the United States, the number of births to older women has increased, among 35- to 39-year-olds, 46.3 births per 1,000 (from 31.7 in 1990), and among 40- to 45-year-olds, 9.1 births per 1,000 (from 5.5 in 1990) reported for 2005 (Jayson 2007). What social expectations regarding age and motherhood do these older women challenge?

Our Aging World

Much has been written about the graying or aging of America, a change in our demographic structure referred to as **population aging** (Clark et al. 2004). One way to confirm population aging is to look at the median age of the U.S. population. (The median age is the age where half the population is older and the other half is younger.) The median age was 17 years in 1820 and 23 years in 1900, and by 2000, it increased to 35 years. By 2030, the median age is predicted to increase to 42 years.

Demography is the study of the size, composition, and distribution of populations, and demographers have identified several reasons for population aging. First, population aging is caused by a decline in birth rates (Moody 2006). With a smaller number of children, the average age of the population increases. In 1900, America was a relatively young population, with children and teenagers making up 40 percent of the population. But by 1990, the proportion of youth dropped to 24 percent.

Population aging can also occur because of improvements in life expectancy as a result of medical and technological advances (Moody 2006). As people live longer, the average age of the population increases. In 1900, life expectancy at birth was 47; according to the Centers for Disease Control, the life expectancy for a child born in 2003 is 77.5 years (National Center for Health Statistics 2006).

Longer life expectancy has also made it necessary to redefine what it means to be old or elderly. Gerontology scholars and researchers now make the distinction between young-old (aged 65 to 75), the old-old (aged 75 to 84), and the oldest-old (aged 85 or older) (Moody 2006). Unless noted otherwise, the use of the term "elderly" in this chapter will refer to those age 65 years or older.

Finally, the process of population aging can be influenced because of birth cohorts (Moody 2006). A **cohort** is a group of people born during a particular period who experience common life events during the same historical period. For example, the Depression of the 1930s produced a small birth cohort that had a minimal impact on the average age of the population. However, the baby boom cohort after World War II is a very large cohort, and its middle-aged baby boomers will contribute to the aging of the U.S. population. When Kathleen Casey-Kirschling, our nation's first baby boomer, born January 1, 1946, applied for her social security benefits in 2007, Social Security Commissioner Michael Astrue said it signaled "America's silver tsunami." An estimated 10,000 people a day will become eligible for Social Security benefits during the next two decades (Ohlemacher 2007). Graphically, the aging of America is displayed in Figure 6.1. Data from the U.S. Census Bureau dramatically show the affect of the baby boom generation on the overall age structure.

Demographers predict that the number of Americans 65 or older will increase over several decades (Figure 6.2). To provide a context for aging in the United States, it is helpful to examine trends in the rest of the world (He et al. 2005). Populations are aging in all countries, though the level and pace vary by geographic region. In 2000, 420 million people (7 percent) in the world were 65 and older; by 2030, the number is projected to increase to 974 million (12 percent). In 2000, China and India had the largest elderly populations, 87.5 million and 46.5 million, respectively. The United States was ranked third in the world, with 35 million (Table 6.1).

Gerontologist Harry Moody warns, "Population aging is a long-range trend that will characterize our society as we continue to into the twenty-first century. It is a force we all will cope with for the rest of our lives" (2006:xxiii). Population aging means an increase in the number or proportion of elderly and signals the need for changes in health care, employment status, living arrangements, and social welfare for the elderly and the rest of society.

The aging of American society is likely to transform state, regional, city, and suburban populations. A 2007 Brookings Institution report predicts the emergence of two major senior populations, each with a set of specific needs and geographic impacts. "Yuppie senior" affluent populations are expected to emerge in the South and West (in cities such as Las Vegas, Denver, Dallas, and Atlanta), increasing their demands for new types of community cultural amenities. Their affluence will continue to support

Photo 6.2

On October 15, 2007, Kathleen Casey-Kirschling signaled the beginning of the senior boom in the United States. Casey-Kirschling is pictured here with Michael Astrue, Commissioner of the Social Security Administration, on the day she filed for her Social Security benefits. Approximately 10,000 people a day will become eligible for Social Security benefits during the next two decades.

The Dramatic Aging of
America, 1900 2030

*Source: U.S. Census
Bureau. Adapted from
Himes (2001).*

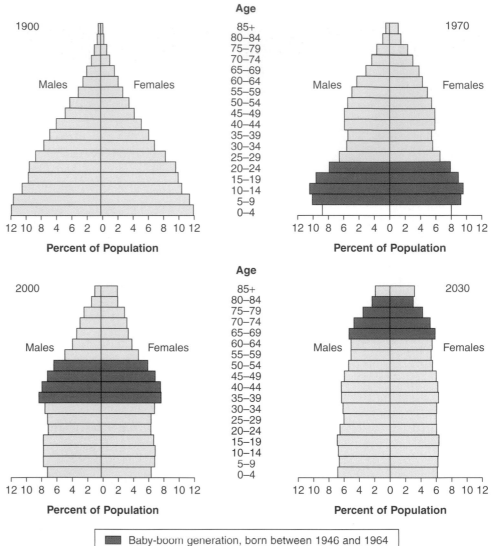

Percentage Aged 65
and Over of the Total
U.S. Population:
2000 to 2050

Source: He et al. 2005.

*Note: The reference
population for these data is
the resident population.*

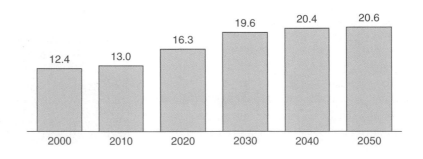

Taking a World View

Aging in China

The growth of the older population in China is expected to accelerate in the next 50 years, surpassing the aging population in many Western European countries and the United States. Charles Kincannon, Wan He, and Loraine West (2005) report that more than a half-century ago, 1 in 25 Chinese people was aged 65 or older, but by the beginning of the twenty-first century, 1 in every 14 was an older Chinese person.

> According to population projects by the U.S. Census Bureau's International Programs Center, China's older population will quadruple. Given China's total population of over 1.2 billion in 2000, this accelerated aging process will involve huge numbers of people. By 2050, it is projected that there will be 349.0 million people 65 and older in China, almost one-fourth more than the total 2000 population of the United States. (Kincannon et al. 2005:245)

(In contrast, the 2050 projection for the number of elderly in the United States is 33.7 million.)

Elderly Chinese rely on a variety of sources for financial support.

> Just over half of the Chinese men aged 65 and over who are no longer working relied primarily on their family for financial support, while four in ten received primary support from a retirement pension in 2000. Older women were far more likely to be dependent on their families for support—82 percent—and much less likely to rely on a retirement pension—13 percent. (Kincannon et al. 2005:250)

Reliance on family support is greatest among the oldest-old Chinese population, largely because they were not eligible to receive benefits under China's pension system (first established in the 1950s and limited to those with at least 20 years of employment) and if they did qualify, pension benefits were insufficient to use as the primary source of financial support. The researchers predict that as China expands its system of social insurance programs, reliance on family support is likely to decline.

Employment continues to serve as a source of support for China's elderly; in fact, a higher percentage of the Chinese older population is employed than is that of the United States. According to the U.S. Census 2000, 18 percent of men and 9 percent of women aged 65 and older were employed in the United States, and China 2000 Census data reveal that nearly 34 percent and 17 percent of similarly aged men and women are employed in China. More than 90 percent of employed Chinese elderly were engaged in agriculture; in China's urban areas, the employment rates of the older population were slightly below those in the United States.

The progressive aging of its older population is a serious issue for China. Better medical care and health conditions in China have led to longer lives for older people in the country. The result, as Kincannon et al. observe, is that,

> Over time, a nation's older population may grow older on average as the larger proportion services to 80 years and beyond (the oldest old). The oldest old and the young old have very different economic, demographic, and health statuses, thus they also have very different needs for health services, old-age care, residential arrangements, or assistance with the requirements of daily life. The oldest old are more likely to be widowed (especially the oldest-old women), to be frail or sick, and to be unemployed and lack financial resources. Their needs may put tremendous pressures on their families as well as on society. The oldest old, therefore, are a group within the older population that warrants special attention. (2005:245)

Table 6.1

Top Three Countries With More Than 2 Million People Aged 65 and Over: 2000 and 2030 (numbers in thousands)

	Rank		Number of People 65 or Older	
	2000	2030	2000	2030
China	1	1	87,538	239,480
India	2	2	46,545	127,429
United States	3	3	35,601	71,453

and encourage the economic and civic growth of their geographic areas. Conversely, cities in the Northeast and Midwest are predicted to have a disproportionately higher population of "mature seniors"—less stable financially or physically able than yuppie seniors. Mature seniors are likely to require more social and public support programs, along with affordable private and institutional housing and accessible health care providers (Frey 2007).

Sociological Perspectives on Age, Aging, and Inequality

Functionalist Perspective

Age helps maintain the stability of society by providing a set of roles and expectations for each particular age group or for a particular life stage. These roles are reinforced by our major social institutions—education, the economy, and family. We assume that children, 18 years or younger, should be in school. After high school graduation, young adults have the choice of entering the workforce or continuing their education (where their student role continues), whereas adulthood is a time set aside to build one's career and to begin a family. Retirement is another important age-related stage, with societal roles and expectations for a retired person.

What Does It Mean to Me?

It is assumed that after college graduation, young adults will live independently from their families of origin. However, the phenomenon of **boomerangers** has been noticed by demographers; boomerangers are young adults who leave home for college, but return after graduation either because of economic constraints (they may be unemployed or underemployed) or personal choice. Surveys of recent college graduates conducted in the early 2000s noted that nearly half of those surveyed expected to live with their families for some period. Is boomeranging a viable transitional stage for recent college graduates? Does this delay adulthood? Explain.

Photo 6.3

The modernization theory of aging suggests that the declining status of the elderly is associated with their decreasing economic and labor contribution. What status do we attribute to the position of store greeter? High or low?

Consider how each age group has its own function or role—the young attend school preparing for their adult lives, adults are employed and building their lives, and the elderly retire. **Disengagement theory** defines aging as a natural process of withdrawal from active participation in social life. Older people disengage from society (from their work and certain parts of their lives), and in turn, society disengages from them (Turner 1996). The theory contends that people enter and exit a set of roles throughout their lives. These transitions are natural and functional for society (Mabry and Bengston 2005; Moody 2006).

This process is portrayed as orderly, timely, and necessary for the well-being of the entire society. For example, in the workplace, older workers must relinquish their jobs to make room for younger workers from the labor market. This process is socially supported through retirement plans, pensions, and Medicare (Mabry and Bengston 2005). Disengagement is portrayed as positive for the elderly because they are now able to participate in activities and a lifestyle that earlier they would have never been able to (more fully engaged in community, family, or leisure activities). The final form of disengagement is death.

However, this perspective fails to acknowledge how vulnerable and powerless adults are in their older years. Is this disengagement natural or forced by society? The next perspective answers this question.

Conflict Perspective

The **modernization theory of aging** suggests that the role and status of the elderly declines with industrialization. Specifically, their power, wealth, and prestige are linked with their labor contribution or their relationship to the means of production. In hunting and gathering societies, the elderly had a low status because they were unable to contribute to the primary means of production. However, their status increased during the time of stable agricultural societies, when older people controlled land ownership. In modern industrial society, life experience is surpassed by technological expertise, thus the status of the elderly declines.

From a conflict perspective, the two groups at odds with one another are the young versus the old. As Donald Cowgill explains, society systematically advantages the young, supported by what he called the "cult of youth"—a value system that glorifies youth "as a symbol of beauty, vigor and progress and discriminates in favor of youth in employment and in the allocation of community resources" (1974:15–16).

Cowgill (1974) identifies how four aspects of modernization lower the status of older people. The first is health technology. Modern health advances improve the population's overall health and longevity. This creates an older and healthier workforce, willing and able to stay in the labor force a bit longer. "As the lives of workers are prolonged, death no longer creates openings in the labor force as it once did" (p. 12). Society then creates a new opening through retirement, forcing people out of their most valued and senior roles in the labor force. Elderly workers are reduced to retirement status with less income and influence.

The next two aspects, education and economic technology, are related to one another. In a modern society, the young have more opportunities to acquire education and training. The status of younger members of society is elevated because they become more literate than their parents. Society relies on their increased literacy in the workplace, creating new information- and technology-based occupations. The people most qualified for these positions are the younger, more literate, workers. Older workers perform more traditional jobs, some that are less valued or that eventually become obsolete.

The final aspect is urbanization. A modern society is more urban, characterized by increased social mobility and migration. Cowgill argues that the young migrate more than the old do. The migration produces a physical and emotion separation between a child and the family of origin, tearing down the bonds of the extended family. Yet, it also promotes the cultural image of the young moving to something better, while the old are left behind.

Feminist Perspective

Women constitute the majority in the U.S. older population. In 2005, there were 21 million women aged 65 years and older compared with 15 million men (Robinson 2007). The standards of our culture create more problems for women than for men as they transition into their middle and later years. Women seem more vulnerable to societal pressure to retain their youth and, consequently, face more questions about their self-worth, which may lead to serious problems ranging from low self-esteem to depression (Saucier 2004). Questions about self-worth and value are also raised in the workplace, where researchers have documented how women experience greater age discrimination during all ages than do men (Duncan and Loretto 2004).

Susan Sontag (1979) notes how society is much more permissive about aging in men. She writes about the **double standard of aging**—men are judged in our culture according to what they can do (their competence, power, and control), but women are judged according to their appearance and beauty. Women's identity is more closely associated with their physical appearance than is that of men. As a result, society considers men "distinguished" in their old age, but women must disguise the fact that they are aging. Sontag argues that because women are unable to maintain their youthful looks as they age, they are pressured to defend themselves against aging at all costs.

Feminist scholars argue that the cosmetic industry focuses on a male and youth standard. Though cosmetic products are advertised for women's use (ever notice that there are no male cosmetic counters at your department store?), feminists assert that the industry is responding to the male-defined standard of female beauty. In addition, the industry is also responding to the image of unattainable youthful beauty upheld by society (Calasanti, Slevin, and King 2006); says Sontag (quoted in Freedman 1986:200), "Women are trained to want to continue looking like girls forever."

What Does It Mean to Me?

For the next few days, note the portrayals of older adults in popular electronic and print media. How would you describe the aging standard for men? For women? Are these portrayals accurate?

Photo 6.4

This fashion model promotes a particular image of beauty and of ideal body shape. How does this image influence the self-image of young girls, adult women, or older women? Are other images of beauty promoted in fashion? in the media?

Interactionist Perspective

Interactionists reveal how our age-related roles are socially defined and expected. Age is tied to a system of matching people and roles (Hagestad and Uhlenberg 2005). What does it mean to be "middle aged"? Middle age is not just measured according to years, but is also associated with a set of role expectations. We share a definition of what it means to be middle aged, and there is an expectation that we need to assume a particular role once we are middle aged.

These role expectations stigmatize particular age groups. **Stigma**, coined by Erving Goffman, is defined as a discrediting attribute. Older adults are discredited in society, stereotyped as less capable, fragile, weak, and frail. **Ageism** or the stereotyping (or discrimination) of older adults can damage the self-concepts of the elderly (Miller, Leyell, and Mazachek 2004) and represents a self-perpetuating cycle of fears that old and younger adults have toward aging in general, disability, death, competition for resources, and the perceived inferiority of particular individuals (Yang and Levkoff 2005). These images may increase social isolation, dependency, and elderly abuse and may become a self-fulfilling prophecy for others (Thornton 2002). However, researchers have found that older adults from cultures with more positive attitudes toward aging than mainstream society may be able to avoid exposure to or internalization of the negative stereotypes of old age (Levy and Langer 1994).

Interactionists examine how the problems associated with aging have been defined and by whom. Society relies on trained experts such as gerontologists, physicians, nurses, and social workers to identify and respond to the problems of aging. Yet, research indicates that this group of professionals is just as likely to be prejudiced against older people as other groups are. A. J. Levenson argues, "Medical students' attitudes have reflected a prejudice against older persons surpassed only by their racial prejudice" (1981:161). He points to medical schools as part of the problem,

Table 6.2

Summary of Sociological Perspectives: Inequalities Based on Age and Aging

	Functional	Conflict/Feminist	Interactionist
Explanations of age and inequality	As we age, we enter and exit age specific roles and statuses. Each status helps maintain social stability. Elderly disengagement is a normal and necessary process.	Inequality is based on one's relationship to the labor force—with the elderly having the least amount of power in society. Women are disadvantaged in the aging process more than men are.	Our age related roles are socially defined and expected.
Questions asked about age and inequality	How do our age related roles and statuses contribute to the stability of society? Is there a way to successfully disengage the elderly from society?	How does modernization contribute to the lowering status of the elderly? What can be done to improve the status of the elderly? How and why does society have two standards for aging—one for men, the other for women? What can be done to reduce the negative impact of the aging standard for women?	How do we learn our age related roles? In what ways do our perceptions and beliefs perpetuate ageism?

putting little value on geriatrics as a specialty. Doctors often think that because aging cannot be stopped, illnesses associated with old age are not that important.

The medical industry has not ignored aging completely. Though controversial, the medical and cosmetic industries actively promote anti-aging vitamins, hormones, surgeries, and pharmaceutical drugs, encouraging wellness to patients and clients while sending a message that they can "beat back old age" (Wilson 2007). Americans spend nearly $50 billion per year in anti-aging vitamins, treatments, hormones, and pharmaceutical drugs (Wilson 2007). Are we responding to a genuine problem or one carefully manufactured by the medical and cosmetic industries?

A summary of all sociological perspectives is presented in Table 6.2.

The Consequences of Age Inequality

Ageism

Ageism is defined by Robert Butler as the "systematic stereotyping of and discrimination against people because they are old, just as racism and sexism accomplish this with skin color and gender" (1969:243). Todd Nelson (2005) described ageism as "prejudice against our feared self." He suggests that age prejudice is one of the most socially condoned and institutionalized forms of prejudice. For example, a standard message in birthday greeting cards is how unfortunate one is to be a year older.

Ageism marks a sharp distinction between "us" and "them." William Bytheway (1995) explains it as, "The issues of these pronouns creates a conceptual map on which groups of people are variously included and excluded. In particular, the old who are discriminated against occupy a different territory on these us/them maps from 'us'" (p. 117).

Older adults tend to be marginalized, institutionalized, and stripped of their responsibility, dignity, and power (Nelson 2002). Dependency is one of the most negative attributes of being identified as "old" in our society (Calasanti and Slevin 2001). Stereotypes about the capacities, activities, and interests of older people reinforce the view that they are incapable of caring for themselves (Pampel 1998). Older adults are not generally disliked, but are likely to be victims of paternalistic prejudice, which stereotypes them as likable, but incompetent (Packer and Chasteen 2006). There is widespread acceptance of negative stereotypes about the elderly regarding their intellectual decline, conservatism, sexual decline, and lack of productivity (Levin and Levin 1980).

In a comparative study of young adults in the United States and Germany, German young adults tended to view aging more negatively than did Americans in the sample (McConatha et al. 2003). Germans were more likely to be pessimistic about the likelihood of finding contentment in old age and did not expect to feel good about life when they were older. The study attributed the differences in aging attitudes to Germany's more prevalent negative stereotypes of older people, a response to the increasing costs of providing extension pension and health benefits to the elderly in Germany. On the other hand, in the United States, effective political advocacy groups, increasingly healthy and influential older adults, and educational aging programs may account for a reduction of ageism. The study also revealed that American and German women were more concerned about age-related physical changes than were men.

What Does It Mean to Me?

Jasim McConatha and her colleagues (2003) also asked young adult Germans and Americans what age they consider as "old." The average age that Germans reported was 64 for men and 60 for women; the U.S. sample reported younger ages—53 for men and 48 for women. What age do you consider as "old"? What expectations do you have about growing older?

Age and Social Class

The most economically vulnerable in our society are very young or old. As discussed in Chapter 2, the rate of child poverty in the United States is one of the highest among Western industrialized countries. According to the U.S. Census, the poverty rate for children was 17.4 percent or 12.8 million in 2006 (DeNavas-Walt, Proctor, and Smith 2007). For a more extensive discussion on children and poverty, refer to Chapter 2, "Social Class and Poverty."

Though most of us envision a life filled with leisure activities, travel, and good living, there is another possibility—that one's retirement can also be a time of serious economic hardship. Although some elderly are economically stable, others may be economically disadvantaged.

In 2005, 3.6 million elderly (or 10 percent) were living in poverty in the United States. Retirement represents a precipitous income drop for most elderly. According to the U.S. Census, the median income by age of householder was highest at $60,242 for those aged 45 to 54 for 2003 (He et al. 2005). Median income begins to decline with the next age group, those 55 to 64, to $49,215. Finally, for those aged 65 and older, their reported median income

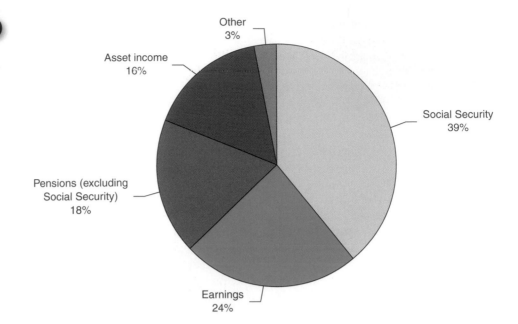

Figure 6.3

Personal Money
Income for the
Population Aged 65
and Older, 2001

Source: He et al. 2005.

*Note: The reference
population for these
data is the civilian
noninstitutionalized
population.*

was $23,787 (He et al. 2005). In 2001, Social Security paid benefits to 91 percent of America's elderly and was the only source of retirement income for many aged 65 and older (He et al. 2006). (Refer to Figure 6.3.)

As we have discussed in the earlier chapters of this section, the inequalities based on race/ethnicity and gender will also determine one's economic status. Poverty rates among the elderly vary by race/ethnicity and gender. As reported in Figure 6.4, older women have higher rates of poverty than older men do. Women are especially susceptible to economic insecurity because of a number of factors. Women have longer life expectancies and are more likely to be widowed and live alone in old age. Because of gender differences in employment and salaries, women are likely to have less retirement income than men do. During 1999, women aged 65 and older received, on average, $8,000 annually as pension income, whereas men received $14,000 annually (He et al. 2005). Non-Hispanic Whites have lower poverty rates than do other reported racial groups. Blacks have the highest poverty rates for both elderly men (17.7) and women (27.4).

In the European Union (EU25), about one in six elderly persons is at risk for living in poverty. In 14 of all EU25 countries, the elderly populations are at higher risk of being poor compared with working-age populations. The risk of poverty is highest in Cyprus, Ireland, and Slovenia. Rates are lowest among the elderly residing in the Czech Republic, Slovakia, Lithuania, Latvia, the Netherlands, and Luxembourg (Zaidi 2006).

Health and Medical Care

Much like their economic vulnerability, the young and the old experience the highest health risks among all age groups. We think of death as something that happens in one's old age; yet, the age group with the highest risk of mortality is newborns and infants within their first year of life. For 2004, the infant mortality rate was 6.79 per 1,000 (Miniño et al. 2005). The three leading causes of death among infants were congenital birth defects, low birth weight, and sudden infant death syndrome. Among the elderly, the majority of deaths are caused by

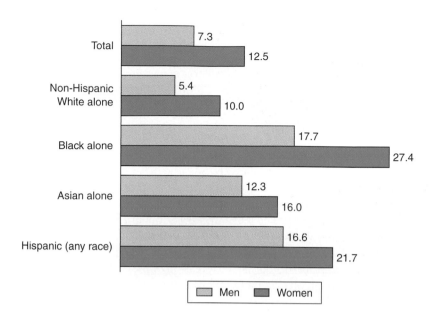

Figure 6.4

Percentage of People Aged 65 and Older in Poverty by Sex, Race, and Hispanic Origin, 2003

Source: He et al. 2005.

Note: The reference population for these data is the civilian noninstitutionalized population.

heart disease, cancer, and stroke. Elderly also experience chronic conditions such as hypertension (high blood pressure) and heart disease that may contribute to fatal disorders. The elderly are more likely to experience chronic illnesses than younger age groups do, 46 percent versus 12 percent (Moody 2006).

Geriatrics is a medical specialty that focuses on diseases of the elderly; some are chronic, but not life threatening, but others will eventually lead to death. The most prevalent chronic disease of the elderly is arthritis, inflammation of the joints and a leading cause of disability in the United States. Other geriatric diseases include osteoporosis (deterioration of bone tissue, prevalent among women), Parkinson's disease (a degenerative neurological disorder), and dementia and Alzheimer's disease (progressive loss of mental abilities and functions). For the last set of diseases, it should be noted that most older adults experience no mental impairment at all (Moody 2006). (A discussion of caregiving and elder abuse is presented in Chapter 7, "Families.")

American elderly, who constitute 13 percent of our population, consume more than 35 percent of total health expenditures—more than 4 times what is spent on younger people (Moody 2006). The United States spends about 16 percent of its gross domestic product (GDP) on health care—the largest expenditure in this category among industrialized countries. During 2005, total health care spending reached $2 trillion, with an average of $6,697 spent per person (Centers for Medicare and Medicaid Services 2006).

According to the Centers for Disease Control (Robinson 2007), older women face distinctly different challenges to maintaining their health compared with older men. The financing and availability of health care is particularly important for older women. Because women make up a higher proportion of the older and frailer population, are less likely to have a spouse to assist them, and need more help with personal care and routine needs, older women use more health care and long-term care services than men do. Most older adults are covered by Medicare, but because older women rely on long-term care (not covered under Medicare), they make higher out-of-pocket payments or rely on Medicaid more often than older men do. Women rely on informal (unpaid) caregivers (adult children, family members, or friends), community-based services (senior centers and convenient transportation), and formal (paid) care services (home health care and nursing home care).

Ageism in the Workplace

In some businesses, age discrimination has an acronym, TFO, "too f——ing old" (Fisher 2004). Although much attention has been given to the marginalization of Black, Latino, or female workers, members of another labor force group—older workers—have experienced their own set of unique problems.

Older workers, those 45 years or older, may find it increasingly difficult to do their work, keep their jobs, or find another position (Swuwade 1996). A survey of senior executives found that 82 percent consider age bias a serious problem in the U.S. workplace. Ninety-four percent of these managers, most of them in their 40s and 50s, said they believed their age had resulted in their not being considered for a particular job (Fisher 2004). In the workplace, older workers have been discriminated against in favor of younger, cheaper, less experienced workers.

It used to be that seniority mattered in the workplace, but in *The Incidence of Job Loss: The Shift from Younger to Older Workers, 1981–1996,* researchers Michele Siegel, Charlotte Muller, and Marjorie Honig (2000) describe how especially vulnerable older workers are in our new economy. The authors explain that during the recession of the early 1980s, men age 45 to 59 were less likely to lose their jobs than were men age 25 to 39. The job loss rate for younger men was 60 percent higher than for men age 45 to 59. It was normal practice that younger, less tenured employees were laid off, maintaining employment for older, more experienced workers. But during the recession of the early 1990s, older men were just as likely to be laid off as were younger men. From 1991 to 1992, the rate of job loss among older workers was identical to the rate for younger workers, and the job loss rate among older college graduates was higher than among younger college graduates (Siegel et al. 2000).

When older workers are laid off, they are more likely to find jobs that are lower paying, temporary, and low-skill in comparison with the jobs they left. Many equate older workers with more experience but also with higher salaries, which discourages potential employers (Fountain 2002). There is also the belief that older workers cannot be (re)trained, hindering their ability to find or retain a new job (Swuwade 1996). Yet, older workers are protected under the Age Discrimination in Employment Act of 1967. The act prohibits employers from discriminating based on age against people 40 to 64 years old.

In their analysis of the effectiveness of similar age discrimination policies in New Zealand, Geoffrey Wood, Mark Harcourt, and Sondra Harcourt (2004) reported widespread noncompliance among employers. Two New Zealand laws, the New Zealand Rights Act of 1993 and the New Zealand Employment Relations Act of 2000, prohibit age discrimination in the workplace. Yet based on job applications from New Zealand businesses, the researchers found that 56 percent of job applications asked one or more unlawful questions about the applicant's age. Wood et al. characterize New Zealand's labor force composition as skewed toward younger workers. Part of the reason is functional (younger workers are cheaper in salaries and pension investments), yet Wood et al. (2004) note that discriminatory managerial practices are so embedded in social and economic structures, they are not easily changed with legislation.

Responding to Age Inequalities

Social Security and Medicare are social insurance programs designed to protect citizens against a specific set of risks. In the case of Social Security, the risk is ensuring that we have sufficient resources at retirement and making sure that these resources will last one's lifetime. How much health care will we need and for how long are the risks addressed by Medicare. Both are proven effective antipoverty programs, supporting elderly financial independence and enabling the elderly to live better and healthier lives (Feder and Friedland 2005). Since their inception, however, both programs have been the subjects of much debate.

Social Security

Social Security was first enacted with the Social Security Act of 1935. A year before, President Franklin D. Roosevelt convened an executive committee to examine economic insecurity, responding to the nation's Great Depression. The committee's recommendation was to create a program to address the long-range problem of the economic security for the aged and poor, focusing not on providing social assistance (such as welfare assistance) but, rather, a social insurance plan against an uncertain future (Social Security Administration and Medicare Board of Trustees 2007). We tend to think of Social Security only as the monthly payments one receives after retiring, but the program also supports unemployment insurance, aid to dependent children, and state grants to provide medical care.

Demographic shifts have lead to new ways to think about aging and economic security throughout the globe. The current public pension system in the United States and in most of Europe is a pay-as-you-go system—current workers support current beneficiaries of the program (Curl and Hokenstad 2006), today's workers pay for today's retirees (Moody 2006). Angela Curl and M. C. Hokenstad (2006) compared the U.S. Social Security system with public pension systems in Sweden and Canada. In Sweden, pensions are based on average life expectancy at the time of retirement. Workers can retire anytime after 61 years of age, but the later they retire, the higher their payments would be. Because of the availability of part-time jobs and the flexible work environment, the program allows Swedes to draw a partial pension for partial retirement. The elderly can mix their employment income with their pension funds. The system is funded by an 18.5 percent payroll tax (the United States collects 12.4 percent). Unlike the U.S. system, Sweden's system is partially privatized. Individuals can invest the money from their pension account or the government invests on their behalf.

The Canadian public pension system is described as combining "income protection for older adults with policies that promote flexibility" (Curl and Hokenstad 2006:95). The Canadian system has three parts: Old Age Security program, the Canada Pension Plan, and private pensions and savings. The minimum age of eligibility for early retirement benefits is 60. Canada collects a 9.9 percent payroll tax to support the program. As in the United States, both Canada and Sweden have residency requirements (requiring that persons must have lived for a certain number of years in the country) and a universal guaranteed monthly minimum benefit. The researchers praised both countries for promoting flexibility in retirement through gradual or partial retirement programs.

In the United States, policy analysts have long warned about the danger of having few workers paying for the benefits for a growing number of retirees. At the time the Social Security system was established, the elderly constituted 5 percent of the population (Moody 2006) and the average life expectancy was about 50 years of age (Curl and Hokenstad 2006). The proportion of the elderly in 2000 grew to 12 percent, projected to increase to 21 percent by 2050 (He et al. 2005). For 2006, Social Security paid benefits of $546 billion to 49 million beneficiaries.

The Social Security Board of Trustees, a nonpartisan panel responsible for reporting on the financial status of the Social Security Trust Funds, projected that tax revenues would fall below actual program costs in 2017. This is the year when the program will spend more than it receives

Photo 6.5

President George W. Bush praises the U.S. Congress for passing the Medicare Reform Law of 2003. The legislation created a prescription drug coverage program for seniors and disabled Americans. Analysts predict that the program will have little effect on slowing down the increasing costs of prescription drugs for the elderly.

through payroll deductions. The trustees predicted that the trust funds would be exhausted in 2041 (OASDI Trustees 2007). Any worker born after 1975 will reach full retirement age after the trust fund is depleted. At the release of their 2007 report, the trustees warned, "The longer we wait to address these challenges, the more limited will be the options available, the greater will be the required adjustments, and the more severe the potential detrimental economic impact on our nation" (Social Security and Medicare Boards of Trustees 2007).

Medicare

The nation's largest health insurance program, Medicare, covers about 43 million elderly or disabled Americans. Medicare has two parts: Part A Hospital Insurance helps pay for care received as an inpatient in critical access hospitals or skilled nursing facilities, as well as some home health care; Part B Physician and Outpatient Coverage pays for medically necessary services and supplies that are not covered under Part A. Each part is financed under a different structure. Part A is offered as an automatic premium because most recipients (and/or their spouses) paid Medicare taxes while they were working. Part A is financed through a payroll tax paid equally by employers and workers. Three-fourths of the financing for Medicare Plan B is received from general tax revenues, and the remaining quarter is financed directly through paid premiums, $96.40 for 2008.

Since its inception, some have been concerned that Medicare spending, especially for Part A, will outpace its revenue sources (Moon 1999). Medicare has been referred to as a "pay as you go" system; every payroll tax dollar that is contributed into the fund is immediately spent by those currently enrolled (Goodman 1998). Because of increases in health care costs, per capita expenditures for Medicare will increase from $5,943 in 1998 to $10,235 in 2025 (Moon 1999). Out-of-pocket expenses are also expected to increase, from $2,508 per beneficiary in 1998 to $4,855 in 2025. One additional projection casts more doubt on the stability of the system. The number of elderly beneficiaries is expected to continually increase. By 2025, the number of eligible elderly is projected to reach 69.3 million, about 21 percent of the U.S. population (Moon 1999).

Political rhetoric continues to swirl around how to preserve the system while trying to expand its services to an ever-increasing population of elderly. In 2003, the U.S. Congress passed the Medicare Reform Law, which included a prescription drug benefit for the first time in the program's history. The bill provides for a prescription drug benefit for older and disabled Americans, offered and managed by private insurers and health plans under contract with the federal government. Critics attacked the plan for its coverage gap, arguing that the bill failed to provide seniors with substantial relief for the cost of prescription drugs. The plan costs about $420 per year and, once covered under the plan, seniors will have to pay a $250 deductible before the coverage begins. Medicare pays 75 percent of annual drug costs between $251 and $2,250, but coverage stops until the senior's annual drug cost reaches $5,100. Once at that point, the plan will cover 95 percent of all additional costs, but by then, seniors would have paid about $3,600 of their own money. Seniors with low incomes will qualify for extra assistance under the bill. The cost of the bill is estimated at more than $900 billion over 10 years. Analysts predict that the reform bill will have little effect on slowing down the increasing costs of prescription drugs and medical services for the elderly.

In 2007, the U.S. House of Representatives passed the Medicare Drug Price Renegotiation Act, giving the Secretary of the U.S. Department of Health and Human Services the authority to negotiate with drug companies regarding the price of prescription drugs for Medicare beneficiaries. In response, the Bush administration stated that it did not believe the act would have any impact on the price of prescription drugs for the elderly and threatened to veto the bill if passed by the Senate.

In Focus

Senior Political Power

The enactment of Social Security, Medicare, and other old age related policies has created a political constituency of older beneficiaries (Campbell 2003) and along with it, the perception of political might. A form of ageism, the stereotype of "greedy" older voters, willing to put their needs (Social Security and Medicare) ahead of other age groups is accepted in many social, political, and media circles (Street and Crossman 2006).

Robert Binstock (2005, 2006–2007) reveals some truths and myths to this perception of senior power. He explains that the senior political power model builds on the fact that older people represent a significant proportion of the electorate. Data in Table 6.3 reveal how the rate of voting among people aged 65 or older was consistently highest among all reported age groups.

Age is associated with voter registration and actual voting, which is also related to the length of residence in one's home, along with one's level of knowledge about political and social issues. Older persons are more likely to have resided in their homes longer than have younger persons, and older people tend to be more knowledgeable about politics and news issues than younger people are. Studies indicate that older persons' high level of interest in politics does not decline even as they reach advanced old age.

Table 6.3

Percentage of Voting-Age People Who Voted in the Presidential Elections, 1980–2004

Age	1980	1984	1988	1992	1996	2000	2004
18–24	40	41	36	43	32	32	42
25–44	59	58	54	58	49	50	52
45–64	69	70	68	70	64	64	67
65+	65	68	69	70	68	67	69

Source: Binstock 2006–2007.

In addition to their higher rates of voting participation, the elderly also have higher rates of participation in other political areas. Elderly make campaign contributions at higher rates than do younger people—about 28 percent of all contributions to the 2000 presidential campaign. Older voters are twice as likely to contact their state and federal representatives about issues that matter to them than younger voters are. Doris Haddock's story as a successful grassroots lobbyist is highlighted in this chapter's Voices in the Community feature.

Photo 6.6

Seniors continue to be perceived as an important voting constituency. Politicians frequently visit senior and retirement centers during election campaigns.

So, what impact does the voting participation by elderly Americans have on elections? One noticeable impact is on how it shapes candidate behavior—candidates actively court the senior vote. Recent presidential candidates have made key appearances at senior centers or promoted their proposals on Social Security and Medicare to selected senior audiences.

But Binstock argues that the senior vote does not have a "distinctive impact on the outcome of elections" (2006–2007:26). He explains that older Americans do not vote cohesively or as a bloc. Their votes are as diverse as that of any other age group, divided along partisan lines, class, gender, and race. Age is just one characteristic among many that influence voting patterns. For example, in the 2004 presidential election, the 65 or older votes were distributed in approximately the same proportions—52 percent for President George W. Bush and 47 percent for Senator John Kerry—almost identical to the general electorate—51 percent for Bush and 48 percent for Kerry.

Research reveals that young and old voters support the Social Security program. Actually, more recently born cohorts (post-Depression and post–New Deal cohorts) support increasing Social Security spending more than do members of birth cohorts currently receiving Social Security benefits. Overall, higher percentages of prime-working-age groups support increased social spending in health, education, and Social Security than do older age groups (Street and Crossman 2006).

Despite what is known about how older voters actually vote, the image of senior power persists because it serves the purposes of several interest groups. First, the leaders of old-age-based organizations (such as AARP) have their own set of incentives to inflate the political importance of the constituency they represent. Second, other political groups use senior power as a "straw man" to argue that more resources should be allocated to their causes. Finally, journalists use senior power "as a tabloid symbol that simplifies the complexities of politics" (Binstock 2005:77).

Voices in the Community

Doris Haddock

"Eight years ago, Doris Haddock, known affectionately as 'Granny D,' embarked on a remarkable journey. She was 89 years old. Despite an arthritic back, she walked—and skied—all the way across America, holding tight to her trademark straw hat while spreading the word about the need for federal campaign finance reform.

"Granny D weighed less than 45 kilograms back then. She's even tinier now. A self-described 'class clown' growing up, she studied elocution in college. But her only speeches at first were to clerks in the shoe-company office that she ran. Doris Haddock soon had her say, though. She and her husband, Jim, shared a passion for social justice, and for decades, they championed many causes together until Jim's death in 1993.

"One day five years later, Granny D's eyes fastened on a lone paragraph in a newspaper article about two U.S. congressmen who had slipped $50 million into an appropriations measure late one night. The money was a subsidy for a tobacco company.

"Furious, Granny D picked up the phone and called the advocacy group Common Cause. 'I said, "But that's corruption!" They said, "Of course it's corruption. You cannot run (for Congress) today without being a multi-millionaire; or you take illegal money from corporations or special interests." I wasn't even conscious of this. I said, "I'm old, I'm tired. What can I do?" And they said, "Try to get the McCain-Feingold Bill passed." So I said, "OK."'

"McCain-Feingold was a bipartisan measure pending in Congress to eliminate unregulated donations, or 'soft money,' from federal election campaigns. But Haddock wondered how one person could help make the bill the law of the land.

"She talked the matter over with her friends in a local discussion group that calls itself the 'Tuesday Morning Academy.' Before long, they were contacting friends, who were contacting THEIR friends all across America, getting signatures on petitions in support of campaign finance reform.

"Granny D says those petitions were sent to the senators. 'And within a couple of months, we got letters back saying, "Dear Little Old Ladies. Don't worry yourselves. We are going to pass the McCain-Feingold Bill." Two months later, they didn't pass it. I went into deep depression. All the ladies who had been helping me left, and I was stuck with egg on my face.'

"Soon afterward, Granny D and her son happened to drive past an old man who was walking down a road, carrying a peace sign. 'What if I did that?' she asked her son. 'My son said, "You have to have a reason." I said, "Oh, I have a reason." He said, "Oh, my God. Campaign finance reform." And I said, "Yes!" He said, "You don't have any money. How are you going to do it? You live on Social Security." I said, "I'll go as a pilgrim. I'll walk until given shelter, fast until given food."'

"And that's exactly what she did for 14 straight months, from Pasadena, California, to Washington, D.C. Along the way, she relied on strangers for food and a place to lay her head at night. 'Granny's Angels,' she called them. Trudging along, she had no idea whether her journey would make any difference. 'I'm walking for future generations,' she told herself. 'For my 16 great-grandchildren.'

"However, she says, 'As I progressed, I became a circus. People were lining up on either side of the road, waiting for me. Or they were telling people ahead, "She's coming! She's coming!"'

"Granny D reached the nation's capital after skiing through the snow on the towpath of an old canal that reaches the city from the Maryland mountains. Only three people greeted her at her last stop, at Arlington National Cemetery.

"But then a Metro subway train pulled up to the cemetery station. 'And suddenly the cars stopped, and people poured out. And the next one came, and they poured out. So I had 2,300 people walking with me. And when we got near, quite a few of the senators and House members walked with me, too. It was grand! And we went down K Street, because that's where the lobbyists are. And girls in these great, high buildings put out signs saying, "Go Granny, Go!" I was in tears to think that people cared.'

"That was in early 2000. The McCain-Feingold bill passed the Senate and the House, and was signed into law by President Bush two years later.

"Granny D went back to New Hampshire, but not to a rocking chair. She spoke at several rallies, and, in 2005, delivered the graduation address at Hampshire College in Amherst, Massachusetts. Showing the effects of emphysema brought on by 50 years of smoking, she told the audience that she had recently been jailed for reading from the U.S. Constitution in the rotunda of the Capitol Building in Washington.

"'As I thought that did violence to my free-speech rights under the Constitution, I went back and read from the [constitutional amendments called the] Bill of Rights,' she told the graduates. 'That landed me in jail, too. I felt freer in that jail, because I had spoken out as a free person, than I have ever felt in the open air. Be involved! Get involved! Stay involved!'"

Source: Quoted material is from Landphair 2007.

Main Points

- Age is both a biological and social classification. We have socially and culturally defined expectations about the meaning of age, our understanding of it, and our responses to it. Age distinguishes acceptable behavior for different social groups.

- Sociologists examine age and the process of aging through a **life course perspective**. This perspective examines the entire course of human life from childhood, adolescence, and adulthood to old age. **Gerontology** is the specific study of aging and the elderly, which is the primary focus of this chapter.

- **Demography** is the study of the changes and trends in the population and demographers have identified several reasons for population aging. First, population aging is caused by a decline in birth rates. Second, population aging can also occur because of an improvement in life expectancy as a result of medical and technological advances. Finally, the process of population aging can be influenced because of birth cohorts.

- Populations are aging in all countries, though the level and pace vary by geographic region. In 2000, 420 million people (7 percent) in the world were 65 and older; by 2030, the number is projected to increase to 974 million (12 percent). In 2000, China and India had the largest elderly populations, 87.5 million and 46.5 million, respectively. The United States was ranked third in the world, with 35 million.

- Functionalists rely on **disengagement theory** to explain aging as a natural process of withdrawal from active participation in social life. The theory contends that people enter and exit a set of roles throughout their lives. Conflict theorists offer the **modernization theory of aging** to explain how the role and status of the elderly declines with industrialization. Their power, wealth, and prestige are linked with their labor contribution or their relationship to the means of production. According to the feminist perspective, the standards of our culture create more problems for women than for men as they transition into their middle and later years. Women are more vulnerable to societal pressure to retain their youth. Interactionists reveal how our age-related roles are socially defined and expected. Age is tied to a system of matching people and roles, some stigmatized.

- Age also serves as a basis for prejudice or discrimination. **Ageism** is defined by Robert Butler as the "systematic stereotyping of and discrimination against people because they are old, just as racism and sexism accomplish this with skin color and gender" (1969:243). Age distinguishes acceptable behavior for different social groups. Research suggests that ageist attitudes may affect a physician's therapeutic decisions toward older patients.

- Social Security and Medicare are U.S. social insurance programs designed to protect citizens against a specific set of risks. Both have proven to be effective antipoverty programs for older Americans, supporting financial independence and enabling the elderly to live better and healthier lives. Since their inception, however, both programs have also been the subject of debate.

On Your Own

Log on to the Web-based student study site at www.pineforge.com/leonguerrero2study for interactive quizzes, e-flashcards, journal articles, Community and Policy Guides, a Service Learning Guide, the end-of-chapter Web exercises, and additional Web resources.

Internet and Community Exercises

1. The Gray Panthers is a national advocacy organization for older and retired adults. Organized in 1970 by Maggie Kuhn and five of her friends, the Gray Panthers' first goal was to combat ageism. The Gray Panthers have also taken a stand on other important social issues: economic justice, medical care, education, and peace. Recent efforts have been directed toward policies ensuring affordable prescription drugs for seniors, children, cancer, and HIV patients. Gray Panthers have more than 50 local chapters. To see if there is a chapter in your state, log on to *Study Site Chapter 6*. Contact the chapter for information about what activities are supported in your area.

2. The National Institute on Aging (NIA), part of the National Institute of Health, leads the country's scientific effort to understand the nature of aging and to extend the healthy, active years of life for Americans. Formed in 1974, the NIA provides leadership in aging research, training, health information dissemination, and other programs relevant to aging and older people. Its Web site includes resource materials on healthy aging, caregiving, medications, dietary supplements, and diseases. Log on to *Study Site Chapter 6* for links to the NIA resource page.

3. The private, not-for-profit Alliance for Aging Research is the nation's leading citizen advocacy organization for improving the health and independence of older Americans. Founded in 1986, the goal of the Alliance for Aging Research is to promote medical and behavioral research into the aging process. The organization also serves as an advocate of consumer health education and public policy. On its Web site, Alliance for Aging Research includes surveys and quizzes pertaining to healthy aging, such as heart disease, general aging, and cognition. Log on to *Study Site Chapter 6* to test your knowledge in these areas.

4. The percent of the population 65 years of age or older is reported in the United Nations *Human Development Report* for the United States, along with other countries. The percent is also projected for 2015. Determine where the United States is ranked among the top-20 listed countries by logging on to *Study Site Chapter 6* for links to the UN report.

PART III

Our Social Institutions

É mile Durkheim first described the importance of social institutions in society. He likened them to organs in a human body. Each organ does one specific thing—a heart is responsible for our circulatory system, the brain is the key to our nervous system, and our lungs regulate our respiratory system. If something happens to the heart, its functions cannot be assumed by other organs and as a result, the body as a whole becomes compromised.

Social institutions are defined as a stable set of statuses, roles, groups, and organizations that provides a foundation for addressing fundamental societal needs (Newman 2006). For our discussion, we will focus on six institutions—family, education, work and the economy, health and medicine, and the media—each with a specific role in society. In the chapters that follow, we will review the importance of each institution, but also highlight the social problems affecting it. The bases of inequality we reviewed in Part II are intricately related to the social problems based in each institution. Each chapter will conclude with a discussion about the policies and community efforts that are being made to address these social problems.

What Is a Family?

We have all grown up with an understanding of what a family is. If asked to describe a "typical" family, most Americans would first think of something like the nuclear family shown in this image: a woman and a man and their children.

And yet, our description could be broader, encompassing grandparents, aunts, uncles, and cousins—in other words, an extended family.

What is it about the two groups of people shown in these images that makes them seem like a "family"?

Now think about your own experience: the family you grew up in and the family you are a part of now. Is your family consistent with the previous images? Does your own family fit the traditional definition of "family"? Or does your family more closely resemble these other images?

▲ The social reality of what a family is has changed considerably in the past few decades. But our social definition of what constitutes a "typical" family is changing more slowly.

What social problems might be caused by the conflict between our mythology—the notion that most families are "traditional" families—and the reality of the growing diversity in family form?

CHAPTER 7

Families

D o you recall the first "crew" you ever hung out with? You know the one—all of its members know each other, speak the same language, laugh at the same jokes, dress alike, and maybe even look alike. Sound familiar? These are the people that you slept, ate, and lived with in your own home: your family.

You may not always think of it this way, but your family is part of the larger social institution of "the family." Consider for a moment that your family was among the 111 million family groups counted by the U.S. Census in 2003 (Fields 2004). What does your family have in common with the other family groups? Your first response might be that your family has nothing in common with the others. No other family has the unique arrangement or history of individuals related through blood or by choice. From this position, any problem experienced by your family, such as a divorce, would be defined as a personal trouble. A divorce is a private family matter kept among immediate family members. It would be none of anyone else's business.

If we use our sociological imagination, however, we can uncover the links between our personal family experiences and our social world. Divorce is not just a family matter but also a public issue. Looking at the recent divorce rate of four divorces per 1,000 people, divorce doesn't occur in just one household, but in millions of U.S. households. It affects the economic and social well-being of millions of women, men, and children. Divorce challenges the fundamental values of home and love and the value of the family itself. Divorce could be everyone's business.

In this chapter, our goal is to explore this private, yet public world of the family. For our discussion, we'll define the **family** as a construct of meaning and relationships both emotional and economic. The family is a social unit based on kinship relations—relations based on blood, and those created by choice, marriage, partnership, or adoption. A **household** is defined as an economic and residential unit. These definitions will allow for the diversity of families that we'll discuss in this chapter, while not presenting one configuration as the standard. As you'll see, the family as we think we know it may not exist at all.

Photo 7.1

Shawn Hornbeck is embraced by his mother Pam Akers during a press conference. Missing since October 2002, Hornbeck was found in 2007 living with another boy and their abductor, Michael Devlin, in St. Louis, Missouri. Devlin was sentenced to multiple life terms for his crimes.

Myths of the Family

The image of the nuclear family—a father, a mother, and biological or adopted children living together—is exalted as the ideal family. Such television families as the 1960s Cleavers, the 1980s Huxtables, and the 1990s Barones reinforce this ideal image of families. Millions of faithful viewers tuned in each week to see how the family handled (and always resolved at the end of the half hour) the current family crisis. Our image of the nuclear family has been transformed over the years. In the beginning, the only image was of a stay-at-home mom like June Cleaver, but the image has been modified to include a working mother like Claire Huxtable or a stay-at-home mom like Debra Barone. We may admire and idolize them, but these television families do not represent most U.S. families.

Jason Fields (2004) identifies several changes in family composition between 1970 and 2003. First, the percentage of families composed of married couples with children declined from 40.3 percent in 1970 to 23.3 percent in 2003. Take note of this—the nuclear family form applies to less than one-quarter of all 2003 U.S. households. The largest family form for 2003 is married couples without children (28.2 percent). Second, the only increase in family groups during that period came in the category of "other family households." Other family households include families whose householder has no spouse present, but such families may include other relatives, as well as children. These family groups made up 16.4 percent of all U.S. households, increasing from 10.6 percent in 1970. Included in this 2003 category are 12 million single-parent families, 10 million headed by mothers and 2 million headed by fathers.

Finally, there has been an increase in the percentage of all nonfamily households, individuals living alone or with non-relatives. (Refer to Table 7.1.) In 2007, U.S. Census data revealed that for the first time more American women were living without a husband than with one. Based on 2005 data, 51 percent of women were living without a spouse, an increase from 35 percent in 1950 and 49 percent in 2000. In contrast, 47 percent of men reported living without a spouse in 2005. The pattern varies by race, with 30 percent of Black women, 49 percent of Hispanic women, 55 percent of non-Hispanic White women, and 60 percent of Asian women reporting that they were living with a spouse. Analysts reported

Table 7.1

Households by Type: 1970 and 2003 (percent distribution)

	1970	2003
Total number of households	63 million	111 million
Nonfamily households		
Men living alone	5.6	11.2
Women living alone	11.5	15.2
Other nonfamily households	1.7	5.6
Family households		
Married couples with children	40.3	23.3
Married couples without children	30.3	28.2
Other family households	10.6	16.4

Source: Fields 2004, Figure 2.

Note: Columns may not add to 100 percent due to rounding.

that several factors contributed to the increase in the number of single women: late marriage for women, women living longer as widows, or after a divorce, and women being more likely than men to delay remarriage (Roberts 2007).

What Does It Mean to Me?

Even though the nuclear family is not the statistical majority in our society, its image as the "perfect" family persists. Why do you think this is the case? How could a different image of the family be accepted by society?

In addition to the false image of the nuclear family, we also embrace other myths about the family. We tend to believe that the families of the past were better and happier than modern families are. We believe that families should be safe havens, protecting their members from harm and danger. And a final myth relates to the topic of this book: We also assume that the family and its failings lead to many of our social problems.

There is a persistent belief that nontraditional families, such as divorced, fatherless, or working-mother families, threaten and erode the integrity of the family as an institution. These "pathological" family forms are blamed for drug abuse, delinquency, illiteracy, and crime. As a group, female-headed households were condemned in the Personal Responsibility and Work Opportunity Act of 1996, also known as the Welfare Reform Act, for their dependency on the public welfare system. During his presidency, George W. Bush advanced several program initiatives intended to strengthen American families.

Sociological Perspectives on the Family

Functionalist Perspective

From a functionalist perspective, the family serves many important functions in society. Some functionalists claim that the family is the most vital social institution. The family serves as a child's primary group, the first group membership we claim. We inherit not only the color of our hair or eyes but also our family's social position. The family confers social status and class. The family helps define who we are and how we find our place in society. And, without the family, who would provide for the essential needs of the child: affection, socialization, and protection?

Social problems emerge as the family struggles to adapt to a modern society. Functionalists have noted how many of the family's original functions have been taken over by organized religion, education, work, and the government in modern society (Lenski and Lenski 1987), but still, the family is expected to provide its remaining functions of raising children and providing affection and companionship for its members (Popenoe 1993). From a functionalist perspective, the family is inextricably linked to the rest of society. The family does not work alone; rather, it functions in concert with the other institutions. Changes in other institutions, such as the economy, politics, or law, contribute to changes and problems in the family.

Consider the education of children. Before the establishment of mandatory public school systems, the family educated its own children. As a formal system, education has become its own institution, taking primary responsibility for educating everyone's children. Functionalists examine how the institutions of the family and education effectively work together. To what extent should parents participate in their children's education, and to what extent is the educational system responsible for raising our children?

Because of this perspective's emphasis on the family and its social and emotional functions, when the family fails—as in the case of divorce or domestic violence—functionalists take these problems seriously. These problems afflict the family and, according to functionalists, can lead to additional problems in the society, such as crime, poverty, or delinquency.

Conflict and Feminist Perspectives

For the conflict theorist, the family is a system of inequality where conflict is normal. Conflict can derive from economic or power inequalities between spouses or family members. From a feminist perspective, inequality emerges from the patriarchal family system, where men control decision making in the family. The persistent social ideals that view the woman as homemaker and the man as breadwinner are problematic from both perspectives. Men's social and economic status increases as their work outside of the home is more visible and rewarded, whereas women's work inside the home remains invisible and uncompensated.

The family structure, as our society has come to define it, upholds a system of male social and economic domination. Theorist Friedrich Engels argued that the family was the chief source of female enslavement. Within the family, Engels believed the husband represented the bourgeois, whereas his wife represented the proletariat (Engels 1902). Just as members of the proletariat were oppressed in the economy, women were oppressed at home.

According to feminist theory, men can maintain their position of power in the family through violence or the threat of violence against women. Feminists argue that domestic violence cannot be solely explained by men's individual attitudes and behavior. Rather, violence against women is linked to larger social structures of male dominance, such as political, economic, and other social institutions like the family. Theorists have encouraged the integration of feminist analyses of gender and power to better comprehend the mechanisms leading to violence, arguing that to stop such violence, structures of gender inequality must change (Brownmiller 1975).

Feminist theorists have also acknowledged how the cultural differences of domestic violence have not been fully recognized. According to Tricia Bent-Goodley (2005:197), "Research has largely focused on White and poor women, despite the fact that domestic violence crosses race, ethnicity, socioeconomic status, religion and sexual orientation." She notes that this narrow research focus has encouraged the perception that women of color, middle- and upper-class women, and women in same sex-relationships do not experience domestic violence, despite evidence to the contrary.

Families are also subject to powerful economic and political interest groups that control family social programs and policies. Conflict arises when the needs of particular family forms are promoted while others are ignored. For example, formal law and social policies recognize heterosexual marriages, but homosexual marriages and families are largely ignored. To resolve social problems in the family, both perspectives suggest the need for structural change.

Interactionist Perspective

Through social interaction, we create and maintain our definition of a family. As we do this, it affects our larger social definition of what everyone's family should be like and how we envision the family that we create for ourselves.

Within our own families, our interaction through words, symbols, and meanings defines our expectation of what the family should be like. How many children in the family? Who does the house cleaning? Who gets to carve the holiday turkey? As a family, we collectively create and maintain a family definition on which members agree. Problems arise when there is conflict about how the family is defined. A couple starting their own family must negotiate their own way of doing things. Two partners may carry definitions and expectations from their families of origin, but together, they create a new family reality.

Problems may also occur when partners' expectations of family or marriage do not match their real lives. In our culture, romantic love is idealized, misleading individuals to believe that they are destined for a fulfilling emotional partnership with one perfect mate. After the realities of life set in, including the first fight, the notion of romantic love is shattered. Couples recognize that it may take more than romantic love to make a relationship work.

Society assigns meaning to particular family groups or relations. More than half a century ago, when a child was born to unmarried parents, it was assumed that the child was unwanted and that the child's future would be less than promising. There was a major social stigma with being referred to as a bastard child. But because currently one in three births involves parents who are not legally married, and most births are wanted or planned, the use of the term "bastard" has disappeared as well as the stigma attached to such a birth (Rutter and Tienda 2005).

Putting all our separate definitions of the family together, we create a portrait of what the American family should be like. But as political and religious forces uphold and encourage a patriarchal nuclear family as the norm, by default, other family forms are considered deviant or against some set of moral codes. Blended families, gay or lesbian families, or single-parent families become social problems based only on how they deviate from the definition of a "normal" family. Refer to Table 7.2 for a summary of all sociological perspectives on the family.

Table 7.2

Summary of Sociological Perspectives: The Family

	Functional	Conflict/Feminist	Interactionist
Explanation of the family and its social problems	Functionalists examine how the family interacts with other social institutions. Functionalists believe that social problems emerge as the family struggles to adapt to a modern society. Functionalists also examine the manifest and latent functions of the family.	Problems emerge from conflict inherent in the family structure. Conflict can derive from economic or power inequalities between family members. From a feminist perspective, the family is a patriarchal system, where men dominate social and economic spheres.	An interactionist focuses on the social meaning and expectations of the "family." Interactionists will also focus on how family members define their own families.
Questions asked about the family	What functions does the family serve? How is the family affected by other social institutions? How does the problem reflect changes between the family and other social institutions?	What is the basis of conflict within the family? How does conflict affect family members and their relationships?	How do we define the "family"? What social forces influence our definition of the family? What are the consequences of these definitions?

Problems in the Family

Divorce

If you do an Internet search on divorce, you might be surprised at your search results. In addition to divorce facts and access to support groups, you'll also find handy guides to complete your own divorce paperwork. Looking to save time, money, and pain? Please try our services. Looking for a divorce lawyer, why not search for one online? And if you'd like to send a divorce greeting card that says, "Happy to be without you," you can find one of those online, too.

Divorce was a rare occurrence until the 1970s. In the 1950s and 1960s, the divorce rate was about 2.2 to 2.6 per 1,000 individuals (U.S. Census Bureau 1999). With the introduction of no-fault divorce laws in the 1970s, the divorce rate began to climb, reaching a high of 5.3 in 1979 and 1981 (U.S. Census Bureau 1999). The increase in divorce rates has been attributed to other factors: the increasing economic independence of women, the transition from extended to nuclear family forms, and the increasing geographic and occupational mobility of families. Furthermore, as our societal and cultural norms about divorce have changed, the stigma attached to divorce has decreased.

In recent years, the divorce rate has remained stable around 4.0 divorces per 1,000 individuals in the total U.S. population. The rate was 4.2 in 2000, declining to 3.6 in 2005 (Centers for Disease Control 2002 and 2006). The U.S. marital rate declined for the same period: 8.7 for 2000 to 7.5 for 2005 (Centers for Disease Control 2002 and 2006). When compared with

European Union (EU) countries, the United States has the higher divorce rate. In 2005, the estimated divorce rate for the European Union was 2.0 per 1,000 (EuroStat 2005a). EU marital rates were also lower for 2005, 4.88 per 1,000 (EuroStat 2005b).

Recent census data on divorce indicate that certain groups are more susceptible to divorce than are others. Based on 2001 Census data, Rose Kreider (2005) reported although one in five U.S. adults has ever been divorced, the percentage of ever divorced was highest among men and women 50 to 59 years of age (41 and 39 percent, respectively). The majority of separated and divorced men and women were between 25 and 44 years of age. The median age of divorce from first marriage was 29.4 years for women and 31.4 years for men. Divorce rates are higher among couples married before 20 years of age, living at 200 percent of the poverty level, with a high school degree or some college, or working full-time (Kreider and Fields 2002).

Sociologists have paid particular attention to immediate and long-term effects of divorce on children. In general, the research indicates that children with divorced parents have moderately poorer life and educational outcomes (emotional well-being, academic achievement, labor force participation, divorce, and teenage childbearing) than do children living with both parents (Amato and Keith 1991). For example, boys living with a divorced mother are four times more likely to display severe delinquency or to engage in early sexual intercourse than are those living in two-parent households (Simons 1996). Some of these effects carry into adolescence and young adulthood (Amato and Keith 1991; Cherlin, Kiernan, and Chase-Lansdale 1995), with more negative outcomes for adult females than for adult males, such as a greater incidence of relationship conflict or difficulty with intimate relationships.

Conversely, research also suggests that marital separation is beneficial to the well-being of children (Videon 2002). Favorable outcomes have included increased maturity, enhanced self-esteem, and increased empathy among children from divorced families (Brooks Conway, Christensen, and Herlihy 2003). Studies have also suggested that parental divorce may shift children from traditional sex-role beliefs and behavior toward more androgynous attitudes and behavior (MacKinnon, Stoneman, and Brody 1984). Research indicates that parent-child relations are important influences on children's well-being, even mediating the effects of marital dissolution (Videon 2002). Divorce is less disruptive if both parents maintain a positive relationship with the child, if parental conflict decreases after separation or divorce, and if the level of socioeconomic resources for the child is not reduced (Amato and Keith 1991).

Research consistently indicates that although men experience minimal economic declines after divorce, most women experience a substantial decline in household income and increased dependence on social welfare (Smock 1994). Data from the 2001 U.S. Census show that although only 15 percent of recently divorced men lived in households where they (or someone they lived with) received noncash public assistance, more than twice as many recently divorced women (or someone they lived with) received noncash public assistance (34 percent) (Kreider 2005). As first reported in Chapter 2, the poverty rate among female-headed households (no male present) is 29 percent compared with 13 percent among male-headed households (no female present).

In his analysis of 14 countries in the European Union, Wilfred Uunk (2004) documented a 24 percent decline in women's median household income after divorce, comparable, he says, to the income change experienced by U.S. women after divorce. His research revealed how median income declines were lower among divorced women from Southern European countries (Greece, Italy, Spain, and Portugal) and Scandinavian countries (Demark and Finland), but higher among women from Austria, France, Luxembourg, and the United Kingdom.

Decline in economic well-being is also true among women from previously cohabitating couples (Avellar and Smock 2005). Cohabiting couples have more precarious financial circumstances, experiencing lower personal and household incomes than married couples

The Family in Sweden

Family sociologist David Popenoe writes,

Many Americans have long had a ready answer to America's family problems: We should be more like Scandinavia. Whether the issues are work-family, teen sex, child poverty, or marital break-up, a range of Scandinavian family and welfare policies is commonly put forth with the assertion that, if only these could be instituted in America, family life in our nation would be improved significantly. (2006:68)

Focusing on Sweden, Scandinavia's largest country, Popenoe argues that there are fundamental differences in the family experience in Sweden and the United States. He concludes that it would be a mistake to think that what works in Sweden could be successfully transferred to the United States.

First, he reminds us that the population of the Scandinavian countries is smaller and more homogenous than that in the United States. The population of Sweden is 9 million, compared with the U.S. population of 300 million.

Second, the marital rate in Sweden is one of the lowest in the world. The number of marriages per 1,000 unmarried women in 2002 was 43.4 in the United States compared with 17.5 in Sweden. Popenoe observes that if this rate holds, only about 60 percent of Swedish women today will ever marry (compared with more than 85 percent in the United States). This is a recent demographic shift—in the 1950s, the marital rate was 95 percent for United States and 91 percent for Sweden.

Popenoe attributes Sweden's low marital rate to Sweden's weak religious structure, a dominant left-wing ideology, and a government that has removed all state benefits for marriage. A secular country, Sweden has no religious pressure to maintain the image and sanctity of marriage. There are no religious or cultural stigmas against cohabiting couples. Popenoe writes that the political left wing in Europe has taken an "antagonistic" stance against strong families—based on feminist concerns about patriarchy, an antipathy toward a bourgeois institution linked to privilege, and the belief that families are an impediment to full equality. All government benefits are given to individuals regardless of their marital or cohabiting status. There are no spousal health care benefits in Sweden as there are in the United States, and individuals are taxed (there is no joint income tax).

As a result, Sweden has a higher rate of nonmarital cohabitation than other Western countries do. Popenoe reports that nearly all Swedish couples live together before marriage compared with two-thirds of U.S. couples. Many Swedish couples don't marry even after children are born. The nonmarital birth rate is higher in Sweden than in the United States, 56 percent versus 35 percent. However, most nonmarital births in Sweden, about 90 percent, are to biological couples living together. In the United States, 60 percent of nonmarital births occur with single non-cohabiting mothers.

Finally, Sweden's divorce rate is about 40 percent compared with 50 percent in the United States. Popenoe attributes the higher U.S. rate to our ethnic, racial, and religious homogeneity and higher residential mobility. In addition, Americans marry at a younger age (another risk factor)—25 years of age for women and 27 years of age for men. In Sweden, the age at first marriage is 31 for women and 33 for men. All Swedish married couples with children 16 years of age or younger, if they want a divorce, need to go through a mandatory six-month waiting period before the divorce is declared final.

have. After the dissolution of a relationship, the level of household income for cohabiting men declines 10 percent but declines 33 percent for cohabitating women. After the dissolution of a cohabiting relationship, women have a higher level of poverty than men do, 30 percent versus 20 percent. Hispanic and African American women are more vulnerable than are White women to experiencing economic decline. Cohabiting relationships may also include children affected by the same economic decline. These children are even more vulnerable because cohabiting mothers have less access to their former partner's income than divorced mothers do.

Violence and Neglect in the Family

Intimate Partner Violence

One of the myths mentioned at the beginning of this chapter was how the family was a safe place for its members. This myth ignores the incidence of violence and abuse in families. Family violence is unique because the aggressor and the victim(s) are part of the same relational unit, with emotional bonds, attachments, and particular power dynamics (Breines and Gordon 1983).

In the United States, nearly 25 percent of surveyed women and 8 percent of surveyed men reported that they were raped or physically assaulted by a current or former intimate partner in their lifetime. Based on these estimates, approximately 1.5 million women and 835,000 men are raped or physically assaulted by an intimate partner annual in the United States (Tjaden and Thoennes 2000). (Refer to Table 7.3 for Bureau of Justice data from 1993 to 2004.) In 2001, 671,710 violent crimes were committed by an intimate partner: current or former spouses, boyfriends, or girlfriends. In about 85 percent of these crimes, the victims were women (Rennison 2003). Data confirm that violence by an intimate partner is a common experience worldwide. A review of more than 50 population-based studies in 35 countries revealed that between 10 and 52 percent of women reported that they had been physically abused by an intimate partner at some point in their lives (Garcia-Moreno et al. 2006).

Table 7.3

Victim-Offender Relationship in Nonfatal Violent Victimization, by Victim and Gender, 1993–2004

| Victim/offender relationship | Average Annual Rate per 1,000 Age 12 or Older | | | |
| | Female | | Male | |
	Rate	Percent 100%	Rate	Percent 100%
Intimates	6.4	22.0	1.1	2.9
Other relatives	2.2	7.7	1.4	3.7
Friend/acquaintances	10.6	36.4	13.3	33.8
Stranger	9.9	33.9	23.5	59.6

Source: U.S. Bureau of Justice 2006.

Research has consistently linked the specific social factors to family violence: low socio-economic status, social and structural stress, and social isolation (Gelles and Maynard 1987). Feminist researchers argue that domestic violence is rooted in gender and represents men's attempts to maintain dominance and control over women (Anderson 1997). Comparative data reveal how the severity and pattern of violence against women is higher in countries with high societal violence and low empowerment of women (Garcia-Moreno et al. 2006).

Studies have also documented how domestic violence is a significant predictor in maternal parenting behavior. Most intimate partner violence involving a female victim is likely to occur in a household where a child is present (refer to Figure 7.1). According to A. Levendosky and S. Graham-Bermann (2000), psychological abuse, rather than physical abuse, is more likely to negatively affect a mother's parenting, which in turn is related to children's behaviors. The researchers report that a mother's experience with psychological abuse was significantly related to a child's antisocial behavior. Children in middle childhood begin to identify with the aggressor and act in emotionally aggressive ways toward their mothers. Yet, female

a. Female Victim Households

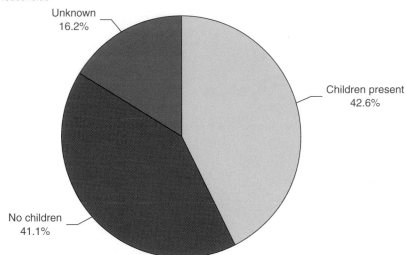

b. Male Victim Households

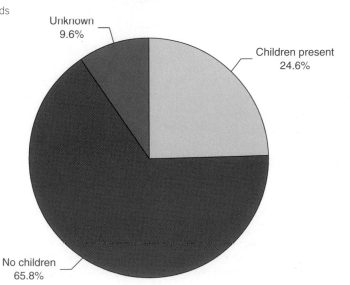

Figure 7.1

Percentage of Household Experiencing Nonfatal Intimate Partner Violence Where Children Under Age 12 Resided, by Gender of Victim, 1994–2004

Source: U.S. Bureau of Justice 2006.

victims identify positive and negative impacts on their parenting because of domestic violence (Levendosky, Lynch, and Graham-Bermann 2000). Battered women report that their emotional feelings or concerns make parenting difficult, noting the reduced amount of quality time or emotional energy they can devote to their children. But these women also report that they had increased empathy and caring toward their children. Researchers suggest that battered women were actively working to protect their children from the effects of violence in their household.

Child Abuse and Neglect

Some children are subject to abuse and neglect in their families. The immediate emotional and behavioral effects of abuse and neglect—isolation, fear, low academic achievement, delinquency—may lead to lifelong consequences, including low self-esteem, depression, criminal behavior, and adult abusive behavior (Child Welfare Information Gateway 2006). In 2004, 872,000 children were victims of child maltreatment. Most child victims, about 62.4 percent, suffered from neglect. The highest rates of victimization (per 1,000 children) were among ethnic minorities: African Americans (19.9 percent), Pacific Islanders (17.6 percent), and American Indians or Alaska Natives (15.5 percent). Asian children had the lowest rate of victimization, 2.9 percent. Tragically, about 1,490 children died as a result of abuse or neglect in 2004 (Children's Bureau 2006).

According to M. McKay (1994), child abuse is 15 times more likely to occur when spousal abuse occurs. Children are 3 times more likely to be abused by their fathers than by their mothers. Poverty is consistently identified as a risk factor for child abuse. It is not clear whether the relationship exists because of the stresses associated with poverty or if reporting is higher because of the constant scrutiny of poor families by social agencies. Poor health care, lack of social and familial support, and fragmented social services have been linked with both poverty and child abuse (Bethea 1999).

While physical, sexual, and emotional abuses are often identified, cases of neglect often go unnoticed. **Neglect** is characterized by a failure to provide for a child's basic needs, and it can be physical (inflicting physical injury), educational (failure to enroll a school-age child in school, allowing chronic truancy), or emotional in nature (spousal abuse in the child's presence, permission for drug or alcohol use by the child, inattention to a child's needs for affection) (National Clearing House on Child Abuse and Neglect Information 2002).

Elder Abuse and Neglect

The elderly are also victims of abuse, usually when in the care of their older children and their families. Elder abuse can also occur within a nursing home or hospital setting. Federal definitions of elder abuse, neglect, and exploitation first appeared in the 1987 Amendments to the Older Americans Act (National Center on Elder Abuse 2002a). Elder abuse can consist of physical, sexual, or psychological abuse, neglect or abandonment, or financial exploitation. **Domestic elder abuse** refers to any form of maltreatment of an older person by someone who has a special relationship with the elder (a spouse, child, friend, or caregiver); **institutional elder abuse** refers to forms of abuse that occur in residential facilities for older people.

The number of U.S. adults age 60 years or older is projected to increase from 35 million in 2000 to more than 72 million by 2030 (He et al. 2005). As the population ages, there will be an increased need for long-term care of the elderly, with spouses and adult children assuming the role of caretaker. An estimated 15 million individuals provide informal care to relatives and friends (Navaie-Waliser et al. 2002). Although caregiving can positively affect the physical and psychological well-being of the care recipients, the added burden of elder care may strain the family's and the caregiver's emotional and financial resources.

Photo 7.2

As the U.S. population ages, there will be an increased need for elderly long-term care. Senior men and women have the option of living alone, with spouses or adult children as caretakers, or in residential care facilities (pictured here).

There were 465,747 reports of elder abuse in domestic settings in 2004 (National Center on Elder Abuse 2006). Men and women 80 years or older were at higher risk for abuse or maltreatment. The majority of reported victims were women (65.7 percent). Among incidents with a known perpetrator, most alleged perpetrators were female and related to the victim (33 percent adult children). Elder abuse, particularly with family perpetrators, has been attributed to social isolation; personal problems such as mental illness, alcohol, or drug abuse; and domestic violence (spouses make up a large percentage of elder abusers) (National Center on Elder Abuse 2002a). A major risk factor is dependency: Abusers tend to be more dependent on the elderly person for housing, money, and transportation than are relatives who do not abuse (Lang 1993).

Teen Pregnancies and Newborn Abandonment

The U.S. teen birthrate, though declining, is still the highest in the developed world. The birthrate for teenagers (age 15–19 years) declined throughout the 1990s, falling from 59.9 live births per 1,000 teenagers in 1990 to 43 in 2002 (Ventura, Abma, Mosher, and Henshaw 2006). (Refer to Figure 7.2.) Despite the overall decline, birthrates for Black and Hispanic teens remain higher than for any other ethnic-racial group (Ventura et al. 2006). The U.S. teen birthrate has been attributed to a range of factors from inadequate sexuality education to

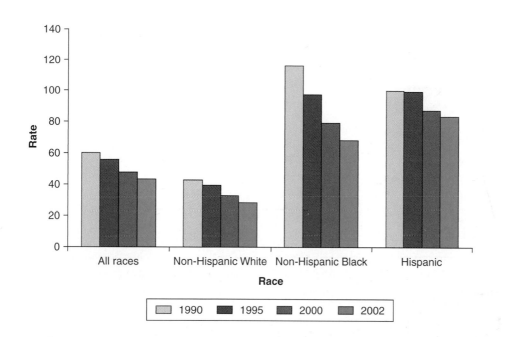

Figure 7.2

Live Birth Rates for Teenagers Aged 15 to 19 Years, by Race and Hispanic Origin, 1990–2002

Source: Ventura et al. 2006.

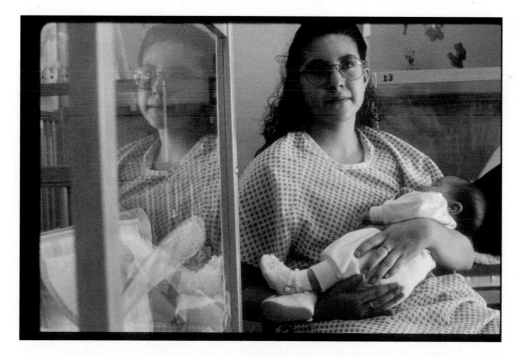

Photo 7.3

A teen mother sits with her baby at the New Futures School in Albuquerque, New Mexico. Programs like New Futures School provide pregnant teens and teen mothers with parenting and child development classes and the opportunity to complete their high school degrees. High school completion and college enrollment rates are lower for teen mothers than for later childbearers.

declining morals (Somers and Fahlman 2001). Research suggests that earlier or more frequent sexual activity among U.S. teens is not the cause of the higher birthrates; rather, sexually active American teens are less likely to use contraceptives than are their European peers (Card 1999). The U.S. teen birth rate is nearly twice the rates in Canada and Great Britain and approximately four times the rates in France and Sweden.

In Focus

Teen Parenting and Education

Pregnant teens or teen parents were routinely expelled from school until Title IX prohibited public schools from discriminating against them. High school graduation rates and college enrollment are lower for teen mothers than for later childbearers, though some longitudinal research reveals that most teen mothers do gain a high school diploma and are employed in stable jobs (Furstenberg, Brooks-Gunn, and Morgan 1987). F. F. Furstenberg et al. (1987) found that teen mothers with the best life outcomes are those who had higher educational aspirations and better-educated and financially stable families.

In her longitudinal study, Lee SmithBattle (2007) examined the impact of parenting on 19 teen mothers' educational goals and school progress. Contrary to popular belief, she observed the teens' renewed commitment to their education. However, based on the accounts of the teen mothers, she discovered that "schools exacerbated their difficulties by failing to provide educational options or by enforcing policies that disregard their complex realities" (2007:366). She writes,

> In addition to work demands, family responsibilities, and transportation difficulties, school policies and practices created additional barriers that undermined teens' aspirations and hindered their school progress … continuing or remaining in school was complicated by cumbersome enrollment processes, stringent attendance policies, lack of educational options, and bureaucratic mismanagement.
>
> Kate's schooling was interrupted for a full year as a result of enrollment difficulties and limited-schooling options. When morning sickness led to many school absences early in her pregnancy, she was referred to the pregnancy school in her urban district. Because home tutoring was not offered as an option, she was forced to withdraw when enrolling in the pregnancy school proved insurmountable …
>
> Kate's mother added that "we just gave up" when Kate's home school failed to transfer her transcript to the pregnancy school in a timely manner. Jenna was also referred to the pregnancy program but refused to be transferred because of its poor academic reputation. She dropped out near the end of her pregnancy for a full year because home schooling was not offered. As she said, "When I was 7 or 8 months pregnant, I stopped going to school. When the new semester started, I withdrew, because after the baby, I would have had to wait a certain amount of weeks to go back and then I'd end up failing. So it was no use."
>
> … [A] cascade of negative events, including inflexible school policies and disciplinary practices, landed Dawn, a suburban student, in educational limbo. After being homeschooled for several months, she returned to her senior year expecting to graduate with her class. She was eventually notified that she would not graduate because she had not completed assignments for one course, presumably because the teacher of that course had not relayed them to her home schooling teacher. Deeply disappointed, Dawn resolved to return to school the following fall and take the few credits she needed to graduate. At the beginning of the fall semester, Dawn was driving her son to day care, going on to school, and leaving mid-morning for work. Her plan to graduate at the end of the fall semester crumbled when the school principal revoked her parking privileges for a declining grade point average after a car accident led to the lack of transportation and several absences. Without a parking pass, Dawn could not drive her son to day care or go to work after school. She was thoroughly discouraged and lacked the skills and family support to appeal to the school board. Her student status was further complicated when she was kicked out of her home and moved in with her grandmother who resided in a different school district.

"I'm tryin' to get into a school here in the city. But they're tellin' me that I have to go two semesters in order to graduate from this new school, when I only would have had to go one semester to graduate from my school. It's just a mess … I actually want to go to school and they won't let me because of something stupid."

At her last interview, she was considering reenrolling in her suburban school so that she could complete the few credits she needed to graduate. Her plans seemed unrealistic because of the lengthy driving time involved (at least 30 to 40 minutes one way) and the complex scheduling that would be required with day care and an employer. (SmithBattle 2007:360–63)

SmithBattle concludes,

The gap between teen mothers' aspirations and the support to achieve them suggests that educators and other professionals are missing a critical opportunity to promote teen mothers' school progress and their long-term educational attainment and success. Schools that cultivate teen mothers' educational ambitions may ultimately contribute to positive chain reactions and the reduction of teen mothers' prior adversity. (p. 369)

Teen mothers, in comparison with their childless peers, are likely to be poorer and less educated, less likely to be married, and more likely to come from families with lower incomes (Hoffman 1998). Their children often lag behind in standards of early development (Hoffman 1998); are less likely to receive proper nutrition, health care, and cognitive stimulation (Annie E. Casey Foundation 1998); and are at greater risk of social behavioral problems and lower intellectual and academic achievement (Maynard 1997). Early childbearing also affects teen fathers. Teen fathers are more likely to engage in delinquent behaviors such as alcohol abuse or drug dealing. In addition, they complete fewer years of school and earn less per year (Annie E. Casey Foundation 1998).

Although teenage childbirth has always been considered a social problem, another phenomenon has captured the public's attention. Reports of abandoned babies found in trash bins, restrooms, parks, and public buildings have become more frequent since the late 1990s. The first case that caught national attention was "Prom Mom" Melissa Drexler. While attending her senior prom in 1997, Drexler gave birth to a full-term baby boy in a bathroom stall. She wrapped her baby in several plastic bags, left him in a garbage can, and returned to her prom. She pleaded guilty to manslaughter. Although there is no comprehensive national reporting system for abandoned babies, data collected by the U.S. Department of Health and Human Services (HHS) indicate increases in the rate of abandonment (D'Agostino 2000). Between 1991 and 1998, the number of infants abandoned in public places rose from 65 to 105 cases. HHS also reports that the number of children abandoned in hospitals grew from 22,000 in 1991 to 31,000 in 1998.

The Problems of Time and Money

In a survey conducted by the Radcliffe Public Policy Center (2000), 79 percent of respondents reported that having a work schedule that enabled them to spend time with their family was a top priority. Nearly everyone reported feeling pressed for more time in their lives, wanting to spend more time with their families, to have more flexible work options, and even just more time to sleep (Radcliffe Public Policy Center 2000).

Researchers have documented the increasing strain on working parents trying to find more time for their family. Bianchi et al. (2006) report that mothers still spend more time with their children, an average of 13 hours per week compared with 7 hours per week for fathers.

Families with children must determine how best to provide financial support to their family, while making sure their children have parental time. Suzanne Bianchi, John Robinson, and Melissa Milkie (2006) report that among families with children about 30 percent of both parents are working full time. Compared with their counterparts in other countries, a higher percentage of American dual-worker couples work long weeks (80 or more hours combined) leaving less family time during the week. Mothers still spend more time with children, an average of 13 hours per week compared with 7 hours per week for fathers. Mothers do most of the routine care (custodial daily care) of children, whereas fathers spend their time with children doing interactive activities (enrichment activities such as talking or reading to them) (Bianchi et al. 2006).

According to Lillian Rubin (1995), economic realities make it especially difficult for working-class parents to maintain their juggling act. In her book, *Families on the Fault Line,* Rubin documented how structural changes in the economy undermine the quality of life among working-class families. This is a functional argument: Because the family is part of our larger social system, what happens at an economic level will inevitably affect the family. The reality of long workdays and weeks does take its toll on families: The loss of intimacy between couples, the lack of time for couples and their children, tense renegotiations over household work, and juggling child care arrangements are just some of the issues that working families face.

Rubin's study shed some light on the condition of working-class families, but often overlooked is the plight of lower-income families—too rich to be classified as living in poverty, but still too poor to be working class. In their two-year study, Lisa Dodson, Tiffany Manuel, and Ellen Bravo (2002) studied lower-income families in Milwaukee, Wisconsin; Denver, Colorado; and Boston, Massachusetts. The researchers concluded that lower-income families deal with basic problems on a daily basis: managing the safety, health, and education of their children while staying employed. Lower-income parents are not "bad" parents; it's just that their parenting may require more time and resources than they have. Among low-income families, there is a higher prevalence of children with chronic health issues or special learning needs; at least two-thirds of the families in the Dodson et al. study reported having a child with special needs. These children require much more time and patience from their parents, sometimes jeopardizing parents' ability to maintain employment and earnings. To support lower-income parents, the authors recommend comprehensive and flexible child care, along with workplace flexibility (taking time off work, adjusting their work schedule) (Dodson et al. 2002).

Community, Policy, and Social Action

Family Medical Leave Act of 1993

The Family Medical Leave Act (FMLA) of 1993 was envisioned as a way to help employees balance the demands of the workplace with the needs of their families. The act provides employees with as many as 12 weeks of unpaid, job-protected leave per year. It also provides for group health benefits during the employee's leave. The FMLA applies to all public agencies and all private employers with 50 or more workers.

As of 2007, 11 states have enacted their own family and medical leave laws, extending the coverage provided by the FMLA or extending coverage to those not eligible under FMLA guidelines. In 2004, California became the first state to enact a law that provides paid family care leave. The California Family Rights Act allows employees to take a paid leave to care for a child, spouse, parent, or domestic partner who has a serious health condition or to bond with a new child. Employees who take such leave can receive 55 percent of their pay up to $780 per week for a maximum of six weeks. The California law applies to all employers, not just those with 50 or more employees (U.S. Department of Labor 2007).

Although there is strong support for the FMLA and its stated goals, the act has also been criticized since its enactment. Almost 41 million workers are not covered by the FMLA because they work for private employers not covered under the law (AFL-CIO 2002) and small businesses employing less than 50 people. Though an estimated 20 million Americans have used the FMLA, nearly two-thirds of eligible workers have not taken advantage of the FMLA because they could not afford the lost wages (AFL-CIO 2002; National Partnership for Women and Families 2002).

Based on their analysis of public policies for working families in more than 170 countries, Jody Heymann, Alison Earle, and Jeffrey Hayes (2007) concluded that U.S. policies lag behind all high-income countries, as well as most middle- and low-income countries in their study. The United States does not offer paid leave for mothers in any segment of the workforce. Only three other nations have no paid leave—Liberia, Papua New Guinea, and Swaziland; 169 countries offer guaranteed leave with income to women. Sixty-six countries offer fathers paid paternity leave or have the right to paid parental leave; the United States offers neither.

Community Responses to Domestic Violence and Neglect

Responses to domestic violence can be characterized as having a distinct community approach. In the area of child abuse, the U.S. Office of Juvenile Justice and Delinquency Prevention established community-based children's advocacy centers to provide coordinated support for victims in the investigation, treatment, prosecution, and prevention of child abuse in all 50 states. Programs at each center are uniquely designed by community professionals and volunteers to best meet their community's needs. One such center is Project Harmony, based in Omaha, Nebraska. Project Harmony provides medical exams, assessment, and referrals. Project Harmony serves children who are victims of abuse and their non-abusing family members. By co-housing representatives from child protective services, law enforcement, and project staff, Project Harmony attempts to improve communication and coordination between all professionals involved in a child's case.

Since its inception in 1995, the Violence Against Women Office (VAW) has handled the U.S. Department of Justice's legal and policy issues regarding violence against women. The office offers a series of program and policy technical papers for individuals, leaders, and communities to support their efforts to end violence against women. These papers highlight some of the best program models and practices and were produced by the Promising Practices Initiative of the STOP Violence Against Women Grants Technical Assistance Project (Little, Malefyt, and Walker 1998). Two featured programs are the following:

- The Minnesota Domestic Abuse Intervention Project (DAIP) was developed in 1980 and serves as a national and international program model. It was the first program of its kind to coordinate the intervention activities of each criminal justice agency in one city. The goals of the program include victim safety, offender accountability, and changes in the climate of tolerance toward violence in the Duluth community. The program also offers a men's nonviolence education program, an advocacy and support group for the men's partners, a class for women who have used violence, and a victim advocacy program for Native Americans through the Mending the Sacred Hoop project.
- The Women's Center and Shelter (WC&S) of Greater Pittsburgh was founded in 1974. WC&S coordinates its program efforts with the medical community, criminal justice agencies, and other organizations. The programs focus on the ability of women to take control of their own lives. WC&S provides comprehensive victim services, which also includes parenting education and employment readiness. WC&S also provides a follow-up program for former shelter residents and child care services for residents, nonresidents, and follow-up participants. WC&S has also created a school-based curriculum called Hands Are Not for Hurting. The curriculum uses age-appropriate lessons that encourage nonviolent conflict resolution and teach youth that they are responsible for the choices they make.

The National Center on Elder Abuse (NCEA) believes that community education and outreach are important in combating the problem of elder abuse and neglect. The NCEA supports community "sentinel" programs, which train and educate professionals and volunteers to identify and refer potential victims of abuse, neglect, or exploitation. In 1999, the NCEA established partnerships with the Human Society of the United States (HSUS), Meals on Wheels Association of America (MOWAA), and the National Association of Retired Senior Volunteer Program Directors. These organizations were selected because of their unique access to isolated elders in their homes. NCEA funded six coalition projects in Arizona, California, New York, North Carolina, and Utah. The program trained more than 1,000 professionals and volunteers to serve as sentinels; as a result, there was an increase in the number of abuse referrals in communities where sentinels were used. Administrators also noticed an increase in the level of satisfaction among volunteers, who as a result of the project were able to assist individuals who they believed might be victims or potential victims (National Center on Elder Abuse 2002b).

Teen Pregnancy and Infant Abandonment

In the 1980s, social and human service programmers defined prevention as an effective way to address the problems of teen pregnancy and parenthood (Card 1999). In the 1990s, several new prevention approaches emerged, particularly after the passage of the Welfare Reform Act of 1996. Under Section 905 of the act, the HHS was mandated to ensure that at least 25 percent of all U.S. communities had teen pregnancy prevention programs in place. Most states identified target goals related to teen birthrates. For example, New Hampshire's goal is to reduce the nonmarital teen rate to 21.0 per 1,000 by 2005; in North Dakota, the goal is to reduce its teen birthrate by 2 percent per year.

Along with funding national educational programs such as the Girl Neighborhood Power: Building Bright Futures for Success program, HHS also funded comprehensive state- and community-wide initiatives through the Abstinence Only Education Program and the Adolescent Family Life Program (administered through the Office of Population Affairs). An example of this approach is School-Linked Reproductive Health Services (The Self Center), a school-neighborhood clinic partnership in Baltimore, Maryland. The program offers education, counseling, and reproductive services to teens. Students at participating schools reported reduced levels of sexual activity, and among those already sexually active, there was more effective contraceptive use. Also reported was a delay in the onset of sexual activity among abstinent youth (Card 1999).

Under welfare reform, two primary provisions affected teen parents and their access to welfare assistance. First, teen parents must stay in school; financial assistance cannot be provided to unmarried, minor, custodial parents who do not have a high school degree or equivalent. Early data on the educational provision indicate that programs offering a range of options—GED programs designed for public assistance recipients, educational activities, and life skills training—proved helpful for teen parents. Second, teen parents must live in an adult-supervised setting (U.S. Department of Health and Human Services 1998).

Ariel Kalil and Sandra Danzinger (2000) studied the effects of these welfare reform rules on low-income teen mothers living in Michigan, most of who were complying with the new welfare requirements and satisfied with their living arrangements. Many were still experiencing child care problems, depression, and domestic violence. Living with the baby's grandmother reduced teen mothers' economic strain; however, it was also related to poorer educational outcomes. Although the teen mothers expressed high educational goals, researchers were uncertain whether the attendance requirement under welfare reform would actually increase overall educational attainment of these young mothers. A study analyzing teen pregnancy policies from 1951 to 1992 in England and Wales concluded that education and job opportunities were more effective in reducing the birth rate than was removing welfare and housing benefits (Selman and Glendinning 1994).

In response to newborn and infant abandonment, 47 states have passed laws since 1999 that offer safe and confidential means to relinquish unwanted newborns without the threat of prosecution for child abandonment. State laws vary according to child's age (72 hours to one year old) and the personnel or places authorized to accept the infant (hospital personnel, emergency rooms, church, and police).

It is unclear how effective these safe-surrender or safe-haven laws have been in reducing infant abandonment or death. New Jersey, home of the first infant abandonment case that gained national attention, passed a safe-haven law in August 2000, modeled after Texas 1999 legislation. Between August 2000 and November 2003, New Jersey officials assisted 17 babies. These children were adopted, placed in foster care, or returned to their mother (New Jersey Safe Haven Protection Act 2007). In 2006, however, six dead newborns were found abandoned in New York, despite the state's safe haven laws. Critics argue that the state's safe haven laws were poorly advertised, with most residents not knowing about the law's provisions. In Illinois, a discussion of the safe haven law is included the high school health curriculum, possibly ensuring more awareness among high-risk youth (Buckley 2007). For more on the Illinois safe haven law and the woman who helped create the legislation, turn to the Voices in the Community feature.

In 2007, a mother abandoned her three-month-old baby boy at a neonatal clinic in Rome, Italy. She was the first to use the clinic's modern founding wheel. The original foundling wheel, used in the Middle Ages, was a revolving wooden barrel built into the church's exterior wall; the mother could deposit her baby, turn the wheel and the baby would be safely protected inside the church. The Rome clinic uses a small modern structure equipped with a heated cradle and a respirator. Also available at other Italian hospitals and churches, these structures are equipped with an alarm to alert medical or church staff when a child is deposited. Foundling wheels are also used in other countries such as Germany, Switzerland, and the Czech Republic (Povoledo 2007).

What Does It Mean to Me?

Investigate whether your state has a safe-haven law. What are the features of the law? What protections does it offer the mother? How many children have been protected under the law?

Voices in the Community

Dawn Geras

In Illinois, the Newborn Abandoned Infant Protection Act was spearheaded by a coalition of community women's groups (Collier 2001). Dawn Geras, a Northbrook businesswoman, started her volunteer group, Save Abandoned Babies, after being moved by several newspaper articles on the issue. A second group of women (from Melrose Park) were inspired to act after watching an *Oprah* show on the topic. A third group of women (from the Rockford area) completed the coalition. One of the Rockford women was Sue Moye, whose daughter, Kelli, abandoned her newborn baby when she was 15. During a two-year period, the coalition encouraged and lobbied legislators to draft a protection bill. The act became law in 2001 (Collier 2001).

Geras currently serves as the president of Save Abandoned Babies Foundation, a nonprofit volunteer organization. The organization, with its motto, "No Shame, No Blame, No Names," is shifting its focus to education and awareness about the Illinois law.

In November 2006, Geras joined adoptive families, adoption agencies, and government representatives to celebrate the five-year anniversary of the Illinois Abandoned Newborn

Infant Protection Act. At the celebration, Geras said, "Even one life saved is a victory for our children and to know that 27 newborns have been safely, legally relinquished and adopted into loving families over these five years validates the importance of this legislation" (Illinois Department of Children and Family Services 2006).

In addition, the original law was amended to allow infants who are seven days old or younger to be accepted (previously the law had set the age limit to three days) and to require that the safe-haven law be taught to students in all school health education classes. Geras explained, "Educating the public about the existence of this life saving law is crucial. If people don't know the law exists, how can they be expected to use it? This law not only saves the lives of these precious babies but also saves the parents from potential criminal charges" (Illinois Department of Children and Family Services 2006).

Supporting Different Family Forms

According to sociologist David Popenoe (1993), there has been a serious decline in the structure and function of the family since the 1960s. Families are not meeting society's needs as they once did and have lost most of their "functions, social power and authority over their members" (Popenoe 1993:527–28). He attributes the weakening of family function to high divorce rates, declining family size, and the growing absence of fathers and mothers in their children's lives. Data reviewed at the beginning of this chapter—the declining percentage of nuclear families, along with the increase in single-parent families and nonfamily household—lend support to his observation.

But critics of Popenoe's position argue that the "family in decline" hypothesis relies too heavily on the definition of family in its nuclear form (Bengston 2001). Sociologist Judith Stacey (1996) agrees that the "family" as defined by a nuclear form of mom, dad, and children is in decline. This family system has been replaced by what Stacey calls a "postmodern family condition," one characterized by diverse family patterns and forms, where no single family form is dominant. There have been significant changes in the traditional family's structure and functions (Bengston 2001); modern families are best characterized by their diversity (Hanson and Lynch 1992).

The proportion of nuclear families is decreasing and being replaced with family structures that include single-parent, blended, adoptive, foster, grandparent, and same-sex partner households (Copeland and White 1991). The increasing diversity of American families requires that we broaden our research and policy agendas beyond traditional family forms (Demo 1992). Perhaps one solution to the "problem" of families is to appreciate and embrace other family forms. Let's examine two family forms: cohabitation and grandparents as primary caregivers to grandchildren.

Cohabitation

Larry Bumpass (1998) argues that the increase in cohabitation reflects and reinforces the declining significance of marriage as a life course marker in our society. Almost half of U.S. young adults report living in a cohabiting union at some point in their lives (Bumpass 1998), a trend reflected in most Western societies. **Cohabiting** is defined as sexual partners, not married to each other, but residing in the same household.

In their *State of Our Unions* 2005 report, Barbara Defoe Whitehead and Popenoe report that from 1960 to 2004, the number of U.S. unmarried couples increased by 1200 percent. In 2004, the U.S. Census reported more than 5 million cohabiting couples. Additionally, about 25 percent of unmarried women aged 25 to 39 years are currently living with a partner and an additional 25 percent report living with a partner some time in the past. Defoe Whitehead and Popenoe report that more than half of all marriages are preceded with a period of living together, compared with virtually none more than 50 years ago.

Researchers increasingly acknowledge that not all cohabitations will eventually lead to marriage and may instead serve as alternate forms of marriage (Manning and Smock 2002). The pattern is particularly true for Black, Hispanic, and mainland Puerto Rican women compared with White women (Bumpass and Lu 2000). Cohabitation rates have increased among Black and White adults, with higher increases observed among Whites. Bumpass and Lu (2000) report that for women aged 19 to 44 years, 59 percent of high school dropouts have cohabited compared with 37 percent of high school graduates.

Defoe Whitehead and Popenoe (2005) cite a recent study that concluded that premarital cohabitation, when limited to a woman's future husband, was not associated with an elevated risk of divorce. Conversely, they observe, "No evidence has yet been found that those who cohabit before marriage have stronger marriages than those who do not."

The increase of cohabitation has coincided with the increase in unmarried childbearing—currently about 40 percent of all cohabiting households contain children (Defoe Whitehead and Popenoe 2005). Though data indicate that about two-fifths of all children will live with their mother and a cohabiting partner and about a third of the time children spend with unmarried mothers is spent in a cohabiting relationship, family scholars acknowledge that more needs to be known about the impact of cohabitation on the family experiences and life outcomes of these children (Bumpass and Lu 2000).

Grandparents as Parents

Fifty-nine-year-old Pat and Ken Owens of Lewiston, Maryland, are the primary caretakers of their grandchildren, Michael and Brandi (Armas 2002). Michael and Brandi are among 3.7 million children living in households where the grandparent is the primary householder (Fields 2003), an increase from 2.2 million grandchildren in 1970 (Casper and Bryson 1998). In 2002, about 1.3 million children lived with their grandparents alone, with neither parent present. The areas with the highest percentage of these households are in the South and rural counties across the Midwest and West. Wyoming and Oklahoma led the country in the percentage of grandparents who are primary caregivers. (See U.S. Data Map 7.1 for the number of grandparents who are primary caregivers for each state.) According to the 2002 U.S. Census, Black children were more likely than were children from other ethnic/racial groups to live in a grandparent's household (Fields 2003).

Grandparents may assume caretaking responsibilities when parents are unable to live with or care for their children because of death, illness, divorce, incarceration, substance abuse, or child abuse or neglect. The Owenses have not heard from Michael and Brandi's mother in the past two years and only recently began receiving financial support from Michael's father. The alternative for the Owenses' grandchildren would have been foster care, something that Pat Owens did not want to happen. "I don't want to make it sound like it's easy because there are some tough, tense times. But I'm very proud of the fact that all the grandchildren still play together and go to school together," said Mrs. Owens (Armas 2002:A6).

Research identifies how grandparents who care for their grandchildren are at high risk for emotional and physical distress. This distress is related to a deficit of social resources, such as marital status, social support, economic resources, and the demands of the caregiving role itself. Grandparent caregivers are more likely to experience depression and suffer from fair to poor physical health and activity limitations than are grandparents in more traditional roles (Chase Goodman and Silverstein 2006).

The sudden responsibility for children leaves many grandparents on fixed incomes with unexpected financial burdens. In one study, children living in a grandparent's household without a parent present were twice as likely to be living below the poverty level as were

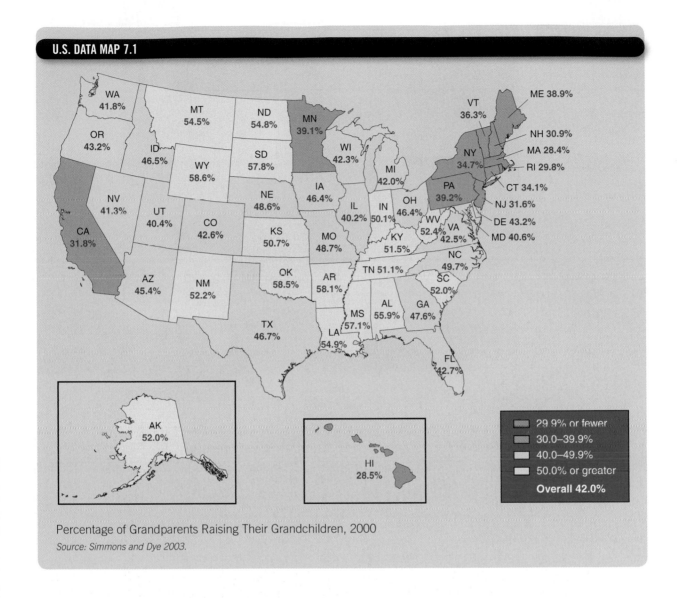

U.S. DATA MAP 7.1

Percentage of Grandparents Raising Their Grandchildren, 2000

Source: Simmons and Dye 2003.

children living with both grandparents and a parent. In addition, children who lived with just their grandparents were also at risk of not being covered by health insurance (Fields 2003). If eligible, grandparents can get assistance through HHS Temporary Assistance for Needy Families (TANF) benefits. Some states, such as Illinois, Pennsylvania, and Wisconsin, offer guardianship or kinship care subsidy programs for grandparents.

Emotional and social support is available for grandparent-headed households. Local grandparent support groups are listed by organizations such as the American Association for Retired Persons (AARP), Generations United, or GrandsPlace. These organizations also provide fact sheets, community links, and suggestions for clothing and school supplies, recipes, and travel and activity guides for grandparents and their grandchildren. AARP (2002) recognized several model support programs such as the Kinship Support Network in San Francisco, California; Project Healthy Grandparents in Atlanta, Georgia; and Grandma's Kids Kinship Support Program in Philadelphia, Pennsylvania.

The U.S. Census reports that 3.7 million children reside in households where the grandparent is the primary householder (Fields 2003). About 1.3 million children lived with their grandparents alone, with neither parent present.

Main Points

- The **family** is defined as a construct of meaning and relationships both emotional and economic. The family is a social unit based on kinship relations—relations based on blood and those created by choice, marriage, partnership, or adoption.
- A **household** is defined as an economic and residential unit.
- From a functionalist perspective, social problems emerge as the family struggles to adapt to a modern society. Functionalists have noted how many of the original family functions have been taken over by organized religion, education, work, and the government in modern society, but still, the family is expected to provide its remaining functions of raising children and providing affection and companionship for its members.

- For the conflict theorist, the family is a system of inequality where conflict is normal. Conflict can derive from economic or power inequalities between spouses or family members. From a feminist perspective, inequality emerges from the patriarchal family system, where men control decision making in the family.
- According to an interactionist, through social interaction, we create and maintain our definition of a family. As we do this, it affects our larger social definition of what everyone's family should be like and the kind of family that we create for ourselves.
- We reviewed several problems based on the family in this chapter:
  - **Divorce**. The U.S. divorce rate has remained stable at around 4.0 divorces per 1,000 individuals in the total U.S. population.

The rate was 4.1 in 1999, 4.2 in 2000, and 3.6 in 2005. Recent census data on divorce indicate that certain groups are more susceptible to divorce than others are.

– **Domestic violence.** Current data suggest that 21 to 34 percent of women will be assaulted by an intimate partner during their lifetime. Research has consistently linked the following social factors to family violence: low socioeconomic status, social and structural stress, and social isolation. Feminist researchers argue that domestic violence is rooted in gender and represents men's attempts to maintain dominance and control over women.

– **Child abuse and neglect.** In 2004, 872,000 children were victims of child maltreatment. Neglect is characterized by a failure to provide for a child's basic needs. It can be physical, educational, or emotional in nature.

– Responses to domestic violence can be characterized as having a distinct community approach. In the area of child abuse, the U.S. Office of Juvenile Justice and Delinquency Prevention established community-based children's advocacy centers to provide coordinated support for victims in the investigation, treatment, prosecution, and prevention of child abuse.

– **Elder abuse and neglect.** Elder abuse can occur in the care of their older children and their families or within a nursing home or hospital setting. **Domestic elder abuse** refers to any form of maltreatment of an older person by someone who has a special relationship with the elder (a spouse, child, friend, or caregiver); **institutional elder abuse** refers to forms of abuse that occur in residential facilities for older people. The National Center on Elder Abuse (NCEA) believes that community education and outreach are important in combating the problem of elder abuse and neglect. The NCEA supports community "sentinel" programs, which train and educate professionals.

– **Teen pregnancy.** The United States has the highest teen birthrate in the developed world. But teenage birthrates have declined. The birthrate for teenagers (age 15–19 years) declined throughout the 1990s, falling from 59.9 live births per 1,000 teenagers in 1990 to 43 in 2002.

– **The problems of time and money.** Economic realities make it especially difficult for working-class parents to maintain their juggling act. The Family Medical Leave Act (FMLA) of 1993 was envisioned as a way to help employees balance the demands of the workplace with the needs of their families. Although there is strong support for the FMLA, the act has also been criticized since its enactment.

■ The proportion of nuclear families is decreasing and is being replaced with family structures that include single-parent, blended, adoptive, foster, grandparent, and same-sex partner households. The increasing diversity of American families requires that we broaden our research and policy agendas beyond traditional family forms. Perhaps one solution to the "problem" of families is to appreciate and embrace other family forms.

On Your Own

Log on to the Web-based student study site at www.pineforge.com/leonguerrero2study for interactive quizzes, e-flashcards, journal articles, Community and Policy Guides, a Service Learning Guide, the end-of-chapter Web exercises, and additional Web resources.

Internet and Community Exercises

1. Interview a student who is also a parent. The student can be from your social problems class or any of your other classes. What challenges do student-parents face? How are their challenges different from those of students who are not parents? Does your school support programs or services for parents? Check with your professor about campus policies regarding research involving human subjects.

2. Contact two domestic violence shelters in your community. What is the mission of each organization? Do different shelters serve different groups? If so, how are services tailored to various groups? Does each program promote violence prevention? Domestic shelters often support student internships for a quarter or semester. If you're interested in learning more about a shelter, ask about its internship program.

3. Identify your school's family friendly policies. Does your school provide child care on site? Has your school provided leaves under the Family and Medical Leave Act? You may have to contact the human resources department in your school for more information.

4. Investigate your state's safe-haven law for abandoned babies. Have there been any recent cases of abandonment?

CHAPTER 8

Education

Education is assumed to be the great equalizer in our society. There are inspirational stories of women and men who, after a tough childhood or adulthood, complete their education, become successful members of society, and are held as role models. Education is presented as an essential part of their success, serving as a cure for personal or situational shortcomings. If you are poor, education can make you rich. If your childhood was less than perfect, a college degree can make up for it. On the occasion of launching the Head Start program in 1965, President Lyndon Johnson is quoted as saying, "If it weren't for education, I'd still be looking at the southern end of a northbound mule" (Zigler and Muenchow 1992).

Yet, along with these images of success, we are also bombarded with images of failure. Media coverage and political rhetoric highlight problems with our educational system, particularly with our public schools. In recent state and national political campaigns, the quality of teaching and preparation of teachers were scrutinized, and school districts with low scores on standardized exams were criticized. U.S. students are said to be falling behind the accelerated pace of higher education internationally. Though the United States still ranks among the top nations in the educational attainment of older adults 35 to 64 years of age (percent holding an associate's degree or higher), it ranks seventh in the educational attainment of younger adults aged 25 to 34 years (Wagner 2006). In 2003, more than 50 percent of Canadian and Japanese 25- to 34-year-olds earned a college degree, compared with 39 percent of American 25- to 34-year-olds (refer to Figure 8.1) (Callan 2006). In 2006, a panel of education, labor, and public policy experts, including two former education secretaries, warned that if the United States does not keep pace with the educational gains made in other countries, our standard of living would be seriously compromised.

So which is it: Is education a key to individual success or an institutional failure? In this chapter, we'll first examine this question by reviewing our educational system from different sociological perspectives. Then we'll explore current social problems in education, along with policy and program responses.

Figure 8.1

Percentage of Adults With an Associate's Degree or Higher, by Age, Top 10 Countries, 2003

Source: Callan 2007.

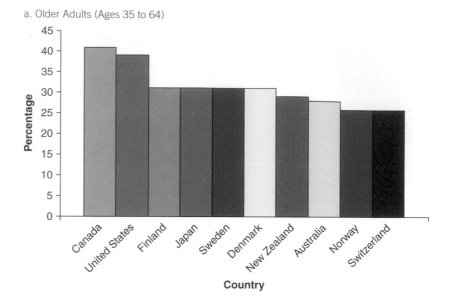

a. Older Adults (Ages 35 to 64)

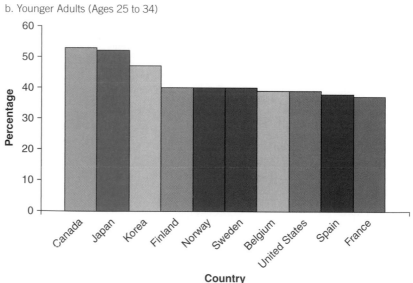

b. Younger Adults (Ages 25 to 34)

The New Educational Standard

We tend to define a high school degree as the educational standard of the past, now replaced with a bachelor's degree. Actually, data from the U.S. Census Bureau confirm that the United States has an increasingly more educated population (see Table 8.1 and U.S. Data Map 8.1).

For 2003, the U.S. Census Bureau (Stoops 2004) reported that an all-time high of 85 percent of adults (25 years and older) completed at least a high school degree, and more than 27 percent of all adults had attained at least a bachelor's degree. Younger Americans are more educated than are older Americans. Based on the data presented in Table 8.1, approximately 70 percent of all Americans 75 years or older attained a high school degree or more compared

Table 8.1

Educational Attainment of the Population, 25 Years and Older, 2005

Characteristics	Percentage With		
	High School Graduate or More	Some College or More	Bachelor's Degree or More
Population aged 25 years and over	85.2	72.3	27.7
By Age Group:			
25–29 years old	86.2	71.2	28.8
30–34 years old	87.3	68	32
35–39 years old	88.1	68.9	31.1
40–44 years old	88.6	71.1	28.9
45–49 years old	89	71.5	28.5
50–54 years old	89.2	69.4	30.6
55–59 years old	87.8	69.9	30.1
60–64 years old	84.5	73.5	26.5
65–69 years old	79	78.9	21.1
70–74 years old	76.1	80.1	19.9
75 years or older	69.9	83	17

Source: U.S. Census Bureau 2006.

Photo 8.1

According to the U.S. Census Bureau (Stoops 2004), in 2003, an all-time high of 85 percent of adults (25 years and older) completed at least a high school degree. More than 27 percent of all adults had attained at least a bachelor's degree.

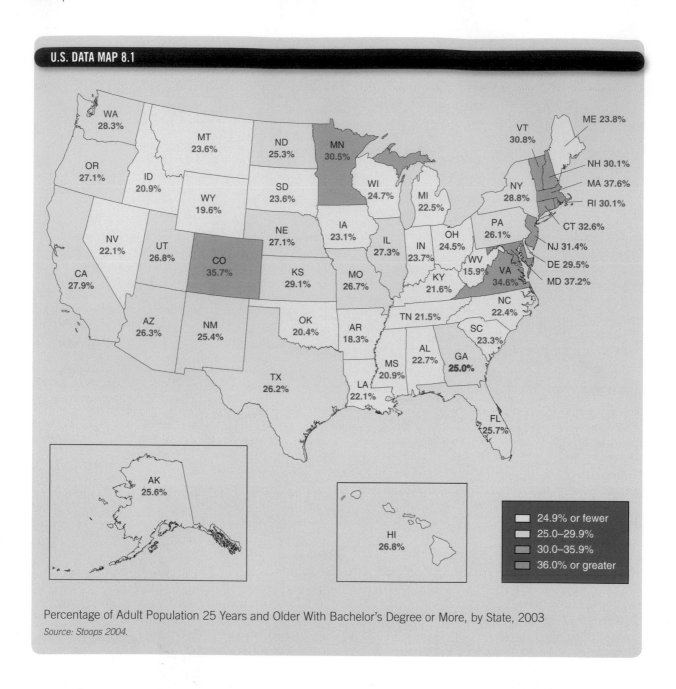

U.S. DATA MAP 8.1

Percentage of Adult Population 25 Years and Older With Bachelor's Degree or More, by State, 2003

Source: Stoops 2004.

with 86 percent of those 25 to 29 years old. The percent attaining a bachelor's degree has also increased, from 17 percent of Americans 75 years or older versus 32 percent of Americans 30 to 34 years of age (the highest among all age groups reported) (U.S. Census 2006). As indicated in the U.S. Data Map 8.1, populations in the Northeast and West had the highest proportions with a bachelor's degree or higher. The state with the highest proportion was Massachusetts (38 percent), and West Virginia had the lowest (15.9 percent). Educational attainment level of adults will continue to rise, as younger, more-educated age groups replace older, less-educated ones.

What Does It Mean to Me?

Compare the educational attainment of three generations of your family: yourself, your parents, and your grandparents. Does your family reflect the same educational attainment pattern as the census data indicate? Are younger generations more educated than older generations in your family? Why or why not?

Sociological Perspectives on Education

Functionalist Perspective

The institution of education has a set of **manifest** and **latent functions**. Manifest functions are intended goals or consequences of the activities within an institution. Education's primary manifest function should come as no surprise: It is to educate! The other manifest functions include socialization, personal development, and employment. Our educational system ensures that each of us will be appropriately socialized and adequately educated to become a contributing member of society. We learn skills and knowledge, as well as about society's norms, values, and beliefs, which are necessary for our survival and, ultimately, the society's survival.

Education's latent or unintended consequences may be less obvious. One unintended function that education serves is as a public babysitter. No other institution can claim such a monopoly over the total number of hours, months, and years of a child's life. From kindergarten through high school, parents can rely on teachers, administrators, and counselors for their child's education and for supervision, socialization, and discipline. In addition, education controls the entry of young women and men into the labor force and the timing of that entry. Consider the surge in employment rates after high school and college graduation. There is always a rush to get a job each summer; employers rely on the temporary labor of high school and college students during busy summer months. Finally, education establishes and protects social networks by ensuring that individuals with similar backgrounds, education, and interests are able to form friendships, partnerships, or romantic bonds.

Functionalists argue that education has been assigned so many additional tasks that it struggles in its primary task to educate the young. In addition to its own main functions, our educational system has taken over functions of other institutions. For example, the educational system provides services to students with family problems, emotional needs, or physical challenges. Schools also provide services for parents in the form of adult education or parenting classes.

Photo 8.2

The title of a popular book claimed, "all I really need to know I learned in kindergarten." What did you learn in kindergarten that prepared you to be part of society?

Conflict Perspective

Conflict theorists do not see education as an equalizer; rather, they consider education a "divider"—dividing the haves from the have-nots in our society. From this perspective, conflict theorists focus on the social and economic inequalities inherent in our educational system and how the system perpetuates these inequalities.

Conflict theorists highlight the socialization function of education as part of the indoctrination of Western bureaucratic ideology. The popular posters and books on "What I learned in kindergarten" could serve as the official list of adult life rules: Share everything, play fair, put things back where you found them. Never mind kindergarten—the indoctrination can begin as early as nursery school. Rosabeth Moss Kanter (1972) describes a child's experience in nursery school as an "organizational experience" creating an **organizational child**. Carefully instructed and supervised by their teachers, students are guided through their day in ordered agendas; they are rewarded for conformity, and any signs of individuality are discouraged. The organizational child is sufficiently prepared for the demands and constraints of a bureaucratic adult world.

Although we consider education as the primary method of achieving equality and mobility in our society, conflict theorists argue that it actually sustains the structure of inequality. As Martin Marger (2008) explains, the relationship between education and socioeconomic status operates in a cycle perpetuated from one generation to the next. It begins with high-income and high-status parents who can ensure greater quality and amount of education for their children. As family income increases, so does the likelihood of college attendance and completion for children. With this higher quality education, their children's income and status will remain high as adults and in turn, they will pass their socioeconomic advantage on to their own children.

Jonathan Kozol, in his books *Savage Inequalities* (1991) and *The Shame of the Nation* (2005), presents a dark portrait of student learning in inner-city schools that are understaffed, undersupplied, and in disrepair. This educational inequality is created by public school systems that rely on property taxes to finance school staffing and operations. As a result, a caste system has emerged in our educational system. The gap between the haves and have-nots is so wide it seems impossible to attain educational equity. As Kozol observes,

> Children in one set of schools are educated to be governors; children in another set of schools are trained to be governed. The former are given the imaginative range to mobilize ideas for economic growth; the latter are provided with the discipline to do the narrow tasks the first group will prescribe. (1991:176)

He introduces us to eight-year-old Alliyah, a third grader he met in 1997–1998. At that time, the New York Board of Education spent $8,000 for her education in a public school in the Bronx. If she had been educated in a typical White suburb of New York, she would have received a public education worth $12,000, and $18,000 if she resided in a wealthy White suburb. According to Kozol, what society chooses to spend on lower-class neighborhoods and schools "surely tells us something about what we think these kids are worth to us in human terms and in the contributions they may someday make to our society" (Kozol 2005:44).

Feminist Perspective

Inequalities are based not just on social class but also on gender. Research reveals the persistent replication of gender relations in schools, evidenced by the privileging of males, their voices, and activities in the classroom, playground, and hallways (Smith 2000).

One of my favorite illustrations of the privilege given to a male voice comes from my own discipline. In sociology, there is a concept called the "definition of the situation," which

Photo 8.3

Are female faculty perceived differently from male faculty? Are there any gender stereotypes associated with female professors?

refers to the phrase: "If men define situations as real, they are real in their consequences" (Thomas and Thomas 1928:572). The concept is an important one to the symbolic interactionist perspective and is often attributed solely to sociologist W. I. Thomas. However, the correct acknowledgment is to Thomas and his wife, Dorothy Swaine Thomas. R. S. Smith (1995) investigated the citation in more than 244 introductory sociology textbooks and found that most attributed the concept solely to W. I. Thomas. One reason for the omission of Swaine Thomas is that she may not have contributed to the phrase (although there is no documented evidence to support this), but Smith (1995) suggests that the omission is because of a professional and structural ideology that historically represented sociology as a "male" domain. Smith notes that the citations began to include Swaine Thomas after the mid 1970s, a time when sociology and introductory texts began to respond to and reflect the changes brought about by a growing women's movement and increasing numbers of female sociologists.

Gender bias and gender stereotypes work to exclude and alienate girls early in their educational experience (AAUW 1992; Sadker and Sadker 1994). Males have favored status in education, particularly in their interactions with their teachers. In the classroom, girls are invisible, often treated as "second-class educational citizens." This is how Myra and David Sadker (1994) explained the subtle yet consequential gender bias in the classrooms they visited. After observing teachers and their interactions with girls and boys in more than 100 classrooms, the Sadkers found that teachers were more responsive to boys and were more likely to teach them more actively. Overall, girls received less attention whereas boys got a double dose, both negative and positive. Boys received more praise, corrections, and feedback, whereas girls received a cursory "OK" response from their teachers. Sadker and Sadker concluded that over time, the unequal distribution of teacher time and attention may take its toll on girls' self-esteem, achievement rates, test scores, and ultimately careers.

Structural factors along with interpersonal dynamics also contribute to the creation and maintenance of gender inequality on college and university campuses (Stombler and Yancey

Martin 1994). Men and women experience college differently and have markedly different outcomes (Jacobs 1996). College women are subjected to male domination through their peer relations (Stombler and Yancey Martin 1994) in the classroom, in romantic involvements (Holland and Eisenhart 1990), and in organized activities. Even activities like fraternity "little sister" programs, which Mindy Stombler and Patricia Yancey Martin (1994) studied, provide the structural and interpersonal dynamics necessary to create an atmosphere conducive to women's subordination.

Interactionist Perspective

In Sadker and Sadker's 1994 study, we identified the differential effects of teacher communication on female and male students. The interaction between teachers and students daily reinforces the structure and inequalities of the classroom and the educational system. From this micro perspective, sociologists focus on how classroom dynamics and practices educate the perfect student and at the same time create the not-so-perfect ones. In what ways does classroom interaction educate and create?

Assessment and testing are standard practices in education. Students are routinely graded and evaluated based on their work and ability. Interactionists would argue that along with assessment come unintended consequences. Based on test results, students may be placed in different ability or occupational tracks. In the practice called **tracking**, advanced learners are separated from regular learners; students are identified as college bound versus work bound.

Advocates of tracking argue that the practice increases educational effectiveness by allowing teachers to target students at their ability level (Hallinan 1994). Yet, placing students in tracks has been controversial because of the presumed negative effects on some students. Opponents argue that labels such as "upper" versus "lower" track or "special" or "slow learners" are used to systematically deny a group of students access to education (Ansalone 2004). In addition to creating unequal learning opportunities (Hallinan 1994), tracking may encourage teachers, parents, and others to view students differently according to their track, and as a result, their true potential may be hindered (Adams and Evans 1996). Although tracking is intended to aid students, it may lead to a self-fulfilling prophecy: Students will fail because they are expected to do so.

Despite the documented negative consequences of tracking, the practice continues in approximately 60 percent of all elementary schools and 80 percent of all secondary schools (Ansalone 2001). Although interactionists do not assess the appropriateness of the label, they would address how the label affects students' identity and educational outcomes. Issues of inequality must also be addressed if the data suggest that students of particular gender or ethnic/racial categories are targeted for tracking. A summary of all sociological perspectives on education is presented in Table 8.2.

What Does It Mean to Me?

Investigate how many female and male students are declared majors in math, engineering, English, nursing, and sociology in your university. Is there a difference in the number of female majors in engineering versus English? Math majors versus nursing majors? What sociological perspective(s) might best explain the gender gap in majors?

Table 8.2

Summary of Sociological Perspectives: Education

	Functional	Conflict/Feminist	Interactionist
Explanation of education	Using a macro perspective, functionalists examine the functions of the educational system. The educational system is strained from performing multiple functions.	Conflict and feminist theorists address how the educational system perpetuates economic, ethnic, and gender inequalities. Students, depending on their social backgrounds, are differentially treated by the educational system.	Using a micro perspective, an interactionist focuses on how the educational experience is created through interaction and shared meanings in the classroom.
Questions asked about education and its social problems	What manifest or latent functions does education serve? How is education affected by other social institutions?	How does education perpetuate social inequalities? Is one group more disadvantaged than another?	How do classroom dynamics determine the educational success of a student? What is the relationship between teacher-student interaction and student success? How does student tracking or labeling impact student achievement?

Problems and Challenges in American Education

The idea that there is a public education crisis is not a new one. A 1918 government report referred to the "erosion of family life, disappearing fathers, working mothers, the decline of religious institutions, changes in the workplace, and the millions of newly arrived immigrants" as potential sources of the public education crisis (Meier 1995:9). At the time, the government's response was the creation of the modern school system with two tracks, one for terminal high school degrees and the other for college-bound students (Meier 1995).

The current call for educational reform was initiated during President Ronald Reagan's administration. In 1983, the National Commission on Excellence in Education released its report, *A Nation at Risk,* a scathing indictment of the education system. The commission was created by Secretary of Education T. H. Bell to respond to what he called the "widespread public perception that something is seriously remiss in our educational system" (National Commission on Excellence in Education 1983). Claiming that we are raising a scientifically and technologically illiterate generation, the commission noted the relatively poor performance of American students in comparison with their international peers, declining standardized test scores, the weaknesses of our school programs and educators, and the lack of a skilled American workforce (National Commission on Excellence in Education 1983).

The educational reform movement marches on, gaining momentum with each elected president. At one time or another, each president after Reagan has referred to himself as the "Education President," declaring an educational crisis and calling for change. Educators and reformers agree that this is an exciting time for American education (Ravitch and Viteritti

Taking a World View

Educational Tracking and Testing in Japan

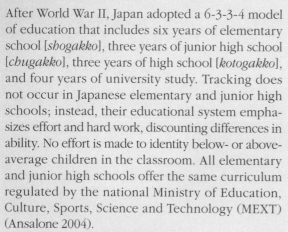

After World War II, Japan adopted a 6-3-3-4 model of education that includes six years of elementary school [*shogakko*], three years of junior high school [*chugakko*], three years of high school [*kotogakko*], and four years of university study. Tracking does not occur in Japanese elementary and junior high schools; instead, their educational system emphasizes effort and hard work, discounting differences in ability. No effort is made to identity below- or above-average children in the classroom. All elementary and junior high schools offer the same curriculum regulated by the national Ministry of Education, Culture, Sports, Science and Technology (MEXT) (Ansalone 2004).

However, a highly competitive form of tracking [*ruikei*] begins at the high school level, separating students into two distinct tracks: general high schools leading to college or vocational high schools leading to jobs. Educational leaders argue that this system is able to accommodate students' different interests and talents and improves students' overall performance on national entrance exams (Ansalone 2004). Entrance examinations, also known as *jyuken higoku* (examination hell), serve as the major sorting mechanism for Japan's high schools and colleges (Bjork and Tsuneyoshi 2005).

Unintentionally, this tracking system has created a two-tier system of schools. Vocational high schools are students' second choices (Ono 2001). In his analysis of vocational high schools in the city of Kobe, Thomas Rohlen (1983) reports that one vocational high school draws students from the lower third of graduating ninth graders in the city. When asked if they had a choice, 80 percent of the students would have rather attended a general high school. Vocational high schools have developed a negative reputation for school violence, smoking, and drug abuse, and their students are considered second-class citizens (Rohlen 1983). In his comparative analysis of educational systems in Japan and the United States, George Ansalone concluded that tracking "promotes differentiation of the curricula, teacher expectations, school misconduct, race, class, gender bias and the development of separate friendship patterns. When tracking is employed, upper track students receive a higher quantity and quality of instruction from more qualified instructors who utilize a greater variety of instructional techniques" (2004:150).

During the last two decades of the twentieth century, several significant changes have occurred in the Japanese educational system. First, government leaders and educational scholars asserted that the emphasis of entrance examinations combined with the demands on students to learn large volumes of content had actually dulled students' interest in learning (Bjork and Tsuneyoshi 2005). Educational reforms began in the 1970s, reducing the amount of material covered by teachers, incorporating more student-centered and integrated learning in the classroom, and reducing the intensity of student learning and testing. MEXT referred to these reform policies as *yutori kyoiku* or reduced intensity reforms. Response to the reforms has been mixed, with some applauding the new student-centered emphasis of the curriculum, but others worry about the impact of a "watered down" curriculum (Bjork and Tsuneyoshi 2005).

Second, admissions into Japan's universities have become less competitive. As its population of 18-year-olds has decreased by more than half a million since 1992, universities have trouble recruiting students. According to Rie Mori (2002), entrance examinations were expected to identify the best students for university education, but this is no longer necessary because there are more universities to accept students who do not score well on the exams. The universalization of higher education in Japan has lead to a greater number of students entering universities who are not as high achieving as students of the past. Japanese higher education must figure out how to educate students with a broader range of learning styles and abilities (Mori 2002).

1997). Under George H. W. Bush's administration, Congress passed America 2000, which was followed by the Goals 2000: Educate America Act in 1994 during Bill Clinton's administration. Congress passed the No Child Left Behind Act of 2001 under George W. Bush's administration. All congressional acts call for coordinated improvements and sweeping reform of our educational system.

David Berliner and Bruce Biddle (1995) contend that the crisis in public education is a manufactured one, constructed by well-meaning or not so well-meaning politicians, educational experts, and business leaders. Berliner and Biddle don't believe that public schools are problem-free; rather, by focusing on the manufactured crisis, they believe we're not addressing the real problems facing our schools, those based in social and economic inequalities. What is the evidence regarding these problems and challenges to our educational system? Let's first examine the basis of education: literacy.

The Problem of Basic Literacy

The United Nations Educational, Scientific and Cultural Organization (UNESCO) estimates that there are more than 770 million illiterate adults in the world (UNESCO 2006). According to the Literacy Volunteers of America (2002), very few U.S. adults are truly illiterate, yet the United States is not a literacy superpower (ProLiteracy Worldwide 2006). What continues to be of concern is the number of adults with low literacy skills who are unable to find and retain employment, support their children's education, and participate in their communities. Basic literacy skills, such as understanding and using information in texts (newspapers, books, a warranty form) or instructional documents (maps, job applications) or completing mathematical operations (filling out an order form, balancing a checkbook) are related to social, educational, and economic outcomes (Sum, Kirsch, and Taggart 2002).

Data from the 2003 National Assessment of Adult Literacy reveal that 30 million U.S. adults demonstrated skills in the "below basic" level (from being non-literate in English to being able to follow written directions to fill out a form). About 63 million adults demonstrated skills in the "basic" level (having basic literary skills to read and understand information in short, simple documents). A total of 43 percent of all Americans are estimated to be in these two levels. In contrast, 57 percent of Americans were categorized in the higher levels—intermediate (able to complete moderately challenging literacy tasks, e.g., refer to a reference document for information) and proficient (able to read and integrate various materials) (ProLiteracy Worldwide 2006).

The U.S. Department of Education reports that individuals at higher levels of literacy are more likely to be employed, to work more weeks per year, and to earn higher wages than are individuals with lower levels of literacy (Kirsch et al. 2002). Education increases an individual's literacy skills, which determine educational success. Basic academic skills influence such educational outcomes as high school completion, college enrollment, persistence in college, field of study, and type of degrees obtained (Sum et al. 2002).

Although the United States spends more per capita on education than other high-income countries, our literacy scores are average in a world comparison. The literacy scores of native-born U.S. adults rank 10th among the 17 high-income nations. The nations that scored higher

Photo 8.4

Unless you're fluent in Portuguese, you would not be able to understand much of the information provided on this directional sign. An estimated 43 percent of the U.S. population has literacy skills at or below the basic level, which includes reading maps or understanding directions to fill out a form.

were Sweden, Norway, Denmark, Finland, the Netherlands, Germany, Canada, Belgium, and Australia. Moreover, the United States ranked first for the largest gap between highly and poorly educated adults, with immigrants and minorities among those poorly educated (Sum et al. 2002).

Inequality in Educational Access and Achievement

Social Class and Education

In October 2003, approximately 3.6 million youth aged 16 to 24 years were not in school and had not earned a high school diploma or GED (Laird et al. 2006). Socioeconomic status is one of the most powerful predictors of student achievement (College Entrance Examination Board 1999). The likelihood of dropping out of high school is five times higher among students from lower-income families than among their peers in high-income families (Laird et al. 2006). Students from lower-income homes or who have parents with little formal education are less likely to be high achievers compared with students from high-income families or students who have parents with a college education (College Entrance Examination Board 1999). A national study revealed that only 5 percent of eighth graders whose parents do not have a high school degree had achievement test scores in the upper quartile; by comparison, more than 50 percent of students who have at least one parent with a graduate degree scored in the top quartile (College Entrance Examination Board 1999).

Poor children are not just poor economically, but also poor educationally. Poor children begin school less prepared and struggle to keep up with their learning and with their classmates (Maruyama 2003). In their analysis of reading and mathematics achievement among kindergarten and first-grade students, K. Denton and J. West (2002) discovered that twice as many first graders from families that were not poor were proficient in understanding words in context and in performing multiplication and division, compared with first graders from poor families. Differences in early school achievement may be attributable to differences in preschool experiences (Maruyama 2003).

Inequalities may emerge before children begin their formal education: in preschool. In 1999, 58 percent of three- or four-year-olds from families earning $40,000 or more attended nursery school compared with 41 percent of children from families with incomes less than $20,000. Although nursery school enrollment has increased during the past few decades, the cost of such programs may prohibit some families from enrolling their children (U.S. Census Bureau 2001), which may negatively affect the children's early school preparation.

Gender and Education

Female college enrollment has steadily increased since the 1970s. In 2000, females were the majority of undergraduate and graduate students (Freeman 2004). Table 8.3 provides a summary of enrollment data from 1970 to 2000. Undergraduate enrollment for women increased from 41.3 percent in 1970 to 54.7 percent in 2000. A similar increase occurred among women enrolled in graduate programs—30.3 percent to 53.6 percent. A more substantial increase occurred among women enrolled in first professional programs (including dentistry, medicine, optometry, and pharmacy). Women constituted 8.5 percent of all those enrolled in these programs in 1970, but increased in 30 years to 46.6 percent. Stoops (2004) reported that for the U.S. population ages 25 to 29, women exceed men in educational attainment—for high school and college completion. About 88 percent of women and 85 percent of men completed high school; 31 percent of women earned a college degree or more compared to 26 percent of men.

Although more women are being awarded bachelor's degrees, fewer women than men complete bachelor's degrees in mathematics, engineering, physics, and other quantitative fields

Table 8.3

Percentage Distribution of Full-Time College Students by Level and Sex, Fall 1970 to Fall 2000

Level	1970 Male	1970 Female	1980 Male	1980 Female	1990 Male	1990 Female	2000 Male	2000 Female
Undergraduate	58.7	41.3	50.7	49.3	47.8	52.2	45.3	54.7
Graduate	69.7	30.3	57.9	42.1	53.6	46.4	46.4	53.6
First-professional	91.7	8.5	71.8	28.2	61.0	39.0	53.4	46.6

Source: Freeman 2004.

(College Entrance Examination Board 1999). By the time they leave high school, males are three times more likely to pursue careers in science, mathematics, or engineering than are females (Strand and Mayfield 2002). Is this because of natural ability? The answer is no, according to the American Association of University Women. The 1992 AAUW study reported that at age 9, there were no differences in math performance between boys and girls, and only minimal differences at age 13. Boys have higher achievement scores in physics, chemistry, earth sciences, and space sciences than girls, and the differences are largest among 17-year-olds.

Differences in mathematical and scientific abilities favoring boys do not appear until the intermediate grades. A study by L. K. Silverman (1986) suggests that females will eventually achieve less than males because they are gradually conditioned by "powerful environmental influences" such as the educational system, peers, and parents to believe that they are less capable than males. A "hidden curriculum" perpetuates gender inequalities in math and science courses. This curriculum takes the form of differential treatment in the classroom, where boys tend to dominate class discussion and monopolize their instructor's time and attention, whereas girls are silenced and their insecurities reinforced (Linn and Kessel 1996). Research suggests that girls, especially gifted ones, fail to achieve their potential because of lower expectations of success, the attribution of any success to chance, and the belief that success will lead to negative social consequences (Silverman 1986).

In 2006, the media declared a "boy crisis" in education when published research revealed how girls had narrowed the gap in their participation and academic achievements in math and science and more women were completing bachelor's degrees than men. Education Sector policy analyst Sara Mead concluded that the concern over boys' academic performance was misguided. Although the percentages of boys and girls taking higher-level mathematics and science courses in high school have increased significantly in the last 20 years, the percentage of girls taking advanced courses has increased more rapidly than has the percentage of boys. Men's higher educational attainment is not declining; rather, it is increasing at a slower rate than that of women. Mead cautioned instead about the growing gaps among students of different races and social classes, especially among Black and Hispanic boys (Perkins-Gough 2006).

Ethnicity/Race and Education

About 3.4 million students entered kindergarten in U.S. public schools last fall and already … researchers foresee widely different futures for them. Whether they are White, Black, Hispanic, Native American or Asian American will, to a large extent, predict their success in school. (Johnson and Viadero 2000:1)

Table 8.4

Differences in Educational Attainment (in percent) by Race and Hispanic Origin for Adults Age 25 to 29 years, 2003

	Non-Hispanic White Alone	Black Alone	Asian Alone	Hispanic (of any race)
High school graduate or more	93.7	87.6	97.1	61.7
Some college or more	65.5	50.2	81.2	31.1
Bachelor's degree or more	34.2	17.2	61.6	10.4

Source: Stoops 2004.

Persistent academic achievement gaps remain between Black, Hispanic, and Native American students and their White and Asian peers. This was the conclusion made by the College Entrance Examination Board's (1999) National Task Force on Minority High Achievement. In the mid-1990s, underrepresented minorities received less than 13 percent of all the bachelor's degrees awarded. The College Entrance Examination Board noted that in the latter half of the 1990s, only small percentages of Black, Hispanic, and Native American high school seniors in the National Assessment Educational Progress test samples had scores "typical" of students who are well prepared for college. Few students in these groups had scores indicating academic skills required for the most selective colleges or universities.

The educational attainment of the young Hispanic population, ages 25 to 29, is lower than for other ethnic/racial groups. Though their high school and college graduation rates have steadily increased during the past decade, Hispanic students continue to trail behind all other groups in educational achievement. As indicated in Table 8.4, at least 87 percent of non-Hispanic Whites, Blacks, and Asians completed high school or more in 2003. In contrast, less than 62 percent of Hispanics (of any race) completed high school or more. At the bachelor's degree level, Asians have the highest proportion of college graduates, 61.6 percent, followed by non-Hispanic Whites, 34.2 percent; Blacks, 17.2 percent; and Hispanics, 10.4 percent.

By 2015, there will be large increases in the number of Latino and Asian American youth, a substantial growth in the number of African American students, and a slight drop in the number of White students. The challenges to our educational system will only increase if demographic predictions hold true. The current educational gaps among racial and ethnic categories have the potential to grow into larger sources of inequality and social conflict (College Entrance Examination Board 1999).

The National Center for Education Statistics (2004) reported that the percentage of males and females who dropped out of high school decreased between 1972 and 2001. However, consistently since the early 1970s, the Hispanic drop-out rate has been the highest. In 2001, 31.6 percent of Hispanic males and 22.1 percent of Hispanic females dropped out of high school. Black students had the next highest drop-out rates: 13 percent for males and 9 percent for females. The lowest drop-out rates were for White males (8 percent) and females (7 percent) (Freeman 2004). Although drop-out rate decreases were noted for White males and females, Black males and females, and Hispanic females, no decrease was detected for Hispanic males.

Ethnicity/race, along with poverty, defines major sources of disadvantage in educational outcomes (Maruyama 2003). Poverty among Latino families produces significant educational disadvantages: Parents may work multiple jobs, may not have the time to spend reading or going over homework with their children, and may not have the skills to read to their children. Economics also play a role in drop-out decisions. To support their families, Latino/a teens may leave school for a paying job. The power of parental and peer influence on Latino/a educational attainment has also been recognized. Parents may have expectations for their children that conflict with school expectations or requirements (AAUW 2001).

According to Claude Steele and Joshua Aronson (1995), the pressure to conform to an image or a stereotype is so strong that it can actually impair intellectual performance. Steele and his colleagues tested the effects of a **stereotype threat** among African American (Steele 1997; Steele and Aronson 1995) and female college students (Spencer, Steele, and Quinn 1999). The stereotype threat is the risk of confirming in oneself a characteristic that is part of a negative stereotype about one's group. The threat is situational, present only when a person can be judged, treated in terms of, or self-fulfill negative stereotypes about group (and self) (Spencer et al. 1999). In their studies, Steele and his colleagues investigated the effect of the stereotype that African Americans and women have lower academic abilities than do White or male students.

It doesn't matter if the individual actually believes the stereotype; if the stereotype demeans something of importance, such as one's intellectual ability, the threat can be disrupting enough to impair intellectual performance (Steele and Aronson 1995). Subjects were compared in test-taking situations using GRE, SAT, or ACT sample questions. In all study conditions where the tests were represented as affected by gender or race, African American and female students underperformed their comparison group. In situations where the stereotype threat was moderated (where subjects were not told that the tests produced gender differences or where subjects were not asked to report their race on the examination form), African American and female students performed as well as White or male students.

Violence and Harassment in Schools

Since a series of deadly school shootings in the late 1990s, part of the focus on education has revolved around safety in schools. School violence can be characterized on a continuum that includes aggressive behavior, harassment, property crimes, threats, and physical assault (Flannery 1997). Victims of school violence may include students, teachers, and staff members. From July 2004 to June 2005, 48 people died in school-related violent incidents. Of these deaths, 21 were homicides and 7 were suicides of school-age youth (ages 5 to 19), a rate of 1 homicide or suicide per 2 million school-aged students during 2004–2005. During the same period, 1,437 children ages 5 to 19 were victims of homicide while away from school. For this period, youth were 50 times more likely to be killed away from school than at school (Dinkes et al. 2006).

Although there were fewer school-associated violent death events in recent years, there were more deaths per event (Anderson et al. 2001). The deadliest U.S. incident took place in April 2007 at Virginia Tech University where 33 students and faculty were killed by a student. The deadliest international incident occurred in 1996 when a gunman killed 16 primary school students and 1 teacher, as well as himself, in Scotland. We spend $200 million annually for school violence prevention and response (Flannery 1997). Though schools have been characterized as "battlegrounds" where both teachers and students fear for their safety (Kingery et al. 1993), nevertheless, schools remain a safe place for students, with the risk of a violent death less than one in two million (Dinkes et al. 2006).

The 2003 National School Based Youth Risk Behavior Survey conducted by the Centers for Disease Control indicated that nationwide about 5 percent of students had missed more than one day of school because they felt unsafe at school or on their way to or from school

The deadliest shooting in U.S. history occurred on the campus of Virginia Tech University (Blacksburg, Virginia) in April 2007. Cho Seung-Hui, a 23-year-old senior, killed 32 students and faculty members before turning the gun on himself. A state panel reviewing the incident concluded that several factors—confusion about state and federal privacy laws, weak enforcement of gun purchasing regulations, and inadequate funding of the state's mental health system—contributed to the tragedy.

(Grunbaum et al. 2004). Hispanic and Black students were more likely than White students to have missed school because they felt unsafe. Among all students, 6 percent said they carried a weapon (a gun, knife, or club) on the school campus, with more male students (8.9 percent) than female students (3.1 percent) doing so. About 9 percent of all students reported being threatened or injured with a weapon on school property, and about 13 percent of students had been in a physical fight on school property (Grunbaum et al. 2004). A 2001 national study reported that 75 percent of students said that they were concerned about a shooting happening in their schools (Bowman 2001). About 87 percent of students believed that shooters want "to get back at those who have hurt them," and 13 percent believed that "nothing could be done to stop school shootings" (Bowman 2001).

Lesbian, gay, bisexual, and transgendered (LGBT) youth are subject to verbal and physical harassment in high schools and middle schools. As reported by the Gay, Lesbian and Straight Education Network's 2005 National School Climate Survey, 64 percent of LGBT youth reported feeling unsafe in school because of their sexual orientation. More than a third experienced some form of physical harassment because of their sexual orientation; 18 percent were physically assaulted for the same reason. More than two-thirds of the surveyed students reported being sexually harassed in school the past year. LGBT students' experience with harassment negatively affected their school attendance, their academic performance, and ultimately, their college aspirations. Students who experienced more frequent harassment, either verbal or physical, were more likely to indicate that they were not planning to go on to college than were students who did not experience the same type of harassment (Kosciw and Diaz 2006).

The extent of sexual harassment in schools has been documented by the American Association of University Women's Educational Foundation (2001). According to its 2001 report, 81 percent of students experience some form of sexual harassment during their school lives. Girls are more likely than are boys to experience sexual harassment. How is sexual harassment defined in schools? The definition offered by the Equal Employment Opportunity Commission (2001) under Title VII of the Civil Rights Act of 1964 reads,

Unwanted sexual advances, requests for sexual favors, and other verbal or physical conduct of a sexual nature constitutes sexual harassment when submission or rejection of this conduct explicitly or implicitly affects an individual's employment, unreasonably interferes with an individual's work performance or creates an intimidating, hostile or offensive work environment.

In schools, sexual harassment can include such behaviors as sexual messages on walls or locker rooms, sexual rumors, being flashed or mooned, being brushed up against in a sexual way, or being shown sexual pictures or material of sexual content (Fineran 2002). Sexual harassment of students has serious consequences, including mental health symptoms (such as loss of appetite, disturbances in sleep, feelings of isolation and sadness) and school performance difficulties (Fineran 2002).

Community, Policy, and Social Action

As a nation, we support the principle of educational excellence and, along with it, the assumption of educational opportunity for all, but in reality, we have an educational system that embraces these ideas, yet fails to achieve them (Ravitch 1997). The educational experiences of poor and minority students fundamentally conflict with the principles of public education, namely that public schools should provide these children with opportunities so that all children can succeed as a result of hard work and talent (Maruyama 2003). Reformers argue that school choice, standardized testing, and school vouchers are improving our educational system. Critics argue that these strategies threaten to erode an already-weak public school structure. There is a deepening chasm between what the American public deems as important in education (safety, skills, discipline) and the goals of the reform movement (access, standardization, multiculturalism) (Finn 1997). Although we have not completely abandoned our public educational system, we still have not found a way to agree on what is appropriate or essential to save it.

Policy Responses—The Basis for Educational Reform

Educate America Act of 1994 and No Child Left Behind Act of 2001

Providing fuel to the reform movement have been congressional acts passed in 1994 and 2001. Although they were adopted under presidents from different political parties, both congressional acts provide strong support for school reform and, along with it, changes to our educational system.

The Goals 2000: Educate America Act introduced the notion of "standards-based reform" at state and community levels. This 1994 act, signed into law by President Bill Clinton, provided the grounds for sweeping reform at all levels and from all angles: curriculum and instruction, professional development, assessment and accountability, school and leadership organization, and parental and community involvement. However, school reform hinged on the use of student performance standards and the creation of the National Education Standards and Improvement Council. A summary of the act reported, "Performance standards clearly define what student work should look like [at] different stages of academic progress and for diverse learners" (Goals 2000 1998:14). The act established performance and content standards in math, English, science, and social studies, and it encouraged participation from the entire community—local officials, educators, parents, and community leaders—in raising academic standards and achievement.

The No Child Left Behind Act (NCLB) of 2001 endorsed the mounting interest in school choice. Some of the major provisions of the bill include new reading and math standard assessments for Grades 3 to 8, more flexibility for states and local school officials with regard to budget spending and program development, the creation of a teaching quality program, consolidation of bilingual and immigrant education programs, and increases in federal funding for President Bush's Reading First plan.

The more controversial elements of the act signed by President George W. Bush include the provision for public school choice and charter schools. The act provided support to permit children in chronically failing public schools to transfer to other schools with better academic records. The bill also provided for annual testing of students in reading and math in the third to eighth grades, which would establish academic records for comparison. If there were no improvement in test results in two years, parents would have the option to move their children to another school. In such an event, the school district must pay for the child's transportation to a better school, and the failing school loses the per-pupil payment. Critics have argued that such school choice provisions will only work if there are schools to choose from within a district and if there is any room in these schools. The law does not provide school leaders with the means to create new slots for students (Schemo 2002). Refer to this chapter's In Focus feature for more about standardized testing.

Promoting Educational Opportunities—Head Start

Called the most popular and most romantic of the War on Poverty efforts (Traub 2000), Head Start remains the largest early childhood program. More than 30 million poor and at-risk preschoolers have been served under Head Start since 1965. Head Start began with a simple model of service: organized preschool centers. At these centers, programs focused on the "whole child," examining and encouraging physical and mental health. Integrating strong parental involvement, Head Start provided a unique program targeting child development and school preparedness. Over the years, the Head Start program expanded to serve school-age children, high school students, pregnant women, and Head Start parents. In 1994, amendments to the Head Start Act established Early Head Start (EHS) services targeting economically disadvantaged families with children three years old or younger. EHS serves both children and their families through a comprehensive service plan that promotes child development and family self-sufficiency (Wall et al. 2000).

The effectiveness of Head Start programming, particularly the educational component, has been the focus of public and government debate (Washington and Oyemade Bailey 1995). Early program research and evaluation efforts were spotty, with the major findings pointing to short-term or "fade out" gains in student learning and testing (Washington and Oyemade Bailey 1995).

Recent research efforts have been more systematic, including longitudinal studies with larger cohorts of Head Start children. Evaluation data have identified a positive relationship between early childhood program enrollment and high school graduation, home ownership, school attendance, and motivation. For example, Diana Slaughter-Defoe and Henry Rubin (2001) noted that among African American middle school students and adolescents, Head Start graduates perceived themselves as more achievement oriented and motivated than did their non–Head Start peers. Head Start parents, teachers, and the family environment were important in sustaining the educational aspirations and expectations of Head Start graduates. Studies consistently indicate the effectiveness of strong parental involvement with the Head Start program.

Head Start parents are more likely to remain involved in their children's education once they begin elementary school, and children whose parents have high participation rates perform better on achievement and development tests (Washington and Oyemade Bailey 1995).

In 1998, the congressional reauthorization of Head Start included an assessment component, requiring the Department of Health and Human Services (HHS) to document the program's effectiveness with low-income three- and four-year-old children. In its 2005 impact study report, DHHS compared the performance of Head Start children with non–Head Start children from fall 2002 through summer 2005. For Head Start children in both age groups, small-to-moderate positive effects were found in pre-reading, pre-writing, and vocabulary skills. Both groups also had significantly better access to health care; in addition, the three-year-olds also reported significantly better overall health than did the non–Head Start group. Positive parenting practices (e.g., reading to one's child) were also documented for children in both Head Start age groups (U.S. Department of Health and Human Services 2005).

Mentoring, Supporting, and Valuing Networks

Women and Girls

In its 1992 report, the American Association of University Women called on local communities and schools to promote programs that encourage and support girls studying science, technology, engineering, and mathematics (STEM). Studies indicate that most girls and women learn best in cooperative, rather than competitive, learning activities. With seed money from the W. K. Kellogg Foundation, the AAUW Educational Foundation initiated the Girls Can! Community Coalitions Project in 1996. The project funded 10 community-based projects that encouraged schools and community groups to improve girls' educational opportunities.

AAUW continues its support of STEM projects through the National Girls Collaborative Project established in 2005. The National Girls Collaborative Project goals include maximizing access to shared resources within projects and with public and private organizations interested in expanding girls' STEM participation, strengthening the capacity of existing and evolving projects by sharing promising research and program models, and using the collaboration of individual girl-serving STEM programs to create the tipping point for gender equity in STEM (National Girls Collaborative 2006).

Six regional collaborative sites operate in California, Florida, Massachusetts, the Mid Atlantic, and the Northwest. Small mini-grants help fund tutoring, career days, field trips, and special events to expose girls and boys to STEM education and careers. One such grant helped fund California State Summer School for Mathematics and Science (COSMOS), a month-long residential academic experience for top California high school students in science and mathematics. Students reside at one of four University of California campuses—Davis, Irvine, San Diego, or Santa Cruz—while taking COSMOS classes in their areas of interest.

Mentoring can also begin in one's own community and among friends. In 1996, Michele Deane noticed that a number of girls in her Boyle Heights (Los Angeles, California) neighborhood did not have anything to do after school. She created a youth organization for local Mexican- and Latin-American girls and women, beginning with a group of her friends, that now serves more than 200 girls and women each year. According to Deane, her organization, Girls Today, Women Tomorrow, created a state of consciousness for thinking "bigger":

> What do I want to do? Who can I become with the support of everyone around me? It's a state of consciousness coming from the world around you instead of seeing only the obstacles. Once they saw other people doing it, they started doing it for their friends, for the younger kids growing up. (Quoted in Wiland and Bell 2006:187)

In Focus

No Child Left Behind and Standardized Testing

These reforms express my deep belief in our public schools and their mission to build the mind and character of every child, from every background, in every part of America.

—George W. Bush, January 2001

When he signed the No Child Left Behind (NCLB) Act, President Bush approved a plan that increased federal pressure on states to pursue a standards-based reform agenda. Under the NCLB, states are required to institute a system of standardized testing for all public school students in Grades 3 to 8 and high school, about 45 million tests annually. All students must be tested in reading and math by 2005 (this was achieved for Grades 4 and 8) and in science by 2007–2008. Each state must have a plan for annual yearly progress toward the goal of academic proficiency for all students (regardless of economic status, ethnicity/race, gender, or disability) by 2014.

Supporters of the act point to the need for increasing accountability of our public schools through standardized testing. The act is based on the premise that test score results are the product of schooling. The idea is that we can give all students the same test, review their test scores, and discover what is "wrong" in their schools. Schools that fail to make annual progress are labeled as "needing improvement" and are likely to face penalties and corrective action, including withdrawal of federal funding.

But NCLB has been called the "most intrusive federal intervention in local schools" (Dillon 2004:A13). According to James Popham (2003), standardized achievement tests measure little of what is taught in school.

> But, by and large, they measure what children bring to school, not what they learn there. They measure the kinds of native smarts that kids walk through the door with. They measure the kinds of experiences the kids have had with their parents.

Education professor David Marshak argues that NCLB supporters did not envision the full impact of the law:

> It [NCLB] puts a standardized test gun to the head of every child, every educator, and every parent in the nation. It guarantees pain and suffering for millions of children and teens whose cognitive and learning styles don't readily fit the narrow structure of standardized testing. It places enormous new demands on most states to pay for the development and administration of new tests. Finally, it seems that the standard for yearly improvement set by the No Child Left Behind Act will be impossible for many schools to meet—certainly many bad schools but also many schools that serve middle-class and upper-middle-class children and are currently held in high esteem by the parents whose children attend them. (2003:230)

Although expressing commitment to the basic intent of NCLB, many state leaders and educators have expressed frustration in implementing the act's requirements and achieving its goals. In particular, school administrators and educators have been critical of a key feature of NCLB: the "one size fits all" accountability standard that assumes that all schools, districts, and groups of students will demonstrate progress according to the standardized measures. The standards, say critics, seriously compromise the abilities of schools to address the unique educational needs of special education students, low-income and minority students, and students with limited English proficiency.

At the fifth anniversary of No Child Left Behind, the Bush administration announced gains in reading and math test scores among 9- and 13-year-olds (earning the highest scores in the history of the test) and improvement in scores among African American and Hispanic students (White House 2006), though minimal progress has been shown in narrowing achievement gaps between White and minority students (Dillon 2007). Nationally, students in 1,800 schools failed to meet state targets for reading and math scores for more than five years (Schemo 2007). The administration called for additional funding for programs to improve teacher quality, improving options for parents of children in struggling schools, and an early testing program for public high school students (White House 2006). President Bush pledged reauthorization of the bill in 2007. Many state and national leaders and educators continue to express dissatisfaction with how the law has been funded, implemented, and applied, calling for major reform before reauthorization.

The volunteer program includes fitness activities, a computer lab, video shooting and editing classes, and a community garden. The garden serves as a connection to the environment and their Latin culture, but also raises girls' awareness about the kind and quality of food they consume. Program graduates return to the program and serve as mentors and volunteers. Ginette Sanchez credits the program for her academic and life successes: "I always thought that I wouldn't have a future. Now that I have positive role models, I'm going to college and I'm being positive by thinking that I'm going to be someone in life as well" (quoted in Wiland and Bell 2006:189).

Voices in the Community

Wendy Kopp

Wendy Kopp's vision for Teach for America began as her senior undergraduate thesis. This Princeton graduate is the youngest and only female to receive the university's Woodrow Wilson Award, the highest honor bestowed to alumni. In the following excerpt from her book, *One Day, All Children ... The Unlikely Triumph of Teach for America and What I Learned Along the Way,* Kopp (2001) explains how Teach for America began with an idea:

Princeton University was not the most likely place to become concerned about what's wrong in education, but it made me aware of students' unequal access to the kind of educational excellence I had previously taken for granted. I got to know students who had attended public schools in urban areas—thoughtful, smart people—as well as students who attended the East Coast prep schools. I saw the first group struggle to meet the academic demands of Princeton and the second group refer to it as a "cake walk." Clearly at Princeton I could not glimpse the depths of educational inequity in our country, but the disparities I did see got me thinking. It's really not fair, I thought, that where you're born in our country plays a role in determining your educational prospects.

In an effort to figure out what could be done about this problem, I organized a conference about the issue. At this time I led an organization called the Foundation for Student Communication ... So in November of my senior year, my colleagues and I gathered together fifty students and business leaders from across the country to propose action plans for improving our educational system....

At one point during a discussion group, after hearing yet another student express interest in teaching, I had a sudden idea: Why didn't this country have a national teacher corps of top recent college graduates who would commit to teach in urban and rural public schools? A teacher corps would provide another option to the two-year corporate training programs and grad schools. It would speak to all college seniors who were searching for something meaningful to do with our lives....

The more I thought about it, the more convinced I became that this simple idea was potentially very powerful. If top recent college graduates devoted two years to teaching in public schools, they could have a real impact on the lives of disadvantaged kids. Because of their energy and commitment, they would be relentless in their efforts to ensure their students achieved. They would throw themselves into their jobs, working investment-banking hours in classrooms instead of skyscrapers on Wall Street. They would question the way things are and fight to do what was right for children.

Beyond influencing children's lives directly, a national teacher corps could produce a change in the very consciousness of our country....

In the end, I produced, "A Plan and Argument for the Creation of a National Teacher Corps," which looked at the educational needs in urban and rural areas, the growing idealism and spirit of service among college students, and the interest of the philanthropic sector in improving education. The thesis presented an ambitious plan: In our first year, the corps would inspire thoughts of graduating college students to apply. We would then select, train and place five hundred of them as teaching in five or six urban and rural areas across the country.

In its first year, Teach for America received 2,500 applications, of which, as Kopp planned, 500 were selected and trained for two years of teaching. Since then, more than 17,000 teachers have been placed or are currently placed in 25 locations throughout the United States. Corp members' salaries and health benefits are paid directly by the school districts they are placed in.

Photo 8.6

Wendy Kopp, founder of Teach for America, is pictured here with Senator Edward Kennedy (D-MA). In 2004, Kopp was the first to receive the John F. Kennedy New Frontier Award, an honor presented to young Americans under the age of 40 for their commitment to public service.

What Does It Mean to Me?

For more information about Teach for America, log on to *Study Site Chapter 8*. The organization's site includes updates on program sites and application deadlines.

LGBT Students

The best estimate of the number of LGBT students is about 5 to 6 percent of the total student population (Human Rights Watch 2001). LGBT youth have been a driving force behind creating change in their schools and communities. Support groups and organized student activities have emerged in states such as California, Illinois, and Washington, providing valuable support to LGBT teens and their friends and families (Bohan and Russell 1999; Human Rights Watch 2001).

One such student group is the Gay Straight Alliance (GSA) in East High School in Salt Lake City, Utah. As J. Bohan and G. M. Russell (1999) chronicle, a group of students proposed creating a student alliance to provide a support network for LGBT students and their heterosexual friends in October 1995. In response to the students' proposal, the school board and the state legislature banned all non-curricular clubs rather than allow the Gay Straight Alliance. The club continued to meet, paying rental and insurance fees for the use of school facilities. According to Bohan and Russell, students indicated how the club had a positive impact on their lives. The alliance served as a safe refuge, decreasing their feelings of isolation and vulnerability, students said, and they reported decreases in substance abuse, depression, suicidal impulses, truancy, and conflict with parents. The straight student members also reported positive effects. In September 2000, Utah's Salt Lake City School Board voted to permit non-curricular student groups to meet on school grounds, reversing its 1995 decision against the GSA (Human Rights Watch 2001). According to the Gay, Lesbian, and Straight Education Network, there are more than 3,000 registered clubs for LGBT students and their friends throughout the United States.

What Does It Mean to Me?

Is there a Gay Straight Alliance organization at your school? If yes, investigate its history. What are its mission, student membership, and activities? If your school does not have an alliance, would your school community support one? Why or why not?

Antiviolence Programs in Schools

As awareness of school violence has increased, so have the calls for effective means of prevention (Aber, Brown, and Henrich 1999). The current focus is less on reacting to school violence and more on promoting school safety through prevention, planning, and preparation (Shaw 2001). The largest and longest-running school program focusing on conflict resolution and intergroup relations is the Resolving Conflict Creatively Program (RCCP). Initiated in 1985 in New York City by the local chapter of Educators for Social Responsibility, the program is a research-based K–12 school program in social and emotional learning. RCCP is in 375 schools nationwide, serving 6,000 teachers and more than 175,000 students.

RCCP begins with the assumption that aggression and violent behavior are learned and therefore can be reduced through education. The program teaches children conflict resolution skills, promotes intercultural understanding, and provides models and opportunities for positive ways of dealing with conflict and differences. For kindergarten students, puppets and other objects are used to illustrate how conflict can be resolved by talking rather than hitting. RCCP includes training for teachers, parents, administrators, and school staff.

An evaluation of the New York City programs indicated that students who received RCCP instruction developed more positively than did students without any RCCP exposure. RCCP students were more prosocial, perceived their world in a less hostile way, saw violence as unacceptable, and chose nonviolent ways to resolve conflict. An additional finding of the evaluation revealed that reading and math scores were higher for RCCP students, especially those who had 25 RCCP lessons over the school year. Evaluators concluded that the RCCP-intensive children were more able to focus on academics when there was less conflict with peers.

With the increase in school violence, more attention has been given to school safety through security screening or police-school liaison projects. Schools are more aware of the links between safety and violence and other student behaviors such as drop-out rates, academic failures, bullying, and suicide (Shaw 2001). Violence prevention programs have become common throughout the country with the primary focus on early education. In addition to the RCCP, other national initiatives include the Safe and Drug Free Schools Initiative and the Safe Schools/Healthy Students Initiative. Regional initiatives include the PeaceBuilders elementary program operating in Arizona, California, Utah, and Ohio; the BrainPower Program in Southern California; Healthy Schools Bullying Prevention Program in South Carolina; and Positive Adolescent Choices Training (PACT) based in Ohio. U.S. and international approaches focus on school safety and less on school violence, use programs to serve students and the entire school population, develop school-community partnerships, and use evaluated program models (Shaw 2001). The most effective school-based violence prevention programs are those that included parental involvement and support, with parents backing school limits and consequences at home (Flannery 1997). Antiviolence programs have also been established on college and university campuses (refer to Table 8.5).

Does Having a Choice Improve Education?

There is a new term now, public school choice. Data issued by the National Center for Education Statistics reveal that more parents are turning away from local public schools to private schools or charter schools (Zernike 2000). The number of students enrolled in public schools of choice increased by 2.5 million from 1993 to 1999, reflecting a 50 percent increase. Today, at least one in four children attends a school other than the one nearest home (Wilgoren 2001). Parents and children have two additional options within the public education system: magnet schools and charter schools.

Magnet schools offer specialized educational programs from elementary school through high school. These schools are organized around a theme such as performing arts, science, technology, or business or around different instructional designs such as free (where students can direct their own education) or open schools (with informal classroom designs). Often, magnet schools are placed in racially isolated schools or neighborhoods to encourage students of other races to enroll. Magnet schools have been criticized for creating a two-tier system of education (Kahlenberg 2002).

Table 8.5

What College and University Campuses Can Do to Make a Difference: Recommendations From the Tool Kit to End Violence Against Women

The National Advisory Council Against Women (2001) released the following list of recommendations as part of its Tool Kit to End Violence Against Women. The list includes recommendations for strengthening prevention efforts and improving services and advocacy for victims on campus:

- Institutionalize a campus-wide response to violence against women.
- Create an interdisciplinary task force to address violence against women.
- Establish a fair campus adjudication process.
- Administer sanctions for perpetrators that convey the seriousness of the offense.
- Invest in comprehensive and accessible on-campus and community services to victims.
- Provide training on violence against women for all campus law enforcement.
- Form partnerships with local victim service programs and criminal justice agencies.
- Highlight men's ability and responsibility to prevent violence against women.
- Enlist men in education efforts.
- Participate in full disclosure of campus crime data reports.

Source: National Advisory Council on Violence Against Women 2001.

Charter schools are nonsectarian public schools of choice that operate free from most state laws and local school board policies that apply to traditional public schools. A charter contract establishes the school's operation, usually limited to three to five years, detailing the school's mission and instructional goals, student population, educational outcomes, and assessment methods, along with a management and financial plan. These schools have grown in popularity since 1991, when Minnesota became the first state to pass an "outcome-based" school law. As of 2005, about 4,600 charter schools were serving more than a million students in 40 states and Washington, D.C. Charter schools are characterized by innovative teaching practices and accountability to students and families. If a school fails to meet its goals, it cannot be renewed under its charter.

School choice has become an even hotter topic with the idea of school vouchers. Simply stated, school vouchers allow the transfer of public school funds to support a student's transfer to a private school, which may include religious institutions. Supporters of school vouchers argue that the system would give parents more choice and freedom in school selection and would create incentives for school improvement (Good and Braden 2000; Kennedy 2001).

Opponents argue that vouchers would siphon money away from public schools, removing any ability to resolve the schools' problems, thus only increasing problems. Others argue that schooling is a public good and must be provided by the government to all children (Good and Braden 2000) equally and fairly. In June 2002, the U.S. Supreme Court ruled that school voucher programs did not violate any church versus state separation and upheld the constitutionality of using public funds to support private school systems (Bumiller 2002).

On the effectiveness of voucher programs and charter or magnet schools, the research remains mixed and ultimately divided according to party lines. The same issues concerning charter schools and voucher systems continue to be the subject of inquiry and debate: defining clear systems of accountability, establishing comparable performance standards, and ensuring the racial and economic integration of students.

In 2006, data analysis based on the 2003 National Assessment of Educational Progress test by the U.S. Department of Education revealed that children in charter schools were performing worse than were their peers in public schools. Fourth graders in traditional public schools scored an average of 4.2 points better in reading and 4.7 points better in math than did comparable students in charter schools. The federal commissioner of education statistics, Mark S. Schneider, said at the release of the report that the department would not put its stamp on research comparing public and charter schools. He admitted, "This is one of the most contentious issues with regard to the charter school research debate" (Schemo 2006).

According to researchers S. Saporito and A. Lareau (1999), if there is one consistent finding on school choice, it is that students from poorer families or with less-educated parents are less likely to apply to or participate in public choice programs than are those from middle-class families. In addition, the researchers raise questions about the school selection process for White and African American families. Although school choice advocates suggest that promoting racial equality is one of the by-products of school choice, Saporito and Lareau (1999) found that White families as a group are more likely to avoid schools with higher percentages of Black students, whereas African American families show no such sensitivity to race. African American families in their study were likely to select schools with lower poverty rates. The researchers concluded that race was a persistent factor in the choice process.

Main Points

- For 2003, the U.S. Census Bureau reported that an all-time high of 85 percent of adults (people 25 years and older) had completed at least a high school degree, and more than 27 percent of all adults had attained at least a bachelor's degree. Younger Americans are more educated than are older Americans. About 70 percent of all Americans 75 years or older attained a high school degree or more, compared with 86 percent of those 25 to 29 years old. The educational attainment level of adults will continue to rise, as younger, more educated age groups replace older, less educated ones.

- The institution of education has a set of **manifest** and **latent functions.** Manifest functions are intended goals or consequences of the activities within an institution. Education's primary manifest function is to educate. The other manifest functions include personal development, proper socialization, and employment. Education's latent or unintended consequences may be less obvious.

- Conflict theorists do not see education as an equalizer; rather, they consider education a divider: dividing the haves from the have-nots in our society. From this perspective, conflict theorists focus on the social and economic inequalities inherent in our educational system and how the system perpetuates these inequalities. Inequalities are based not just on social class but also on gender. Research reveals the persistent replication of gender relations in schools, evidenced by the privileging of males, their voices, and their activities in the classroom, playground, and hallways.

- From an interactionist perspective, the interaction between teachers and students reinforces the structure and inequalities of the classroom and the educational system. Sociologists focus on how classroom dynamics and tracking practices educate the perfect student and at the same time create the not-so-perfect ones.

- Although educational attainment is increasing, it is still unevenly distributed among social groups.

Socioeconomic status is one of the most powerful predictors of student achievement. The likelihood of dropping out of high school is higher among students from lower-income families.

■ Census data indicate slight differences in educational attainment for men and women. A "hidden curriculum" perpetuates gender inequalities in math and science courses. This curriculum takes the form of differential treatment in the classroom, where boys tend to dominate class discussion and monopolize their instructor's time and attention, whereas girls are silenced and their insecurities reinforced.

■ Persistent academic achievement gaps remain between Black, Hispanic, and Native American students and their White and Asian peers. In the mid-1990s, underrepresented minorities received less than 13 percent of all the bachelor's degrees awarded. The College Entrance Examination Board noted that in the latter half of the 1990s, only small percentages of Black, Hispanic, and Native American high school seniors in the National Assessment Educational Progress test samples had scores typical of students who are well prepared for college. Latino/a students are at greater risk of not finishing school than any other ethnic-racial group.

■ Part of the focus on education has revolved around safety in schools. School violence can be characterized on a continuum that includes aggressive behavior, harassment, threats, property crimes, and physical assault.

■ Recent acts and initiatives to reform education include the Educate America Act of 1994 and the No Child Left Behind Act of 2001. Several education programs promote educational opportunities. Probably the most well-known early childhood program is Head Start.

■ Programs such as the National Girls Collaborative Project support girls studying mathematics and science. Lesbian, gay, bisexual, and transgendered (LGBT) youth have been a driving force behind creating change in their schools and communities. Support groups and organized student activities have emerged in states such as California, Illinois, and Washington, providing valuable support to LGBT teens and their friends and families.

■ There is a new term now: *public school choice.* School vouchers allow the use of public school funds to support student transfers to private schools, which may include religious institutions. Supporters of school vouchers argue that the system would give parents more choice and freedom in school selection and would create incentives for school improvement. Opponents argue that vouchers would siphon money away from public schools, removing any ability to resolve the schools' problems, thus only increasing problems.

On Your Own

Log on to the Web-based student study site at www.pineforge.com/leonguerrero2study for interactive quizzes, e-flashcards, journal articles, Community and Policy Guides, a Service Learning Guide, the end-of-chapter Web exercises, and additional Web resources.

Internet and Community Exercises

1. Contact your local ProLiteracy Worldwide volunteer program or search their national Web site (log on to *Study Site Chapter 8*). Click on "find a program" and select your state. If a program does not exist in your community, find the one nearest you (your local library may have information on literacy programs). What services or activities do they provide students? Does the program have data regarding the number and types of students they have served? On the effectiveness of their program? What skills are necessary to become a literacy tutor?

2. Interview a student who is a first-generation college student. The student can be from your social problems class or any of your other classes. What challenges do first-generation college students face? Do you believe their challenges are different from those of students who are second- or third-generation college students? Does your college provide programs for first-generation students? If not, what type of services or support might be valuable for this group of students?

3. Investigate whether your local school district supports educational outreach programs for girls or minority students. Select one program and answer the following: What group does the program serve? What educational "gaps" does the program address and how? How effective is the program? Contact the local school district, the YWCA, or the American Association of University Women (log on to *Study Site Chapter 8*) for more information.

4. In addition to public school choice, homeschooling has become an option for many families. It was estimated that 1.1 million children were homeschooled in 2003, about 1.7 percent of all students age 5 to 17 years (National Center for Educational Statistics 2004). This figure includes students who were homeschooled only and students who were homeschooled and enrolled in school for 25 hours or less per week. Parents offered a variety of reasons for homeschooling their children: belief that they can give their child a better education at home, religious reasons, the poor learning environment at school, and family reasons. Homeschoolers are more likely to be located in rural and suburban areas of the Western United States (Bauman 2001). Home School World provides a listing of all U.S. national and state homeschool organizations, including a listing of international organizations (under "Homeschool Organizations"). Log on to *Study Site Chapter 8*.

5. Log on the United Nations Human Development Report Web site and compare global literacy rates for males and females (ages 15 years and older). In general, is the literacy rate higher for men or women? Identify the countries with the lowest literacy rates for each group. Log on to *Study Site Chapter 8* for more information.

CHAPTER 9

Work and the Economy

W hen C. Wright Mills first wrote about the "sociological imagination" in 1959, he identified unemployment as a public issue. Public issues, like social problems, are matters that transcend the individual and have much more to do with the social organization of our lives. Mills writes,

When, in a city of 100,000, only one man is unemployed, that is his personal trouble, and for its relief we properly look to the character of the man, his skills and his immediate opportunities. But when in a nation of 50 million employees, 15 million are unemployed, that is an issue, and we may not hope to find its solution within the range of opportunities open to any one individual. The very structure of opportunities has collapsed. (1959/2000:14)

Almost 50 years after Mills wrote these words, unemployment remains a public issue. According to the U.S. Bureau of Labor Statistics (2007a), 6.8 million women and men were unemployed in December 2006, an overall rate of 4.5 percent. For the European Union as a whole, unemployment is around 8 percent, with rates highest in France, Germany, Italy, and Spain (Blanchard 2004). But unemployment isn't the only problem facing workers. For some, work is a dangerous place, leading to injury or death. Others are victims of discrimination or harassment at the workplace. And for many working men and women, their paychecks do not provide a livable wage.

Work isn't just what we do; work is a basic and important social institution. It fuels our economy and provides economic support for individuals and families. In December 2006, more than 145.9 million women and men were employed in the United States, about 63 percent of the population 16 years or older (U.S. Bureau of Labor Statistics 2007a). But work is also important for our social and psychological well-being. Individuals find a sense of fulfillment and happiness, and for most, work provides a self-identity. Much of our social status is conferred through our occupation or the type of work we do. Because of the importance of work, problems related to work become categorized as social problems, as everyone's problem. In this chapter, we will examine the work that we do, the social organization of work itself, and the social problems associated with our work and economy. We will begin first with a review of the changing nature of work.

The Changing Nature of Work

During the late eighteenth century and early nineteenth century, the means of production shifted from agricultural to industrial. In agrarian societies, economic production was very simple, based primarily on family agriculture and hunting or gathering activities. Each family provided for its own food, shelter, and clothing. This changed during the **Industrial Revolution**, an economic shift in how people worked and how they earned a living. Family production was replaced with market production, in which capitalist owners paid workers wages to produce goods (Reskin and Padavic 1994).

As a whole, we don't produce goods anymore, we provide services. Since the late 1960s, the U.S. economy has shifted from a manufacturing to service-based economy (Brady and Wallace 2001). In 1950, manufacturing accounted for 33.7 percent of all nonfarm jobs; by 2000, manufacturing's share had dropped to 14 percent (Pollina 2003). The **service revolution** is an economy dominated by service and information occupations. Examine Table 9.1, which lists the fastest-growing occupations in 2004 and 2014. Notice that none of the occupations on the list involve manufacturing; instead, jobs in retail, food, and customer service dominate the list.

Table 9.1

Ten Occupations With the Largest Projected Job Growth (in thousands), 2000–2014

Industry Description	2004	2014
Retail salespersons	4,256	4,992
Registered nurses	2,394	3,096
Janitors and cleaners, except maids and housekeeping cleaners	2,374	2,813
Waiters and waitresses	2,252	2,627
Combined food preparation and serving workers, including fast food	2,150	2,516
Customer service representatives	2,063	2,534
General and operations managers	1,807	2,115
Postsecondary teachers	1,628	2,153
Nursing aides, orderlies, and attendants	1,455	1,781
Home health aides	624	974

Source: U.S. Bureau of Labor Statistics 2005.

This shift has been referred to as **deindustrialization**, a widespread, systematic disinvestment in our nation's manufacturing and production capacities (Bluestone and Harrison 1982). Less manufacturing takes place in the United States because most jobs and plants have been transferred to other countries. The expansion of industrialization in other regions such as Asia, the Caribbean, Eastern Europe, and Africa has encouraged managers to reduce their manufacturing costs by exporting manufacturing jobs there (Sullivan 2004). In addition, U.S. factories have closed as a result of mergers or acquisitions as well as poor business. And thanks to technological advances, it takes fewer people to produce the same amount of goods. (Refer to this chapter's In Focus feature for more on the effects of deindustrialization.)

Job losses in manufacturing have been particularly acute since early 2000: U.S. manufacturers have lost 2.7 million jobs since July 2000. Affected areas have experienced devastating social and economic losses, turning some into ghost towns (Brady and Wallace 2001) or leading cities to the brink of bankruptcy—the plight of Cleveland, Ohio, and Detroit, Michigan, in the mid-1980s and early 1990s. The phenomenon was also observed in Great Britain in the 1970s and 1980s in cities such as Birmingham and Manchester (Carley 2000). Lost manufacturing jobs are often replaced with unstable, low-paying service jobs, or no jobs at all. As a result, cities may experience a significant loss of revenue to support basic public services such as police, fire protection, and schools (Bluestone and Harrison 1982).

In addition to the transformation in the type of work we do, there has also been a transformation in who is doing the work. The first significant workforce change began in World War II with the entry of record numbers of women into the workplace. In 1940, the majority of 11.5 million employed women were working as blue-collar, domestic, or service workers out of economic necessity (Gluck 1987). White and Black women's entry into defense jobs signaled a major breakthrough. One-fourth of all White women and nearly 40 percent of Black women were wage earners who previously worked in lower-paid clerical, service, or manufacturing jobs. By 1944, 16 percent of working women held jobs in war industries. At the height of wartime production, the number of married women in the workplace outnumbered single working women for the first time in U.S. labor history. Almost one in three women defense workers were former full-time homemakers. In Los Angeles, women made up 40 percent of the aircraft production workforce.

These heavy industry jobs may have paid better, but the jobs held an important symbolic value: These jobs were men's jobs. After the war, although the proportion of women workers in durable manufacturing increased in many cities, many women were forced back into low-paying, female-dominated occupations (Gluck 1987) or back to their homes.

Labor force participation rates have steadily increased for White, Black, and Hispanic women since World War II (see Table 9.2). In 2005, 69,288,000 women 18 years or older were employed, accounting for 46 percent of the U.S. labor force (U.S. Department of Labor, Women's Bureau 2005). Women dominated secretarial, receptionist, nursing, and bookkeeping occupations during that year (U.S. Department of Labor, Women's Bureau 2005).

Photo 9.1

Coal mining is an inherently dangerous occupation. The 2006 explosion at the Sago Mine (West Virginia) and the 2007 collapse of the Crandall Canyon Mine (Utah) renewed industry focus on worker safety and emergency response procedures. Pictured here are rescue workers setting up their drill rig at the Crandall Canyon Mine. Six miners and three rescue workers were killed at the mine.

In Focus

The Economic and Social Impacts of Deindustrialization

Timothy Minchin (2006) documents the 2000 closing of the International Paper (IP) mill in Mobile, Alabama, in 2000. The papermaker closed the 70-year-old plant, shifting its manufacturing focus to newer, more efficient plants. Minchin identifies economic, social, and emotional effects on the plant's 800 workers, their families, and communities.

In the U.S. South, where wages have lagged behind the rest of the country, IP workers were earning almost twice as much as the average manufacturing employee in the area. IP jobs had the reputation of being among the best-paying jobs in the area, but after the closure, though most IP workers found new jobs, they are earning less than what they made at IP.

International Paper was one among many Alabama companies forced to close its doors largely because of increased overseas competition. According to Minchin (2006), employment in the paper industry slumped from 647,000 in 1990 to 499,000 in 2004. More than 48,000 Alabama jobs were lost between January 1999 and February 2003 because of plant closures and layoffs. He writes,

> The loss of manufacturing jobs hit Alabama hard because the state remained more dependent on manufacturing than many others; in 2000, 19.3 percent of Alabama's jobs were in manufacturing, compared to 14.2 percent for the nation as a whole. Between 2000 and 2002, Alabama lost 12.5 percent of its manufacturing jobs, while the nation in general saw 9.5 percent of these positions disappear. (Minchin 2006:48)

Minchin documents the personal toll on IP's workers, observing how workers and supervisors felt angry and betrayed about how the company could treat such "dedicated employees." The closure affected a male-dominated operation—about 80 percent of IP's Alabama plant employees were male. Many reported the loss of marriages, homes, and automobiles. Drug use and increased drinking was also reported among the laid off workers. Factory men, who for all their lives were their family's primary income earners, were unhappy with their wives being forced to seek work.

Minchin says that plant closures had a particularly harsh effect on African American communities. Black employees reported their disappointment in being "last hired, first fired" at the IP plant. Through legislative and legal efforts, Black employees were beginning to move into higher-paying production positions at the time the factory was closed. African American workers rebounded slower than did their White counterparts. According to Minchin, Blacks were more likely to live and work in urban areas (where plant closings have been particularly high) rather than in suburban areas where job opportunities were opening up.

What Does It Mean to Me?

Just for one week, count the number of times you hand cash, a credit card, or a check to someone for payment. How many times do you hand it to a woman? M. F. Fox and S. Hesse-Biber (1984) offer the following description for the typical working woman—a working mother who attended high school but had little or no college experience, working in retail, clerical, or service occupations. For the current list of the 20 leading occupations for employed women, go to the U.S. Department of Labor's Women's Bureau (log on to *Study Site Chapter 9*). Has "women's work" changed? Why or why not?

Table 9.2

Civilian Labor Force Participation Rates for White, Black, and Hispanic Women, 1930–2010 (Projected)

	All Women	White Women	Black Women	Hispanic Women
1930	24.3	n.a.	n.a.	n.a.
1940	25.4	n.a.	n.a.	n.a.
1950	31.4	n.a.	n.a.	n.a.
1960	37.1	36.2	n.a.	n.a.
1970	41.6	42.6	49.5	n.a.
1980	51.5	51.2	53.1	47.4
1990	57.5	57.4	58.3	53.1
2000	60.2	59.8	63.2	56.9
2010 Projected	62.2	61.6	66.2	59.4

Source: U.S. Bureau of Labor Statistics 2003b, 2004; U.S. Census Bureau 1951, 1960, 1966.

The second workforce change has been the record numbers of elderly Americans returning to work. Since the mid 1980s, the labor force participation rates of older Americans have consistently increased (Toossi 2005). In 2005, 23.4 million Americans aged 55 or older were employed in the labor force, reflecting a 37.2 percent participation rate (Rix 2006). The percent of Americans 55 years or older staying employed or going back to work after retirement is projected to increase by 2014, more than any other age group (refer to Figure 9.1).

Although Americans are working longer partly because they are living longer, additional factors contribute to the increase in 65-plus employment. Government policies have eliminated mandatory retirements and outlawed age discrimination in the workplace. In 2000, older Americans were also encouraged to go back to work with the removal of age restrictions and taxes on their earned wages (Toossi 2005). At a time when most elderly Americans are considering retirement, some cannot afford to live on their retirement income, forcing them back to the workplace. Seniors are a valuable commodity in the workplace: Employers are grateful for their skills and work ethic. And many seniors are working because they want to do so.

The final workforce shift began with the latest immigration boom between 1996 and 2000 from Latin America and Asia (Mosisa 2002). The foreign-born U.S. population rose from a low of 9.6 million in 1970 to 14.1 million in 1980, 19.8 million in 1990 (Gibson and Lennon 1999), and 28.4 million in 2000 (Lollock 2001). In 1960, 1 of 17 U.S. workers was foreign born, compared with 1 of 8 today. More than 80 percent of the labor force increase among workers between 35 and 44 years old can be attributed to the increase in foreign-born workers. Ronald Pollina (2003) predicts that, as our native population continues to age, the U.S. workforce will become increasingly dependent on foreign-born workers.

Photo 9.2

Women accounted for more than 45 percent of the U.S. labor force in 2005. The leading occupations for U.S. women continue to be in retail, administrative support, or service.

Figure 9.1

Projected Percentage
Change in Number
of Labor Force
Participants by Age
Group, 2004–2014

Source: Toossi 2005.

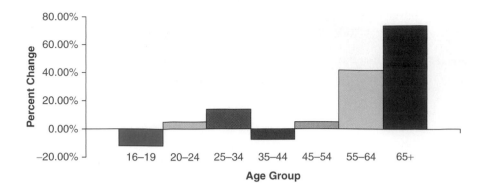

Foreign-born workers account for half or more of the increase in the job categories of administrative support, services, precision production, craft and repair, operators, fabricators, and laborers. Foreign-born workers are overrepresented in low-paying occupations that do not require high school degrees. During 2000, about 60 percent of all foreign-born workers were living in four states—California, New York, Florida, and Texas. Immigrants settle in regions with perceived economic opportunities, seeking established ethnic enclaves providing interpersonal and job support (Mosisa 2002).

Sociological Perspectives on Work

Functionalist Perspective

According to the functionalist perspective, work serves specific functions in society. Our work provides us with some predictability about our life experiences. We can expect to begin paid employment around the age of 18 or after high school graduation (or we could delay it for four or five more years by attending college and graduate school). Your work may determine when you get married, when you have your first child, and when you purchase your home. Work serves as an important social structure as we become stratified according to our occupations and our income. Finally, even for the most independent among us, the way we live depends on the work of thousands—for our food, clothing, safety, education, and health. Our lives are bound to the products and activities of the labor force (Hall 1994).

Recall that from this perspective, work can also produce a set of dysfunctions that can lead to social problems (or that may be problems themselves). Employers encourage workers to become involved with their work, hoping to increase their productivity as well as their quality of work (a function). However, getting too involved in one's work may lead to job stress, overwork, and job dissatisfaction for workers (all are dysfunctional). Although technology improves the speed and quality of work for some (a function), as machines replace human laborers, technology can also lead to job and wage losses (dysfunctions).

As some researchers have focused on the functions and dysfunctions of work, others have tried to understand the nature of work itself. Frederick W. Taylor, a mechanical engineer, offered an analysis that revolutionized nineteenth- and twentieth-century industrial work. Using what he called **scientific management**, Taylor broke down the functional elements of work, identifying the most efficient, fastest, best way to complete a task. In one of his first research projects, Taylor determined the best shovel design for shoveling coal. Taylor believed that with the right tools and the perfect system, any worker could improve his or her work productivity, all for the benefit of the company. In his book, *The Principles of Scientific Management,* Taylor wrote, "In the past the man has been first; in the future the system must be first" (1911:5).

Although scientific management in its pure form has rarely been implemented, Taylor's principles continue to serve as the foundation for modern management ideology and technologies of work organization (Bahnisch 2000). Beyond simply changing how work was organized, Taylor also offered his ideas about the organization of work: the need for defining a clear authority structure, separating planning from operational groups, providing bonuses for workers, and insisting on task specialization. Taylor's model shifted power to management, forcing skilled workers to give up control of their own work (Hirschhorn 1984), which was a cause for concern expressed by theorists from the next theoretical perspective.

Conflict Perspective

Power, explained Karl Marx, is determined according to one's relationship to the means of production. Owners of the means of production possess all the power in the system, he believed, with little (probably nothing) left for workers. As workers labor only to make products and profits for owners, workers' energies are consumed in the production of things over which they have no real power, control, or ownership (Zeitlin 1997). According to Marx, man's labor becomes a means to an end; we work only to earn money. Marx predicted that eventually we would become alienated or separated from our labor, from what we produce, from our fellow workers, and from our human potential. Instead of work providing a transformation and fulfillment of our human potential, work would become the place where we felt least human (Ritzer 2000).

Modern systems of work continue to erode workers' power over their labor. Deskilling refers to the systematic reconstruction of jobs so that they require fewer skills, and ultimately, management can have more control over workers (Hall 1994). Although Taylor (1911) proposed scientific management as a means to improve production, sociologist Harry Braverman (1974) argued that by altering production systems, capitalists and management increase their control over workers. Once dependent on the workers' abilities, the nature of work shifts to managerial priorities. Management, according to Braverman (1974:119), "controls each step of the labor process and its mode of execution." Workers lose their power when their skilled labor is taken away.

Although Marx predicted that capitalism would disappear, capitalism has grown stronger, and at the same time, the social and economic inequalities in U.S. society have increased. Capitalism has become more than just an economic system; it is an entire political, cultural, and social order (Parenti 1988). Modern capitalism includes the rise and domination of corporations, large business enterprises with U.S. and global interests. Conflict theorists argue that capitalist and corporate leaders maintain their power and economic advantage at the expense of their workers and the general public.

In the 1990s and early 2000s, energy trader Enron was the epitome of corporate power and success. The company bought and sold natural gas, power facilities, telecommunications, and other energy-related businesses. In 2000, Enron operated in 30 countries, employing 18,000 women and men (Enron Corporation 2000). Fortune magazine named Enron as the "Most Innovative Company in America" for six consecutive years (Enron Corporation 2001), and it was reportedly worth $70 billion at its peak (National Public Radio 2003). A great company to work for, a powerful and successful company, Enron was unstoppable—or was it?

Through a complex web of tax-sheltering partnerships, Enron hid millions of dollars in debts and company losses. Enron's accounting firm, Arthur Andersen, was a partner in the deception and was later convicted of obstruction of justice in the government's investigation. The company's losses were first made public in October 2001, when the Securities and Exchange Commission launched a formal investigation into the company's dealings (Fowler 2001). On December 2, 2001, Enron declared bankruptcy, the largest in U.S. history. Enron executives, including Chairman Kenneth Lay, left the company in disgrace.

Enron's employees lost their jobs and more than $1 billion in pension holdings. Several class action lawsuits have been filed against Enron executives, directors, and consultants, alleging that they knew about Enron's financial troubles but chose not to inform its employees or stockholders. Enron was not the only company involved in erroneous claims. In the 1990s, more than 700 companies were forced to correct misleading financial statements as a result of accounting failures or fraud (Frontline 2002). According to Lynn Turner, chief accountant of the Securities and Exchange Commission from 1998 to 2001 (Frontline 2002), Enron's demise is a symptom of something larger in our economic system.

It's beyond Enron … It's embedded in the system at this time…. There's been a change in culture that arose out of the go-go times of the 1990s. Some people call it greed. But I think it is an issue where we got a lot of financial conflicts built into the system, and people forgot, quite frankly, about the investors.

Feminist Perspective

From a feminist perspective, work is a gendered institution. Through the actions, beliefs, and interactions of workers and their employers, as well as the policies and practices of the workplace (Reskin and Padavic 1994), men's and women's identities as workers are created, reproduced, then solidified in the everyday routines of informal work groups and formal workers' organizations (Brenner 1998). We already discussed the importance of World War II for women's employment; but recall that after the war, there was pressure on women to resume their roles as housewives or to assume more appropriate occupations. As a gendered institution, work defined the roles appropriate for World War II women and for women today. The workplace does not treat women and men equally. Women are concentrated in different—and lower ranking—occupations than men are, and women are paid less than men are (Reskin and Padavic 1994).

A fundamental feature of work is the **sexual division of labor**: the assignment of different tasks and work to men and women. This division of labor leads to a devaluing of female workers and their work, providing some justification for the differential compensation between men and women (Reskin and Padavic 1994). In the United States, as in most other countries, women earn less than men (England and Browne 1992). There is no country in the world where women make the same or more than men do (refer to Figure 9.2). In the early 1960s, U.S. women earned about 59 cents for every dollar earned by men (Armas 2004). By 2005, for every dollar a man earns, a woman makes 81 cents (U.S. Department of Labor, Women's Bureau 2005). (Refer to Figure 9.3 for a comparison of women's median weekly earnings as a percent of men's for 1979 and 2005. Note how gains have been made in almost every age group presented.) Pay equity is greater for men and women working as elementary school teachers (a ratio of 94.9), accountants (93.7), or general office clerks (96), but there is no single occupation where U.S. women make the same amount of money as men do (U.S. Department of Labor, Women's Bureau 2005).

In pay, compared with men, women are disadvantaged because they are in lower-paying feminized jobs or because they are paid less for the same work (Budig 2002). Sociologists and feminist scholars insist that no natural differences between men and women would lead to

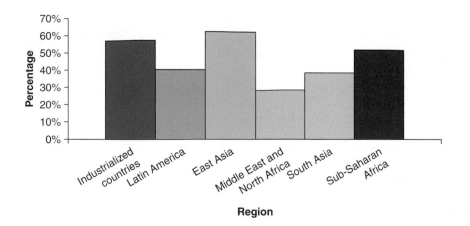

Figure 9.2

Women's Earnings as
a Percentage of Men's
Earnings, in U.S.
Dollars, 2003

Source: UNICEF 2007.

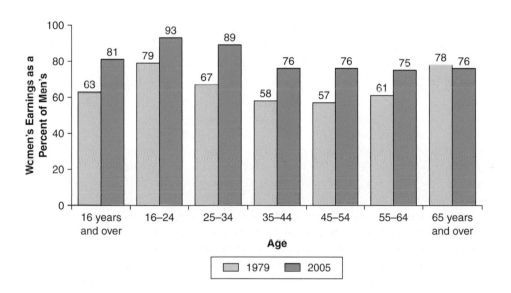

Figure 9.3

U.S. Women's to Men's
Earnings Ratio by Age
Groups, 1979 and
2005

*Source: U.S. Bureau of
Labor Statistics 2006a.*

*Note: Earnings are median
usual weekly earnings of
full-time wage and salary
workers.*

this. Instead, these researchers offer several structural explanations for the differences: differential socialization (women are socialized to pursue careers that traditionally pay less or are lower in status), differential training (men are better educated, so should be rewarded with higher pay), and workplace discrimination (Reskin and Padavic 1994). For additional discussion of these perspectives, refer to Chapter 4, "Gender."

One workplace problem feminist advocates and scholars have focused on are the exclusionary and discriminatory practices in the global field of information technologies (IT) (Rosser 2005). The IT workforce represents a vertically and horizontally gender-stratified labor market. Women are concentrated in the lowest paid positions—performing the tedious, hands on, eye straining work of making hardware and software products on assembly lines. Their work is dangerous, is menial, and offers little economic or geographic mobility. Women of color are disproportionately represented in these jobs in China, India, Mexico, and Taiwan. Conversely, men dominate IT positions with more creative and financial control such as venture capitalists, computer scientists, or engineers. Men are economically rewarded for their positions of power in the industry.

What Does It Mean to Me?

Though more women are earning MBAs, the number of women female chief executives continues to be low. In 2005, only 16 percent of Fortune 500 corporate officers were women, and there were only nine female CEOs (1.6 percent) (Creswell 2006). What would take to increase the number of female chief executives?

Interactionist Perspective

The sociologies of work and symbolic interaction were developed side by side in the 1920s and 1930s at the University of Chicago. There are strong similarities between the two sociological perspectives: In the same way that symbolic interactionists are interested in how individuals negotiate their social order, the sociologists of work are interested in the negotiated order of work (Ritzer 1989). Interactionists address norms in the workplace, how workers interact with their peers, how workers deal with stress, and how workers find meaning in the work they do. This perspective allows us to understand the process by which individuals understand, interpret, and create their work.

For example, Geraldine Byrne and Robert Heyman (1997) interviewed nurses in the United Kingdom, investigating the relationship between their perceptions of work and patients and how it influenced their communication with patients. The researchers noted how nurses distinguished their work with "major trauma" and "minor" patients. They defined their work as more valuable and satisfying in trauma cases, giving them an opportunity to feel technically expert and rewardingly useful. Their time with minor patients was described as boring or repetitive and a small part of their daily work. Although nurses felt that all patients experience some anxiety in the hospital, those with more serious illnesses or injuries would be more anxious than others. As a result, nurses spent more time with the seriously ill or injured patients, but made sure to "pop-in" with all patients as a way of demonstrating that they had not forgotten about them and providing some nursing contact.

According to symbolic interactionists, we attach labels and meanings to an individual's work (and major). If you meet a fellow student for the first time, one of the first questions you may ask is, "What's your major?" Why ask about a major? Think of it as a shortcut for who you are. Based simply on whether you are a sociology major or a physics major, people make assumptions about how much you study or your academic quality. This is no different from asking someone what he or she does for a living. These social constructs create an order to our work and our lives, but they can also create social problems.

What Does It Mean to Me?

What assumptions are made about majors on your campus? Which major is the "party" major? Which major is the "serious" major? Where do these assumptions come from? Are they true? How could they be changed?

Problems arise when these social constructs serve as the basis of job discrimination. A recent study by researchers from the University of Chicago and Massachusetts Institute

Table 9.3

Summary of Sociological Perspectives: Work and the Economy

	Functional	Conflict/Feminist	Interactionist
Explanation of work and the economy and its social problems	Sociologists use this perspective to examine the functions and dysfunctions of work and employment. Functionalists also analyze the functional elements of work itself.	Conflict or feminist theorists focus on how economic, ethnic, and gender inequalities are perpetuated in the economy and the workplace.	From this perspective, sociologists investigate how our work experiences are created through interaction and shared meanings. Social problems emerge from the meanings we associate with our work.
Questions asked about work and the economy	How does the institution of work help preserve the social order? How is economic and social stability maintained by the institution of work? How do other institutions affect our work? our economy?	What social inequalities are present in the institution of work? How do we become alienated from our work?	How is our work and workplace socially defined? How do we behave based on our meaning of work? Are there positive and negative meanings of work?

of Technology revealed discrimination in the recruiting process based only on what was perceived about someone's first name (Bertrand and Mullainathan 2003). Researchers sent 5,000 resumes in response to job advertisements in the *Boston Globe* and *Chicago Tribune*. First names were selected, based on a review of local birth certificates. Fictional applicants with "White" first names—Neil, Brett, Emily, and Jill—received one callback for every 10 resumes mailed out. In contrast, equivalent "Black" applicants—with names such as Aisha, Rasheed, Kareem, and Tamika—received one response for every 15 resumes sent. Other aspects of discrimination were revealed in the study. If the resume indicated that the applicant lived in wealthier, more educated, or more-White neighborhoods, the rate of callbacks increased. This effect did not vary by race. See Table 9.3 for a summary of all perspectives.

Problems in Work and the Economy

Unemployment and Underemployment

According to the U.S. Bureau of Labor Statistics (2007a), about 145 million Americans were employed and 6.8 million were unemployed in December 2006. Compared with the unemployment rate of Whites, 4.0 percent, the rate was 8.4 percent for African Americans and 4.9 percent for Hispanics or Latinos. Nearly 1.2 million individuals were unemployed for 27 weeks or more (U.S. Bureau of Labor Statistics 2007a).

In addition to unemployment, we should be aware of another rate, **underemployment**. Underemployment is defined as the number of employed individuals who are working in a job that underpays them, is not equal to their skill level, or involves fewer working hours than they would prefer (taking a part-time job when a full-time job is not available). In February 2003, the number of people working part-time because of cutbacks or because they were unable to find a full-time job was 4.8 million (U.S. Bureau of Labor Statistics 2003a).

There is significant variation in unemployment and underemployment rates. People who are young, non-college educated, and ethnic/racial minorities have higher underemployment rates (Bernstein 1997). Minority group underemployment is significantly higher than underemployment among non-Hispanic Whites. Min Zhou (1993) reports at least 40 percent of the members of each minority group he analyzed (Puerto Ricans, Blacks, Mexicans, Cubans, Chinese, and Japanese) were underemployed. In particular, Blacks and Puerto Ricans have the highest rates of labor force nonparticipation (were not in the labor force and had not worked in the last two years) and joblessness. Joblessness includes subemployment (individuals who were not in the labor force but worked within the last two years) and underemployment rates (either based on low wage or occupational mismatch). Recent immigrants, a large portion of the Asian and Hispanic minority groups, may have difficulty in securing employment because of lack of job skills or language proficiency and, as a result, are more likely underemployed than are native-born and non-Hispanic White workers (DeJong and Madamba 2001).

Scholars have documented the destructive effects of joblessness on overall health (Rodriguez 2001) and emotional well-being (Darity 2003). Unemployment has been consistently linked with higher levels of alienation, anxiety, and depression (Rodriguez 2001) and a lower sense of overall health (Darity 2003). Cross-national data reveal how long periods of unemployment are related to increased rates of suicide and spousal abuse (Darity 2003). Among Blacks and non-Hispanic Whites, long-term exposure to unemployment produces a "scarred worker effect." The experience of unemployment undermines the worker's will to perform, leading that person to become less productive and less employable in the future (Rodriguez 2001).

The Problem of Globalization

Globalization is a process whereby goods, information, people, communication, and forms of culture move across national boundaries. Though we tend to think of globalization as an economic phenomenon, we should not lose sight of its political, social, and cultural implications (Eitzen and Zinn 2006).

Globalization has transformed the nature of economic activity (Eitzen and Zinn 2006). It has been credited with bringing the world together—creating a world market where all businesses, employers, and employees must compete. This competition keeps corporations focused on innovation, quality, and production. Increasing productivity and output creates more jobs and stimulates economic growth (Weidenbaum 2006), creating a new middle class and reducing poverty in many countries (Yergin 2006). The United States has benefited from globalization, experiencing a doubling in foreign trade during the 1990s, which led to the creation of more than 17 million new jobs (Yergin 2006).

Yet, globalization has its dark side (Weidenbaum 2006). Foremost, worker security has declined everywhere, including the United States. Skilled workers are threatened by the unfair competition of low-cost sweatshops, of cheaper labor to be found in other nations. More than 3 million U.S. manufacturing jobs were lost from 2001 to 2004 (Dobbs 2004). These high-paying jobs ($16 per hour) migrated to other countries, in a process characterized as the "race to the bottom." Corporations are moving to lower wage economies, shifting their production from the United States to Mexico to China (where the average manufacturing wage is 61 cents per hour) (Eitzen and Zinn 2006).

Furthermore, progress in poverty reduction has been limited and geographically isolated (Weller and Hersh 2006). Globalization has brought no real gains in income among the poor; the number of poor people actually increased from 1987 to 1998 in several countries. The world's poorest 10 percent, or about 400 million people, lived on 72 cents per day in 1980. The same number of people lived on 79 cents in 1990 and 78 cents in 1999.

Women are particularly vulnerable in the new global economy. The global assembly line is filled with girls and women engaged in work that is low-wage, temporary, part-time, or home-based and usually under unsafe working conditions. More than half the world's legal and illegal immigrants are women—Third World women moving to postindustrial societies for jobs as nannies, maids, and sex workers (Eitzen and Zinn 2006).

Contingent or Temporary Workers

Contributing to the increase in unemployment and underemployment is the use of **contingent workers**. A contingent workforce is composed of full-time or part-time temporary workers. There were about 5.7 million contingent workers in 2005 (U.S. Bureau of Labor Statistics 2005), with more than half (55 percent) reporting that they would have preferred to have a permanent job.

Companies rely on temporary workers to handle short-term projects and work overloads or to fill in for employees who are vacationing, sick, or on family leave (Saftner 1998). Companies have also used temporary workers to eliminate positions and to reduce costs (Davidson 1999). Overall, temp workers are less likely to have health insurance or pension coverage. In addition, they are not covered by health and safety regulations and may not be qualified for workers' compensation if injured on the job (Davidson 1999). Many temps feel like second-class corporate citizens (Eisenberg 1999).

The term **outsourcing** refers to a practice by businesses of hiring external contractors to do the jobs that a regular work staff once completed. One common task that is outsourced is accounting. In the past, businesses would have an in-house accountant, someone who did the books and took care of accounts receivable or payable. But businesses began to contract with external accountants or accounting firms to handle their accounts. By paying accountants as contract workers, companies avoid paying them employee or medical benefits.

U.S. companies also outsource internationally, a strategy referred to as **offshore outsourcing** or **offshoring**. By 2015, more than 3.3 million U.S. jobs will be sent overseas, about 2.5 percent of the total U.S. employment. European companies, motivated by their own need to reduce costs, are also practicing offshoring, sending work to China and parts of Central and Eastern Europe. It has been estimated that more than a million European Union jobs would be sent outside of Europe by 2015 (Parker 2004).

India, with its large number of English-speaking college graduates, is expected to receive more than half of the U.S. and European jobs. Indian labor is used for what is referred to as "back office" work—call center support, research and development, preparing tax returns, processing health insurance claims, and transcribing medical notes (Waldman 2003)—usually at wages substantially lower than their U.S. counterparts receive. In 2005, Indian outsourcing businesses generated more than $17.7 billion in software and IT services exports (King 2006). Back office locations have sprung up in China, Morocco, Brazil, Chile, and Mexico—some established by the same Indian outsource companies who started it all.

But offshore outsourcing isn't the perfect solution. Employees in India work 10- to 12-hour night shifts to be at their desks at the same time as colleagues in the United States (India

Photo 9.3

By 2015, more than 3.3 million U.S. jobs will be sent overseas, about 2.5 percent of the total U.S. employment. India is expected to receive most of the U.S. jobs, characterized as "back office" work, which includes call center support, processing and transcription, and research and development.

is 10.5 hours ahead of U.S. Eastern Standard Time). Doctors in India report high levels of substance abuse and relationship breakups among its outsourcing workers. India's outsourcing industry has a rate of 60 percent employee turnover per year (Thottam 2004).

What Does It Mean to Me?

Between 600,000 and 800,000 individuals are globally trafficked per year. Human trafficking includes the involuntary movement of individuals within one country or from one country to another for exploitation. Exploitation may include physical labor or sexual exploitation (UN Anti-Human Trafficking Unit 2006). In 2007, slave labor and illegal detention atrocities were exposed in the Shanxi providence of China. Owners and managers of brick making kilns in the northern providence were found guilty for their roles in the enslaving adult and child workers. Government raids freed more than 850 adults and children from the area. The United States is reported to be a high destination country where victims are brought to be exploited, yet not much is discussed about our country's role in human trafficking. Is trafficking a U.S. social problem? Why or why not?

A Livable Wage

Despite record increases in employment, some Americans still needed two jobs or more to make a decent living. In 1990, the federal minimum wage was $3.80 per hour; in 1997, the minimum wage was increased to $5.15 per hour. Individuals earning poverty-level wages, about $8.47 per hour in 2000, are characterized as low-wage workers (AFL-CIO 2003). Data indicate that low-wage workers are likely to be minority, female, non-college educated, and non-union, working in low-end sales and service occupations (Bernstein 1997; Bernstein, Hartmann, and Schmitt 1999; U.S. Department of Labor 2002).

U.S. Data Map 9.1 identifies the percentage of workers who were paid $5.15 or less within each state in 2005. The green shaded states are those above the U.S. average of 2.5 percent,

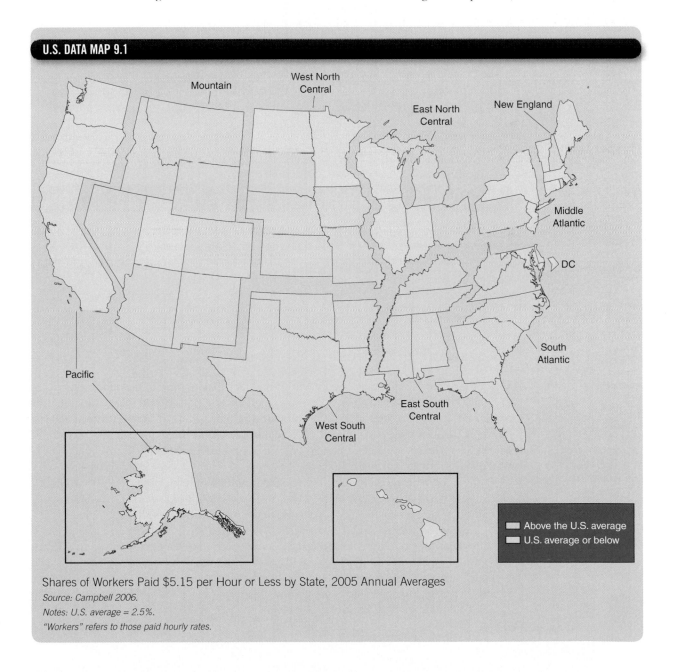

U.S. DATA MAP 9.1

Legend: Above the U.S. average / U.S. average or below

Shares of Workers Paid $5.15 per Hour or Less by State, 2005 Annual Averages

Source: Campbell 2006.

Notes: U.S. average = 2.5%.

"Workers" refers to those paid hourly rates.

indicating a higher percentage of workers are paid $5.15 or less in the state. States with the highest percentages were in the South—Oklahoma (4.3) and West Virginia (4.3). The states with the lowest percentages were Alaska (.5) and Washington (.6) (Campbell 2006).

In her 2001 book, *Nickeled and Dimed: On (Not) Getting By in America,* Barbara Ehrenreich explored life and work on minimum wage in three states: Florida, Maine, and Minnesota. Working as a hotel maid, a nursing home aide, a sales clerk, a waitress, and a cleaning woman, Ehrenreich rated her work performance as a B or maybe even a B+. In each new job, Ehrenreich had to master new terms, new skills, and new tools (and not as quickly as she thought she would be able to master them). How did Ehrenreich survive on minimum wage? She discovered that she needed to work two jobs or seven days a week to achieve a "decent fit" between her income and her expenses. She describes getting her meals down to a "science": chopped meat, beans, cheese, and noodles when she had a kitchen in which to cook; if not, fast food at about $9 per day. For housing, she shuffled between motel rooms and apartments, moving to a trailer park at one point. Ehrenreich concluded,

> Something is wrong, very wrong, when a single person in good health, a person who in addition possesses a working car, can barely support herself by the sweat of her brow. You don't need a degree in economics to see that wages are too low and rents too high. (2001:199)

What Does It Mean to Me?

The median income for 2005 was $46,326 (DeNavas-Walt, Proctor, and Lee 2006). The median is the exact point where 50 percent of all incomes are above and 50 percent are below. If this is the middle—or "average"—household income, could you raise a family of four on this income? Determine a monthly budget for your family, including rent/mortgage, food, entertainment, car expenses (gas/maintenance/insurance), clothing expenses, and savings account.

A Hazardous and Stressful Workplace

According to the U.S. Bureau of Labor Statistics (2006b), there were 5,702 fatal work injuries in 2005. The majority of fatalities occurred to men, mostly caused by the type of work they do. Operators, fabricators, and laborers accounted for more than one of every three fatalities. The most fatalities, about one-quarter of all fatal incidents, occurred from highway accidents. Workplace violence—including assaults and suicides—accounted for 13 percent of all work-related fatal occupational injuries in 2005.

In the same year, a total of 4.2 million nonfatal injuries and illnesses were reported in private industry workplaces, a rate of about 4.8 cases per 100 full-time workers. Approximately 4.0 million were injuries, the most occurring in the manufacturing sector (20 percent). The manufacturing sector also had the largest share of occupational illnesses, about 39 percent. The U.S. Department of Labor monitors illnesses such as skin diseases, respiratory conditions, and poisonings. New reported workplace illnesses were related directly to work activity, such as contact dermatitis or carpal tunnel syndrome. Some conditions, such as long-term illnesses related to exposure to carcinogens, are usually underreported and not adequately recognized (U.S. Bureau of Labor Statistics 2006c).

Taking a World View

Mexico's Maquiladoras

Maquiladoras are textile, electronics, furniture, chemical, processed food, or machinery assembly factories where workers assemble imported materials for export (Abell 1999; Lindquist 2001). The maquiladoras program allows imported U.S. materials to enter Mexico without tariffs; when the finished goods are sent back to the United States, the shipper pays duties only on the value added by the manufacturer in Mexico (Abell 1999; Gruben 2001). The program began in 1965 as an employment alternative for Mexican agricultural workers. Drawn to Mexico because of its proximity to U.S. borders and by low labor costs, nearly every large U.S. manufacturer has a maquiladora location. Several Asian and European companies like Sony, Sanyo, Samsung, Hitachi, and Phillips also have maquiladora locations (Lindquist 2001). There are an estimated 4,000 export manufacturers along Mexico's border with the United States (Abell 1999; Smith 2004), employing more than one million people (Sowinski 2000). These factories account for almost half of Mexico's total exports (Smith 2004).

The maquiladora program became controversial as soon as it appeared (Gruben 2001). Supporters of the maquiladora program argue that if these plants had not located in Mexico, they would have gone to other low-wage countries. Opponents argue that the program helped U.S. firms and others take advantage of the low-wage Mexican labor force. The maquiladoras have been criticized for their treatment of female workers, dangerous work conditions, and impact on the physical and social environment of border towns.

Most maquiladora workers are women. Yolanda is a worker from Piedras Negras.

As the sun rises, Yolanda is already awake and working—carrying water from a nearby well, cooking breakfast over an open fire, and cleaning the one-room home that she and her husband built out of cardboard, wood, and tin. She puts on her blue company jacket and boards the school bus that will take her and her neighbors across Piedras Negras to a large assembly plant. Yolanda and 800 co-workers each earn US$25–35 a week for 48 hours' work, sewing clothing for a New York-based corporation that subcontracts for Eddie Bauer, Joe Boxer, and other U.S. brands. These wages will buy less than half of their families' basic needs. (Abell 1999:595)

Yolanda's job provides a wage, but not a livable one to support herself or her family. The daily salary for maquiladora women is about $4.67, described not as a standard of living but, rather, as a standard for survival (Moffatt 2005). According to Elizabeth Fussell (2000), early maquiladora factories attracted the "elite" of the Mexican female labor force: young, childless, educated women. Current maquiladora laborers are likely to be the least-skilled Mexican women: slightly older, poorly educated women with young children (Fussell 2000).

Yolanda's town of Piedras Negras is no different from other maquiladora towns such as Tijuana and Matamoros. Once quaint border towns have been transformed by maquiladora activity. Despite their profits, companies do not invest in the physical and social infrastructure of these border towns. As a result, most factory neighborhoods lack basic health and public services such as clean drinking water or sewage systems, electricity, schools, health facilities, and adequate housing (Abell 1999).

Sexual harassment is often used as a method of intimidation in the maquiladora. Supervisors taunt female workers and proposition them by offering lighter workloads in exchange for dates and sexual favors. Supervisors have also sexually assaulted female workers. Women, on and off their jobs, are subject to intimidation and violence. For example, in the town of Ciudad Juarez, since 1993, approximately 300 women have been found murdered. Although it is unclear how many of these women were maquiladora workers, all these women have been identified as victims of an unsafe environment, easy targets for serial killers, drug traffickers, and even police officials (Moffett 2005).

Under the 1994 North American Free Trade Agreement (NAFTA), tariff breaks formerly limited to all imported parts, supplies, and equipment used by Mexican maquiladoras now also apply to manufacturers in Canada and the United States (Lindquist 2001). Although there has been a rapid growth in the number of maquiladoras and their workers, the program was strained in the late 1990s because of cost increases (added taxes, NAFTA administration, and rising wages) and labor problems (high turnover and worker shortages) (Lindquist 2001). By 2004, however, there was a maquiladora resurgence, with factories exporting goods worth more than $7 billion and the addition of 55,000 more positions (Smith 2004).

During recent national debate about immigration, NAFTA and the maquiladora program were blamed for failing to improve Mexico's economic and business infrastructure and thus failing to reduce illegal immigration to the United States.

The tragedies of the 2006 explosion at the Sago Mine in West Virginia and the 2007 collapse of the Crandall Canyon Mine in Utah renewed focus on mine worker safety. In West Virginia, 13 were killed in a methane gas explosion. Six miners were entombed in the Crandall Canyon Mine in a tunnel collapse and 3 others killed attempting to rescue the trapped workers. These incidents led to an ongoing reevaluation of mine regulation, employee training and safety, and the development of emergency response teams, along with improvements in underground communications technology. Coal mining is an inherently dangerous occupation with deaths and injuries documented worldwide. China and Russia have the largest number of coal mine fatalities annually.

The National Institute for Occupational Safety and Health (NIOSH) (2003) defines job stress as the harmful emotional or physical response that occurs when a job's characteristics do not match the capabilities, resources, or needs of the worker. Certain job conditions are likely to lead to job stress: a heavy workload, little sense of worker control, a poor social environment, uncertain job expectations, or job insecurity. Eventually, job stress can lead to illness, injury, or job failure. Studies have analyzed the impact of stress on our physical health, noting the relationship of stress with sleep disturbances, ulcers, headaches, or strained relationships with family or friends. Recent evidence suggests that stress also plays a role in chronic diseases such as a cardiovascular disease, musculoskeletal disorders, and psychological disorders (NIOSH 2003).

Discrimination at the Workplace

The Equal Employment Opportunity Commission (EEOC) was established in 1964 by Title VII of the Civil Rights Act. The EEOC monitors and enforces several federal statutes regarding employment discrimination. Under Title VII of the Civil Rights Act of 1964, employment discrimination based on race, religion, sex, or national origin is prohibited. Discrimination based on age was added to the prohibited activities in 1967 with the passage of the Age Discrimination in Employment Act, and employment discrimination against individuals with disabilities was prohibited in Title I and Title V of the Americans with Disabilities Act of 1990. Under current federal law, discrimination based on sexual orientation is not prohibited. The EEOC has received between 72,000 and 88,000 charges annually since 1992 (EEOC 2003).

Of the 84,000 cases that were filed during 2002, about 11 percent were based on national origin discrimination. Cases in this category and those involving religious discrimination increased after September 11, 2001, whereas discrimination charges based on race and sex declined. Complaints of discrimination based on national origin have increased 20 percent during the last eight years (McDonough 2003). According to EEOC spokesman David Grinberg, "Most people think about race and gender discrimination—national origin discrimination doesn't come to mind, but it's having a greater impact on the workplace" (McDonough 2003). Many incidents of national origin discrimination may still be unreported because of fear of retaliation or lack of awareness about EEOC laws.

U.S. Garment Industry and Sweatshop Labor

Look at the labels on your shirts and jeans. Where were they made? We know that most of our clothing is manufactured in countries like China, Hong Kong, India, Thailand, or Vietnam. But an estimated 873,000 textile, apparel, and furnishing workers were employed in the United States in 2006; most were production workers or sewing machine operators (233,000) in the garment industry. These manufacturing jobs are located in California, North Carolina, Georgia, New York, Texas, and South Carolina (U.S. Bureau of Labor Statistics 2007b).

All manufacturers must follow the Fair Labor Standards Act (FLSA), which establishes federal minimum wage, overtime, child labor, and industrial homework standards. The Department of Labor's Wage and Hour Division makes routine enforcement sweeps in major garment centers, fining businesses that are in violation of the FLSA. In its 1994 report, the General Accounting Office (GAO) concluded that "sweatshop working conditions" remain a major problem in the U.S. garment industry. Actually, "the description of today's sweatshop differs little from that at the turn of the century" (GAO 1994:1).

According to Sweatshop Watch (2003), there is no legal definition of a "sweatshop." The GAO (1994) defines a **sweatshop** as a workplace that violates more than one federal or state labor law. The term has come to include exploitation of workers, for example, with no livable wages or benefits, poor and hazardous working conditions, and possible verbal or physical abuse (Sweatshop Watch 2003); employers who fail to treat workers with dignity and violate basic human rights (Co-Op America 2003); and businesses that violate wage or child labor laws and safety or health regulations (Foo 1994). The term sweatshop was first used in the nineteenth century to describe a subcontracting system in which the contractors earned profits from the margin between the amount they received for a contract and the amount paid to their workers. The margin was "sweated" from the workers because they received minimal wages for long hours in unsafe working conditions (Sweatshop Watch 2003).

A random sample of apparel manufacturers in Southern California in 1996 revealed that 43 percent failed to pay their workers the minimum wage, 55 percent had overtime liabilities, and one-third were not registered with the state (U.S. Department of Labor 1996). According to its October 2000–December 2000 *Garment Enforcement Report*, the Department of Labor recovered $519,666 in back wages for 712 garment workers. During the quarter, the department's Wage and Hour Division investigated 67 employers nationwide and found 48 percent (28 cases) in violation (U.S. Department of Labor 2000). In December 2006, the U.S. Department of Labor announced

Photo 9.4

A California garment factory worker wears a face guard to protect her from dust while she sews. An estimated 23,000 apparel or textile businesses operated in the United States in 2000.

that $2.9 million in back wages was collected for garment workers (U.S. Department of Labor 2006).

Some have argued that sweatshops continue to exist because of unscrupulous manufacturers and increased competition from low-wage workers (U.S. Department of Labor 1996). Because garment production is labor intensive, manufacturers will subcontract to decrease their overhead, primarily saving on labor costs. Through the use of sweatshop labor, manufacturers shift much of their costs, risks, and responsibilities onto subcontractors (Foo 1994).

Community, Policy, and Social Action

Federal Policies

When President William Howard Taft signed Public Law 426–62 in March 1913, he created the U.S. Department of Labor. From the beginning, the department was intended to foster and promote the welfare of U.S. wage earners, to improve working conditions, and to advance opportunities for profitable employment. In its current mission statement, the department included improving working conditions, advancing opportunities for profitable employment, protecting retirement and health care benefits, helping employers find workers, and strengthening fee collective bargaining as part of its charge. The department administers and enforces more than 180 federal laws that regulate workplace activities for about 10 million employers and 125 million workers.

In addition to the FLSA, the Department of Labor enforces several statutes applicable to most workplaces. It regulates the Employee Retirement Income Security Act (pension and welfare benefit plans), the Occupational Safety and Health Act (ensuring work and a workplace free from serious hazards), the Family and Medical Leave Act (granting eligible employees as many as 12 weeks of unpaid leave for family care or medical leave), and several acts that cover workers' compensation for illness, disability, or death resulting from work performance.

Two labor issues continue to be debated in Congress. The first is raising the minimum wage. In 2000, Congress failed to pass legislation that would have increased the minimum wage from $5.15 to $6.15. In 2001, Senator Ted Kennedy of Massachusetts and Representative David Bonior of Michigan, both Democrats, introduced legislation that proposed a $1.50 raise in the minimum wage over three years. Efforts to increase the minimum wage are supported by unions and poverty organizations that argue that doing so will help the nation's working poor and low-income families. Opponents, who include members of the business community and the U.S. Chamber of Commerce, argue that increasing the minimum wage would put an unnecessary stress on medium-size and small businesses but would not decrease poverty. Some predict that businesses would be forced to eliminate jobs, reduce work hours, or be put out of business. Results from policy analyses and academic research have not provided conclusive evidence for either argument (Information for Decision Making 2000). In 2007, the U.S. Congress passed legislation that would increase the minimum wage to $7.25 over two years. States, unwilling to wait for congressional action, have passed higher minimum wage laws during the past decade. Today, 29 states have minimum wages ranging from $6.15 to $7.63 per hour (Uchitelle 2006). Maryland became the first state to require a living wage. Although the state minimum wage is $6.15 an hour, under the new 2007 law, Maryland employers with state contracts will have to pay workers a minimum of $11.30 an hour in the Baltimore-Washington areas and $8.50 an hour for workers in rural counties (Greenhouse 2007a).

What Does It Mean to Me?

The Association of Community Organizations for Reform Now (ACORN) is the nation's oldest and largest grassroots organization. ACORN organizes low- and moderate-income people in more than 850 chapters in more than 100 U.S. cities under a campaign for livable wages. A livable wage is defined as a wage that would allow a full-time primary worker with three dependents to earn just above the poverty line. Go to *Study Site Chapter 9* to determine the living wage for your state or area.

The second issue involves protection against workplace discrimination based on sexual orientation. Several congressional actions provide protection against forms of discrimination: Title VII of the Civil Rights Act of 1964 (race, color, gender, national origin, or religion), the Age Discrimination in Employment Act (1967), the Vocational Rehabilitation Act (1973), and the Americans with Disabilities Act (1973). The Employment Non-Discrimination Act, which prohibits employment discrimination based on sexual orientation, was introduced in Congress in 1994 but has not yet received enough votes to pass (Human Rights Campaign 2003; Kovach and Millspaugh 1996).

Expanding Employment Opportunities for Women

In 1992, Congress passed the Nontraditional Employment for Women (NEW) Act to broaden the range of Job Training Placement Program (JTPA) efforts on behalf of women (U.S. Department of Labor, Women's Bureau 2000). Also enacted in 1992 was the Women in Apprenticeship and Nontraditional Occupations (WANTO) Act, designed to provide technical assistance to employers and labor groups. Through both programs, the federal government has provided support to more than 20 state and 32 community-based initiatives to provide technical assistance to employers and labor organizations, to improve JTPA programs, to promote apprenticeships in nontraditional occupations, and to provide information through community workshops, seminars, and outreach. According to the Women's Bureau (U.S. Department of Labor, Women's Bureau 2000), almost 5,000 women participated in training or job placement programs; 30,000 women were reached through community programs; and 3,000 employers and labor groups were provided technical assistance through NEW and WANTO initiatives. Funding for WANTO ended in 2003.

Hard Hatted Women, based in Cleveland, Ohio, received a WANTO 2002–2003 grant to increase the employment of women in a variety of blue-collar jobs in building and manufacturing trades. With WANTO funding, the program provided technical assistance to employers and labor unions, on-site counseling for women, a job search networking club, and training (U.S. Department of Labor 2003). Hard Hatted Women continues as the only community nonprofit organization dedicated to supporting women in high-wage, nontraditional, blue-collar careers. The program maintains two areas of focus: education and training and supporting economic equity policy initiatives. Between 1992 and 2005, in its Pre-Apprentice Training Program, Hard Hatted Women graduated 441 women, placing almost 60 percent of graduates in jobs (Hard Hatted Women 2006).

Funded by the Women's Foundation of California (a 501 (c) (3) private organization that funds community-based organizations), La Cocina is a nonprofit kitchen incubator for women entrepreneurs. La Cocina provides its commercial kitchen facilities to women, about half Latina, to produce specialty foods for their businesses. According to Valier Perez Ferreiro, participants are selected because "we are finding amazing entrepreneurs who are already cooking or have a product that is so promising that it deserves to be seen in the market and that we think has a chance for success" (Novak 2007, p.B4). For $10 per hour or less, women can use La Cocina's facilities and equipment. There are 19 kitchen incubators throughout the United States, some based in such urban areas as Denver, New York City, and Minneapolis. La Cocina also provides training from local experts in business operations and management, banking, and food science. Jill Litwin brought her Peas of Mind line of organic foods to La Cocina, which introduced her to a buyer for Whole Foods Market. As of 2007, her Peas of Mind products were in 80 California stores (20 of them Whole Food Markets).

Worker-Friendly Businesses— Conducting Business a Different Way

Each year *Fortune* magazine releases a list of the "100 Best Companies to Work For." For 2007, the magazine wrote,

> Ten years ago, when we began compiling this list, the idea that your employer would deliver your groceries (a new perk at Microsoft) or allow you to do your laundry at work (Google) might have seemed crazy … Indeed, much has changed in the American workplace over the past decade. (Levering and Moskowitz 2007:94)

No. 1 on Fortune's list was Google. This employer features 11 free gourmet cafeterias in its Mountain View, California, headquarters. But Google was named number one for more than its food. This business rewards its 6,500 employees with benefits that include competitive salaries and stock options, free onsite washers and dryers (including free detergent), a $5,00 subsidy to purchase a hybrid car, onsite language classes, car wash and oil change services, and an annual ski trip (all expenses paid). Google also provides free work transportation for about 1,200 of its employees aboard 32 luxury shuttle buses. The shuttle route includes 40 pickup and drop-off locations, covering more than 4,000 miles a day (Helft 2007).

There is no big secret to creating worker-friendly organizations, although some organizations are slow to learn their values. Beyond the standard employee benefit package of vacations, health care, retirement plans, and life insurance, innovative employers have used dependent care, flexibility work options, expanded leave time, and enhanced traditional benefits to attract and retain employees. As a result, these employers have reduced employee turnover, increased employee satisfaction, and improved worker productivity (Schmidt and Duenas 2002).

Even during tough economic times, several U.S. businesses have been holding firm to a no-layoff policy. Nucor is the nation's most profitable steelmaker and has not laid off an employee in 33 years (Clark 2002). Southwest Airlines is the most consistently profitable airline and hasn't had a single layoff in its 31-year history. And that AFLAC duck has something else to quack about: His company hasn't laid off a full-time worker in 47 years. Employers avoid layoffs by asking staff to put in overtime during busy periods and cutting hours during the slow ones. Temporary workers are also part of the equation. Companies like SC Johnson protect their full-time workforce by using and cutting temporary positions. Some companies keep their employees at work during slow periods—just doing other things. At Nucor, when there isn't enough business to support their $25-per-hour welding duties, workers are put on a factory maintenance detail at $10 per hour. In turn, these companies have been rewarded with loyal and productive employees (Clark 2002).

Voices in the Community

Judy Wicks

Businesses have also found a way to give back to their communities, to combine their work with their social activism. Judy Wicks has been leading the way as the owner of Philadelphia's White Dog Cafe. In an interview with Maryann Gorman (2001), Wicks examines how her business and activist philosophies have merged at the White Dog.

A person could go to the White Dog Café just to eat. Many customers do—at least for the first time.... But many of the diners who come just for the food end up staying for the activism. Wicks tells of a salesman who sold the restaurant its insurance, then attended a Table Talk on the School of the Americas and became a regular at the annual protests held at the School. Wicks jokes, "I use food to lure innocent customers into social activism."

The Dog offers diners an array of learning experiences in lecture or hands-on formats as well as field trips to other countries. There are Table Talks on a variety of topics: the American war on drugs, the Supreme Court's decision in the 2000 election, racism. And the White Dog often organizes rides to rallies and marches, like the demonstrations against Clinton's impeachment that Jesse Jackson called for, the Million Mom March, Stand for Children and more....

Wicks' White Dog venture began when she sensed the dissonance between her profession and her activism. Her energies drained by splitting her values into the commercial and nonprofit worlds, she sought a simpler solution, which first led her to managing someone else's restaurant. "I realized with the nonprofit and the for-profit that in order to really be effective, I needed to focus on one organization," Wicks says, "So I eventually abandoned the publishing work and focused on the restaurant. But it wasn't all my restaurant, and when I started experimenting with bringing my values to work, like having a breakfast for Salvadoran refugees, I had to part ways with my partner there. This idea of compartmentalizing your life, and having certain values in one area and other values in another, has never worked for me."

The first step, Wicks says, was simply to change her life so that she lived where she worked. "I think that society teach[es] us, 'Separate work from home; don't mix them together. That'll be too stressful.' And I've always worked against that." So the White Dog, located in a block of row houses near the University of Pennsylvania, became both her home and her business. "I'm kind of a holistic person," she says, "I live where I work. I live 'above the shop,' in the old-fashioned way of doing business." ...

Not only does she not have to commute, ... but she doesn't have to go food shopping or do the dishes. "But it's more than that," she says, "It's about energy and focus. And relationships. Being able to foster all these relationships."

The relationships she speaks of are those with her staff. One of the White Dog's missions is "Service to Each Other," an in-house goal of creating a tolerant and fair workplace.... Moving past standard business boundaries in this way has informed Wicks' creative development of the White Dog's missions. From encouraging her employees to re-create their job descriptions to fostering a sense of community among other Philadelphia restaurants, Wicks has shown that profitable businesses don't have to be stark-raving competitive.

In 1990, Wicks began the Sister Restaurant program, promoting small minority-owned restaurants in Philadelphia. Currently, the program includes five such restaurants. The White Dog Web site features information about the cafe and also updated schedules for the Table Talks, community events, films, and special events. Log on to *Study Site Chapter 9* for more information.

Organized and Fighting Back

In 2005, union membership among manufacturing workers dropped below the percentage of all American workers for the first time (Greenhouse 2007b). According to the Bureau of Labor Statistics, 12 percent of all employees were union members, but in the heart of U.S. organized labor, only 11.7 percent of manufacturing employees were members (down from 13 percent in 2005). The decline was attributed to large-scale layoffs and buyouts in the automobile industry, along with the labor movement's increasing difficulty in organizing nonunion workers (Greenhouse 2007b).

Historically, labor unions have served as bargaining agents for workers, fighting for fair wages, safe work environments, and benefits from employers. Many of the worker benefits advocated by early labor unions are now mandated through federal, state, or local labor laws. In our changing and global economy, unions cannot sustain themselves by simply negotiating pay, benefits, and working rights. To remain vital and relevant, unions need to develop multilevel strategies and use new skills (Lazes and Savage 2000).

In the 1990s, following years of declining membership, U.S. unions shifted their membership strategy (Fantasia and Voss 2004). Changing their focus from the declining U.S. manufacturing sector, union organizers began to successfully rebuild their membership in the service and public sectors dominated by female, immigrant, and minority workers. Unions are presented as organizational vehicles of social solidarity, "emphasizing direct [worker] action as an important source of collective power" (Fantasia and Voss 2004:128). In addition, these unions have a strong orientation toward social justice that appeals to their new members, connecting the labor movement with the movement for social citizenship and universal civil rights (Fantasia and Voss 2004).

Ehrenreich and Thomas Geoghegan (2002) called for "a new approach to rebuilding unions—and to labor law reform … The first step toward the revitalization of American unions should be to create a form of membership accessible to any worker." Ehrenreich founded United Professionals, a nonprofit nonpartisan organization for white-collar workers, regardless of their occupation or employment status in 2006. The organization's mission is to "protect and preserve the American middle class, now under attack from so many directions, from downsizing and outsourcing to the steady erosion of health and pension benefits. We believe that education, skills and experience should be rewarded with appropriate jobs, livable incomes, benefits and social supports" (United Professionals 2006). In addition to providing a way to connect with other white-collar workers, members are able to receive insurance support, job assistance, and legal counseling. The organization also serves as an advocacy group, addressing national, state, and local issues. The organization hopes to establish state chapters in the future.

In recent years, innovative collaborations have been forged between unions, workers, and college students. One example of union innovation is the Union of Needletrades, Industrial, and Textile Employees (UNITE). In 1996, it launched a "Stop Sweatshops" campaign linking union, consumer, student, civil rights, and women's groups in the fight against sweatshops.

UNITE helped form United Students Against Sweatshops (USAS) in 1998, bringing together a coalition of student groups to raise awareness about the problem of sweatshop labor in the manufacturing of collegiate clothing (caps, shirts, and sweatshirts sold in campus stores). In March 1998, Duke University adopted the nation's first Code of Conduct for University Trademark Licensees. Under the code, any clothing with the Duke logo would be subject to labor and human rights standards. The student group Duke Students Against Sweatshops played a key role in shaping the anti-sweatshop code (Sweatshop Watch 2000). In 2004, UNITE announced that it was merging with the Hotel Employees and Restaurant Employees International Union. The new union, called UNITE HERE, represents 440,000 active members and more than 400,000 retirees (UNITE 2004).

Also working with USAS is the Student Labor Action Project (SLAP). This campus-based coalition supports the growing economic justice movement among college students by "making links between campus and community organizing, providing skills training to build lasting student organizations, and developing campaigns that win concrete victories for working families" (Student Labor Action Project 2007). Living wage campaign organizations exist at such schools as Georgetown University, Harvard University, University of Miami, and Vanderbilt University. SLAP provides training and support for campus-based campaigns for living wages, union organizing, and union contract negotiations through a series of on-campus workshops.

Photo 9.5

Why is there poverty at the world's richest university? In 2001, Harvard University students demanded answers to this question, staging a 21-day sit-in at the president's office. Students demanded a better living wage for lower paid university workers. While achieving salary increases for workers, the Harvard Living Wage Campaign continues to advocate for better labor practices and salaries for the university's service, janitorial, and dining staff.

Main Points

- In December 2006, about 63 percent of the U.S. population 16 years of age or older were employed. Work isn't just what we do; work is a basic and important social institution. Because of the importance of work, problems related to work quickly become categorized as social problems—as everyone's problem.

- During the late eighteenth century and early nineteenth century, the means of production shifted from agricultural to industrial. In agrarian societies, economic production was very simple, based primarily on family agriculture and hunting or gathering activities. During the **Industrial Revolution**, an economic shift occurred in how people worked and how they earned a living. Family production was replaced with market production, in which capitalist owners paid workers wages to produce goods.

- Since the 1960s, there has been another shift, referred to as **deindustrialization**, a widespread, systematic disinvestment in our nation's manufacturing and production capacities. Most manufacturing jobs and plants have been transferred to other countries. Local U.S. factories have closed as a result of globalization, mergers, or acquisitions and poor business.

- At the same time, more women are working, and more elderly people are returning to work. A very recent shift includes an immigration boom and the presence of more foreign born workers, often working in low-paying jobs.

- According to the functionalist perspective, work serves specific functions in society. Our work provides us with some predictability about our life experiences.

- Conflict theorists argue that capitalist and corporate leaders maintain their power and economic advantage at the expense of their workers and the general public.

- From a feminist perspective, work is a gendered institution. Through the actions, beliefs, and interactions of workers and their employers, as well as the policies and practices of the workplace, men's and women's identities as workers are created, reproduced, then solidified in the everyday routines of informal work groups and formal workers' organizations.

- According to symbolic interactionists, we attach labels and meanings to an individual's work. These social constructs create an order to our work and our lives but can also create social problems.

- Problems related to work include unemployment and **underemployment** (the number of employed individuals who are working in a job that underpays them, is not equal to their skill level, or involves fewer working hours than they would prefer), **globalization**, **contingent** (temporary) work, **outsourcing** (including **offshore outsourcing**), **sweatshop** labor, discrimination in the workplace, and hazardous work.

- Several federal policies and institutions have been created with regard to the workforce. This includes the U.S. Department of Labor, which oversees several laws and organizations relevant to the workforce. Two labor issues continue to be debated in Congress: raising the minimum wage and workplace discrimination.

- In addition to government efforts, some companies have independently tried to make the atmosphere at work more worker-friendly through such areas as enhanced traditional benefits, flexible time, and expanded leave. Others try to follow a no-layoff policy.

- Labor unions, which historically have attempted to defend worker rights and whose efforts have led to many federal laws, still attempt to help workers, often in innovative ways. Student groups have provided strong support for unions and living wage organizations on campus.

On Your Own

Log on to the Web-based student study site at www.pineforge.com/leonguerrero2study for interactive quizzes, e-flashcards, journal articles, Community and Policy Guides, a Service Learning Guide, the end-of-chapter Web exercises, and additional Web resources.

Internet and Community Exercises

1. Explore the *Occupational Outlook Handbook,* published by the U.S. Bureau of Labor Statistics (log on to *Study Site Chapter 9*). The online handbook provides career descriptions, earning information, and job prospects for a range of occupational groups. For information about what sociologists and other social scientists do, click on "Professional and related occupations" on the handbook's main page, then find and click on "Social Scientists, other." Research other occupations that might be of interest to you.

2. Do you know what the fourth Thursday in April is? It is "Take Our Daughters and Sons to Work Day." The program is sponsored by the Ms. Foundation for Women, which first created the "Take Our Daughters to Work Day" in 1993. "Take Our Daughters and Sons to Work" is a program that explores career opportunities and work/life issues with girls and boys. To find out more about the national program and related activities, log on to *Study Site Chapter 9*. Does your school or workplace support this program? Why or why not? Do you believe that these or similar activities are effective in changing girls' and boys' definitions of work and family life? Why or why not?

3. The Kneel Center at the Cornell University Library presents an on-site historical exhibit of the Triangle Factory Fire. On March 25, 1911, a fire at the Triangle Waist Company in New York City killed 146 immigrant workers. In 2003, the site was recognized as an official landmark by New York's Landmarks Preservation Commission. This tragedy led to the creation of local and federal policies prohibiting sweatshop conditions and ensuring worker health and safety. Log on to *Study Site Chapter 9* for more information. The Web site provides a history of early industrial sweatshops, along with a detailed narrative of how the community, workers, and unions responded to the tragedy. The Web site includes photographs, interviews, and documents from the period.

4. The United Nation's *Human Development Report 2006* tracks unemployment rates for more than 20 nations. Log on to *Study Site Chapter 9* to compare the U.S. rate with the other listed countries. How does our unemployment rate compare with other developed countries? Which countries have lower rates of unemployment?

5. Investigate whether there is a local ACORN chapter in your city or state. Identify the local initiatives for the chapter. For more information about the New Orleans ACORN chapter (featured in Chapter 2, "Social Class and Poverty"), log on to neworleans.acorn.org.

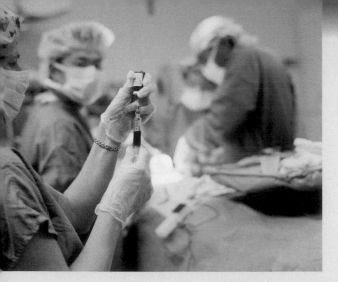

CHAPTER 10

Health and Medicine

I n 2006, the Centers for Disease Control and Prevention (CDC) released a report of health
trends for the year. In "Health, United States, 2006" (National Center for Health Statistics
2006), the CDC reported that life expectancy reached a record high of 77.9 years, with
gaps between Blacks and Whites and men and women narrowing (see Table 10.1). In
addition, the CDC reported improvements in many other aspects of our health:

- Heart disease remains the leading killer, but deaths from heart disease fell 16 percent
  between 2000 and 2004, and deaths from cancer—the number 2 killer—dropped
  8 percent.
- Infant mortality fell to 6.8 deaths per 1,000 live births in 2004, down from 6.9 deaths
  per 1,000 live births in 2003.
- Americans are increasingly using many types of preventive or early detection health
  services.

Yet, the study concludes,

Even as progress is made improving life expectancy, increased longevity is accompanied
by increased prevalence of chronic conditions and their associate pain and disability. In
recent years, progress in some arenas—declines in infant and cause-specific mortality,
morbidity from chronic conditions, reduction in prevalence of risk factors including smoking and lack of exercise—has not been as rapid as in earlier years or trends have been moving in the wrong direction … improvements have not been equally distributed by income,
race, ethnicity, education and geography. (National Center for Health Statistics 2006:3)

Table 10.1

Life Expectancy at Birth in Years According to Race and Sex in the United States, Selected Years

	All Races		White		Black	
	Male	Female	Male	Female	Male	Female
1900	46.3	48.3	46.6	48.7	32.5	33.5
1950	65.6	71.1	66.5	72.2	59.1	62.9
1960	66.6	73.1	67.4	74.1	61.1	66.3
1970	67.1	74.7	70.7	78.1	60.0	68.3
1980	70.0	77.4	70.7	78.1	63.8	72.5
1990	71.8	78.8	72.7	79.4	64.5	73.6
2000	74.3	79.7	74.9	80.1	68.3	75.2
2003	74.8	80.1	75.3	80.5	69.0	76.1

Source: National Center for Health Statistics 2006.

Before we discuss health inequalities and challenges, let's use our sociological imagination to better understand the relationship between health, illness, and society.

Sociological Perspectives on Health, Illness, and Medicine

If you are thinking that this is going to be a discussion about human physiology, theories about germs and viruses, full of medical terms, you'd be wrong. Although medicine can identify the biological pathways to disease (Wilkinson 1996), we will need a sociological perspective to address the social determinants of health. "Health is a result of an individual's genetic makeup, income and educational status, health behaviors, communities in which the individual lives, and the environments to which he or she is exposed" (Lurie and Dubowitz 2007:1119).

Consider the gross inequalities in health between and within countries. For example, life expectancy at birth is highest for a child born in Japan (81.9 years) and lowest for one born in Sierra Leone (34 years). Within the United States, there is a 20 year gap in life expectancy between most and least advantaged populations (Marmot 2005). There is no biological reason why life expectancy should be 48 years longer in Japan than Sierra Leone or why there is a gap in life expectancy about the rich and poor in America. To better understand the connection between our social structure and health, we must investigate how our political economy, our corporate structure, and the distribution of resources and power influence health and illness (Conrad 2001a).

The sociology of health and illness includes the field of epidemiology. **Epidemiology** is the study of the patterns in the distribution and frequency of sickness, injury, and death and the social factors that shape them. Epidemiologists are like detectives, investigating how and why groups of individuals become sick or injured (Cockerman and Glasser 2001). They don't focus on individuals; rather, epidemiologists focus on communities and populations, addressing how health and illness experiences are based on social factors such as gender, age, race, social class, or behavior (Cockerman and Glasser 2001). Epidemiology has successfully increased public awareness about the risk factors associated with disease and illness, leading many to quit smoking, to participate in more physical exercise, and to eat healthier diets (Link and Phelan 2001). For example, Type 2 diabetes, the most common form of the disease, occurs when the body does not produce enough insulin or the cells ignore the insulin. An estimated 16 million Americans have Type 2 diabetes. However, the disease can be effectively managed with healthy behaviors like meal planning, exercise, and weight management (American Diabetes Association 2003). Modernization, fast foods, and physical inactivity have led to significant increases in the number of Type 2 diabetes cases in countries such as Brazil and India. Indian public health officials project that in 20 years, there may be 78 million diabetics in their country. A disease that usually affects the old is affecting the younger Indian population primarily because they have adopted a modern life-style and diet (Kleinfield 2006).

Photo 10.1

Ogimi, a village on Japan's southern island of Okinawa, is known as "longevity village" because of the long life span of residents. Pictured here is a 105-year-old female resident. As a whole, Japan has one of the world's oldest populations.

What Does It Mean to Me?

All of us practice healthy behaviors we believe or were told can prevent or cure illness or disease. Brushing our teeth is one practice that we routinely do without really remembering why we do it. What other healthy behaviors do you practice? Why do you do them? Where did you learn them?

Epidemiologists use three primary measures of health status: fertility, mortality, and morbidity. These data are routinely collected by the National Center for Health Statistics, Centers for Disease Control. **Fertility** is the level of childbearing for an individual or population. The basic measure of fertility is the crude birthrate, the number of live births per 1,000 women ages 15 to 44 in a population. The U.S. crude birthrate for 2005 was 14 births per 1,000 women (Hamilton, Martin, and Ventura 2006). Related to this is the measure of **fecundity**, the maximum number of children that could be born (based on the number of women of childbearing age in the population).

In the early 1900s, a woman could expect to give birth to about four children, whereas a woman during the Great Depression of the 1930s could expect to have only two (U.S. Census Bureau 2002). The lowest number of births per woman was 1.8 children in the mid-1970s. Since then, the rate has averaged around two births per woman (U.S. Census Bureau 2002). Fertility is determined by a set of biological factors, such as the health and nutrition of childbearing women. But innovations in medicine, in the form of infertility treatments, have also made childbirth possible for women who once considered it impossible. Fertility is also determined by social factors, such as our social values and definitions of the role of women, the ideal family size, and the timing of childbirth.

Mortality is the incidence of death in a population. The basic measure of mortality is the crude death rate, the number of deaths per 100,000 people in a population in a given year. For 2004, the U.S. death rate was 800 deaths per 100,000 (Miniño et al. 2006). In the United States, it is unlikely that we'll die from acute infectious diseases, such as an intestinal infection or measles. Rather, the leading causes of death are attributed to chronic conditions such as coronary heart disease, cancer, stroke, or chronic lower respiratory disease, all of which have been linked to heredity, diet, stress, and exercise. The leading causes of death vary considerably by age. The leading cause of death of college-age Americans is unintentional injuries, followed by homicide and suicide. Among the elderly, mortality caused by chronic diseases (heart disease, cancer, chronic bronchitis, diabetes) is more prevalent.

Infant mortality is the rate of infant death per 1,000 live births. For 2004, the infant mortality rate was 6.79 per 1,000 (Miniño et al. 2005). The three leading causes of death among U.S. infants were congenital birth defects, low birth weight, and sudden infant death syndrome. Infant mortality is considered a basic indicator of the well-being of a population, reflecting the social, economic, health, and environmental conditions in which children and their mothers live. Though infant mortality rates in the United States have declined, rates are disproportionately higher for minority children. In 2003, infant mortality rates were highest for infants of non-Hispanic Black mothers (13.6 deaths per 1,000 live births), American Indian mothers (8.7 per 1,000), and Puerto Rican mothers (8.2 per 1,000). Infant mortality rates were lowest for infants of Cuban mothers (4.6 per 1,000) and Asian or Pacific Islander mothers (4.8 per 1,000) (National Center for Health Statistics 2006).

The U.S. infant mortality rate is historically higher than other industrial societies. In its 2007 report, the Save the Children organization identified Iceland as the country with the lowest infant mortality rate with a rate of 2 deaths per 1,000 births. (Refer to Table 10.2 for a complete list of the top 24 countries.) When comparing U.S. infant mortality rates to those of other countries, researchers noted that maternal health care is more widely and uniformly available in other countries with national health care programs.

Table 10.2

Newborn Mortality by Country, 2006

Country	Newborn Mortality Rate (per 1,000)
Iceland	2
Finland, Japan, Sweden	3
Czechoslovakia, Denmark, Germany Norway, Iceland	4
Australia, Austria, Canada, Cyprus, Israel, Luxembourg, Malta, Netherlands, Republic of Korea, Switzerland, United Kingdom	5
Croatia, Slovakia	6
United Arab Emirates, United States	7

Source: Save the Children 2007.

Morbidity is the study of illnesses and disease. Illness refers to the social experience and consequences of having a disease, whereas disease refers to a biological or physiological problem that affects the human body (Weitz 2001). Epidemiologists track the **incidence rate**, the number of new cases within a population during a specific period, along with the **prevalence rate**, the total number of cases involving a specific health problem during a specific period (Weitz 2001). For example, the 2005 incidence rate for diabetes was 1.5 million people age 20 years or older; the prevalence rate was 20.8 million or 7 percent of the population (Centers for Disease Control 2006). Incidence rates help measure the spread of **acute illnesses**, which strike suddenly and disappear quickly, like chicken pox or the flu. The prevalence rate measures the frequency of long-term or **chronic illnesses**, such as diabetes, asthma, or HIV (Weitz 2001). The National Center for Health Statistics publishes the *Morbidity and Mortality Weekly Report,* a weekly summary of surveillance information on reported diseases/deaths.

In addition to epidemiological analyses, sociologists have also applied theoretical perspectives to better explain the social problems of health and illness.

What Does It Mean to Me?

Based on data from the National Center for Health Statistics, determine the current rates of fertility and mortality (adult and infant) in your state. Log on to *Study Site Chapter 10*. How do these rates compare with overall national figures (as reported in earlier paragraphs) or with neighboring states?

Functionalist Perspective

Émile Durkheim conducted the first empirical analysis of suicide in the late 1800s. Before Durkheim's work, scientists attributed suicide primarily to psychological or individual factors. However, Durkheim treated suicide as a social fact and identified the relationship between suicide and the level of social attachment or regulation between an individual and society. His research is the first true epidemiological analysis, but most importantly revealed the relationship between illness and the larger social structure.

The stability of society is paramount from a functionalist's perspective. Consider for a moment what happens when you become sick. When are you sick enough not to attend class? How do others begin to treat you? According to the functionalist perspective, illness has a legitimate place in society. The first sociological theory of illness was offered by Talcott Parsons (1951), addressing how individuals are expected to act and to be treated while sick (Weitz 2001). This set of behaviors is part of Parsons' theory of the **sick role**.

The sick role has four parts. In the first, sick people are excused from fulfilling their normal social role. Illness allows them to be excused from work, from chores around the house, or even from attending class! Second, sick people are not held responsible for the illness. The flu that's going around is no one's fault, so you aren't personally blamed if you catch it. (Although your roommates may blame you if they catch what you have.) Third, sick people must try to get well. Illness is considered a temporary condition, and sick people are expected to take care of themselves with appropriate measures. In relation to this, Parsons offers the last part, that sick people are expected to visit medical authorities and to follow their advice.

Although Parsons legitimized the social role of illness, he also identified a critical source of the problem in health care today. In the fourth element, Parsons identified the authority and control of the physician. Even though you're the one who is sick, the doctor has the ultimate power to diagnose your condition and tell you that you're "really" sick. Doctors play a prominent role in managing our illnesses, but they don't do it alone. Doctors, along with nurses, pharmaceutical corporations, hospitals, and health insurers, form a powerful medical industry. The medical industry has served us well with its technological and scientific advances, offering a wider array of medical services and treatment options. However, this industry has also created a set of problems, or dysfunctions, as functionalists like to refer to them. Medicine has shifted from a general practitioner model (a family doctor who took care of all your needs) to a specialist model (where one doctor treats you for a specific ailment). You are receiving quality care, but at a price (and you are paying to be treated by many different doctors, instead of just one). As a result, health care costs have become less affordable, leaving many without adequate coverage and care. The system intended to heal us does not treat everyone fairly. We will explore this further in the next perspective.

Photo 10.2

In Kampala, Uganda, a nurse (left) provides counseling to a patient (right) who is affected by cancer as a result of HIV. The nurse is a volunteer trained by The Aids Support Organization (TASO). Uganda has been called a pioneer in the battle against AIDS in Africa, reducing the prevalence of the disease in the country from a high of 18.3 to 6.2 percent in 2005.

Conflict Perspective

According to conflict theorists, patterns of health and illness are not accidental or solely the result of an individual's actions. Conflict theorists identify how these patterns are related to systematic inequalities based on ethnicity/race or gender and on differences in power, values, and interests.

The experience of AIDS treatment in Africa highlights the inequality of health care delivery and access. The World Health Organization (WHO) reports that though antiretroviral treatments for AIDS are available, most HIV-positive children from poor countries are not receiving these drugs. An estimated 2.3 million children under the age of 15 are infected with HIV and 800,000 of them need antiretroviral drugs to stay alive, yet only 60,000 to 100,000 are being treated. Fewer than 10 percent of pregnant women with HIV in poor or middle-income countries are receiving medication to prevent the transmission of the virus to their newborns. So why have rich(er) countries been able to virtually eliminate pediatric AIDS? Observers note that the disease would and could have been eradicated earlier if there were stronger health systems in African countries, a stronger commitment from the global community, and more money to be made helping these poorer countries (Altman 2006).

Conflict theorists may take a traditional Marxist position and argue that our medical industry is based on a capitalist system, founded not on the value of human life, but pure profit motive. Studies consistently identify that those in upper social classes have better health, health insurance, and medical access than men and women of lower socioeconomic status. A conflict theorist argues that instead of defining health care as a right, our capitalistic system treats health care as a valuable commodity dispensed to the highest bidder. The alternative would be a dramatic change in the medical system, ensuring that health care is provided to all regardless of their race, class, or gender.

Instead, what we have in place is a medical system responsive to middle- or upper-class patients and their needs. According to Ken Silverstein (1999), 6.1 million people died world-wide of malaria or acute lower-respiratory infections because there were no drugs available to treat these illnesses. Silverstein notes that pharmaceutical companies have pursued drugs that maximize their profitability, focusing less on diseases of the poor or drugs that are commercially unviable. As he explains, the interest is in lifestyle drugs, "remedies that may one day free the world from the scourge of toenail fungus, obesity, baldness, face wrinkles, and impotence" (p. 14).

The medical system itself ensures that those already in charge maintain power. In health care, no other group has greater power than medical physicians and their professional organization, the American Medical Association (AMA), established in 1847. On its Web site, the AMA identifies itself as "the nation's most influential medical organization." In his book, *The Social Transformation of American Medicine,* Paul Starr (1982) explains how the AMA's authority over the medical profession and education was secured in the early 1900s, with a series of events that culminated with the Flexner Report. The 1910 report was written by Abraham Flexner and was commissioned by the Carnegie Foundation and supported by the AMA. Through the report, Flexner and the AMA were able to pass judgment on the quality of each medical school, based on an assessment of its curriculum, facilities, faculty, admission requirements, and state licensing record. This report eventually led to strict licensing criteria for all medical schools, which led to the closure of schools that could not meet the new standards. Starr (1982) reveals that although the increased standards and school closures may have improved the quality of medical training and care, they also increased the homogeneity and cohesiveness of the profession. From 162 schools in 1906, the number of medical schools dropped to 81 by 1922. Some of the closed schools were exclusively for African Americans and women. According to Rose Weitz (2001), with the increasing cost of education and higher educational prerequisites, fewer minorities, women, immigrants, or poor students could meet the requirements. As a result,

> Fewer doctors were available who would practice in minority communities and who understood the special concerns of minority or female patients. At the same time, simply because doctors were now more homogenously White, male, and upper class, their status grew, encouraging more hierarchical relationships between doctors and patients. (Weitz 2001:327)

The AMA (2003) continues its lobbying and legislative efforts today, pursuing several legislative goals including medical liability reforms (asking for limits on punitive or noneconomic damages), preserving Medicare physician payments, expanding health insurance options, and obtaining regulatory relief from Medicare administration.

What Does It Mean to Me?

The consumer movement has shifted some of the power in the doctor-patient relationship to the patient. Drug ads, previously reserved for professional medical journals, are now commonly featured in popular magazines and in television advertisements. Pharmaceutical companies routinely take two or three full-page ads, featuring drug warnings, side effects, and precautions, along with a description of their drug and its benefits. It sometimes is difficult to figure out what the drug is for. How do you think this popular diffusion of pharmaceutical information has redefined the relationship between doctor and patient? Between your doctor and you?

Feminist Perspective

According to Peter Conrad (2001b), illness and how we treat it can reflect cultural assumptions and biases about a particular group. Take, for example, the case of women and their medical care. Conrad explains that throughout history, there are examples of medical and scientific explanations for women's health and illnesses that reflect dominant and often negative conceptions of women. Since the 1930s, women's natural physical conditions and experiences, such as childbirth, menopause, premenstrual syndrome, and menstruation, have been medicalized. **Medicalization** refers to the process through which a condition or behavior becomes defined as a medical problem (Weitz 2001). Although the medicalization of these conditions may have been effective in treating women, various feminist theorists see it as an extension of medicine's control of women (Conrad 2001b), specifically normal female experiences linked with the female reproductive system (Markens 1996), inappropriately emphasizing the psychological, biomedical, or sociocultural origins (Hamilton 1994). Once a condition is defined as a medical problem, medicine, rather than the woman herself, gains control of its diagnosis and treatment.

Menopause, a natural physiological event for women, was defined in the medical community as a "deficiency disease" in the 1960s when commercial production of estrogen replacement therapy became available (Conrad 2001b; Lock 1993). Although a few medical writers refer to menopause as a natural process, many continue to describe it as a "hormonal imbalance" that leads to a "menopausal syndrome" (Lock 1993). Although estrogen replacement treatment was presented as a means for women to retain their femininity and to maintain good health, feminists argued that menopause was not an illness; actually, estrogen therapy may not be necessary and may actually be dangerous (Conrad 2001b). A recent study indicated that although estrogen is an effective short-term treatment for hot flashes or night sweats, estrogen does little to improve the quality of older women's lives (Haney 2003).

Studies have suggested that the meanings and experiences of menopause may also be bound by cultural definitions. In North America, where women are defined by their youth and beauty, aging women are set up as a target for medicalization. In Japan, however, public attention focuses on a woman's life course experience. For a middle-aged Japanese woman, what matters is how well she fulfills her social and familial duties, especially the care of elderly family members, rather than her physical or medical experiences. The Japanese medical community has a different perspective on menopause than their American colleagues do: Most doctors in Japan define menopause as natural and an inevitable part of the aging process (Lock 1993).

What Does It Mean to Me?

Do you know what a "mommy job" is? Mentioned on ABC's *Brothers and Sisters* program and featured on print and television advertisements, a mommy job is a cosmetic surgical procedure that may include a breast lift, a tummy tuck, and liposuction to reduce the stretch marks, slackened skin, and excess fat that result from pregnancy and child birth. Targeting women of childbearing age, the marketing of mommy makeovers has been described as an attempt to "pathologize the postpartum body, characterizing pregnancy and child birth as maladies with disfiguring aftereffects that can be repaired with the help of scalpels ..." (Singer 2007:E3). From a sociological perspective, is a mommy job a surgical necessity or invention? What do you think?

Interactionist Perspective

From the interactionist perspective, health, illness, and medical responses are socially constructed and maintained. In the previous sections, we discussed how health issues are defined by powerful interest or political groups. We just reviewed how the medicalization of women's conditions reflects our cultural assumptions or biases about women. Each example demonstrates how social, political, and cultural meanings affect our definition and response to health and illness.

A patient's experience with the medical system can be disempowering (Goffman 1961), but the experience can be mediated by social meaning and interpretations (Lambert et al. 1997). According to S. Peterson, M. Heesacker, and R. Schwartz (2001), when people contract a disease, they define their illness according to a socially constructed definition of the disease, which includes a set of images, beliefs, and perceptions. Patients use these definitions to create a personal meaning for their diagnosis and to determine their subsequent behavior. The authors argue that these social constructs have a greater influence on the patient's actions and decisions about his or her health than recommendations from health professionals do.

Sociologists also examine how the relationship between doctors and their patients is created and maintained through interaction. In particular, sociologists focus on how medical professionals use their expertise and knowledge to maintain control over patients. Research indicates that doctors' power depends on their cultural authority, economic independence, cultural differences between patients and doctors, and doctors' assumed superiority to patients (Weitz 2001). Studies consistently demonstrate the systematic differences in the level of information provided by physicians to their patients. Although differences might be attributed to the doctor responding to a patient's particular communication style, researchers argue that information varies according to the doctor's impressions of a patient (e.g., intelligence) or according to subjective judgments about what information the patient needs (Street 1991). Educated and younger patients tend to receive more diagnostic information, as do patients who ask more questions and express more concerns; doctors are likely to communicate as equals with their educated, older male patients (Street 1991). African Americans, Asian Americans, and Hispanics are more likely than are Whites to experience difficulties in communicating with their doctors. The difficulties include not understanding their doctor, not feeling that the doctor listened to them, or having questions for their doctor that they did not ask (Collins et al. 2002).

Interactionists and social constructionists also investigate how a disease is socially constructed. This doesn't mean that disease and illness do not exist. Rather, the focus is on how illness is created and sustained according to a set of shared social beliefs or definitions. In his essay, "The Myth of Mental Illness," Thomas Szasz (1960) argues that mental disorders are not actually illnesses. He considered mental illness a convenient myth to cover up the "everyday fact that life for most people is a continuous struggle" (p. 118). The disease of mental illness is constructed and maintained through a set of medical, legal, and social definitions. The social construction of disease has also been applied to anorexia nervosa (Brumberg 1988), black lung disease (Smith 1987), and chronic fatigue syndrome (Richman and Jason 2001).

For a summary of sociological perspectives on disease and illness, see Table 10.3.

What Does It Mean to Me?

What are the social constructs for cancer? for AIDS? for diabetes? How would your social definitions of a disease affect your experience of the disease?

Table 10.3

Summary of Sociological Perspectives: Health and Medicine

	Functional	Conflict/Feminist	Interactionist
Explanation of the social problems of health and medicine	Although illness may threaten the social order, it does have a legitimate place in society. This perspective also addresses the functions and dysfunctions of the medical industry.	Patterns of health and illness reflect systematic inequalities based on ethnicity/race or gender, and also differences in power, values, and interests. Conflict theorists examine the power of the medical industry and its consequences. Feminist theorists examine medicine's control of women, specifically of normal female experiences linked with the female reproductive system.	Acknowledges how illness is created and sustained according to a set of shared social beliefs or definitions. Theorists in this perspective address how social, political, and cultural meanings affect our definition and response to health and illness.
Questions asked about health and medicine	What is the role of illness in our society? What functions and dysfunctions does the medical industry provide?	How does the medical industry exert control over its patients? Is everyone treated fairly and equally by the medical system?	Who or what defines what it means to be "sick" in our society? How do our definitions of disease and illness shape our beliefs and behaviors?

Health Inequalities

Gender

As noted in Table 10.1, women live about five years longer than men. The three leading causes of death for males and females are identical: heart disease, cancer, and stroke. Although women live longer than men, women experience higher rates of nonfatal chronic conditions (Waldron 2001; Weitz 2001). Men experience higher rates of fatal illness, dying more quickly than women when illness occurs (Waldron 2001; Weitz 2001).

These differences in mortality have been attributed to three factors: genetics, risk taking, and health care (Waldron 2001). Biological differences seem to favor women; more females than males survive at every age (Weitz 2001). Because of differences in gender roles, men are more likely to engage in risk-taking behaviors or potentially dangerous activities: driving too fast or incautiously, using legal or illegal drugs, or participating in dangerous sports (Waldron 2001). The workplace offers more dangers for men. More men than women are employed, and men's jobs tend to be more hazardous (Waldron 2001); about 9 of every 10 fatal workplace accidents occur to men (Men's Health Network 2002). Finally, because women obtain more routine health examinations than men do, their health problems are identified early enough for effective intervention (Weitz 2001). Typically, women eat healthier diets and smoke and drink less alcohol than men do (Calnan 1987).

According to the Men's Health Network (2002), "No effective program exists which is devoted to awareness and prevention of the leading killers of men." Although men die of cancer at twice the rate of women by the age of 75, there is little education for men in cancer self-detection and prevention. Whereas there is a popular national campaign for breast cancer, there is no national educational campaign teaching men how to self-examine for testicular cancer, a leading killer of men from 15 to 40 years of age. In addition, there are no quality educational programs regarding prostate cancer, a cancer that strikes one in five men (Men's Health Network 2002).

Social Class

Regardless of the country where a person lives, social class is a major determinant of one's health and life expectancy (Braveman and Tarimo 2002). The link between class and health has been confirmed in studies conducted in Australia, Canada, Great Britain, the United States, and Western Europe (Cockerman 2004). Although no factor has been singled out as the primary link between socioeconomic position and health, scholars have offered many factors—standard of living, work conditions, housing conditions, and the social and psychological connections with others at work, home, or the community—to explain the relationship (Krieger, Williams, and Moss 1997).

Weitz (2001) offers several explanations for the unhealthy relationship between poverty and illness. The type of work available to poorly educated people can cause illness or death by exposing them to hazardous conditions. Poor and middle-class individuals who live in poor neighborhoods are exposed to air, noise, water, and chemical pollution that can increase rates of morbidity and mortality. Inadequate and unsafe housing increases the risk of injury, infections, and illnesses, including lead poisoning when children eat peeling paint. The diet of the poor increases the risk of illness. The poor have little time or opportunity to practice healthy activities like exercise, and because of life stresses, they may also be encouraged to adopt behaviors that might further endanger their health. Finally, poverty limits individual access to preventative and therapeutic health care.

The relationship between health and social class afflicts those most vulnerable, the young. Children in poor or near-poor families are two to three times more likely not to have a usual source of health care than are children in non-poor families (Federal Interagency Forum on Child and Family Statistics 2007). Access to a regular doctor or care facility for physical examinations, preventative care, screening, and immunizations can facilitate the timely and appropriate use of pediatric services for youth. Even children on public insurance (which includes Medicaid and the State Children's Health Insurance Program) were more likely not to have a usual source of care than were children with private insurance. Children in families below the poverty level had lower rates of immunization and yearly dental check ups (both basic preventative care practices) than did children at or above the poverty level (Federal Interagency Forum on Child and Family Statistics 2007). More than 9 million U.S. low-income children are uninsured (Kaiser Family Foundation 2007a).

Education

A similar relationship has been documented between education and health—the higher your education, the better your health (no matter how it is measured—mortality, morbidity, or other general health measures). Schooling might be a more important correlate to good health than is one's occupation or income (Grossman and Kaestner 1997).

Recent studies on the effects of compulsory education in Sweden, Denmark, England, and Wales consistently identify that a longer educational experience leads to better health (Kolata 2007). Michael Murphy and his colleagues (2006) identified mortality trends by educational level for Russian men and women between 1980 and 2001. Murphy et al. concluded that better educated men and women had a significant mortality advantage over less educated men and women. In 1980, life expectancy at age 20 for university educated men was 3 years greater than for men with only an elementary education. By 2001, however, the gap between university and elementary educated men had increased to 11 years. Similar differentials were also noted among Russian women.

Researchers suggest that education helps individuals choose and practice a healthier lifestyle regarding diet, exercise, and other health choices. Highly educated men and women are likely to visit their primary physicians more often and regularly and may be more willing to use new medical technologies or medicines. Knowledge about the health consequences of smoking and drinking has been shown to decrease smoking and excessive alcohol consumption. Educated parents will also transmit their healthier lifestyle to their children (Grossman and Kaestner 1997).

Researchers have demonstrated the link between education and future orientation. Future-oriented individuals attend school for longer periods. Educated individuals are able to link their current actions to their future, not only for their education, but also for preventative health care practices. For example, a future-oriented person will say, "I'm going to college now so that I can have a good job when I graduate." Applied to health behaviors, the same person will say, "I won't start smoking because I know there are long-term health consequences of smoking." Studies have shown that men and women who discount the future are more likely to become addicted to alcohol or other drugs (Becker and Mulligan 1994).

Other Health Care Problems

The Rising Cost of Health Care

The United States spends about 15 percent of its gross domestic product (GDP) on health care—the largest expenditure in this category among industrialized countries. (A GDP comparison with other countries is presented in Figure 10.1.) In 2005, total health care spending reached $2 trillion, with an average of $6,697 spent per person in health expenses (Centers for Medicare and Medicaid Services 2006). Analysts note that for 2005, health care spending grew at the slowest pace in six years, partly because of the increasing use of generic drugs. Overall health care spending increased 6.9 percent in 2005, lower than the 7.2 percent increase in 2004. In 2005, home health care spending increased more than any other category. More than 30 percent of health care spending went to hospitals in the same year. (Refer to Table 10.4 for a summary of health insurance plans.)

Even though the U.S. health system is the most expensive in the world, "Comparative analyses consistently show the United States underperforms relative to other countries on most dimensions of [health] performance" (Davis et al. 2007:viii). The United States remains the only major industrial country without some form of universal health coverage. In their analyses of health care systems and outcomes in Australia, Canada, Germany, New Zealand, the United Kingdom, and the United States, Karen Davis and her colleagues (2007) concluded that the United States failed to achieve better health outcomes and scored last on the dimensions of

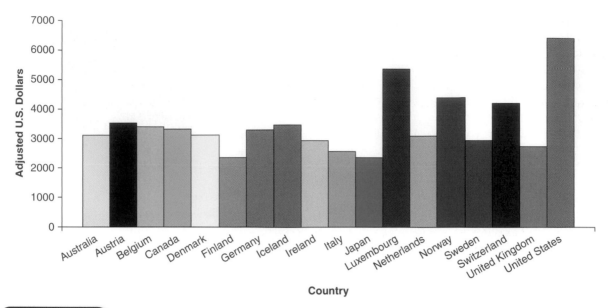

Figure 10.1

Total Health
Expenditures per
Capita, United
States and Selected
Countries, 2005

*Source: Organisation for
Economic Development
and Co-Operation 2007.*

Table 10.4

U.S. Health Insurance and Health Care Delivery Systems

Most Americans receive health insurance from their employers. This type of insurance is referred to as group insurance or as employment-based private insurance. Employers buy into a health insurance program, paying for part or all of the cost of the insurance premiums. A premium is a monthly fee to maintain your health coverage.

Most college students are probably covered by their parents' insurance plan. That handy insurance card in your wallet identifies your insurance provider, the amount of your deductible (payment due at the time of service), and the amount of coverage for prescription drugs or emergency services. There are many different types of group insurance programs:

- *Fee for service plan.* Under this plan, also known as an indemnity health plan, insurance companies pay fees for services provided to the people covered by the policy. This type of program emphasizes patient choice and immediate patient care.
- *Health maintenance organization (HMO).* These organizations operate as prepaid health plans. For your premium, the HMO provides you and your family comprehensive care. This plan is also known as managed health care, a plan that controls costs by controlling access to care. You'll be assigned to a primary care provider who will provide most of your medical care, but if necessary, that doctor will refer you to specialists within the HMO practice or to providers contracted by the HMO. Under the plan, there is limited coverage for any treatment outside the HMO network.
- *Preferred provider organization (PPO).* A PPO is a combination of the fee for service and HMO plan. With a PPO, you can manage your own health care needs by selecting your own doctors. These specialists will be on a preferred provider list supported by the PPO plan. If you use a provider outside of your plan, you may have to pay a larger percentage of your health care expenses.
- *Federal health plans.* Medicare is available to Americans 65 years or older or those with disabilities, whereas Medicaid pays for medical and long-term care for the poor; low-income children, pregnant women, and elderly; the medically needy; and people requiring institutional care.

access, patient safety, efficiency, and equity despite spending the most per capita on health care. Given our lack of universal health care coverage, when compared with these other nations, more Americans are uninsured or underinsured and are more unlikely to seek necessary care because of costs. Germany ranked first on access to health care and the United Kingdom ranked first for health care equity. The United States ranked last on both measures.

Policymakers and consumers have been keeping an eye on the cost of prescription drugs, one of the fastest-growing sectors of medical care. Increases in drug costs are expected to outstrip the overall growth in health care spending for the next 10 years. Spending for prescription drugs in 2004 totaled $188.8 billion, about 4.5 times more than the $40.3 billion spent in 1990 (Kaiser Family Foundation 2006).

The cost of prescription drugs remains a significant burden for elderly Americans. The American Association of Retired People (AARP) annually examines prices for 193 brand-name prescription drugs used by Americans 50 years of age or older. AARP reported that from 2005 to 2006, manufacturer prices increased 6.2 percent (an increase more than one and half times the rate of general inflation for the same period, 3.7 percent). Eighty-one drugs most commonly used by older Americans had price increases of more than 5 percent during the first nine months of 2006. In 2006, the typical older American, who takes four prescription drugs daily, spent $270.24 more for brand-name (non-generic) drugs than in the previous year. For 2005, the cost increase was lower at $189.72 (Binder et al. 2007).

Inequalities in Health Insurance

Access to health care is unevenly distributed across the U.S. population (Conrad and Leiter 2003). Data from the U.S. Census Bureau reveal that 46.6 million or 15.9 percent of Americans had no health insurance at anytime during 2005, an increase of 1.3 million from 2004 (DeNavas-Walt, Proctor, and Lee 2006). (Refer to U.S. Data Map 10.1.) Although most Americans receive health insurance through their employers, many uninsured individuals are either employed or are dependents of employed people (House Energy and Commerce Subcommittee on Health and Environment 1993). Nearly four in five of those without health insurance were in the labor force or had at least one parent who was employed (Families USA 2003). From 2000 to 2006, the number of companies offering health insurance fell from 69 to 61 percent (Kaiser Family Foundation 2007b). Most of the working poor are not eligible for public assistance medical programs (e.g., Medicaid) even though their employers do not provide health insurance (Seccombe and Amey 2001). Among households with incomes less than $25,000, only 75.6 percent received health insurance (DeNavas-Walt et al. 2006).

Several social factors are related to health coverage. About 11.2 percent or 8.3 million children under the age of 18 did not have health insurance (DeNavas-Walt et al. 2006). When parents lose jobs, children lose health insurance coverage. Children in poverty, with an uninsured rate of 19.0 percent, were more likely to be uninsured than was the population of all children. Children 12 to 17 years old were more likely to be uninsured than were those younger than 12 years old (DeNavas-Walt et al. 2006).

Among all minority groups, Hispanic Americans are most likely not to have insurance. Nearly half (45 percent) of Hispanics younger than age 65 and two-thirds

Photo 10.3

The fastest growing sector of medical care is prescription drug spending. In 2004, more than $188 billion was spent on prescription drugs. The cost of drugs remains a significant burden for elderly Americans.

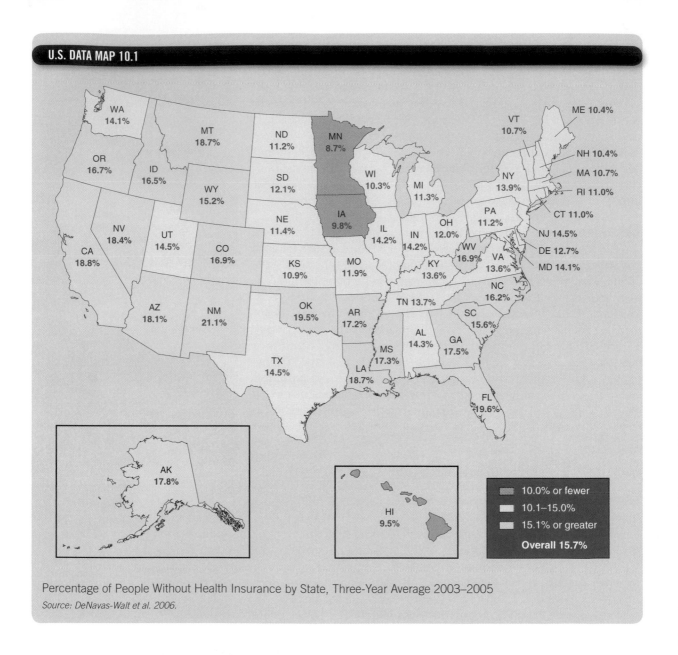

U.S. DATA MAP 10.1

Percentage of People Without Health Insurance by State, Three-Year Average 2003–2005

Source: DeNavas-Walt et al. 2006.

(65 percent) of working-age Hispanics with low incomes were uninsured for all or part of 2000 (Commonwealth Fund 2003a). Twenty-two percent of Hispanic children did not have any health insurance compared with 7.2 percent for non-Hispanic White children (DeNavas-Walt et al. 2006). Hispanic American women, compared with White women, are three times more likely to lack health insurance (NWHIC 2002). Most Hispanic women obtain health insurance through their employers; however, lower-income or part-time workers are less likely to be offered health coverage through their employment (NWHIC 2002). According to Michelle Doty (Commonwealth Fund 2003a), "Lack of insurance, unstable coverage, language barriers, and low income all contribute to the growing health care crisis among Hispanics."

However, losing health insurance is also becoming a middle-class issue, according to the Employee Benefit Research Institute. In 2001, about 800,000 people had incomes in excess of

$75,000 but no health insurance (Broder, Pear, and Freudenheim 2002). More than one-third of those uninsured in 2007, about 17 million Americans, have family incomes of $40,000 or more (Pear 2007a). They either lost their jobs in high-wage industries, were employed but unable to afford health insurance premiums (Broder et al. 2002), work as independent contractors (e.g., real estate agents), or would not be covered by insurers because of a preexisting health condition (Pear 2007). Advocates for the uninsured are hopeful that policymakers will respond to a more effective lobbying coalition of high-wage workers, the unemployed, the poor, and minorities (Broder et al. 2002). Annual premiums for employer-sponsored health insurance for 2006 averaged $4,242 for single coverage and $11,480 for family coverage (Kaiser Family Foundation 2007b).

What Does It Mean to Me?

Policy analysts, advocates, and health care professionals sharply criticized President George W. Bush for stating in 2007, "People have access to health care in America. After all, you just go to an emergency room." From a sociological perspective, what is wrong with the president's statement? Does his statement address access as discussed in the previous section? Why or why not?

Community, Policy, and Social Action

Health Care Reform

During his first administration, President Bill Clinton said problems connected with the U.S. health care system were the most pressing in the United States. In 1993, Clinton pushed for passage of the Health Security Act, an attempt at comprehensive health care reform. The act would have required all employers to provide health insurance to their employees and gave small businesses and unemployed Americans subsidies to purchase insurance. After Congress rejected Clinton's health care plan, Americans looked to the private market to restrain health care costs and to enhance patient care and choice. U.S. medicine moved aggressively toward managed care arrangements, HMOs, and for-profit health plans (Oberlander 2002).

According to Stuart Altman and Uwe Reinhardt (1996), although Americans in general are pleased with the quality of health care they receive, they are troubled by other aspects of health care. Americans are upset primarily about the cost of health care, both out-of-pocket costs and the cost of the health care system as a whole. Second, the structure of the health insurance system is viewed as having a number of shortcomings, foremost among them being the lack of universal coverage. Finally, although the quality of health care is high, people believe it is not uniformly so.

The U.S. health care system is identified as a private health care system, but in reality, it is a mixed system of public and private insurance (Oberlander 2002). Health care reform did not die with the defeat of the Health Care Act. Pick up any newspaper or listen to the evening news and you'll hear stories about insurance reform, prescription drugs, or quality care. The issue was debated among 2008 presidential candidates and received a boost with the release of *Sicko,* Michael Moore's documentary film comparing health care in Canada, Cuba, Great Britain, and the United States. Although there have been no other presidential proposals for universal health coverage, political attention on the issue has been more limited in scope and at a slower pace (Begley et al. 2002), focused on improving the medical experiences of those already insured by regulating managed care and expanding existing programs (Oberlander 2002).

State Health Care Reforms

The federal government has failed to reach a compromise on comprehensive health coverage, so the burden of health care reform falls to the states (Beatrice 1996). Several states have aggressively moved forward on health reform, and several—Florida, Hawaii, Massachusetts, Minnesota, Oregon, and Washington—are committed to providing health coverage for all of their citizens. Following are summaries of three state plans.

Hawaii was one of the first states to act on health care reform. In 1974, the state passed the Hawaii Prepaid Health Care Act, requiring employers to provide health insurance for all employees working more than 20 hours per week and to pay at least 50 percent of the cost. Hawaii is the only state that requires employer payments to medical insurance under a congressional exemption of the Employee Retirement and Income Security Act (ERISA). ERISA bars states from requiring all employers to offer health insurance, from regulating or taxing self-insured plans, and from mandating the specific benefits to be covered by employer health plans (Beatrice 1996). Hawaii's plan also limits employees' share of the insurance premium expenses to no more than 1.5 percent of their income. Recently, there has been a call to repeal or at least revise the Prepaid Health Care Act because of increasing health care costs. The percent of uninsured Hawaiians is 9.5 percent (DeNavas-Walt et al. 2006).

The MinnesotaCare Act became law in Minnesota in 1992. Also known as the HealthRight Act, the legislation included a variety of laws aimed at reducing costs and expanding access to health care for the uninsured (Beatrice 1996). MinnesotaCare is funded through a tax on health care providers and through enrollee premiums (based on family size, number of people covered, and income) (Sacks, Kutyla, and Silow-Carroll 2002). The act set price controls for health care spending (repealed in 1997), set statewide managed care guidelines, initially mandated that all non-HMO physicians follow a state fee structure (repealed in 1995), placed all HMOs under the regulation of the Commission of Health, and mandated that HMOs be nonprofit (Citizens Council on Health Care 2003). The act also subsidized health insurance to low- and middle-income uninsured families and individuals. Minnesota expanded Medicaid eligibility to 275 percent of the federal poverty guidelines (e.g., a family of four with $45,000 income would be eligible) and placed all MinnesotaCare recipients into HMOs (Beatrice 1996; Citizens Council on Health Care 2003). A small number of Minnesotans, about 9 percent, are uninsured (DeNavas-Walt et al. 2006).

Massachusetts became the first state to provide universal health care coverage to all its residents in 2006. The plan is estimated to insure half a million people within three years, about 95 percent of the state's uninsured population. Called a "moderate plan," the plan will allow the state to provide sliding scale coverage, low-cost coverage, and free insurance coverage to uninsured residents depending on their income, age, or employment status. Individuals who can afford health insurance will be penalized on their state income taxes if they do not purchase it. The plan will cost $1.2 billion over three years, using a combination of federal and state (existing and new) monies.

What Does It Mean to Me?

Does your state have a comprehensive health care policy? If the information is available, trace the history of the policy from legislation to implementation. What benefits and problems exist within the system? If your state does not have a comprehensive health policy, what roadblocks exist to adopting such a policy?

State Children's Health Insurance Program

The State Children's Health Insurance Program (SCHIP) was adopted in 1997 as an amendment to the Social Security Act, Title XXI. The program is administered under the Centers for Medicare and Medicaid Services (CMS). SCHIP enables states to implement their own children's health insurance programs for uninsured low-income children 18 years old or younger and targets the children of working parents or grandparents. For example, a family of four that earns as much as $34,100 a year is eligible. The insurance plan would pay for regular checkups, immunizations, prescription medicines, and hospitalizations. SCHIP uses comprehensive outreach materials and educational programs to recruit eligible children and their families, especially through elementary and secondary schools. In many states, as SCHIP enrollments began, so did Medicaid enrollments. By June 2005, the number of children enrolled in SCHIP had reached a record high of more than 4 million. Enrollment growth exceeded 22 percent in four states—Delaware, Oregon, Virginia, and Wyoming. State officials expected SCHIP enrollment increases to continue.

During its 2007 meeting, the National Governors Association asked the Bush administration to provide more federal funds to support the SCHIP program. Responding to increasing need, many states have expanded program coverage and eligibility. Governors reported that they were running out of federal funds to support the popular program. States would be short $13 billion to run the program if they continued with their current program eligibility rules and benefits through 2012 (Pear 2007b). Michael O. Leavitt, Secretary of Health and Human Services, responded by saying that the administration would work with Congress to find a short-term solution to their funding problem, but also recommended that states should better manage their SCHIP programs (Pear 2007c). Based on his concern that SCHIP expansion would be too costly and would be a step toward universal health coverage, in fall 2007, President Bush vetoed the House bill that proposed coverage for more than 10 million children as well as expansion of coverage to include dental services, mental illness, and pregnant women with low incomes. SCHIP advocates accused the president and his congressional supporters of misplacing their concerns for socialized medicine over the necessary expansion of health coverage for children. With the House unable to override the president's veto, the administration promised to work with Congress on a bill compromise.

State Prescription Drug Plans

In an effort to control drug costs for their residents, several states have offered innovative cost-control models. Pennsylvania is second to Florida in the proportion of its population that is 65 years or older (Pear 2002). More than 300,000 people in Pennsylvania are enrolled in Pharmaceutical Assistance Contract for the Elderly (PACE). Men and women 65 years or older can enroll in the program if they have annual incomes less than $14,000 for an individual or

Photo 10.4

Gemma Frost (age 9) is pictured with her mother Bonnie during a news conference promoting the State Children's Health Insurance Program (SCHIP), along with Representative Steny Hoyer (D-MD) and House Speaker Nancy Pelosi (D-CA). Gemma received health care benefits through the program, after suffering injuries in a car accident. In 2007, President George W. Bush vetoed a bill that would have reauthorized and expanded the SCHIP program.

$17,200 per couple. The program costs patients a $6 co-payment per filled prescription and is financed largely from state lottery proceeds. The program requires the use of low-cost generic drugs, which account for about 45 percent of all filled prescriptions. PACE was established in 1984 with strong bipartisan support, but with the rising cost of drugs and the increasing number of patients, lawmakers are looking for more cost cutting strategies (Pear 2002). Pennsylvania also supports PACENET, an assistance program available for elderly individuals with household incomes between $14,500 and $23,500 or between $17,700 and $31,500 per couple (with a $40 deductible and an $8 co-payment).

In 2003, Pennsylvania joined eight other states and the District of Columbia to form a nonprofit consortium to buy drugs in bulk, passing on the savings to their citizens (Freudenheim 2003). The consortium will include a drug benefit manager who will help states maximize their drug benefits by receiving full price discounts and rebates, determining the most cost-effective and appropriate drugs, and including coverage for mail order prescriptions and for importing drugs from Canada. The program is supported by the Heinz Family Philanthropies, a charitable organization.

U.S. laws prohibit the importation of Canadian drugs into the United States, unless their safety is certified by the U.S. Department of Health and Human Services. Despite this law, Boston, Massachusetts, and the state of New Hampshire announced in 2003 that they would begin buying prescription drugs from Canada. Springfield, Massachusetts, was the first city

Taking a World View

Health Care in Canada

Imagine a world where you never see a doctor's bill, an insurance statement, or any other paperwork related to health care (Weitz 2001). That place exists: Canada. Canada has a publicly financed, privately delivered health care system known as Medicare. Their national health insurance system provides all Canadians access to universal, comprehensive coverage for medically necessary hospital, inpatient, and outpatient care.

Although we tend to think of the Canadian insurance program as one program, there are actually 13 federally supported programs, each administered by one of its 10 provinces and 3 territories (Taylor 1990). The legislative foundation of Canada's national health insurance program comes from the 1957 Hospital Insurance Act and the 1966 Medical Care Insurance Act. In 1984, both acts were consolidated into the Canada Health Act, which in addition outlawed extra billing by physicians and user fees by facilities and guaranteed a one-tier system of health insurance (Livingston 1998). Although the federal government is responsible for setting national standards of health care, each province or territory is responsible for the management and delivery of health care services and some aspects of prescription care and public health (Health Canada 2003). The federal government supports a province's health insurance program as long as it is universal (covering all citizens), comprehensive, accessible (with no limits on services), portable (each province must recognize each other's coverage), and publicly administered (under the control of a public nonprofit organization) (Marmor and Mashaw 2001).

The insurance system is funded by a progressive federal tax: Those who earn more money pay a higher proportion of their income in taxes (Weitz 2001). Overall, Canada spent less per capita on health care than the United States did. For 2005, Canada spent $3,326 compared with $6,401 in the United States (refer to Figure 10.1). The total budgets of hospitals and the level of physicians' fees are determined by annual negotiations between provinces and the health care providers (Marmor and Mashaw 2001). Budgets are adjusted each year, accounting for inflation, new programs, and changes in service volumes (Marmor and Mashaw 2001). Most Canadian physicians are paid according to a fee-for service plan (Taylor 1990).

Administrative costs are much lower in Canada than in the United States. Because doctors and hospitals receive their payments from one source, they do not have to keep track of multiple insurance plans or file for insurance reimbursement (Marmor and Mashaw 2001). This benefit is also passed on to patients: Canadians do not have to file claim forms or pay out of pocket for uncovered expenses. In addition, Canadian doctors have lower malpractice insurance costs. The Canadian Medical Association formed the nonprofit Canadian Protective Association in 1912. All member physicians were first charged a flat fee for malpractice insurance; currently, there are six fee categories organized according to medical specialty (Taylor 1990).

According to Clifford Krauss (2003), growing complaints about health care services have begun to erode public confidence in Canada's health care system. Although Canadians continue to support their health system, many worry about its effectiveness, particularly its service delivery. Krauss cites many Canadian studies that reveal how the health system is overworked and understaffed and that patients wait impatiently for their health services. A recent Canadian government study reported 4.3 million Canadian adults or 18 percent of those who saw a doctor in 2001 had difficulty in seeing a doctor or getting a test completed in a timely fashion (Krauss 2003).

A report prepared by Canada's Fraser Institute revealed that in 2006, Canadian patients experienced an average waiting time of 17.8 weeks between receiving a general practitioner's referral and undergoing treatment and 9.0 weeks between specialist consultation and treatment. Compared with 1993, overall waiting time is about 91 percent longer (Esmail and Walker 2006). Such long wait periods are considered symptomatic of a failing and overburdened health care system as well as a serious threat to patient health. The authors of the report, Nadeem Esmail and Michael Walker, conclude, "The promise of the Canadian health care system is not being realized. On the contrary, a profusion of research reveals that cardiovascular surgery queues are routinely jumped by the famous and politically connected, that suburban and rural residents confront barriers to access not encountered by their urban counterparts and that low-income Canadians have less access to specialists…" (2006:5).

In Focus

Bird Flu Pandemic

Human cases of the bird flu or avian influenza were first recorded in 1997. The disease is an infection that occurs naturally among birds. Many wild birds carry the virus, but remain unaffected by it. However, the virus can cause disease or death among domesticated birds, such as chickens, turkeys, and ducks.

The influenza virus can be spread to humans from contact with domesticated infected poultry or surfaces contaminated with secretions from infected birds. The specific virus that has been found in human patients is called Avian Influenza A (H5N1). Illness resulting from human to human infection has been reported but is rare, according to the Centers for Disease Control.

Outbreaks of H5N1 have been reported in the Republic of Korea, Vietnam, Japan, Thailand, Cambodia, the Lao People's Democratic Republic, Indonesia, China, and Malaysia. Of these, Japan, the Republic of Korea, and Malaysia have controlled their outbreaks and are now considered free of the disease according to the World Health Organization (WHO). Since 2003, 186 (of 307) deaths have been documented as a result of the disease. There have been no reported cases of the bird flu in the United States.

Scientists and medical experts have been concerned about the potential for a bird flu pandemic (world wide spread of the virus), primarily because of the uncontrollable aspect of the primary carriers of the virus, wild birds. Migratory birds visit lands and waters also used by domesticated animals and their humans. WHO explains that poverty in these countries exacerbates the problem, for example, families will consume poultry even when death or signs of illness appear in the flock. Though ill or dying, these birds are the primary source of food for many poor people. Butchering and food preparation potentially can expose more humans to the disease. Additionally, it is difficult for rural farmers to accurately assess the source of illness among their poultry, delaying reporting to health authorities.

Second, health officials are concerned about H5N1 mutating to enable human to human infection. All flu viruses change (consider how each year there is a new flu shot to be given during flu season), so according to the CDC, if H5N1 can be spread by person-to-person contact, because the bird flu virus does not commonly infect humans, we have little or no immune protection. In April 2007, the Federal Drug Administration approved the first human H5N1 vaccine.

to import Canadian drugs for city employees in 2003. Boston plans to do the same for its city employees and retirees, estimating a savings of about $1 million annually. For example, 90 pills of Lipitor, a popular cholesterol drug, cost $183.97 in the United States, but the Canadian price is $136.70, a savings of $47.27 per prescription (Testa 2003). The Nevada State Board of Pharmacy approved regulations in 2006 that would allow its residents to purchase online prescriptions filled by state-approved Canadian pharmacies.

In 2004, the Health and Human Service Taskforce on Drug Importation concluded that savings on foreign drugs were not as much as consumers would expect and warned about significant risks to consumers purchasing imported drugs. The report questioned the safety and effectiveness of foreign made drugs. At the release of the report, consumer and health advocates weighed in, criticizing the taskforce for failing to address the fundamental problem of providing affordable prescriptions drugs to those who cannot afford it.

Voices in the Community

Victoria Hale

In 2005, Victoria Hale was named by *Esquire* magazine as its business woman of the year. She is the director and founder of OneWorld Health, the first nonprofit pharmaceutical company in the United States.

Hale worked as an analyst for the U.S. Food and Drug Administration's Center for Drug Evaluation and Research and then at Genentech (a biotechnology firm) when she first envisioned her company. Her position in the pharmaceutical industry allowed her to see how drugs were being set aside simply because they were not making enough money. The industry dedicates less than 10 percent of its total research and development budget to eradicate diseases of the developing world, which account for 90 percent of the world's total infections (Heffernan 2005). Of the more than 1,500 drugs marketed worldwide between 1974 and 2004, only 21 or 1.3 percent of these drugs were used to treat diseases of the developing world (Buse 2006).

She established OneWorld Health to accomplish what she believed the industry should be doing—investing in the development and distribution of drugs that could be used to eradicate diseases in the developing world. Says Hale, "We deliberately chose neglected diseases that others were not working on.... There has been little research done on these diseases, and limited money for research or development. We don't choose projects because money is available, we choose projects and then we go find funding" (Roth 2006:2). OneWorld Health is funded by charitable contributions.

Hale's strategy was simple. She searched for drugs whose patents had expired or were not being used because of low profit margins. The first drug OneWorld Health invested in was paramomycin, an antibiotic that cures a parasitic disease called visceral leishmaniasis, also known as black fever or Kala Azar. The disease afflicts a half a million people annually worldwide, particularly in Bangladesh, Brazil, India, Nepal, and Sudan. Hale discovered that the development of paramomycin had been shelved before completing its clinical trials. After continuing and completing clinical trials with the drug, Hale found a company, Gland Parma, based in Hyderabad, India, that agreed to produce paromomycin and sell it for $10 per full course of treatment, affordable for the poor people of India.

The Institute of OneWorld Health's CEO and founder Victoria Hale is photographed with the Institute's Chief Medical Officer Ahvie Herskowitz.

The use of paromomycin was approved in India in late 2006. The government publicly announced its goal of eradicating the disease by 2010. Eradication dates have also been set in Bangladesh and Nepal for 2015. After dealing with the initial skepticism of the pharmaceutical industry and its executives, Hale's OneWorld Health has been heralded as an innovative socially minded organization. Hale's work will continue, "You can't take care of all two billion of the world's poorest poor at one time. But you can go disease by disease and determine which one you can succeed with" (Roth 2006:2). OneWorld Health is working on three additional drugs—for malaria (the most severe parasitic disease), diarrhea (the number two killer of children in the developing world), and Chagas disease (a parasitic illness of Central and South America).

For more information about OneWorld Health, go to *Study Site Chapter 10.*

Work-Based Health Clinics

In an effort to manage and control the cost of their health care expenses, more U.S. corporations are providing on-site medical care. Since 2005, Cigna, Sprint Nextel, and Credit Suisse have opened on-site health clinics. More than 100 of the nation's largest 1,000 employers now provide primary care clinics or preventative health services on-site for their employees (Freudenheim 2007). One such company, Pepsi Bottling Group, operates 11 employee clinics in the United States with plans to add more in the future. Toyota's San Antonio, Texas, primary care center is open for workers and their families, complete with two full-time doctors, a blood test lab, and an X-ray center.

Photo 10.6

Community health centers bring health care to the people. This remote area medical (RAM) clinic is an annual medical clinic set up at the Wise County Fairgrounds in Wise, Virginia. Over three days, the clinic will assist more than 7,000 rural residents. All services, equipment, and supplies are donated.

On-site clinics are a "modern model that is ... proving to be cost-effective ... Not only does it pick up health issues earlier, but it doesn't require time away from work and at the same time creates a culture of caring," says Sean Sullivan, president of the Institute for Health and Productivity Management (Wells 2006:48). The health care savings from these clinics is estimated between 5 and 20 percent. The range of services includes routine physical examinations, allergy and flu shots, routine monitoring for chronic diseases, prescriptions, and advice on weight loss or smoking cessation (Freudenheim 2007; Wells 2006). On-site medical facilities and staff can be outsourced or company managed and hired; both models allow companies to customize their clinics and range of services to meet the needs of their workforce (Wells 2006).

Community-Based Health Care for Minorities

Regardless of their health insurance status, minority Americans are more likely than Whites to be disconnected from the health care system and a regular doctor (Commonwealth Fund 2003b). Unique community-based health care approaches have emerged to serve these groups.

Community health centers (CHCs) were based on neighborhood health clinics first established during the War on Poverty in the 1960s. CHCs are operated by a variety of non-profit organizations, health departments, religious or faith-based organizations, or medical organizations or schools. Costs are covered through a variety of sources, ranging from private insurance to government contracts or grants. These centers have been called the most effective tool to reduce health disparities and can increase access to health care to an estimated 14 million individuals, two-thirds of whom are minority group members (Hargreaves, Arnold, and Blot 2006).

One example of a community health center is the Project Brotherhood Black Men's Clinic, located on Chicago's South Side. The clinic was created to address the disproportionate disease burden and shorter life expectancy for Black men (Tanner 2003). Eric Whitaker, cofounder of the clinic, explained how many men avoid traditional health care providers because they never find doctors who look and talk like them. Focus groups indicated that Black men need a reason other than their health to go to the clinic, so program administrators latched on to a cornerstone of Black cultural life: the barbershop.

Project Brotherhood opened a barbershop alongside the community health clinic, where men are able to get free haircuts and listen to informal presentations by the clinic staff on HIV, heart disease, and cancer. In addition, the clinic offers job search and resume writing information and parenting classes. As Whitaker describes it, "It's a place where information is exchanged, it's a place of familiarity.... We just transported that idea to the clinic setting" (Tanner 2003:A12). He explains, "We want to learn how to listen to our community and discover its own perception of needs and assets" (Phalen 2000). In 1999, the clinic averaged about 4 medical visits per week; by 2005, the average number of medical visits increased to 27 per week (Project Brotherhood 2006).

Using the same strategy of bringing care to community members, the American Diabetes Association (ADA) has enlisted the help of African American churches to educate their parishioners about the disease (American Diabetes Association 2003). Recognizing the importance of the African American church as a source of community support and vehicle for communication, "Diabetes Sunday" began in 1996. Diabetes Sunday is often held during Black History Month. Church pastors discuss diabetes and distribute education materials. Certified diabetes educators are on hand to answer questions after the service. Similar community-based programs target Latino, Asian Pacific Islander, and immigrant populations.

Main Points

- Although medicine can identify the biological pathways to disease, we need a sociological perspective to address the social determinants of health. Research continues to demonstrate the relationship between the individual and society and the structural effects on health: how our health is affected by our social position, work, families, education, and wealth and poverty.

- **Epidemiology** is the study of the patterns in the distribution and frequency of sickness, injury, and death and the social factors that shape them. Epidemiologists focus on communities and populations, addressing how health and illness experiences are based on social factors such as gender, age, race, social class, or behavior.

- According to the functionalist perspective, illness has a legitimate place in society.

- Conflict theorists believe that patterns of health and illness are not accidental or solely the result of individuals' actions. Theorists identify how these patterns reflect systematic inequalities based on ethnicity/race or gender and differences in power, values, and interests.

- Although the **medicalization** of such conditions as premenstrual syndrome or menopause may have been effective in treating women, various feminist theorists see this trend as an extension of medicine's control of women.

- From an interactionist's perspective, health, illness, and medical responses are socially constructed and maintained.

- Women live longer than men, but women experience higher rates of nonfatal chronic conditions. Men experience higher rates of fatal illness. These differences in **mortality** have been attributed to three factors: genetics, risk taking, and health care.

- Data consistently support the notion that those with higher education, income, or occupational prestige have lower rates of **morbidity** and mortality. No factor has been singled out as the primary link between socioeconomic position and health; however, scholars have offered many factors—standard of living, work conditions, housing conditions, and the social and psychological connections with others at work, home, or the community—to explain the relationship.

- Another problem with health is access to health care and insurance; many groups do not have insurance or access to it, and now the middle class is also experiencing problems in retaining health insurance. Illegal immigrants also do not have insurance and often use the emergency room as their place of primary care.

- The United States spends about 15 percent of its gross domestic product (GDP) on health care—the largest expenditure in this category among industrialized countries. The rising cost of health care has been attributed to various factors: increases in the application of high technology for medical treatment and diagnosis, the aging population of the United States, the overall demand for health care, the amount of uncompensated care, and the cost of prescription drugs (the fastest-growing spending category and a particular problem for the elderly).

- Health care reform has been a topic of debate in Congress and the White House for several years. Recent federal health care initiatives include Medicare reform, the State Children's Health Insurance Program, and the Patient's Bill of Rights. But as health care continues to be debated at the federal level, some states are taking action to make reforms. Communities and organizations are also becoming involved.

On Your Own

Log on to the Web-based student study site at www.pineforge.com/leonguerrero2study for interactive quizzes, e-flashcards, journal articles, Community and Policy Guides, a Service Learning Guide, the end-of-chapter Web exercises, and additional Web resources.

Internet and Community Exercises

1. Based on the American Medical Association's Web site, review the organization's mission statement, history, and legislative initiatives. Log on to *Study Site Chapter 10*. Is there evidence to demonstrate how the AMA maintains its power and influence on the medical profession?

2. Select a specific disease or illness that you believe affects college-age men and women. Identify several Internet Web sites and support groups related to the disease or illness. How is the disease defined? Are there objective and subjective aspects of the disease? Does it vary by gender? by ethnicity/race?

3. Select one local hospital or health care system in your area or state. Through the Internet (or by visiting the hospital), identify the organization's mission statement, its patient bill of rights, and community-based programs. How does this organization define *care*? Does its definition appear to be consistent with the community's population and needs? Why or why not?

4. The United Nations' Human Development Report tracks several health indicators: life expectancy at birth, infant mortality rates, and health expenditure per capita. Log on to *Study Site Chapter 10* for the UN link and compare the United States with Canada, Germany, Mexico, Sweden, and the United Kingdom.

CHAPTER 11

The Media

Imagine that you wake up tomorrow in a sort of "Twilight Zone" parallel society where everything is the same except that media do not exist: no television, no movies, no radio, no recorded music, no computers, no Internet, no books or magazines or newspapers.

—Croteau and Hoynes (2000:5)

W hat would your life be like without the media? Without communication, and the media to communicate with, there would be no society. The term media is the plural of medium, derived from the Latin word medius, which means middle. A medium is a method of communication—television, telephone, cable, Internet, radio, or print—between (or in the middle of) a sender and a receiver. But taken all together, **the media** are the "different technological processes that facilitate communication between the sender of the message and the receiver of that message," as defined by David Croteau and William Hoynes (2000:7).

Communication is a basic social activity (Seymour-Ure 1974). It is impossible not to communicate and not to come in contact with the media. According to the U.S. Census Bureau (2007), in 2004, the average U.S. adult watched 1,546 hours of television, listened to 185 hours of recorded music, and watched 67 hours of home videos. The average U.S. adult also spent 188 hours reading the daily newspaper, 124 hours reading magazines, and 176 hours on the Internet.

Photo 11.1

In 2007, for the first time, Americans spent more time on their cell phones than on landlines and pay phones. People are using their cell phones in various ways—phone calls, text messaging, video messages, e-mail, and Web access.

What Does It Mean to Me?

For at least a week, monitor your own media usage. How many hours of television do you watch? How many hours are you on the Internet? On your phone? Would it be possible for you to live a day without media?

The media reflect "the evolution of a nation that has increasingly seized on the need and desire for more leisure time" (Alexander and Hanson 1995:i). Technological developments have increased our range of media choices, from the growing number of broadcast and cable channels to the ever-increasing number of Internet Web sites. New technologies have also increased our viewing control of and access to the media. For example, technology allows us to choose where and when we want to see a recent film. Are you taking a long road trip? You can pack your portable DVD player and watch your favorite movie on the way. Need to access your e-mail? You can check e-mail with a wireless connection in your classroom or with your iPhone from nearly any location. In fact, u cn comnC8 w/yr F W txt msgN (translation: You can communicate with your friends with text messaging). As Croteau and Hoynes observe, "We navigate through a vast mass media environment unprecedented in human history" (2001:3).

Yet, the media have been blamed for creating and promoting social problems and have been accused of being a problem themselves. The most commonly identified problem is the media's content: controversial programming that features violence, racism, and sexism. Media critics have expressed concern about the highly controlled process by which the images that we see are conceived, produced, and disseminated by media conglomerates. Social researchers and policymakers have identified the unequal advantage some social groups have over others in our increasingly high-tech media environment. Before we review the media and their related social problems, we will first examine the media from a sociological perspective.

Sociological Perspectives on the Media

Functionalist Perspective

Functionalists examine the structural relationship between the media and other social institutions. Even before the content is created, political, economic, and social realities set the stage for media content. The media are shaped by the social and economic conditions of American life and by society's beliefs in the nature of men and women and the nature of society (Peterson 1981). The first American printing press arrived in Boston along with a group of Puritans fleeing England in 1638. The press became an instrument of religion and government, used to print a freeman's oath that presented the conditions of citizenship in this new country, as well as an almanac and a book of hymns (Peterson 1981).

Through electronic and print messages, the media continue to frame our understandings about our lives, our nation, and our world. The media serve as a link between individuals, communities, and nations. They help create a collective consciousness, a term used by Émile Durkheim to describe the set of shared norms and beliefs in a society. The mass media provide people with a sense of connection that few other institutions can offer. Live media events, such as the Olympics, the Super Bowl, or the Oscars, are set off from other media programs. People gather in groups to watch, they talk about what they see, and they share the sense that they are watching something special (Schudson 1986). News events captured by the media, such as the 1995 Oklahoma City bombing or the terrorist attacks of September 11, 2001, connect a nation and even the world.

What Does It Mean to Me?

The digital equivalent of hanging out can be found on social networking sites such as MySpace and Facebook. Aimed at teenagers and young adults, these sites provide opportunities for personal expression and connection. How does MySpace or Facebook provide a way to connect with others? If you have your own page, how does it work for you? Are there any dysfunctions related to these social networking sites?

In particular, television has contributed to a corresponding nationalization of politics and issues, taking local or regional events and turning them into national debates. Socially and politically, the media make our world smaller. In 1992, when the city of Los Angeles experienced three days of violence and civil unrest, it made us more aware of the racial tensions and inequalities, not only in Los Angeles, but also in our own cities. In 2004, the entire nation watched as gay and lesbian couples were allowed to marry in San Francisco. Other mayors and cities began to follow their lead, and the number of gay and lesbian wedding ceremonies increased across the country, along with the growing debate about how same-sex marriages would affect the institution of marriage.

The media have been accused of creating serious dysfunctions and social problems in society. Research has documented the link between viewing of media violence and the development of aggression, particularly among children who watch dramatic violence on television and film. Television has been called the "other parent" or the "black box," accused of draining the life and intelligence out of its young viewers. Popular media culture has been accused of undermining our educational system, subverting traditional literacy (Postman 1989). Public health studies have documented the link between television viewing and poor physical health among children and adults. The one thing television viewers seem to do while watching television is eat, and the danger is in what they choose to eat and drink—unhealthy snack foods or high calorie drinks (Van Den Bulck 2000). We will examine these dysfunctions more closely in the section, "Do the Media Control Our Lives?"

Conflict Perspective

The media, says Noam Chomsky (1989), are like any other businesses. The fundamental principle in American media is to attract an audience to sell to advertisers. Yes, you read that correctly. Commercial television and radio programming depend on advertising revenue, and in turn, the networks promise that you, the consumer audience, will buy the advertisers' products. "The market model of the media is based on the ability of a network to deliver audiences to these advertisers" (Croteau and Hoynes 2001:6).

In the United States, media organizations are likely to be part of larger conglomerates where profit making is the most important goal (Ball-Rokeach and Cantor 1986). Since the very beginning of mass communications, ideas, information, and profit have mixed. The first books printed in the colonies may have been devoted to religion, but the printers made money (Porter 1981). The media, according to conflict theorists, can only be fully understood when we learn who controls them.

One of the clearest and some say most problematic trends in the media is the increasing consolidation of ownership. The corporate media play a major role in managing consumer demand, producing messages that support corporate capitalism, and creating a sense of political events and social issues (Kellner 1995). In 1984, more than 50 corporations controlled most of our newspapers, magazines, broadcasting, books, and movies. By 1997, Ben Bagdikian reported that there were 10 media giants. Bagdikian declared, "Media power is political power" (1997:xiii). In 2002, media scholar Mark Crispin Miller reported on the new 10 multinational conglomerates controlling the media, listed in Table 11.1. Four of them—Disney, AOL Time Warner, Viacom, and News Corporation Limited—are truly multimedia corporations, producing movies, books, magazines, newspapers, television programming, music, videos, toys, and theme parks in the global marketplace.

Miller (2002) warned that the most corrosive influence of these 10 media conglomerates was their impact on journalism. Journalism has traditionally been referred to as the fourth estate, an independent institutional source of political and social power that monitors the actions of other powerful institutions such as politics, economics, and religion. However, conflict theorists remind us that someone is in charge of the fourth estate. Those who control the media are able to manipulate what we see, read, and hear. The media, serving the interest of interlocking state and corporate powers, frame messages in a way that supports

Table 11.1

Ten Largest Media Companies

1. AOL Time Warner (United States)
2. Disney (United States)
3. General Electric (United States)
4. News Corporation Limited (Australia)
5. Viacom (United States)
6. Vivendi (France)
7. Sony (Japan)
8. Bertelsmann (Germany)
9. AT&T (United States)
10. Liberty Media (United States)

Note: For more information about these media companies, visit the Web site, "Who Owns What," sponsored by the Columbia Journalism Review. Log on to Study Site Chapter 11 for a link.

the ruling elite and limits the variety of messages that we read, see, and hear (Chomsky 1989).

E. J. Epstein reveals that what we consider news is not the product of chance events, "it is the result of decisions made within a news organization" (1981:119). He explains that the crucial decisions on what constitutes news—what will and what won't be covered—are made not by the journalists but by executives of the news organization. Although the public expects news reporters to act like independent fair-minded professionals, reporters are employees of corporations that control their hiring, firing, and daily management (Bagdikian 1997). News executives are in control of the selection and deployment of specific reporters, the expenditure of time and resources for gathering the news, and the allocation of space for the presentation of news (Epstein 1981). The news divisions of the media cartel are motivated to work on behalf of their parent companies and their advertisers, even when that means working against the public interest (Miller 2002). Bagdikian (1997) asserts that the integrity of much of the country's professional news has become more ambiguous than ever.

A 2002 study released by Fairness and Accuracy in Reporting (2004) documented that network news programs rely on White (95 percent), male (85 percent), and Republican (75 percent) sources for their news. Women made up only 15 percent of all news sources and were featured rarely as experts but most often as "ordinary Americans" in news stories (Howard 2002).

Feminist Perspective

Douglas Kellner says that the media represent "a contested terrain, reproducing on the cultural level the fundamental conflicts within society" (1995:101). Feminist theorists attempt to understand how the media represent and devalue women and minorities. This perspective examines how the

media either use stereotypes disparaging women and minorities or completely exclude them from media images (Eschholz, Bufkin, and Long 2002).

One of the most important lessons young children learn is expected gender roles, learning masculine versus feminine behaviors. Although these lessons are taught by parents and teachers, a significant source of cultured gendered messages is television programs (Powell and Abels 2002). As L. R. VandeBerg explains, television programs are "designed to … evoke, activate, reference, and occasionally challenge mainstream social myths, policies and beliefs, including those concerning gender" (1991:106).

In his analysis of children's programming, M. Barner (1999) found that women are typically portrayed in passive roles as housewives, waitresses, and secretaries, whereas men are seen in active roles as construction workers or doctors. In their study of children's shows *Barney and Friends* and *Teletubbies,* K. Powell and L. Abels (2002) reported that neither popular program showed men and women in nonstereotypical roles. The female characters were followers most of the time, were underrepresented in a variety of occupations, and played feminine roles. The male characters on these children's programs were leaders or directors of action and were larger and stereotypically male in appearance.

Photo 11.2

According to a 2007 Pew Survey, viewers of *The Daily Show with Jon Stewart* (and *The Colbert Report*) were the most well-informed about current events. When asked to name national and international political leaders or about current news stories, more *Daily Show* and *Colbert Report* viewers were able to correctly answer the questions than viewers who relied on broadcast television (network and local news) or those who obtained their news from the Internet or newspapers.

What Does It Mean to Me?

Research suggests that women's magazines also promote a traditional gender ideology. Review four women's magazines, *Ms., More, New Woman,* and *Essence,* and identify the images of women presented. Through the articles and advertisements, how do these magazines promote feminism, independence, beauty, or social class?

In recent years, the media have adopted more feminized forms of pop culture. "Girl power" programming features a female action heroine, as in the UPN program *Buffy the Vampire Slayer* or the WB's *Charmed,* that is both feminist and feminine (Corliss and McDowell 2001). But M. G. Durham points out that although these new "kick ass" heroines are powerful and strong, the girls' bodies "conform to the social ideals of slenderness and voluptuousness that epitomize current dominant definitions of beauty" (2003:26). The new girl heroines meet every standard for conventionally, yet unrealistically defined beauty: They are White, fine-featured, thin, heterosexual, middle class, and usually blonde. Their performances are an important part of the interdependent cycle of media profits, advertising, and girls' purchasing power. These heroines and their beauty images are linked to the advertisers who underwrite their shows, modeling for cosmetics, soft drinks, or clothing. According to Nancy Signiorelli, "The majority of women in television have not really changed during the past two decades; society has undergone numerous changes, and while things are not perfect, they are greatly improved" (1990:80).

Interactionist Perspective

In what they tell us and what they choose not to tell us, the media define our social world (McNair 1998). The interactionist perspective focuses on the symbols and messages of the media and how the media come to define our "reality." It might be best to view the media, as Michael Gurevitch and Mark Levy suggest, as "a site on which various social groups, institutions, and ideologies struggle over the definition and construction of social reality" (1985:9).

The mass media become the authority at any given moment for "what is true and what is false, what is reality and what is fantasy, what is important and what is trivial" (Bagdikian 1997:xliv). The mass media define what events are newsworthy. The first criterion is proximity; events happening close are more newsworthy than those happening at a greater distance. Second, deviation is an important criterion. Events that can be reported as disruptions (natural disasters, unexpected deaths, murders), deviations from cultural or social norms of behavior (especially sexual, e.g., philandering clergy or politicians), and lifestyle deviance (alternative lifestyle report) make the news (Galtung and Ruge 1973).

Sociologist Herbert Gans (1979) explained that the news isn't about just anyone, it is usually about "knowns." Women and men identified by their position in government or their fame and fortune are automatically newsworthy. Knowns are elites in all walks of life, but especially from politics, the entertainment industry, and sports. Incumbent presidents appear in the news most often. The president is the only individual whose routine activities are noteworthy. (When was the last time that a news crew followed your every move?) In recent years, "celebutantes" (daughters from wealthy or famous families), such as Paris Hilton and Nicole Richie, have been deemed newsworthy by the media with their every move and fumble chronicled.

One of the biggest news stories of the late twentieth century was the death of Princess Diana. The facts and circumstances of her 1997 death and the fairytale life that preceded it conformed to all the criteria of newsworthiness. Although it occurred in France, her death involved cultural proximity: Princess Diana was a worldwide celebrity. The nonstop coverage of her life and death began as a story of celebrity fascination but ended as an international tragedy (McNair 1998). Shortly after Princess Diana's death, Mother Teresa, a renowned humanitarian, died. Although Mother Teresa's death was a news story, it never got the same attention and coverage as the death of Princess Diana did. As journalist Daniel Schorr says, the difference between the two women's lives was "the difference between a noble life well lived and a media image well cultivated.... Mother Teresa was celebrated, but was not a celebrity" (1998:15).

Photo 11.3

Female action heroines are both feminist and feminine, combining beauty with strength. Michelle Ryan portrayed Jamie Sommers in NBC's remake of the Bionic Woman series.

What Does It Mean to Me?

"The death of Anna Nicole Smith is the number one story around the world tonight," declared CNN's Larry King on the evening of February 8, 2007. The day was described as a feeding frenzy for our national media with attacks on U.S. troops and the death of four Marines in Iraq taking a back seat to the model's sudden death in Florida. Wolf Blitzer, King's CNN colleague, described Smith's story as "tabloid gold," with the public riveted to continuing coverage concerning her death and burial and the paternity of her baby. Why was Smith's life and death newsworthy?

Taking a World View

Women's Feature Service, New Delhi, India

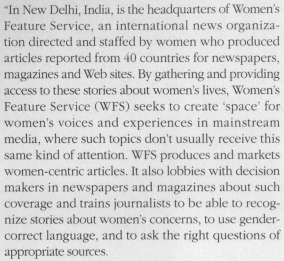

"In New Delhi, India, is the headquarters of Women's Feature Service, an international news organization directed and staffed by women who produced articles reported from 40 countries for newspapers, magazines and Web sites. By gathering and providing access to these stories about women's lives, Women's Feature Service (WFS) seeks to create 'space' for women's voices and experiences in mainstream media, where such topics don't usually receive this same kind of attention. WFS produces and markets women-centric articles. It also lobbies with decision makers in newspapers and magazines about such coverage and trains journalists to be able to recognize stories about women's concerns, to use gender-correct language, and to ask the right questions of appropriate sources.

"WFS exists because of the felt need for a gender balance in news coverage and because of dissatisfaction about the ways in which news organizations—in India and elsewhere—treat news coverage about women. Often, the media either ignore important stories altogether, relegate reporting to obscure places in the newspaper, or sensationalize incidents without examining the underlying context or causes. The media tend to focus on women only when it comes to 'women's issues,' forgetting that women also have an equal stake in so-called 'male concerns' such as the budget, economy, globalization, agriculture, and conflict resolution....

"Despite the presence of women journalists on the crime beat, incidents of rape and dowry deaths (shockingly regular occurrences in the Indian subcontinent) are usually reported in a routine manner, with the police being the sole source of information. Deadline pressure is one reason, but the other is that editors rarely insist that reporters get more information from other sources. Nor is there any follow-up to an incident. When it comes to issues that impact most directly on women, news that should cause concern and lead to analytical articles that examine a particular issue in depth, is often dismissed in a couple of paragraphs on an inside page.

"The other drawback is that there are only a few women writers—and fewer men—who can give a fresh perspective or insights into issues that concern women. Many women journalists have been conditioned (both socially and through the competitiveness of this profession) to adopt masculine attitudes and values. For instance, for a month after the United States declared its 'war on terror' and began bombing Afghanistan, none of the leading newspapers in India wrote on its editorial page about the women in the conflict.

"In India, there is a glass ceiling that women journalists have yet to break. Not a single mainline newspaper in India has a woman chief editor. One reason, of course, is that women joined the profession late—the first batch of women entered the profession in the 1960s—and took to covering politics even later. But it must be pointed out that there are male chief editors who are much younger than many senior women journalists are. Though women journalists have proven as competent, if not more, than men, they still lag behind in the power game. Two women who are at the top—Shobhana Bhartia (managing editor of *The Hindustan Times*) and Malini Parthasarathy (executive editor of *The Hindu*)—both belong to the families that own the newspapers. They've had to work hard to prove themselves and overcome some amount of intra-office opposition, but the fact remains they would not have risen this far but for their family connections. Interestingly, both do not have brothers, giving rise to the question: Would they have been given these opportunities had there been male siblings?"

Source: Quoted material is from Parekh 2001.

Photo 11.4

Broward Country
Chief Medical Examiner
Dr. Joshua Perper
(center) addresses
the media at a press
conference regarding
the death of Anna
Nicole Smith on
February 9, 2007,
in Fort Lauderdale,
Florida.

The mass media play a large role in shaping public agendas by influencing what people think about (Shaw and McCombs 1997) and, ultimately, what people consider a social problem (Altheide 1997). David Altheide (1997) describes the news media as part of the "problem-generating machine" produced by an entertainment-oriented media industry. The news informs the public, but its message is also intended to serve as entertainment, voyeurism, and a "quick fix" rather than providing an understanding of the underlying social causes of the problem.

Altheide (1997) argues that the fear pervasive in American society is mostly produced through messages presented by the news media. According to Barry Glassner (1997), the disproportionate coverage of crime and violence in the news media affects readers and viewers. Despite evidence that Americans have a comparative advantage relative to diseases, accidents, nutrition, medical care, and life expectancy, American women and men perceive themselves to be at greater risk than do their counterparts elsewhere and express fears about this (Altheide 1997). In a national poll, respondents were asked why they believe the country has a serious crime problem. About 76 percent said they had seen serious crime in the media whereas only 22 percent said they had a personal experience with crime (Glassner 1997). In a study of 56 local news programs, crime was the most prominently featured subject, accounting for more than 75 percent of all news coverage in some cities (Klite, Bardwell, and Salzman 1997).

For a summary of sociological perspectives, see Table 11.2.

Table 11.2

Summary of Sociological Perspectives: The Media

	Functional	Conflict/Feminist	Interactionist
Explanation of the media and social problems	Functionalists examine the structural relationship between the media and other social institutions. Sociologists using this perspective also examine the functions and dysfunctions of the media.	Conflict theorists focus on the media and how their messages are controlled by an elite group. Feminist theorists address how the media use stereotypes disparaging women and minorities or completely exclude them from media images.	From this perspective, sociologists investigate how the media define our social reality.
Questions asked about the media	How do other institutions affect the media and their content? What functions do the media serve in society? What are their dysfunctions?	Who owns the media? How are the media's messages manipulated? What images of women and minorities are presented by the media?	How do the media define what is newsworthy? How do the media define the public agenda? How do the media define what we believe is a social problem?

The Media and Social Problems

The Digital Haves and Have-Nots

The term **digital divide** was first used in the mid-1990s by policy leaders and social scientists concerned about the emerging split between those with and those without access to the computer and the Internet. The term refers to the gap separating individuals who have access to new forms of technology from those who do not. In addition, others have identified a gap between those who can effectively use new information and communication tools and those who cannot (Gunkel 2003). Despite the increasing diffusion of computers and an overall increase in Internet use, a deep divide remains "between those who possess the resources, education and skills to reap the benefits from the technology and those who do not" (Servon 2002).

The divide is also a global phenomenon. Less than 10 percent of the world's population uses the Internet (Gullién and Suárez 2005). According to data collected by the Central Intelligence Agency (2007), the European Union and the United States rank the highest with Internet users—each has more than 200 million users. Developing non-Western countries such as those in Africa, South America, and South Asia have less than a million Internet users (Refer to Table 11.3).

The Internet has been described as both empowering and discriminating, enabling residents in some countries to pursue a better life, but others are left behind (Gullién and Suárez 2005). Cross-national research has linked the diffusion of Internet technology with developed service sector economies, an educated population, economic development and significant research and development investment. Users in less developed countries have basic access

Table 11.3

Top Ten Internet Users, by Country

1.	European Union	247,000,000
2.	United States	205,327,000
3.	China	123,000,000
4.	Japan	86,300,000
5.	India	60,000,000
6.	Germany	50,616,000
7.	United Kingdom	37,600,000
8.	South Korea	33,900,000
9.	France	29,945,000
10.	Italy	28,870,000

Source: Central Intelligence Agency 2007.

problems: economic (cost of basic necessities versus the cost of Internet access), technological (varying ability of local networks), and geographical (limited access outside urban areas) aspects (Vartanova 2002). Mainly because of income differentials, the Internet is beyond the reach of global citizens. For example, in countries with low human development, the average annual income is about $1,200 (in U.S. dollars); the cost of owning a cheap personal computer would take more than half of that income, about $700 (Drori 2004; United Nations Development Programme 2001).

The digital divide implies a chain of causality. Access and ability to use the computer and Internet technology help improve one's social and economic well-being; lack of access to computers and the Internet harms one's life chances. But it is also true that those who are already marginalized in society will have fewer opportunities to access and use computers and the Internet (Warschauer 2003). The digital divide is a symptom of a larger social problem in the United States: social inequality based on income, educational attainment, and ethnicity/race.

Data from the National Center for Education Statistics (DeBell and Chapman 2006) revealed how computer and Internet use was divided along demographic and socioeconomic lines. Use of both technologies was higher among White students than among Black and Hispanic students. For example, two of every three White students (67 percent) use the Internet, but less than half of Black students (47 percent) and Hispanic students (44 percent) do. Higher computer and Internet use was positively associated with students living with more highly educated parents and those living in households with higher family incomes. (Refer to Table 11.4 for a summary of the NCES data. U.S. Data Map 11.1 reports the percent of households with Internet access per state.)

Internet access is not the only issue facing underserved communities. Wendy Lazarus and Francisco Mora (2000) identified four online content barriers. The greatest barrier is the lack of locally relevant information. They discovered that low-income users seek practical and relevant information that affects their daily lives, topics such as education (adult high school degree programs), family (low-cost child care), finances (news on public benefits, consumer information), health (local clinics, low-cost insurance resources), and personal enrichment

Table 11.4

Percentage of Children in Nursery School and Students in Grades K–12 Who Use Computers and the Internet, by Student and Household Characteristics, 2003

Characteristics	Percent Using Computers	Percent Using the Internet
Student Race/Ethnicity		
White	93	67
Black	86	47
Asian or Pacific Islander	91	58
Hispanic	85	44
Parent Educational Attainment		
Less than high school	82	37
High school graduate	89	54
Some college	93	63
Bachelor's degree	92	67
Graduate	95	73
Household Income		
Under $20,000	85	41
$20,000–34,999	87	50
$35,000–49,999	93	62
$50,000–74,999	93	66
$75,000 or more	95	74

Source: DeBell and Chapman 2006.

(foreign-language newspapers). In some instances, information may be available in printed documents, but these may be difficult to locate or obtain. General information may exist online, but it might not be suitable to low-income audiences. For example, online housing services might list high-end rental units rather than lower-rent housing.

The second barrier identified by Lazarus and Mora (2000) was the lack of information at a basic literacy level. According to the authors, a number of online tutorials that review computer program and Internet skills are written at a higher level of literacy. The third barrier was the need for content for non-English speakers. There is little government material (for example, on voting, Medicare, or taxes) translated in Spanish. Last, there is a need for more Web sites that reflect diverse cultural heritages and practices.

What Does It Mean to Me?

Access the Web site for your city or state government. Determine whether links and information are available in different languages. How many different language translations are provided? For what specific material or information?

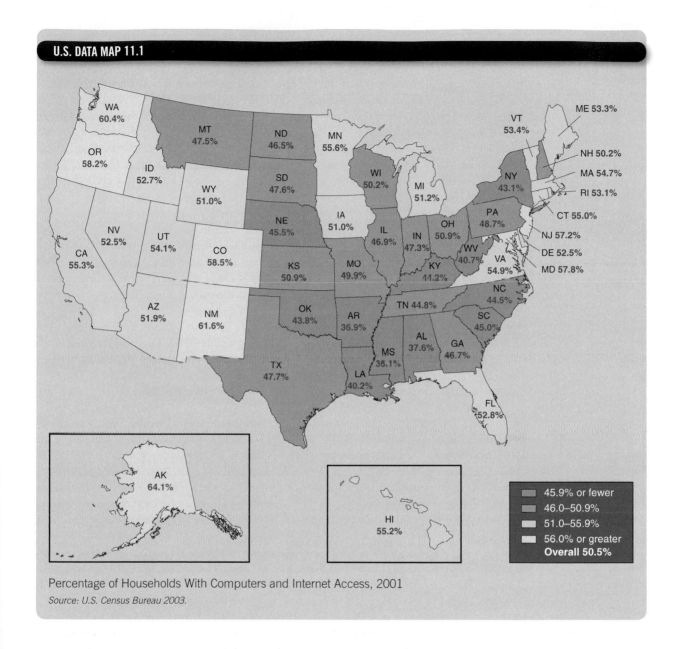

U.S. DATA MAP 11.1

Percentage of Households With Computers and Internet Access, 2001

Source: U.S. Census Bureau 2003.

Legend:
- 45.9% or fewer
- 46.0–50.9%
- 51.0–55.9%
- 56.0% or greater
- **Overall 50.5%**

Media Indecency

Although the media are owned and operated by private organizations, their content is regulated by federal standards. The Federal Communications Commission (FCC) enforces federal law and commission rules regarding indecency, defined as language that depicts or describes sexual or excretory organs or activities in language that is offensive, based on community standards. The FCC regulates program content between 6 a.m. and 10 p.m., when children are most likely to be watching television or listening to the radio (FCC 2004b). After an incident during the 2004 Super Bowl, the FCC began to impose huge fines on, among others,

Viacom and Clear Channel Communications for broadcasting shock-jock Howard Stern's radio program. In response, Clear Channel Communications declared a zero tolerance policy for indecency and stopped broadcasting Stern's program.

In a study of 400 hours of television programming, the Parents Television Council (2003) found that foul language increased during every time slot between 1998 and 2002. The council reviewed programs for all instances of foul language including curses (*hell* and *damn*), offensive epithets (*bitch, bastard*), scatological language (*ass* or *asshole*), sexually suggestive or indecent language (*screw*), and censored language (*shit* or *fuck*). Foul language during the family hour (8 to 9 p.m. in most areas of the country) increased by 94.8 percent between 1998 and 2002, the council found. A 2003 study by the Kaiser Family Foundation found that 64 percent of all shows and 71 percent of all prime-time shows have at least some sexual content. Only 15 percent of all sexual references or actions were deemed "responsible," suggesting responsibility or consequences (Kunkel et al. 2003).

Although many people were outraged by Janet Jackson's exposed breast during the 2004 Super Bowl halftime show, others suggested that earlier portions of the program were also inappropriate. Commercial advertisements featured during the game included Budweiser ads showing a sleigh ride interrupted by horse flatulence and a dog biting a man's crotch. Comedian Bill Cosby explains,

> What needs to be discussed prior to commenting on Miss Jackson's performance is what came before it. Look at the rest of the halftime show and the ads. Didn't I witness a couple of hundred situations where the female was being exploited or her body parts were being exploited? (quoted in Kelly, Clark, and Kulman 2004:50)

In 1997, the FCC implemented the television rating system for cable and broadcast television. This system is used with the V-chip control device (the V stands for violence) in your television, which can block programming by age-based categories or content labels. The TV parental guidelines were established by the National Association of Broadcasters, the National Cable Television Association, and the Motion Picture Association of America. The guidelines include seven ratings, reported in Table 11.5. All shows except news, commercials, and sports

Table 11.5

FCC Television Rating System

TV-Y	Programming appropriate for all children, including children ages 2 to 6.
TV-Y7	Programming for children age 7 and older. Programming in this category may include mild fantasy violence or comedic violence.
TV-Y7FV	Programs where fantasy violence may be more intense.
TV-G	Programs are suitable for all ages, containing little or no violence, no strong language, and little or no sexual dialogue or situations.
TV-PG	Programs contain material that parents may find unsuitable for young children, for example, one or more of the following: moderate violence, sexual situations, suggestive dialogue, or infrequent coarse language.
TV-14	Programming that many parents would find unsuitable for children under 14 years of age because it is likely to contain intense violence, sexual situations, strong language, or suggestive dialogue.
TV-M	Programming to be viewed by adults and unsuitable for children under 17 because it may contain graphic violence, explicit sexual activity, or crude indecent language.

Source: FCC 2004c.

programming (including the Super Bowl) begin with a rating in the upper left-hand corner. Critics argue that the rating system is ineffective because show producers rate their own programs, and many parents remain confused by the rating system.

Do the Media Control Our Lives?

Internet Abuse and Addiction

Researchers define **Internet addiction** as the excessive use and nonproductive use of the Internet. The literature suggests that addiction does exist but that it affects very few users (Griffiths 2003). Internet addiction is a broad term that covers a variety of behaviors and impulse controls (Griffiths 2002). Five types of addiction have been identified: cybersexual addiction (compulsive use of adult Web sites for cybersex and cyberporn), cyberrelationship addiction (over-involvement in online relationships), net compulsions (online gambling, shopping, or trading), information overload (compulsive Web searching), and computer addiction (computer game playing) (Griffiths 2002). (Refer to this chapter's In Focus feature for a discussion of cyberporn and online predators.)

Internet use, including Web surfing, instant messaging, shopping, and visiting chat rooms, can become a problem when it begins to interfere with one's job or social life ("Doctors Diagnose" 2003). Online computer games, such as Sims Online, EverQuest, or Ultima Online, attract an estimated hundreds of thousands of players each day. Aaron Hazell (quoted in Snider 2003) admits to playing Ultima Online for as many as 18 hours per day, even skipping his job as a software developer a few times so that he could continue playing the game. According to Hazell, "These games have an amazing amount of depth. You can buy a boat and sail to another continent. Or you can buy a house and then chop wood and make furniture for your home. The possibilities are endless" (p. 23).

Employers have identified Internet abuse as a serious work problem. According to a recent SurfControl study, office workers who spend one hour a day at work on nonwork Internet activities (shopping online, booking vacations) could cost businesses as much as $35 million per year. The study found that 5 percent of the Internet use at the office was not work-related (Griffiths 2003). Unlike Internet addiction, Internet abuse has no negative effects for the user, except for a decrease in work productivity if occurring at the workplace (Griffiths 2003).

A survey of human resource management directors revealed that 83 percent of the companies have Internet access policies regulating how much Internet time employees can have during their workday. Despite the existence of such policies, 64 percent of companies disciplined employees and more than 30 percent terminated employees for inappropriate use of the Internet. The leading causes for disciplinary action or termination included accessing pornography, online chatting, gaming, sports, investing, and shopping at work (Case and Young 2002).

The Problem With Television

In his 1932 book, *Brave New World,* Aldous Huxley writes about a society in which the pain of living is eased by the "television box." In Huxley's world, the television was constantly on, like a "running tap." Although TV may not be on 24/7, individuals in the industrialized world devote an average of three hours per day to television viewing. About 40 percent of families always eat dinner while watching television (TV Turnoff Network 2004). In 2006, Nielsen Media Research reported that the average American home had more television sets than people—2.73 TV sets versus 2.55 people (Bauder 2006).

In Focus

Moral Panic, Cyberporn, and Online Predators

Cyberporn or Internet based pornography has been called the newest danger to the innocence of childhood (Potter and Potter 2001). V. Lo and R. Wei (2002:13–14) argue that Internet pornography is different from traditional pornography (e.g., magazines and videos). First, it is widely available via the World Wide Web, e-mail, and real-time data feeds. Second, it is both an active and inactive format through digitized moving images, animated sequences, hot chats, and interactive sexual games. Finally, there has been an explosion of cyberporn since consumers have also become producers of their own pornographic material. According to K. Podlas, "Would-be Larry Flynts now require only a web-cam, video cameras to upload real-time images to a Web site, maybe a digital camera, and software ranging from shareware to a $100 commercial package" (2000:849). The moral panic over cyberporn, according to J. Li (2000), was set off by the June 1995 issue of *Time* magazine, which cited a study (later refuted) that concluded that 83 percent of online materials constituted porn.

A moral panic, defined by Stanley Cohen, is a period in which "a condition, episode, person or group of persons emerges to become defined as a threat to societal values and interests, its nature is presented in a stylized and stereotypical fashion by the mass media" (2002:9). According to Giselinde Kuipers, in the United States "the public discussion of Internet pornography has evolved into a true moral panic: highly emotional and polarized debates, sustained media attention, the founding of organizations of distressed citizens, skewed and exaggerated representation of the nature and amount of pornography and sex on the Internet and numerous attempts at government regulation" (2006:390).

Kuipers (2006) argues that our concern over Internet pornography is socially constructed and turns to other countries to see how it can be viewed differently. In the Netherlands, for example, the public discourse on Internet porn has focused on individual responsibility. The Netherlands is consistent with other European Union countries where ethnic and racial hatred is considered a more serious threat than sexual Internet content. European governments are working toward legislation against online racism, while U.S. lawmakers continue to focus on the regulation of pornographic sites.

Actually, there is no conclusive count of the number of pornographic Web sites, though some estimate that the number of pornographic Web sites has grown in the past few years, increasing by nearly 300 sites per day (Chen 1999) and generating more than $700 million per year (Hapgood 1996). The literature identifies the number of sites as being from 150,000 to one million. The Henry J. Kaiser Family Foundation (2001) reported that 70 percent of 15- to 17-year-olds accidentally stumbled across pornography online, with 9 percent saying that this happens very often. Among those who were exposed to pornography, 45 percent said they were upset by the experience, but more than half (55 percent) said they were not too upset or not at all upset at viewing this material.

In 1998, the U.S. Congress adopted the Children's Online Privacy Protection Act (COPPA), which makes it a crime for commercial Web site operators to post any material that is harmful to minors. Minors were defined as those 17 years of age or younger. In 2004, the U.S. Supreme Court voted 5 to 4 to prevent the enforcement of COPPA. The American Civil Liberties Union and online publishers challenged COPPA, believing that it violates free speech rights. The Supreme Court argued that it would be difficult to enforce COPPA restrictions. In the case of protecting minors from pornography, the justices suggested the use of filtering software on home computers.

Despite the popularity of NBC's *To Catch a Predator* and other media reports about sexual predators using the Internet to lure children or teens for sexual exploitation, there is little empirical research on the topic (Potter and Potter 2001). R. H. Potter and L. A. Potter (2001) highlight several national studies in the United States and Australia. In Australia, the first systematic study of Internet content was conducted in 1995, cataloging the different types of sexual content on line. The Australian Broadcasting Association (ABA) concluded that the likelihood of involuntary exposure to sexually explicit or hard core materials was low. In a follow up study, ABA researchers asked children about their perceptions of Internet harm. Children in the 2000 study were characterized as blasé about the sexual material they encountered on the Internet, saying that though they were appalled by some of the material, they did not think it constituted harm. In addition, the children reported that they did not need to be supervised by adults or did not require any filtering software to prohibit or limit their access to sexual material on the Internet. The first U.S. national study of online victimization of minors by David Finkelhor, Kimberly Mitchell, and Janis Wolak (2000) found that 19 percent of surveyed youth received a sexual solicitation or approach via the Internet. In 10 percent of these cases, the perpetrator asked for a face-to-face meeting; in most instances, however, the youth terminated contact and never met with the perpetrator.

In 2007, MySpace reported that it identified and removed more than 29,000 registered sex offenders who had posted personal profiles, along with information about where they live, on its Web site. Facebook has come under similar scrutiny for its accessibility to predators or convicted sex offenders. There is no empirical research to confirm how many children, if any, were victims of these sex offenders. State lawmakers are suggesting stricter regulations on profile registration for teens (requiring parental consent before creating one's profile) and Internet access to sex offenders (requiring registration of email addresses).

The average child or adolescent watches nearly three hours of television per day, not including time spent watching videotapes or playing video games. A recent study indicated that children spend more than six hours a day with various media combined (American Academy of Pediatrics 2001a). By the time the average young person graduates from high school, he or she will have spent about 12,000 hours in the classroom and about 18,000 hours watching television (Greeson and Williams 1986).

A large body of scholarly literature has consistently confirmed the negative impact of television on children and adolescents, who are particularly vulnerable to the messages conveyed through television (American Academy of Pediatrics 2001a). Exposure to violence in the media, including television, poses a significant risk to the health of children and adolescents (American Academy of Pediatrics 2001b). It is estimated that by age 18, the average young person will have viewed more than 200,000 acts of violence on television (American Academy of Pediatrics 2001b). Girls and boys who are exposed to violent behavior on film or TV behave more aggressively immediately afterward, are more tolerant of aggressive behavior (Huesmann et al. 2003), and are desensitized to the pain of others (American Academy of Pediatrics 2001b). Researchers have also linked high levels of exposure to violent television programs to aggression in later childhood, adolescence, and young adulthood (Anderson et al. 2003).

Exposure to violent video games has been linked with heightened levels of aggression in young adults and children (Anderson and Bushman 2001). Interactive video games are the preferred leisure activity for children and adolescents, yet about 80 percent of today's most popular games contain violence (Vessey and Lee 2000). Playing violent video games has been found to account for a 13 to 22 percent increase in violent adolescent behavior (American Academy of Pediatrics 2001b).

Simply stated, television may be bad for your physical health. Elevated television viewing and physical inactivity have been found to promote obesity in children (Faith et al. 2001). In addition to the unhealthy snacking that may occur while watching television, when people watch television, they burn fewer calories (Van den Bulck 2000). TV watching reduces the amount of time spent on more physical activities (Dietz and Gortmaker 1985). Television

viewing by children is also correlated with between-meal snacking, consumption of foods advertised on television, and children's attempts to influence their mother's food purchases (Dietz and Gortmaker 1985). The Kaiser Family Foundation study (Gantz et al. 2007) of advertising time during children's programming found that 50 percent of television ads were devoted to food; 72 percent were for candy, snacks, cereals, or fast food. The food industry has attempted to self-regulate the content of its commercials aimed at young audiences. The Kellogg Company, makers of breakfast cereals such as Froot Loops and Apple Jacks, announced in 2007 that it would phase out advertising products to children under 12 years of age unless the food products met specific nutrition guidelines for calories, sugar, fat, and sodium (Martin 2007). Sweden, France, and the Quebec providence of Canada have banned the advertising of food to children; the need for similar federal regulations has been suggested in the United States (Martin 2007).

Do You Trust the News Media?

Whether we are watching CNN, ABC, NBC, or Fox, we rely on reporters for most if not all of our information about our community and our world. According to the Project for Excellence in Journalism (2004), public attitudes about the press have been growing less positive for about 20 years. In its report, *The State of the News Media 2004,* the Project for Excellence in Journalism identified the disconnection between the public and the news media over motive as the fundamental reason for the decline in public support. Whereas journalists believe they are working in the public interest, the public believes that news organizations are working primarily for profit and that journalists are motivated by professional ambition. The public debate about whether media are liberal or conservative also sensitizes the public to how the media could be manipulated or subjective in content. In addition, people are increasingly distrustful of the large multinational corporations that own and control most of the news media (Project for Excellence in Journalism 2004).

These findings are confirmed by the Pew Research Center for the People and the Press (2005). More than 70 percent of the public feels that news organizations are influenced by powerful people and organizations rather than being independent. Sixty percent of Americans consider news organizations politically biased. In addition, more than 70 percent believe news organizations tend to favor one side, rather than treat all sides fairly.

In addition, the Pew Research Center examined how Americans regard their news organizations. The center documents how, after the media's focus shifted away from terrorism, Americans grew increasingly skeptical. In August 2002, 49 percent of respondents thought that news organizations were highly professional, down from 73 percent in November 2001. About 35 percent felt that the news media usually get their facts straight, and 56 percent believed that news organizations usually provide inaccurate reports. And if the news organizations do make a mistake, 67 percent of respondents believed that the news organizations were likely to try to cover up their mistakes; only 23 percent believed that media are willing to admit their mistakes (Pew Research Center for the People and the Press 2002). Recent accounts exposing untrue or falsified reporting have contributed to the public's mistrust of the media.

During the early months of the 2008 presidential campaign, a news report claimed that that Democratic candidate Senator Barack Obama attended a madrassah (a Muslim religious school) while living in Indonesia. The allegation was first raised in *Insight,* a conservative Internet magazine owned by the *Washington Post,* and was repeated by other news services. The magazine alleged that Obama was raised as a Muslim and that the sources of their information were researchers close to rival Senator Hillary Rodham Clinton's campaign. The deputy headmaster of the school Obama attended stated that the school is a public one that does not focus on religion. Obama's campaign staff maintained that the story was "completely ludicrous" and identified the candidate as a committed Christian. Clinton's staff claimed that they had nothing to do with the story.

Community, Policy, and Social Action

Federal Communications Commission and the Telecommunications Act of 1996

The FCC was established by the Communications Act of 1934 and is charged with regulating interstate and international communications by radio, television, wire, satellite, and cable. As an independent agency, the FCC oversees violations of federal law and policies and reports directly to the U.S. Congress (FCC 2004a). Under current FCC rules, radio stations and broadcast television channels cannot air indecent language or material that shows or describes sexual or excretory functions between 6 a.m. and 10 p.m., when children may be watching. In response to a violation, the FCC may issue a warning, revoke a station's license, or impose a monetary fine. In the entire history of the FCC, the commission has fined two television stations for indecency (Kelly et al. 2004).

Introduced a week before the Super Bowl in 2004, the Broadcast Decency Enforcement Act was designed to amend the Communications Act of 1934. The bipartisan bill increased penalties to $500,000 for violating FCC regulations. Broadcasters could also lose their licenses if they violate indecency standards three times. The act would not apply to cable television programming. In the case of the 2004 Super Bowl halftime show, each CBS station that aired the program could be fined as much as $27,500. After the 2004 Super Bowl halftime show, television broadcasters began to police their programming more closely. The 2004 Grammy and Oscar award telecasts each included a broadcast delay (five seconds to five minutes) to ensure that no inappropriate language or situations would be aired. The act was signed into law in 2005.

Through the FCC, the public can file complaints on a range of issues: billing disputes, wireless questions, telephone company advertising practices, telephone slamming (switching a consumer's telephone service without permission), unsolicited telephone marketing calls, and indecency and obscenity complaints. Along with the Federal Trade Commission (FTC), the FCC is enforcing the National Do-Not-Call Registry, which went into effect on October 2002. The FCC and FTC report that more than 55 million people have signed up for the registry. Its success has led lawmakers to consider implementing a do-not-spam list. However, in June 2004, the FTC rejected the idea of a do-not-spam list, arguing that creating such a list would not help decrease the amount of spam. Actually, such lists might be used by spammers to send more unwanted commercial e-mail.

The first major overhaul of the original 1934 act, the Telecommunications Act of 1996, was seen as a way to encourage competition in the communications industry. The law specified how local telephone carriers may compete, how and under what circumstances local exchange carriers can provide long-distance services, and ways to deregulate cable television rates. Also included in the act were provisions to make telecommunications more accessible to disabled Americans; a Decency Act makes it a crime to knowingly convey pornography over the Internet on a Web site accessible to children.

Although the act was presented as an opportunity to encourage competition and break down media monopolies, industry watchers noted that the 1996 law swept away the minimal consumer and diversity protections of the original 1934 act. The 1996 act reduced competition and allowed more cooperation between media giants. Data we reviewed in an earlier section of this chapter seem to support this. The new law permitted some of the largest industries—those not active in creating media content, such as telephone companies—to enter the television, radio, and cable industry. New industries joined older media companies to form interlocking partnerships, rather than become independent competitors as the act had predicted. For example, U.S. West, one of the largest telephone companies in the country, acquired Continental Cablevision, the third-largest cable system. Sprint, the long-distance telephone company, formed a joint venture with TCI, a cable company. The merger between Disney, ABC Broadcasting, and CAP Cities was made possible only through the passage of the 1996 law (Bagdikian 1997).

What Does It Mean to Me?

File-sharing is restricted by U.S. copyright law. Under this law, copyright owners have the right to control copies and distribution of the original material. Throughout the United States, colleges and universities have implemented policies for P2P (person to person) file sharing including documents, software, music, and movies. What is your university's policy regarding file sharing? How is it enforced? Do you believe it is an effective policy? Why or why not?

Who Is Watching the Media?

Numerous organizations have emerged as watch groups monitoring media accuracy and content. Groups such as Accuracy in the Media (AIM) and Fairness and Accuracy in Reporting (FAIR) are nonprofit, grassroots organizations that attempt to expose biased and inaccurate news coverage, while encouraging members of the media to report the news fairly and objectively. AIM publishes a newsletter and weekly newspaper column, broadcasts a daily radio commentary, and promotes a speaker's bureau to expose faulty reporting. On its Web site, current news stories are posted, along with AIM's analysis of the story's accuracy. In addition to a weekly program and a magazine, FAIR also operates research and advocacy desks that work with activists and professionals on women's issues; analyze the effects of sexism, racism, and homophobia in the media; and monitor the media's marginalization, misrepresentation, and exclusion of people of color in the news and in the newsroom (FAIR 2004).

Several organizations attempt to improve and protect the integrity of journalists in print, electronic, and Internet media. The Project for Excellence in Journalism (2004) began as an initiative by journalists to clarify and raise the standards of American journalism. The project serves as a research organization, conducting an annual review of local television news, producing a series of content studies on press performance, and offering educational programs for journalists. The project also provides information to the public about what to expect from the press, how to write a letter to the editor, and how to talk to the news media. Another organization, the Committee to Protect Journalists, is an independent, nonprofit organization promoting press freedom worldwide by defending the rights of journalists to report the news without fear of reprisal. The organization documents attacks on the press world wide, including the number of journalists killed, missing, or imprisoned. For 2007, the committee reported that 65 journalists were killed, 32 in Iraq (Committee to Protect Journalists 2008).

Voices in the Community

Geena Davis

Responding to the lack of female characters in television and motion picture programming, Geena Davis founded the See Jane Program to promote and advance gender balance in media for children 11 years of age or younger. Davis (2006) explains that after she gave birth to her daughter and began to watch preschool programs, she did a study of her own, examining the characters in her daughter's videos, and realized that the programming was dominated by male characters.

Where the Girls Aren't is the first research project sponsored by the program, using content analysis to examine 101 top grossing G-rated animated and live action films from 1990 to 2004 (Kelly and Smith 2005). In these films, researcher Stacy Smith found that there were three male characters for every one female character. Twenty-three percent of speaking characters were female and more than 80 percent of film narrators were male. This gender imbalance can affect children's gender development, reinforcing children's stereotypical attitudes and beliefs about gender. According to Davis, "By making it common for our youngest children to see everywhere a balance of active and complex male and female characters, girls and boys will grow up to empathize with and care more about each others' stories" (quoted in Kelly and Smith 2005:9–10).

Certainly Davis' experience on the film *Thelma and Louise* helped her recognize the impact of a strong female character on women and men. She acknowledges the affect of media images on adults, but her concern is on the impact on children. "We know that kids learn their value by seeing themselves reflected in the culture. They say, 'I see myself! I must matter. I must count.'" Her goal is for media portrayals to be "normal and natural for children to see worlds and characters—be they Martians or dinosaurs or talking toaster ovens—that are roughly half female and male. Just like the real world that our kids live in" (Davis 2006).

In addition to ongoing research, the See Jane Program continues to educate education professionals, parents, and child professionals about the importance of gender equity in the media. The program's goal is to see females in half of all characters in media made for young children.

Media Literacy and Awareness

Evidence shows that media education can help mitigate the harmful effects of the media. Parents can mount the most effective intervention to reduce the effects of media violence on children. A. I. Nathanson (1999) reports that parental co-viewing of and commenting on the programs seems to reduce the effects of TV violence on the child, probably because it reduces the child's perception that the violence is real and reduces the likelihood that the child will act out the violence. L. Huesmann et al. (1983) also reported on the effectiveness of a school-based intervention that teaches children violence on television is not real and should not be imitated.

According to the Media Awareness Network (2004), **media literacy** is the ability to sift through and analyze the messages that inform, entertain, and sell to us every day. It is about shifting from the role of passive receiver of media to an active critical receiver of it. Media literacy includes asking several questions: For whom is this media message intended? Who wants to reach this audience and why? From whose perspective is this story being told? Whose voices are being heard and whose are absent? (Media Awareness Network 2004). The Action Coalition for Media Education offers 10 basic principles for media literacy, which are presented in Table 11.6.

Media awareness organizations seek to increase awareness and control of the media by key groups, often targeting children and youth. Media Activities and Good Ideas by, with, and for Children (MAGIC) is sponsored by the United Nations Children's Fund. The program "calls on children and young people to learn as much about the media as they can to help them in their choice of media and make sure they benefit from the media" (MAGIC 2004). MAGIC sponsors international conferences and meetings that celebrate children's excellence and creativity in film, video, and Internet programming. Listen Up! Youth Media Network is a coalition of youth media groups, professional media organizations, nonprofit groups, and foundations that help youth be heard in the mass media. Listen Up! partners with youth, creating media in high schools, media art centers, and nonprofits throughout the country. Partner programs include 911 Media Arts Center (Seattle, Washington), Vid-Kid Productions and Teen Vision Productions (White Sulphur Springs, Montana), Appalshop (Whitesburg, Kentucky), and North East School of the Arts (San Antonio, Texas).

Table 11.6

The 10 Basic Principles of Media Literacy Education

1. *Medium.* The form of communication that transmits messages, tells stories, structures learning, and constructs a "reality" about the world.

2. *Media literacy.* An educational approach that seeks to give media users greater freedom and choice by teaching them how to access, analyze, evaluate, and produce media.

3. *Construction of reality.* Media construct our culture, which involves trade-offs. Ask yourself: What are the trade-offs in this media experience? Who produced this media? What kind of reality does this media create? How accurate is this "reality"? What stories are NOT being told and why?

4. *Production techniques.* Media use identifiable production techniques (camera angles, editing, sound effects, colors and symbols, etc.). Ask: What kinds of production techniques does this media employ?

5. *Value messages.* Media contain ideological and value messages. Ask: What kinds of value messages does this media promote?

6. *Commercial motives.* Media are commercial and business interests. Ask: What are the commercial motives behind this media? Who or what paid for this media and why? Who or what owns this media product?

7. *Individual meanings.* Individuals construct their own meaning of media. Ask: What meanings do YOU find in this media? What different meanings might other individuals or groups find?

8. *Emotional transfer.* Commercials and other multimedia experiences operate primarily at an emotional level and are usually designed to transfer the emotion from one symbol or lifestyle onto another. Ask: What emotions does this media tap? What might we consider if we think more deeply about this media?

9. *Pacing.* TV runs at 30 frames per second; movies at 24 frames per second. The conscious mind can process about 8 frames per second; hence, television and movies tend to keep us from conscious analysis and reflection about individual messages and larger industry contexts. Ask: What do you observe about this media upon reflection?

10. *Symbolic rhetoric/techniques of persuasion.* Symbols, flattery, repetition, fear, humor, words, and sexual images are common and effective techniques of media persuasion. Ask: What persuasive techniques is this media using?

Source: Action Coalition for Media Education 2004. Reprinted with permission of ACME. The Action Coalition for Media Education (ACME) is an independently funded coalition that champions critical media literacy education, independent media production, and democratic media reform and justice initiatives. Find out more at http://www.acmecoalition.org.

Main Points

- **Media** are the technological processes facilitating communication, and they have increased to include more broadcast and cable channels and Internet Web sites. Yet, the media have been blamed for directly creating and promoting social problems and have been accused of being a problem themselves through controversial programming and manipulating how we see events.

- Functionalists examine the structural relationship between the media and other social institutions that affect content, including the social and economic conditions of American life and society's beliefs in the nature of men, women, and society. Research has documented the link between viewing media violence and the development of aggression, particularly among children who watch dramatic violence on television and film.

- Conflict theorists believe that the media can be fully understood only when we learn who controls them. The media, serving the interest of interlocking state and corporate powers, frame messages in a way that supports the ruling elite and limits the variety of messages that we read, see, and hear.

- Feminist theorists attempt to understand how the media represent and devalue women and minorities. They examine how the media either use stereotypes disparaging women and minorities or completely exclude them. One of the most important lessons young children learn is expected gender roles.

- The interactionist perspective focuses on the symbols and messages of the media and how the media come to define our "reality." The mass media define what events are newsworthy, using proximity (nearness of events) and deviation (something different from what we are accustomed to) as criteria. The mass media play a large role in shaping public agendas by influencing what people think about and, ultimately, what people consider a social problem.

- The term **digital divide** refers to the gap separating individuals who have access to and understanding of new forms of technology from those who do not. This divide is a symptom of a larger social problem: social inequality based on income, educational attainment, and ethnicity/race. Internet access is not the only issue facing underserved communities. A lack of locally relevant information, a lack of information at a basic literacy level, and English-only content are also barriers.

- Although the media are owned and operated by private organizations, their content is regulated by federal standards enforced by the Federal Communications Commission (FCC). The FCC enforces federal law and commission rules regarding indecency.

- Other social problems include **Internet addiction**, a broad term that covers a variety of behaviors and impulse controls. Internet use can become a problem when it begins to interfere with one's job or social life. Employers have identified Internet abuse as a serious work problem.

- The average child or adolescent watches nearly three hours of television per day. Research confirms the negative impact of television on children and adolescents, including more aggression and physical problems such as health and obesity.

- Public attitudes about the press have been growing less positive for about 20 years. The public believes that news organizations are working primarily for profit and that journalists are motivated by professional ambition; doubts about the political bent of the media and the motives of the multinational corporations that own and control them also lead to distrust.

- Aside from the FCC, numerous organizations have emerged as watch groups monitoring media accuracy and content, including the Project for Excellence in Journalism and the Committee to Protect Journalists.

- Evidence shows that media education can help mitigate the harmful effects of the media. Parents can provide the most effective intervention to reduce the effects of media violence on children, and school-based intervention also seems to help. The goal is **media literacy**, shifting the viewer from the role of passive receiver to active critical receiver.

On Your Own

Log on to the Web-based student study site at www.pineforge.com/leonguerrero2study for interactive quizzes, e-flashcards, journal articles, Community and Policy Guides, a Service Learning Guide, the end-of-chapter Web exercises, and additional Web resources.

Internet and Community Exercises

1. Interview the editorial staff of your school newspaper. Ask the editor how decisions are made regarding the leading stories in the paper. What qualifies as news? Who makes decisions about the content of the paper? About what appears as the lead story on the front page? Is it done by the entire news staff or by a small group?

2. Media watchdog groups such as Fairness and Media in Reporting (FAIR) and Accuracy in the Media (AIM) track major news stories, checking reporters' facts or exposing their biases. Select a current news story and examine how one or both of these groups examines it. Do they reveal any different or new information about the story?

3. Investigate whether there are any youth media literacy organizations in your city or state. The Listen Up! Network (www.listenup.org) identifies current projects by state and region. You may also contact local television or radio stations for additional community contacts. Who is the target audience (at risk youth, ethnic youth, or youth with creative talent) for these programs? What services or programs are provided?

4. Project Censored is a media research group based at Sonoma State University. The group of faculty, students, and community members publishes an annual list of 25 news stories of social significance that have been overlooked, underreported, or self-censored by the major national news media. Go to the Project Censored Web site (www.projectcensored.org) to review the most recent list of news stories. What social problems are identified by these stories? What might be done to promote these stories in mainstream media?

PART IV

Our Social and Physical Worlds

D rug abuse and criminal behavior are usually perceived as personal troubles. Both are considered
deviant behavior, deviating from the normal expectation of not abusing drugs or engaging in criminal
activity. We tend to believe that someone turns to drug use or
criminal activity because of some personal defect, individual
failure, or weakness. Yet both can be defined as public issues,
emerging from the social structure and threatening the quality
of human life.

Drug abuse is a serious personal and social problem. It can kill. It can impair an abuser's ability to effectively function at the workplace, among friends, and at home. Treatment and prevention of drug abuse costs society billions of dollars and diverts needed funds and support from other social problems. Drug activity in a neighborhood can increase violence and decrease personal safety for all residents, even those not directly involved in drugs. The quality of human life, beyond the abuser, is threatened by or damaged by the use of drugs (Goode 2004).

What Does It Mean to Me?

Like drug abuse, criminal activity also has collateral effects, unintended effects on offenders, their families, and their communities (Tonry 2004). What collateral effects of crime can you identify?

Drug abuse and crime are the subject and focus of extensive scholarly research, attempting to understand the extent and origins of both. Both problems also receive global attention, including from governments and public agencies monitoring and combating drug abuse and crime in their part of the world. These issues will be discussed in Chapters 12 and 13.

In the last three chapters of Part IV, we will review social problems that affect our physical and natural worlds—problems related to urbanization, the environment, and war and terrorism. Sociologists acknowledge that our physical and natural environments influence and are influenced themselves by human societies and behavior.

Yet some of you may be wondering why a discussion on urban sprawl or global warming should be included in this textbook. Though they deal with our physical and natural worlds, both have definite human connections—humans cause these problems or experience consequences as a result of them. Environmentalist Paul Hawken explains, "Human activity *is* part of the natural world, in the largest sense, but human activity ignores the means-and-ends, give-and-take factors that are inherent in any maturing ecosystem" (1993:26). The industrial revolution brought great human growth and prosperity at an immense cost to the earth (Hawken 1993). Almost every living system on the planet has lost some capacity to sustain itself. We consume our natural commodities, such as water, air, soil, and minerals, without much thought to their true cost.

However, our understanding of the relationship between our actions and our natural and physical worlds has considerably improved. Former Vice President Al Gore increased public awareness about global warming through his 2006 documentary, *An Inconvenient Truth*. Gore and others from the scientific community have documented how environmental problems are caused by human action (or inaction), our social policies, and priorities. We have also been cautioned about how urbanization and population growth have fueled our consumption of natural resources, particularly in developing parts of the world. Scientists warn us that we are about to run out of these natural commodities leading to irrevocable global damage. The consequences of human action on our urban landscape and environment will be discussed in Chapters 14 and 15.

The terrorist attacks on September 11, 2001, were the result of human action. Since then, the Iraq War and the Bush administration's War on Terror have been part of our nation's psyche, significantly affecting many aspects of our lives—socially, economically, and politically. Our country's response to terrorism has been accused of creating and exacerbating problems here and abroad. War and terrorism as a social problem will be discussed in Chapter 16.

Drug Legacy

◀ Drug abuse is related to many social factors, including hopelessness, poverty, and violence. In her own Brooklyn neighborhood, photographer Brenda Ann Kenneally documented the legacy of drug abuse passed down from parent to child and its effects on their community. While his mom is taking a hit from her crack pipe, Andy "boxes" with the mailboxes in the hallway. The electricity for the family's apartment has been turned off.

▶ Kenneally also documents common violence and self-destructive behavior in her neighborhood. Fay, a crack dealer and user, tries to avoid being hurt by the man she sells crack for. She has smoked all the profits from her drug sales. She has also left her child with one of her customers, hiding him from authorities who would place him in foster care.

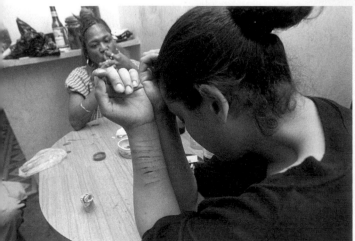

◀ Meanwhile, Lisa reveals the injury she inflicts on herself. Lisa has attempted many times, unsuccessfully, to quit using drugs; out of frustration, she began cutting her left forearm with a knife.

▶ These three women, all family members, are being held in Riker's Island in Queens for drug related charges. In 1987, drug arrests constituted 7 percent of the total of all arrests reported to the FBI. By 2005, drug arrests increased to 13 percent.

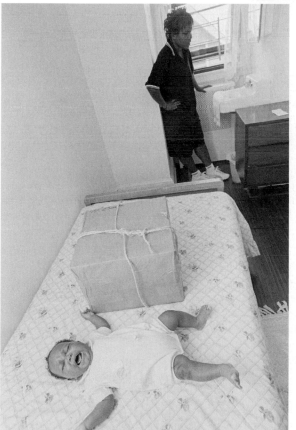

◀ The vicious cycle of drug abuse can be broken, however. One young woman, Moya, learned from her mother, Theresa, to be a cocaine smoker and dealer. Theresa spent eight years in recovery from her own addiction before dying of a weakened heart. At the time, Moya was out of prison on parole. Three weeks after the funeral, Moya was caught dealing again and sent back to prison. She bore a child while in prison but had to find a temporary home for her daughter while she finished her sentence. When released, all the resources Moya had to start over were her baby, named Theresa after her mother, and a single box of belongings.

Moya and Theresa found a temporary home in Hour Children, a convent-run support community. Hour Children operates five residences, offering a safe home environment for formerly incarcerated mothers and their children. As a condition of her stay, Moya is required to get a job or attend school. Perhaps baby Theresa's life may turn out differently from Moya's.

Which do you believe is more effective in breaking the cycle of drug abuse—punishment through incarceration or rehabilitation through community programs like Hour Children?

CHAPTER 12

Alcohol and Drug Abuse

A s a nation, we have been in the War on Drugs for the past 35 years. Some refer to it as a war with "no rules, no boundaries, no end" (PBS 2000). Since the mid-1980s, the United States has adopted a series of aggressive law enforcement strategies and criminal justice policies aimed at reducing and punishing drug abuse (Fellner 2000). Changes in federal law require all sentenced federal offenders to serve at least 87 percent of their court-imposed sentences. Many drug offenders are subject to mandatory minimum sentences based on the type and quantity of drugs involved in their arrest (Scalia 2001). According to the Uniform Crime Report, 1,846,400 drug arrests were made in 2005, an increase from 1,579,600 in 2000 (Bureau of Justice Statistics 2006). Although some consider the increase in drug arrests a good sign, critics charge that mandatory sentencing denies drug users what they really need, access to treatment. Tougher sentencing has failed to decrease the availability of drugs and has failed to reduce illicit drug use. In addition, some argue that the focus on drug-related crimes has distracted law enforcement from monitoring more serious crimes.

There seems to be no argument about the seriousness of the drug problem in the United States and worldwide. According to a 2005 National Survey on Drug Use and Health, 19.7 million Americans ages 12 and older reportedly were current illicit drug users (Substance Abuse and Mental Health Services Administration [SAMHSA] 2005). (Refer to Figure 12.1 for illicit drug use by age and U.S. Data Map 12.1 for the percentage of illicit drug use by state.) Nearly 18 million Americans are alcoholics (National Institute on Alcohol Abuse and Alcoholism [NIAAA] 2004). Globally, more than 76 million individuals have diagnosable drinking problems (World

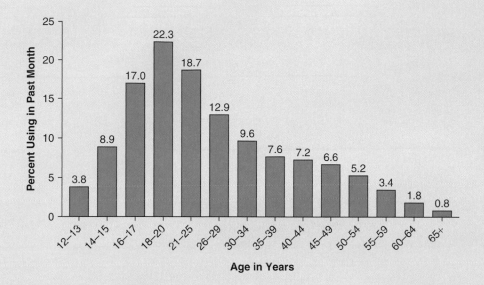

Figure 12.1

Past Month Illicit Drug
Use Among Persons
Aged 12 or Older, by
Age, 2005

Source: SAMHSA 2005.

Health Organization 2004) and about 15 million people have drug use disorders (World Health Organization 2006).

Although we might focus first on one drug user and his or her personal trouble with drugs, it doesn't take long to recognize how drug use affects the user's family and friends, workplace or school, and neighbors and community. Throughout this chapter, we will examine the social problem of drug abuse, reviewing its extent, its social consequences, and our solutions. We begin first with a look at how the sociological perspectives address the problem of drug abuse.

Sociological Perspectives on Drug Abuse

Biological and psychological theories attempt to explain how drug abuse is based in the individual. Both perspectives assume that there is little a person can do to escape from their abuse: Their abuse is genetic or inherited. Abuse may emerge from a biological or chemical predisposition or from a personality or behavioral disorder. Such explanations also have consequences for treatment. Programs focus on the individual, arguing that the abuser needs to be "fixed." Although both perspectives have been important in shaping our understanding of drug abuse, these perspectives cannot explain the social or structural determinants of drug abuse. In this next section, we will examine how sociological perspectives address the problems of drug abuse.

Functionalist Perspective

Functionalists argue that society provides us with norms or guidelines on drug use. Cross-cultural studies reveal that there is variation in the way people expect to behave when they drink. For example, violent behavior is associated with alcohol consumption in the United States, Great Britain, and Australia; yet, drinking behavior is described as "peaceful and harmonious" in Mediterranean and South African countries (Social Issues Research Center 1998).

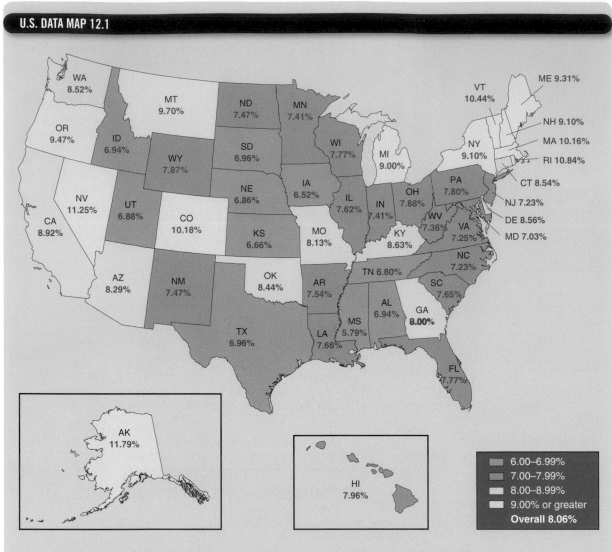

U.S. DATA MAP 12.1

Any Illicit Drug Use in the Past Month, Percentage by State. Annual Averages Based on 2003 and 2004 National Household Survey of Drug Abuse.

Source: SAMHSA 2004.

A set of social norms identifies the appropriate use of drugs and alcohol. The use of prescription drugs, as directed by a physician, is considered acceptable behavior. Prescription drugs alleviate pain, reduce fevers, and curb infections. Alcohol in moderation may be routinely consumed with meals, for celebration, or for health benefits. At least one glass of red wine a day has been shown to reduce one's risk of heart disease.

Yet society also provides norms regarding the excessive use of drugs. For example, college students share the perception that excessive college drinking is a cultural norm (Butler 1993); this perception is enforced by the media and advertisers (Lederman et al. 2003). Aaron Brower (2002) argues that binge drinking is determined by and is a product of the college environment.

Photo 12.1

Social norms identify the appropriate uses of drugs and alcohol. Alcohol consumption with meals or in celebration may be appropriate for adults, but not for young children.

Unlike alcoholics, college students are able to turn their willingness to binge-drink on and off depending on their circumstances (e.g., whether they have to study for an exam).

To explain drug abuse, functionalists rely on Émile Durkheim's theory of **anomie**. According to this perspective, drug abuse can also occur when society is unable to provide guidelines for our behavior. Durkheim believed that under conditions of rapid cultural change, there would be an absence of common social norms and controls, a state he called anomie. If people lack norms to control their behavior, they are likely to pursue self-destructive behaviors such as alcohol abuse (Caetano, Clark, and Tam 1998). During periods when individuals are socially isolated (such as moving to a new neighborhood, experiencing a divorce, or starting a new school year), they may experience high levels of stress or anxiety, which may lead to deviant behaviors, including drug abuse.

Conflict Perspective

Although many drugs can be abused, conflict theorists argue that intentional decisions have been made over which drugs are illegal and which ones are not. Powerful political and business interest groups are able to manipulate our images of drugs and their users.

Heroin, opium, and marijuana were considered legal substances in the late eighteenth and early nineteenth centuries but public opinion and law changed when their use was linked to ethnic minorities and crime. Opium smoking, most associated with Chinese immigrants brought here to work on the railroads, became the subject of intense anti-drug efforts after completion of the railroad system. At the same time, the oral consumption of opium, more widespread among Whites, was never considered as problematic (Reinarman and Levine 1997).

Katherine Beckett (1995) and Dorothy Roberts (1991) describe how women of color have been unfairly targeted in the war on drugs. As crack cocaine use spread throughout the inner cities in the 1980s, prosecutors shifted their attention to drug use among pregnant women, making drug and alcohol abuse during pregnancy a crime. The approach treated pregnant drug users as criminals and was "aimed at punishing rather than empowering women who use drugs during their pregnancy" (Beckett 1995:589). Beckett explains, "Prosecutions of women for prenatal conduct thus create a gender specific system of punishment and obscure the fact that male behavior, socio-economic conditions, and environmental pollutants may also affect fetal health" (1995:588).

Roberts (1991) argues that poor Black women were the primary targets for prosecutors. Research indicates that African American women are about 10 times more likely than are other women to be reported to civil authorities for drug use. Public health facilities and private doctors are more inclined to turn in pregnant Black women than pregnant White women who use drugs. Are Black women being prosecuted for their drug use or for something else? Roberts states, "Society is much more willing to condone the punishment of poor women of color who fail to meet the middle-class ideal of motherhood" (1991:1436).

Feminist Perspective

Theorists and practitioners in the field of alcohol and drug abuse have ignored the experiences unique to women, ethnic groups, gay and lesbian populations, and other marginalized groups. Women face unique social stigmatization as a result of their drug use and may also experience discrimination as they attempt to receive treatment (Drug Policy Alliance 2003b). The scientific literature did not address women's addiction until the 1970s.

Specifically, there has been a lack of sensitivity to the range of drug abuse experiences, beyond the male or White perspective. Early prevention and treatment models treated female abusers no differently than men were treated. However, there is increasing recognition of the importance of gender-specific and gender-sensitive treatment models, including the development of separate women's treatment programs. Female users have a variety of different treatment and psychosocial needs, influenced by their backgrounds, experiences, and drug problems. For example, single career-oriented women without children will have different treatment needs and priorities than will single mothers or married mothers (National Clearinghouse for Alcohol and Drug Information [NCADI] 2003a).

Gail Unterberger (1989) offers a feminist revision of the traditional 12-step statement used by Alcoholics Anonymous. As originally written, the 12 steps send a negative message for women, reinforcing feelings of powerlessness and hopelessness during recovery. Unterberger believes that alcoholic women are more likely to suffer from depression than their male counterparts are, and unlike men, women alcoholics may turn their anger on themselves rather than others. Unterberger's revised 12-step statement is presented in Table 12.1.

Interactionist Perspective

Sociologists Edwin Sutherland and Howard Becker argue that deviant behavior, such as drug abuse, is learned through others. Sutherland (1939) proposed the theory of **differential association** to explain how we learn specific behaviors and norms from the groups we have contact with. Deviance, explained Sutherland, is learned from people who engage

---

Table 12.1

The 12 Steps for Women Alcoholics

1. We have a drinking problem that once had us.

2. We realize that we need to turn to others for help.

3. We turn to our community of sisters and our spiritual resources to validate ourselves as worthwhile people, capable of creativity, care, and responsibility.

4. We have taken a hard look at our patriarchal society and acknowledge those ways in which we have participated in our own oppression, particularly the ways we have devalued or escaped from our own feelings and needs for community affirmation.

5. We realize that our high expectations for ourselves have led us either to avoid responsibility and/or to overinvest ourselves in others' needs. We ask our sisters to help us discern how and when this happens.

6. Life can be wondrous or ordinary, enjoyable or traumatic, danced with or fought with, and survived. In our community we seek to live in the present with its wonder and hope.

7. The more we value ourselves, the more we can trust others and accept how that helps us. We are discerning and caring.

8. We affirm our gifts and strengths and acknowledge our weaknesses. We are especially aware of those who depend on us and our influence on them.

9. We will discuss our illness with our children, family, friends, and colleagues. We will make it clear to them (particularly our children) that what our alcoholism caused in the past was not their fault.

10. As we are learning to trust our feelings and perceptions, we will continue to check them carefully with our community, which we will ask to help us discern the problems we may not yet be aware of. We celebrate our progress toward wholeness individually and in community.

11. Drawing upon the resources of our faith, we affirm our competence and confidence. We seek to follow through on our positive convictions with the support of our community and the love of God.

12. Having had a spiritual awakening as a result of these steps, we are more able to draw upon the wisdom inherent in us, knowing we are competent women who have much to offer others.

Source: Unterberger 1989.

---

in deviant behavior. In his study, "Becoming a Marijuana User," Becker (1963) demonstrated how a novice user is introduced to smoking marijuana by more experienced users. Learning is the key in Becker's study:

> No one becomes a user without (1) learning to smoke the drug in a way which will produce real effects; (2) learning to recognize the effects and connect them with drug use … ; and (3) learning to enjoy the sensations he perceives. (1963:58)

This perspective also addresses how individuals or groups are labeled "abusers" and how society responds to them. For example, consider alcohol abuse among the Native American population. Alcohol abuse and alcoholism are leading causes of mortality among American Indians, and there are disproportionately higher rates of alcohol-related crimes among American Indians. Yet, Malcolm Holmes and Judith Antell (2001) argue that alcohol abuse and its related problems are not entirely objective phenomena; they also involve interpretation and stigmatization of deviant behavior. One persistent societal myth maintains that as a group, American Indians have problems handling alcohol. However, research indicates that factors such as demography (a young population) and geography

Table 12.2

Summary of Sociological Perspectives: Drug Abuse

	Functional	Conflict/Feminist	Interactionist
Explanation of drug abuse	Drug abuse is likely to occur when society is unable to control or regulate our behavior.	Powerful groups decide which drugs are illegal. Certain social groups are singled out for their drug abuse. There has been a lack of sensitivity to the range of drug abuse experiences.	Drug abuse is learned through interaction with others. The perspective also focuses on society's reaction to drug abuse, noting that certain individuals are more likely to be labeled as drug users than others.
Questions asked about drug abuse	What rules exist to control or encourage drug abuse? Are some groups or individuals more vulnerable to drug abuse than others?	What groups are able to enforce their definitions about the legality or illegality of drug use? How are they able to enforce their definitions? How are the experiences of women and minority drug users different from those of White males?	How is drug abuse learned through interaction? How are drug users labeled by society? Why are specific groups targeted?

(rural Western environment) may explain high rates of alcohol-related problems in Indian populations.

The authors highlight the considerable variation in drinking patterns within and between tribal communities; in other words, not all American Indians have drinking problems. The social construction of the "drunken Indian" stereotype links alcohol abuse to the perceived "weaker" cultural and individual characteristics of American Indians. Holmes and Antell explain, "The persistence of such myths in the symbolic-moral universe of the dominant White culture, despite evidence to the contrary, suggests that alcohol use by American Indians still documents allegations of weak will and moral degeneracy" (2001:154). For a summary of these four sociological perspectives, see Table 12.2.

What Is Drug Abuse?

Drug abuse is the use of any drug or medication for a reason other than the one it was intended to serve or in a manner or in quantities other than directed, which can lead to clinically significant impairment or distress. **Drug addiction** refers to physical or psychological dependence on the drug or medication. Although many drugs can be abused, five drugs will be reviewed in the following section: alcohol, nicotine, marijuana, methamphetamine, and cocaine. Marijuana, methamphetamine, and cocaine are considered illicit (illegal) drugs. Most of the information presented in this section is based on data from the National Institute on Drug Abuse (NIDA) and the Office of National Drug Control Policy (ONDCP). For more information, log on to *Study Site Chapter 12.*

Alcohol

Alcohol is the most abused drug in the United States. Although the consumption of alcohol by itself is not a social problem, the continuous and excessive use of alcohol can become problematic. Four symptoms are associated with alcohol dependence or **alcoholism**: craving (a strong need to drink), loss of control (not being able to stop drinking once drinking begins), physical dependence (experiencing withdrawal symptoms), and tolerance (the need to drink greater amounts of alcohol to get "high") (NIAAA 2003c). Alcohol use is related to a wide range of adverse health and social consequences, both acute (traffic deaths or other injuries) and chronic (stroke, alcohol dependence, liver damage) (NIAAA 2005).

Current drinking (12 or more drinks in the past year) and heavy drinking (five drinks on a single day at least once a month for adults) among adults is highest for American Indians and Alaska Natives, followed by Native Hawaiians. Prevalence of deaths from chronic liver disease and cirrhosis is about four times higher, and fatal car accidents caused by alcohol three times higher, among American Indians and Alaska Natives than among the rest of the U.S. population. Adult drinking is lowest among Asian Americans and Pacific Islanders, but alcohol use is increasing significantly in this group.

Among adolescent minorities, African Americans have the lowest rates of drinking and the lowest frequency of being drunk. Hispanic adolescents have the highest rates of heavy drinking, followed by White adolescents. Decline in alcohol abuse with increased age is called "aging out" or simply part of the maturation process. Although studies suggest that White adolescents drink alcohol more heavily and frequently than other ethnic/racial groups do, White adolescents are also more likely to age out. During early to middle adulthood, the frequency of heavy drinking stabilizes among Whites, increases among African Americans, and declines but remains high for Hispanics (Caetano and Kaskutas 1995; Chen and Kandel 1995).

Alcohol researchers have begun to identify the importance of individual attributes, cultural factors, and structural factors in minority drinking. Studies suggest that ethnic/racial groups have different sets of norms and values regulating drinking. For example, some groups exhibit low rates of problem drinking because their culture associates the use of alcohol primarily with eating, social occasions, or rituals (Herd and Grube 1996). However, other ethnic/racial groups may consider drinking as an activity separate from eating or ritual celebrations, leading to higher rates of problem drinking. Researchers have attributed alcoholism among ethnic minorities to three different sources of stress: **acculturative stress**, experienced by most immigrants who are faced with leaving their homeland and adapting to a new country; **socioeconomic stress**, experienced by ethnic minorities who feel disempowered because of social and economic inequalities in U.S. society; and **minority stress**, which refers to the tension that minorities encounter because of racism (Caetano et al. 1998).

Although social class, occupational and social roles, and family history of alcohol all play a role in determining the drinking patterns of people in general, specific factors put women particularly at risk (Collins and McNair 2003). Research indicates that a woman's risk for drinking increases with the experience of negative affective states, such as depression (Hesselbrock and Hesselbrock 1997) or loneliness, and negative life events, such as physical or sexual abuse during childhood or adulthood (Wilsnack et al. 1997). Other factors decrease women's chances of developing alcohol problems. Traditionally, women are socialized to abstain from alcohol use or to drink less than men (Filmore et al. 1997). Women who do not participate in the labor force may have less access to alcohol than men do (Wilsnack and Wilsnack 1992), and women's roles as wife and mother may also discourage alcohol intake (Leonard and Rothbard 1999).

Research confirms how drinking accelerates during adolescence, peaking at young adulthood, and then decelerating to moderate levels in adulthood (Muthen and Muthen 2000). People who begin drinking before age 15 are four times more likely to develop alcohol

Taking a World View

The Scottish Drinking Problem

Some have noted, "Scotland has a special relationship with alcohol" (Ritson 2002). Scotland is known for its Scotch whisky, a national beverage that is more than just a drink or a means of getting drunk. According to Sharon MacDonald, whiskey is "a symbolic distillation of many images of Scottishness, especially hospitality, camaraderie, joviality, and masculinity" (1994:125).

Alcohol abuse in Scotland is increasing because of excessive drinking levels among adults and the frequency and level of drinking among teenagers. In recent studies, 27 percent of Scots men and 14 percent of women between 16 and 64 years old were drinking more than the recommended weekly limits of less than 21 units for men and 14 units for women (a unit of alcohol is eight grams by weight) (Alcohol Information Scotland 2007). Younger Scots, between the ages of 16 and 24, were more likely to exceed the weekly limits—62 percent of males and 56 percent of females. Cirrhosis mortality rates in Scotland were among the highest in Western Europe. In 2002, the rate per 100,000 per year was 45.2 for men and 19.9 for women. Between 1987 and 1991 and between 1997 and 2001, cirrhosis mortality increased among Scottish men by 69 percent and increased among women by 46 percent (Leon and McCambridge 2006).

Men's and women's drinking are not viewed as equally problematic by Highland residents or the Scottish population in general. According to MacDonald (1994), in the Highlands, although more men abuse alcohol, their drinking is viewed as a lesser problem than that of women who drink. Drunkenness is accepted, even expected, of Scottish males, but a drunken woman is considered unrespectable, slovenly, and loose. The drunk Scottish woman is a pathetic figure rather than the humorous figure of a drunk Scot male. It is likely, says MacDonald, that the number of women who abuse alcohol is underestimated because of the added stigma associated with women's drinking. In her research, MacDonald was told of female alcoholics in "hushed, conspiratorial" tones. Women were desperate to conceal their drinking problem. Women's drinking was more likely to be confined to home than men's drinking was (MacDonald 1994).

In January 2002, the Scottish Government launched a "Plan for Action on Alcohol Problems," a national and local program aimed at reducing alcohol-related harm (Ritson 2002). The plan focused on correcting harmful drinking patterns and influencing the habits of children and young people through prevention and education programs. The plan promised to achieve a "cultural change by an immediate investment [in] a national communication strategy that will challenge current stereotypes of binge drinking" (Ritson 2002:218).

---

dependence at some time in their lives compared with people who have their first drink at age 20 or older (NIAAA 2003b). O'Malley, Johnston, and Bachman (1998) report that adolescents who use alcohol are at higher risk for social, medical, and legal problems, such as poor school performance; interpersonal problems with friends, family, and others; physical and psychological impairment; drunk driving; and death. The rate of fatal crashes related to alcohol among drivers ages 16 to 20 is more than twice the rate among drivers age 21 or older (NIAAA 2003b). The most common alcohol-related problem reported by adolescent drinkers is that alcohol use causes them to behave in ways that they later regret (O'Malley et al. 1998). Underage use of alcohol is more likely to kill young people than all illegal drugs combined (NIAAA 2003b).

Tobacco and Nicotine

Tobacco is the world's number one drug problem, killing more people than all other drugs combined (Goode 2004). According to the World Health Organization (2007a), if current smoking patterns continue, by 2010 10 million people per year worldwide will die of diseases caused by cigarette smoking. Cigarette smoking is the most prevalent form of nicotine addiction, the most frequently used addictive drug in the United States (NCADI 2003b). Nicotine is both a stimulant and a sedative to the central nervous system. An average cigarette contains about 10 milligrams of nicotine. Through inhaling the cigarette smoke, the smoker takes in 1 to 2 milligrams of nicotine per cigarette. In 2004, 29 percent or 70.3 million Americans reported current use of a tobacco product (NIDA 2006).

In the United States, the prevalence of smoking is highest among Native Americans/Alaska Natives (33.4 percent), followed by Whites (22.2 percent), Blacks (20.2 percent), Hispanics (15 percent), and Asians and Pacific Islanders (11.3 percent). In 2004, 22 percent of high school students were current smokers (American Lung Association 2006).

Cigarette smoking is the most important preventable cause of cancer in the United States. It has been linked to most cases of lung cancer in Europe and the United States (Crispo et al. 2004). Smoking has also been linked to other lung diseases, such as chronic bronchitis and emphysema, and to cancers of the mouth, stomach, kidney, bladder, cervix, pancreas, and larynx. The overall death rates from cancer are twice as high among smokers as nonsmokers (NIDA 1998). It is estimated that 438,000 annual deaths are attributable to cigarette smoking in the United States (American Lung Association 2006).

Passive or secondhand smoke is a major source of indoor air contaminants. Nonsmokers exposed to secondhand smoke at home or work increase their risk of developing heart disease by 25 to 30 percent and lung cancer by 20 to 30 percent (NIDA 2006). Secondhand smoke is estimated to cause about 3,000 lung cancer deaths per year and may contribute to as many as 40,000 deaths related to cardiovascular disease in the United States (NIDA 1998). More than 79,000 adults die each year as a result of passive smoking in the European Union countries (European Lung Association 2006).

Despite the persistent public health message that smoking is bad for your health, smoking among teenagers has been on the rise since 1991 (Lewinsohn et al. 2000). In a study comparing adolescent smokers with nonsmokers, adolescent smokers were found to have more stressful environments, more academic problems, and poorer coping skills than nonsmokers have. Adolescent smoking has also been associated with a number of environmental factors, such as disruptive home environment, parental and peer smoking, low social support from family and friends, conflict with parents, and stressful life events (Lewinsohn et al. 2000).

Data indicate that the use of smokeless chewing tobacco products (referred to as "snuff," "dip," or "chew") occurs at a significantly younger age than cigarette smoking does. Smokeless tobacco products are consumed orally, with packets of the tobacco tucked in a front lip or cheek. An average size chew kept in the mouth for 30 minutes provides the same amount of nicotine as three cigarettes (National Cancer Institute 2003). Smokeless tobacco may cause permanent gum recession, mouth sores, lesions, and cancers of the mouth and throat. Most adolescent users are White, male, and from a rural area (Newman 1999). Experimentation with chewing tobacco has been associated with personal factors (such as risk taking, physical fighting, and low satisfaction in school) and weak family structure or support (e.g., one-parent family). Regular use of chew has been linked to adult modeling, peer influences, and certain sports teams' membership (Campbell-Grossman, Hudson, and Fleck 2004).

Marijuana

Marijuana is the most commonly used illicit drug, widely used by adolescents and young adults (NIDA 2002a), with an estimated 141 million people or 2.45 percent of the world's population consuming it (UNODC 2006a). It is a favorite drug among youth and adolescents, with use (at least once in a lifetime) estimated as high as 37 percent in some countries (UNODC 2006a). The major active chemical in marijuana is THC or delta-9-tertrahydrocannainol, which causes the mind-altering effects of the drug. THC is also the main active ingredient in oral medications used to treat nausea in chemotherapy patients and to stimulate appetite in AIDS patients (ONDCP 2003a).

Among Americans 12 years or older, approximately 40 percent (more than 97 million) are estimated to have used the drug at least once in their lifetime (ONDCP 2006a). According to the Centers for Disease Control, 38.4 percent of surveyed high school students and 49 percent of college students reported lifetime use of marijuana. Longitudinal data show increases in marijuana use during the 1960s and 1970s, declines in the 1980s, with increasing use since the 1990s (NIDA 2002a).

Acute marijuana use can impair short-term memory, judgment, and other cognitive functions as well as a person's coordination and balance, and it can increase heart rate. Chronic abuse of the drug can lead to addiction, as well as increased risk of chronic cough, bronchitis, or emphysema. Addictive use of the drug may interfere with family, school, or work activities. Smoking marijuana increases the risk of lung cancer and cancer in other parts of the respiratory tract more than smoking tobacco does (NIDA 2002a). Marijuana smoke contains 50 percent to 70 percent more carcinogenic hydrocarbons than tobacco smoke does (ONDCP 2006a). Because marijuana users inhale more deeply and hold their breath longer than cigarette smokers do, they are exposed to more carcinogenic smoke than cigarette smokers are. In 2004, marijuana use was a contributing factor in more than 200,000 emergency room visits (ONDCP 2006a).

Methamphetamine

Methamphetamine or "meth" is a highly addictive central nervous system stimulant that can be injected, snorted, smoked, or ingested orally. A derivative of amphetamine, methamphetamine was therapeutically used in the 1930s to treat asthma and narcolepsy (sleeping disorder) (Pennell et al. 1999). It is the most prevalent synthetic drug manufactured in the United States. The drug is commonly referred to as "speed," "crystal," "crank," "go," and "ice" (a smokable form). More than 10 million people have tried methamphetamine at least once in their lifetime (ONDCP 2006b).

Chronic methamphetamine use can cause violent behavior, anxiety, confusion, and insomnia. Users may also exhibit psychotic delusions, including homicidal or suicidal thoughts. Long-term use of the drug can lead to brain damage, similar to damage associated with Alzheimer's disease, stroke, or epilepsy (ONDCP 2006b).

The increase in methamphetamine use has been attributed to the ease of manufacturing the drug and to its highly addictive nature (ONDCP 2003b). In the early 1990s, the primary sources of methamphetamines were super laboratories in California and Mexico. Super labs are able to produce 10 pounds of meth in a 24-hour production cycle.

Certain aspects of the manufacturing and use of methamphetamines, compared with other illegal drugs, have different consequences. Unlike other drugs, meth is easy to make with common chemicals that are easy to obtain (Pennell et al. 1999). The drug can be manufactured illicitly in laboratories set up in homes, motels, trailers, cars, or public storage lockers operated by independent "cooks." Meth produced in these labs is primarily for personal use or limited distribution. Of the 32 chemicals that are used to make or "cook" meth, about one third of the chemicals are toxic (Snell 2001). The waste and residue remaining from meth cooking can contaminate water supplies, soil, and air, causing danger to people, animals, and plant life in the area. Many of the chemicals are explosive, flammable, and corrosive.

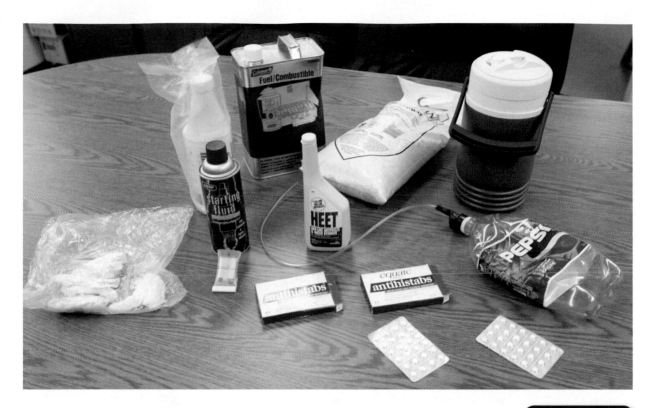

Photo 12.2

Items used in the production of methamphetamine are displayed on a table. The increase in methamphetamine use has been attributed to how easy the drug is to make with common chemicals and household items.

Production in these small-scale labs peaked in the late 1990s and early 2000; in 2006, however, DEA officials reported that the number of small lab seizures had declined in recent years. The DEA credits efforts by states and drug store chains to limit access to the ingredients used to make the drug. Meanwhile, meth trafficking from California and Mexico has been increasing.

Cocaine

Cocaine is one of the oldest known drugs, derived from the leaves of the coca bush. Cocaine is listed as a Schedule II drug, a drug with a high potential for abuse. The drug is a strong central nervous system stimulant and can be snorted, smoked, or injected. Crack is the street name for cocaine that has been processed from cocaine hydrochloride into a smokable substance. Because crack is smoked, the user experiences a high in less than 10 seconds.

Prevalence rates for lifetime use of cocaine are between 1 and 3 percent in developed countries. In the United States, 2.4 million Americans were current cocaine users in 2005 (World Health Organization 2007b). Adults 18 to 25 years old have a higher rate of use than any other age group. Overall, men have higher rates of use than women do. Cocaine initiation is more likely to occur among adults rather than youths under 18. In 1968, the average age of a new user was 18.6 years; it was 23.8 years in 1990 and 21 years from 1995 to 2001 (SAMHSA 2003a). For 2005, the average age of a new user was reported at 19.7 years of age (NIDA 2007).

Some of the most common complications of the drug include cardiovascular disease (disturbances in heart rhythm and heart attacks), respiratory effects (chest pain and respiratory failure), neurological effects (strokes, seizures), and gastrointestinal complications (NIDA 2002b).

The full effect of prenatal drug exposure is not completely known. Babies born to mothers who abuse cocaine are often premature, have low birth weights, and are often shorter in length. It has been predicted that "crack babies" will suffer severe irreversible damage. However, it appears that most crack babies recover, although there is indication of some learning deficits, such as the child's inability to block distractions or to concentrate for long periods of time (NIDA 2002c).

The Problems of Drug Abuse

The Relationship With Crime and Violence

National Crime Victim Surveys indicate that the rate of alcohol-involved violent crimes (crimes in which the offender has been drinking, as perceived by victims) has decreased 34 percent from 1993 to 1998, a shift from 2.1 million incidents in 1993 to 1.4 million in 1998. Alcohol abuse is more often suspected in crimes than abuse of any other drug. However, the number of violent offenses in which the offender was believed to be using other drugs (illicit drugs) increased 19 percent during the same period (443,426 in 1993 to 526,522 in 1998). For 1998, 41 percent of probationers, 41 percent of jail inmates, 38 percent of state prisoners, and 20 percent of federal prisoners reported that they were drinking at the time of the offenses for which they were convicted. Nearly one half of the violent victimizations that involved alcohol occurred in a residence, with more than 20 percent occurring in the victim's home. About one-third of the alcohol-involved victimizations resulted in an injury to a victim. It has been estimated that the loss per victim of alcohol-involved violence was about $1,016 or an estimated annual loss of $400 million per year (Greenfeld and Henneberg 2001).

Drinking- and alcohol-related problems have been associated with intimate partner violence among White, Black, and Hispanic couples. This does not mean that violence can only occur when drinking is involved or that alcohol is the prime cause of the violence. Rather, some people may consciously use alcohol as an excuse for violent behavior. Also, alcohol may be related to violence because heavier drinking and violence have common predictors, such as impulsive personalities (Caetano, Schafer, and Cunradi 2001).

Alcohol use has also been associated with child abuse as both a cause and a consequence (Widom and Hiller-Sturmhofel 2001). Parental alcohol abuse may increase a child's risk of experiencing physical or sexual abuse, either by a family member or another person. Parental alcohol abuse may also lead to child neglect. Studies indicate that girls who were abused or neglected are more likely to have alcohol problems as adults than are other women (Widom and Hiller-Sturmhofel 2001).

Drug Use at the Workplace

Employers have always been concerned about the impact of substance abuse on their workers and their businesses because drug use may undermine employee productivity, safety, and health (Frone 2004). About 8 percent of full-time workers between the ages of 18 to 64 years used an illicit drug in the past month (Larson et al. 2007). Illicit drug use was higher among White workers, male workers, workers with less than a college degree, and workers with lower family incomes (Larson et al. 2007). Drug abuse cost American businesses $81 billion in lost productivity, $37 billion because of premature death, and $44 billion because of illness in 2002. Alcohol abuse contributed to about 86 percent of the costs (U.S. Department of Labor 2003).

In their examination of occupational risk factors for drug abuse, Scott MacDonald, Samantha Wells, and T. Cameron Wild (1999) found that problem drinking or drug use was linked to the quality and organization of work, drinking subcultures at work, and the safety of the workplace. Respondents reporting alcohol problems were more likely to have jobs involving repetitive tasks and dangerous working conditions. Respondents with alcohol problems were also more likely to drink with coworkers and experience some social pressure to drink. The same pattern was also true for workers with drug problems: They considered their jobs "boring" or repetitive, they identified their job as dangerous, they experienced stress at work, or they were likely to be part of a drinking subculture at work. Among all factors they identified, the presence of a drinking subculture at work was the strongest risk factor for alcohol and drug abuse.

By occupation, the highest rates of current illicit drug use and heavy drinking were reported by food preparation workers, waiters/waitresses, and bartenders (19 percent); construction workers (14 percent); service occupations (13 percent); and transportation and material moving workers (10 percent) (U.S. Department of Labor 2003). Among employed adults, White, non-Hispanic males between the ages of 18 and 25 who have less than a high school education are likely to report the highest rates of heavy drinking and illicit drug use (U.S. Department of Labor 2003).

Problem Drinking Among Teens and Young Adults

For 2005, the National Survey on Drug Use and Health (NSDUH) reported that the highest prevalence of binge drinking (drinking five or more drinks within a few hours or within one sitting) was for young adults ages 18 to 25 (42 percent), with the peak rate occurring at age 21 (50 percent). Heavy drinking (five or more drinks on the same occasion on at least five different days in the past 30 days) was reported by 15.3 percent of those 18 to 25 years of age. Binge and heavy drinking were lowest for people age 65 or older, with reported rates of 8.3 percent and 1.7 percent, respectively (SAMHSA 2005). See Figure 12.2 for a summary of 2005 alcohol use by age.

An international comparative study on drinking trends among 15- and 16-year-olds revealed that teens in the United Kingdom were among those most likely to drink heavily and to experience intoxication. In 2003, 52 percent of boys and 56 percent of girls reported binge drinking in the past 30 days. From 1995 to 2003, although there was no significant increase in the proportion of boys engaging in binge drinking, researchers observed a sharp increase in the proportion of girls binge drinking since 1999. The rise in binge drinking among girls is consistent with recent reports documenting the increase in heavy episodic or binge drinking among young women in Britain. The majority of surveyed teens, 75 percent, reported that they had at some time "been drunk" in 2003. Researchers concluded that U.K. teenagers drink in ways that are potentially harmful and that binge drinking among U.K. teens is a matter of real concern (Plant, Miller, and Plant 2005).

By the time they reach the eighth grade, nearly 50 percent of U.S. adolescents report having had at least one drink, and more than 20 percent report having been drunk (NIAAA 2003c). Underage drinkers account for nearly 20 percent of the alcohol consumed in the United States (Tanner 2003). In 2005, among youth ages 12 to 17, 16.5 percent used alcohol in the month before the National Household Survey on Drug Use, lower than the rate of youth alcohol use reported in 2004, 17.6 percent. Among all youth, 9.9 percent were binge drinkers, and 2.4 percent were heavy drinkers (SAMHSA 2005). Adolescent binge drinking is one of the stronger predictors of bingeing through college (Wechsler et al. 1995).

Binge drinking among college students has been called a major U.S. public health concern (Clapp, Shillington, and Segars 2000). (Refer to Figure 12.3 for a comparison of heavy drinking among those enrolled full time in college and those who are not. Drinking rates are consistently higher for those enrolled full time.) Henry Wechsler (1996) reported results from the 1996 Harvard School of Public Health College Alcohol Study, highlighting how

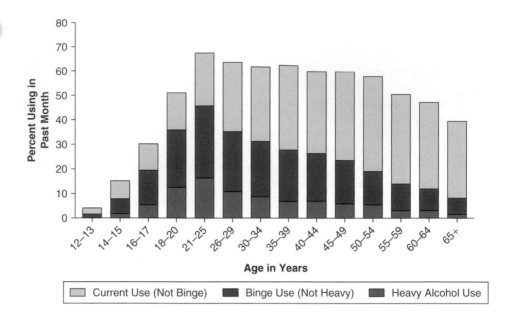

Figure 12.2

Current, Binge, and Heavy Alcohol Use Among Persons Aged 12 or Older, by Age, 2005

Source: SAMHSA 2005.

binge drinking has become widespread among college students. In the Wechsler study, binge drinking was defined as five or more drinks in a row one or more times during a two-week period for men and four or more drinks in a row one or more times during a two-week period for women. The author explains that men, students younger than 24, fraternity and sorority residents, Whites, students in athletics, and students who socialize more are most likely to binge drink. On average, students who engaged in high-risk behaviors such as illicit drug use, unsafe sexual activity, and cigarette smoking were more likely to be binge drinkers. In contrast, students who were involved in community service, the arts, or studying were less likely to be binge drinkers (Wechsler 1996).

Access to alcohol is also related to problem drinking. Weitzman et al. (2003) reported a positive relationship between alcohol outlet density (number of bars and liquor stores near campus) and frequent drinking (drinking on 10 or more occasions in the past 30 days), heavy drinking (five or more drinks at an off-campus party), and drinking problems (self-reported).

The Task Force of the National Advisory Council on Alcohol Abuse and Alcoholism (2002) concluded that 1,400 college students between the ages of 18 and 24 die each year from alcohol-related unintentional injuries, including motor vehicle crashes. About half a million students between the ages of 18 and 24 are unintentionally injured while under the influence of alcohol, and more than 600,000 students are assaulted by another student who has been drinking. In addition, the task force reports that 25 percent of college students report academic consequences (poor grades, poor performance, missing classes) as a result of their drinking, and more than 150,000 develop an alcohol-related health problem. Based on self-reports about their drinking, 31 percent of college students met the criteria for alcohol abuse, and 6 percent met the criteria for alcohol dependence (Task Force of the National Advisory Council on Alcohol Abuse and Alcoholism 2002).

Brower (2002) explains that there is no evidence that drinking in college leads to later-life alcoholism or long-term alcohol abuse. He writes, "Real life is a strong disincentive for the kind of binge drinking that college students do" (p. 255). He suggests using the term **episodic high-risk drinking** to describe more accurately how college students drink: infrequently drinking a large quantity of alcohol in a short period.

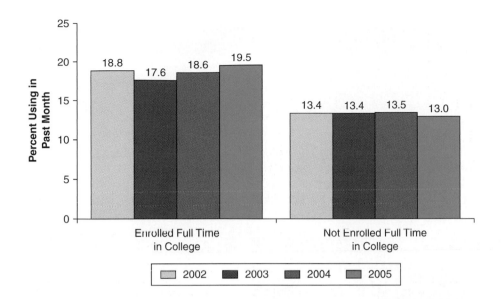

Figure 12.3

Heavy Alcohol
Use Among Adults
Aged 18 to 22, by
College Enrollment:
2002–2005

Source: SAMHSA 2005.

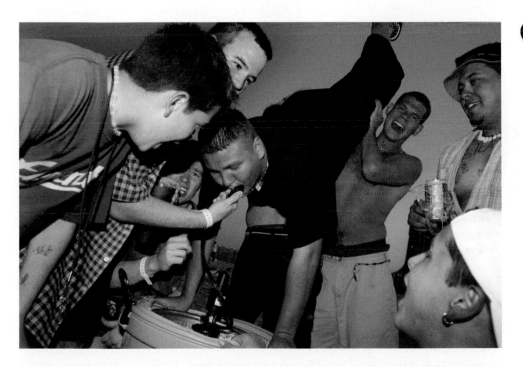

Photo 12.3

Binge drinking among
college students has
been called a major
U.S. public health
concern. According
to the Task Force of
the National Advisory
Council on Alcohol
Abuse and Alcoholism
(2002), 1,400 college
students between
the ages of 18 and
24 die each year
from alcohol-related
unintentional injuries,
including motor
vehicle crashes.

What Does It Mean to Me?

What is the drinking policy on your campus? What educational or service programs are provided for students who abuse alcohol?

In Focus

Alcopops

Flavored alcoholic beverages, sugary fruit-flavored alcoholic beverages such as Mike's Hard Lemonade, Bacardi Silver, and Zima, have become popular with young drinkers. Though the alcohol content must be included in the packaging, some packaging makes these beverages appear more like non-alcoholic soft drinks or energy drinks rather than an alcoholic beverage containing 5 to 7 percent alcohol per volume. State lawmakers and parents have increasingly grown concerned about the popularity of these drinks, believing that these beverages contribute to under-age drinking (Marshall 2007), referring to these beverages as "gateway" drugs (leading youth to more traditional alcohol beverages).

Photo 12.4

Girls are more likely to consume alcopops than boys. The American Medical Association's 2004 survey revealed that a third of all teen girls older than age 12 have tried alcopops. Pictured here are two girls drinking alcopops in a club in Great Britain.

A 2001 national poll revealed that 41 percent of all teens ages 14 to 18 have tried an alcopop, with twice as many 14- to 16-year-olds preferring them to beer or mixed drinks. When asked why they would choose these beverages over beer, wine, or liquor, teens reported that they liked the sweet taste of the drinks and the disguised taste of alcohol. Most surveyed teens believed the products were marketed to their age group (Alcohol Policies Project 2001).

However, there appears to be more concern about the growing consumption of these beverages among young girls. The American Medical Association released 2004 poll results that indicated that a third of all teen girls older than age 12 have tried alcopops (American Medical Association 2004). As part of an effort to urge Governor Arnold Schwarzenegger to reclassify these beverages as hard liquor, Cinthya Luius, a high school senior and member of the San Diego Youth Council, explained, "Kids all over the state and the nation call these products 'cheerleader beer' and 'girlie beer' because they are so popular with underage girls—their sweet taste is designed to appeal to young drinkers." Schwarzenegger vetoed the bill in 2005, but two new bills were introduced in February 2007 (Marshall 2007).

In addition to California, several other states such as Arkansas, Illinois, and Nebraska have considered reclassifying these beverages as hard liquor. Critics argue that these beverages are no different than beer, which contains 4 to 6 percent alcohol per volume. Maine has already reclassified these beverages as hard liquor.

The Increase in Club Drugs

MDMA (3–4 methylenedioxymethamphetamine) is a synthetic psychoactive drug with stimulant and hallucinogenic properties. The pill—popularly known as Ecstasy, Adam, X, XTC, hug, beans, and love drug—first gained popularity at dance clubs, raves, and college scenes.

The 2005 National Survey on Drug Use and Health reported that 4.7 percent or 11.5 million Americans have used MDMA at least once during their lifetime (ONDCP, 2006c). MDMA is usually taken in pill form at the cost of about $25 per tablet, but the drug can also be snorted, injected, or used in a suppository.

Rohypnol, GHB, and Ketamine are other drugs commonly used in club and rave scenes. All three are also known as "date rape" drugs. Previously confined to club or rave subculture, Ecstasy has become a mainstream drug (NIDA 2003a), second only to marijuana as the most frequently used illicit drug among young adults (Johnston, O'Malley, and Bachman 2001).

From 2002 to 2004, a public service announcement sponsored by the Partnership for a Drug Free America featured Jim and Elsa Heird. The Nevada couple's 21-year-old daughter, Danielle, took Ecstasy on three occasions; the last time it resulted in her death. After taking one or one and a half pills, Danielle began to feel ill and decided to stay home to rest. A few hours later when her friends came home, they found her dead. There were no other controlled substances or alcohol in her body at the time of her death (Vaughn 2002). Ecstasy-related deaths like Danielle Heird's are rare. During the third and fourth quarters of 2003, MDMA use was reported in more than 2,000 drug abuse–related emergency room cases (ONDCP 2006d).

Even when they are not fatal, Ecstasy and other related drugs, known collectively as methylated amphetamines, are not harmless drugs. Ecstasy produces an intense release of serotonin in a user's brain, which can cause irreparable damage to the brain and memory functions. Research indicates that long-term brain damage, especially to the parts of the brain critical to thought and memory, may result from its use. Users may also experience psychological difficulties (such as confusion, depression, and sleeping problems) while using the drug and sometimes for weeks after. As a result of using the drugs, individuals can also experience increases in heart rate and blood pressure and physical symptoms such as nausea, blurred vision, or faintness (NIDA 2003a). When users overdose, they can experience rapid heartbeat, high blood pressure, faintness, panic attacks, and even loss of consciousness (Vaughn 2002).

In a study of undergraduates at a large midwestern university, Boyd, McCabe, and d'Arcy (2003) found that men and women were equally likely to have used Ecstasy, and several factors predicted its use. White students were more likely to report lifetime Ecstasy use than were African American or Asian students. According to the researchers, sexual orientation was also related to Ecstasy use: Those who identified themselves as gay, lesbian, or bisexual were more likely to report lifetime, annual, or past month Ecstasy use than were heterosexual students. Students with a GPA of 3.5 or higher were consistently less likely to have used Ecstasy in the past year or their lifetime than were students with GPAs below 2.5. Students who reported binge drinking within the past two weeks were also more likely to report past-month Ecstasy use.

Policy and Social Action

Federal Programs

Throughout the first part of this chapter, I have already referred to three U.S. offices: the NIDA, the ONDCP, and the NIAAA. All three programs are federally funded.

The NIAAA was established after the passage of the Comprehensive Alcohol Abuse and Alcoholism Prevention, Treatment, and Rehabilitation Act of 1970. Signed into law by President Richard Nixon, the legislation acknowledged alcohol abuse and alcoholism as major public health concerns. The law instructed the NIAAA to "develop and conduct comprehensive health, education, research, and planning programs for the prevention and treatment of alcohol abuse and alcoholism and for the rehabilitation of alcohol abusers and alcoholics" (NIAAA

2003b). Since then, the NIAAA's mission has been revised to include support and implementation of biomedical and behavioral research, policy studies, and research in a range of scientific areas to address the causes, consequences, treatment, and prevention of alcoholism and alcohol-related problems (NIAAA 2003c).

NIDA was established in 1974 as the federal office for research, treatment, prevention, training services, and data collection on the nature and extent of drug abuse. Like NIAAA, NIDA is part of the National Institutes of Health, the federal biomedical and behavioral research agency. NIDA's stated mission is to bring "the power of science to bear on drug abuse and addiction" (NIDA 2003b). NIDA supports more than 85 percent of the world's research on the health aspects of drug abuse and addiction.

The ONDCP is the newest federal drug program, operated through the White House. Established in 1988 through the Anti-Drug Abuse Act, the ONDCP's mission was to set national priorities, design comprehensive research-based strategies, and certify federal drug control budgets. According to the act, the purpose of the office was to prevent young people from using illegal drugs, reduce the number of drug users, and decrease the availability of drugs (ONDCP 2003c). Ten years later, ONDCP's mission was expanded under the Reauthorization Act of 1998. Some of the legislative requirements included a commitment to a five-year national drug control program budget, the establishment of a parents' advisory council on drug abuse, development of a long-term national drug strategy, and increased reporting to Congress on drug control activities (ONDCP 2003c). The act also provided support for the High Intensity Drug Trafficking Areas (HIDTA) program, coordinating local, state, and federal law enforcement drug control efforts.

Extensive use of illegal drugs continues despite the efforts of these three lead agencies. And the War on Drugs comes with huge economic cost, $12.9 billion requested by the Bush administration for FY2008 to support 11 federal drug agencies, including ONDCP (White House 2007). Although most advocates support prevention and law enforcement efforts, some have attempted to explore alternative strategies to the problem of drug abuse.

Drug Legalization

The contemporary debate about the legalization of drugs emerged in 1988 during a meeting of the U.S. Conference of Mayors. Baltimore's Kurt L. Schmoke called for a national debate on drug control policies and the potential benefits of legalizing marijuana and other illicit substances (Inciardi 1999). Proponents present several arguments for the legalization of drugs: Current drug laws and law enforcement initiatives have failed to eradicate the drug problem, arresting and incarcerating individuals for drug offenses does nothing to alleviate the drug problem, drug crimes are actually victimless crimes, legalization will lead to a reduction in drug-related crimes and violence and improve the quality of life in inner cities, and legalization will also eliminate serious health risks by providing clean and high-quality substances (Cussen and Block 2000; Silbering 2001). Many supporters of legalization argue that drugs should be legalized based on the libertarian legal code (Trevino and Richard 2002), namely that the legalization of drugs would give a basic civil liberty back to citizens by granting them control over their own bodies (Cussen and Block 2000).

The term legalization is often used interchangeably with another term, decriminalization. The terms vary in the extent to which the law can regulate the distribution and consumption of drugs. In general, **decriminalization** means keeping criminal penalties but reducing their severity or removing some kinds of behavior from inclusion under the law (e.g., eliminating bans on the use of drug paraphernalia). Some would support regulating drugs in the same way alcohol and tobacco are regulated, whereas others would argue for no restrictions at all. **Legalization** suggests removing drugs from the control of the law entirely (Weisheit and Johnson 1992).

In the Netherlands, though possession and sale of marijuana are not legal, coffee shops are permitted to sell limited amounts of the drug to adults (18 years or older) for personal use. Smoking marijuana is permitted in these shops.

Several European countries including Portugal, the United Kingdom, and Switzerland have moved toward drug legalization or decriminalization in some form. The country most identified with liberal drug policies is the Netherlands. Since the mid-1970s, Netherlands' coffee shops have been allowed to sell marijuana products. Though possession and sale of marijuana are not legal, in practice, sales through these shops are not prosecuted and buyers (18 years or older) are not prosecuted for possession of small (personal use) amounts.

In the United States, drug legalization is generally opposed by the medical and public health community (Trevino and Richard 2002). The American Medical Association has consistently opposed the legalization of all illegal drugs, arguing that most research shows drugs, particularly cocaine, heroin, and methamphetamines, are harmful to an individual's health. Opponents charge that drug use is a significant factor in the spread of sexually transmitted diseases such as HIV, and drug users are more likely to engage in risky behaviors and in criminal activity (Trevino and Richard 2002). The Drug Enforcement Administration has also been clear about its opposition to drug legalization, citing concerns about potential increases in drug use and addiction, drug-related crimes, and costs related to drug treatment and criminal justice.

In the 1990s, the drug debate began to change, with legalization proponents advocating a "harm reduction" approach. Many opposed to legalization began to accept aspects of the harm reduction approach. Harm reduction is a principle suggesting that "managing drug misuse is more appropriate than attempting to stop it all together" (Inciardi 1999:3). Proponents acknowledge that current drug policies are not working, but they are still not in favor of full decriminalization (McBride, Terry, and Inciardi 1999). The harm reduction approach emphasizes treatment, rehabilitation, and education (McBride et al. 1999), including advocacy for changes in drug policies (such as legalization), HIV/AIDS-related interventions, broader drug treatment options, counseling and clinical case management for those who want to continue using drugs, and ancillary interventions (housing, healing centers, advocacy groups) (Inciardi 1999).

What Does It Mean to Me?

The Drug Policy Alliance (2003a) is an organization "working to broaden the public debate on drug policy and to promote the realistic alternatives to the war on drugs based on science, compassion, health, and human rights." Since 1996, 46 states have enacted more than 150 drug policy reforms. The reforms usually target drug sentencing and the legalization of medical marijuana. For more information on drug policies in your state, go to *Study Site Chapter 12*.

Punishment or Treatment?

Stricter federal policies have increased the number of men and women serving jail or prison time for drug-related offenses. As conflict and symbolic interaction theories suggest, drug laws are not enforced equally, with certain groups being singled out. Although most illicit drug users are White, Blacks constitute about 80 to 90 percent of all people sent to prison on drug charges. Nationwide, Black men are sent to state prison on drug charges at 13 times the rate of White men (Fellner 2000). Drug enforcement usually targets urban and poor neighborhoods while ignoring drug use among middle- or upper-class people. Whereas our society treats middle- or upper-class drug use as a personal crisis, lower-class drug use is defined as criminal. Consider, for example, how the media reported the recent drug treatment of celebrities Lindsey Lohan and Britney Spears—were either referred to as a criminal?

Sasha Abramsky (2003) explains that with tougher drug laws, the U.S. drug war was taken away from the public health and medical officials and placed into the hands of law enforcement and courts. The notion that drug abuse is a disease was replaced with the idea that drug abuse is a crime. In contrast, the Netherlands defines drug abuse as a public health issue and has implemented harm reduction strategies, placing the priority on drug education and treatment, rather than on punishment.

However, as overall crime rates began to decline, public support for the get-tough-on-drugs policy began to wane. Research conducted by the Pew Research Center in 2001 found that 73 percent of Americans favored permitting medical marijuana prescriptions, 47 percent favored rolling back mandatory minimum sentences for nonviolent drug offenders, and 52 percent believed drug use should be treated as a disease rather than as a crime. Although federal policy seems unlikely to change in the near future, several states are reexamining the way they deal with drug offenders.

Abramsky (2003) identifies key legislative changes in several states. Arizona and California passed recent legislation that diverted thousands of drug offenders into treatment programs instead of prisons. In 1998, Michigan repealed its mandatory life sentence law for those caught in the possession of more than 650 grams of certain narcotics. In 2002, Michigan Governor John Engler signed legislation that rolled back the state's tough mandatory-minimum drug sentences. The Kansas Sentencing Commission proposed reforms of the state's mandatory sentencing codes, along with expansion of treatment programs. The reforms were accepted in March 2003. Abramsky explains, "Increasingly impatient with the costly combination of policing and prosecution, voters, along with a growing number of state and local elected officials, have abandoned their support for incarceration-based anti-drug strategies and have forced significant policy shifts" (2003:26).

Drug Treatment and Prevention Programs

Individual Approaches

Drug addiction is a "treatable disorder" (NIDA 2003c). Traditional treatment programs focus on treating the individual and his or her addiction. The ultimate goal of treatment is to enable users to achieve lasting abstinence from the drug, but the immediate treatment goals are to reduce drug use, improve the users' ability to function, and minimize their medical and social complications from drug use.

Treatment may come in two forms: Behavioral treatment includes counseling, support groups, family therapy, or psychotherapy; medication therapy, such as maintenance treatment for heroin addicts, may be used to suppress drug withdrawal symptoms and craving. Short-term treatment programs can include residential treatment, medication therapy, or drug-free outpatient therapy. Long-term programs (longer than six months) may include highly structured residential therapeutic community treatment or, in the case of heroin users, methadone maintenance outpatient treatment. During the past 25 years, research indicates that treatment does work to reduce drug intake and drug-related crimes. Patients who stay in treatment longer than three months have better outcomes than do people who undergo shorter treatments (NIDA 2003c).

Workplace Strategies

Certain employers, such as employers in the transportation industry or organizations with federal contracts in excess of $100,000, are required by law to have drug-free workplace programs. The federal government, through the Drug Free Workplace Program, also encourages private employers to implement such programs in an effort to reduce and eliminate the negative effects of alcohol and drug use at the workplace (SAMHSA 2003b). The American Management Association reported that the percentage of companies that test employees for drugs increased from 22 percent in 1987 to more than 81 percent in 1997 (Hoffman and Larison 1999). After implementing a drug-free workplace program, employers, unions, and employees are likely to see a decrease in administrative work losses (sick leave abuse, health insurance claims, disability payments, and accident costs), hidden losses (poor performance, material waste, turnover, and premature death), legal losses (grievances, threat to public safety, worksite security), and costs of health and mental health care services (SAMHSA 2003c).

Drug-testing programs have been subject to lawsuits during the past decade for challenging the employees' right to privacy and their constitutional freedom from unreasonable searches by the government (SAMHSA 2003b). There have also been challenges to the accuracy of drug tests. Critics have asserted a positive test does not always correlate with poor job performance, a criterion for assessing the adverse effects of drugs (Klingner, Roberts, and Patterson 1998). Consistent with conflict theories on drug use, some have argued that drug testing promotes various political agendas and reflects the manipulation of interest groups that market and sell drug testing and security services (Klingner et al. 1998). Yet, many U.S. companies consider drug-testing programs part of an effective policy against substance abuse among workers (Hoffman and Larison 1999).

Paul Roman and Terry Blum (2002) report that employee assistance programs (or EAPs) are the most common intervention used in the workplace to prevent and treat alcohol and other drug abuse among employees. The primary goal for many of these programs is to ensure that employees maintain their employment, productivity, and careers. These EAPs usually include health promotion, education, and referral to abuse treatment as needed. Most of these programs do not target the general workplace population; rather, services are directed to those already affected by a problem or in the early stages of their abuse. There is some evidence of the effectiveness of these programs, returning substantial proportions of employees with alcohol problems to their jobs (Roman and Blum 2002).

Campus Programs

The U.S. Supreme Court ruled that drug testing in schools is legal for student athletes (1993) and for students in other extracurricular activities (2002). In both rulings, the Court stated that drug screenings play an important role in deterring student drug use.

However, a national study of 76,000 high school students reported no significant difference between drug use among students in schools with testing versus students in schools without testing. Researchers Ryoko Yamaguchi, Lloyd Johnston, and Patrick O'Malley (2003) reported that 37 percent of 12th graders in schools that test for drugs said that they had smoked marijuana in the previous year, compared with 36 percent of 12th graders in schools that did not test. In addition, 21 percent of 12th graders in schools with testing reported that they had used illicit drugs (cocaine or heroin) in the previous year compared with 19 percent of 12th graders in schools without drug screenings. The study found that only 18 percent of schools did any kind of drug screening between 1998 and 2001. Large schools (22.6 percent) reported more testing than did smaller schools (14.2 percent). Most drug tests were conducted in high schools. The study did not compare schools that conducted intensive regular screenings with those that occasionally tested for drugs. The study indicated that education, rather than testing, may be the most effective weapon against abuse (Winter 2003).

In their review of 94 college drug prevention programs, Andris Ziemelis, Ronald Buckman, and Abdulaziz Elfessi (2002) identified three prevention models that produced the most favorable outcomes in binge-drinking prevention efforts. The first model includes student participation and involvement, such as volunteer services, advisory boards, or task forces, to discourage alcohol or other drug use or abuse. The researchers documented how these activities reinforce students' beliefs that they are in control of the outcomes in their lives and that their efforts and contributions are valued. This model encourages student ownership and development of the program. The second model includes educational and informational processes, such as instruction in classes, bulletin boards and displays, and resource centers. The most effective informational strategies were those that avoided coercive approaches but instead encouraged interactive communication between students and professionals on campus. The last model includes efforts directed at the larger structural environment, changing the campus regulatory environment and developing free alternative programming, such as providing alcohol-free residence halls or mandatory alcohol and drug abuse classes as part of campus intervention. In general, models that discourage or deglamorize alcohol and drug use were associated with better outcomes than were those that merely banned or restricted substance use (Ziemelis et al. 2002).

Voices in the Community

Jill Ingram

This article about Jill Ingram, of the National College Commission of Mothers Against Drunk Driving (MADD), was taken from MADD's Web site (Glenn 2000).

… Ingram became involved in underage drinking prevention after her brother was hit by a drunk driver in 1996.

Ingram's brother, Dan, and his date were on their way home from a college sorority formal when the drunk driver—who [was] driving her van, headlights off, on the wrong side of a divided highway—hit their car head-on. Dan sustained multiple injuries to his ankle, knee, and wrist in the crash.

"There were several witnesses who saw [the drunk driver] and honked to get her attention, but she was too drunk to notice," said Ingram. "Later it was determined that her blood alcohol content was two times over the legal limit."

"I can't even bear to think about what my brother must have gone through during and in the moments that followed the crash," Ingram said. "But I do know what my family went through in the aftermath of the crash. It was then that I knew I had to do something to prevent another family from going through that kind of pain."

Ingram responded by taking action. Her first step was to attend the 1997 MADD National Youth Summit to Prevent Underage Drinking as a youth delegate. While Ingram was at the Summit, a cheerleading squad teammate back home nearly lost her life in an impaired driving crash. This second alcohol-related crash fueled Ingram's fire for preventing the senseless tragedy from happening to anyone else.

Armed with the knowledge and driven by the passion of a true activist, Ingram set out to make a difference. Working in conjunction with the MADD Northern Virginia chapter, she began speaking at high schools and local community colleges to educate students about the dangers of alcohol and impaired driving. She also co-founded the student-led Alcohol and Drug Abuse Prevention Team (ADAPT) on her university's campus.

"Underaged students like myself who choose not to drink didn't seem to have the same options to have fun and get together with friends as the students who do drink," Ingram said of her impression upon arriving at [the University of Virginia] her first year. "The drinkers and partyers at the university had an outlet that was not available to students like me on campus. Students need an outlet for their energies and their leadership abilities," she continued. "Our campus alcohol awareness program is 10 years old, but student involvement this year is the highest it has ever been because we have created new and different options for alcohol-free lifestyles."

Community Approaches

In 1997, the Drug-Free Communities Act became law. The act was intended to increase community participation in substance abuse reduction among youth. The program is currently directed by the White House's ONDCP. The program supports more than 700 coalitions of youth, parents, law enforcement, schools, state, local, and tribal agencies, health care professionals, faith-based organizations, and other community representatives. The coalitions rely on mentoring, parental involvement, community education, and school-based programs for drug prevention and intervention, much like Project Northland.

Based in northern Minnesota, Project Northland was the largest community trial in the United States to address the prevention of alcohol use and alcohol-related problems among adolescents (Williams and Perry 1998). Adopting a holistic approach, the project assumed that prevention efforts should be directed at adolescents and their immediate social environment (family, peers, friends) and should include larger peer groups (teachers, coaches, religious advisers) as well as the broader community of businesses and political leaders. The project was recognized for its programming by the SAMHSA, U.S. Department of Health and Human Services, and the U.S. Department of Education.

Project Northland included youth participation and leadership, parental involvement and education, community organizing and task forces, media campaigns, and school curriculum as part of its strategies for alcohol prevention. The program included two phases. Phase 1 focused on strategies to encourage adolescents not to use alcohol. Phase 2 emphasized changing community norms about alcohol use, reducing the availability of alcohol among high school students, and adopting a functionalist approach in reinforcing community norms and boundaries. Community strategies included making compliance checks of age-of-sale laws (coordinated through local police departments), holding training sessions for responsible beverage servers at retail outlets and bars, and encouraging businesses to adopt "gold card" programs where discounts are provided to students who pledge to remain free of alcohol. At the end of Phase 2, significant differences between the intervention and comparison group were observed. The rates of increase in underage drinking were lower among the intervention students (Perry, Williams, and Komro 2002). In 2002, Project Northland was implemented internationally. First-year program data from primary and secondary schools in Croatia reveal that Project Northland was effective in increasing dialog between Croatian students, parents, and teachers about students' actual use of alcohol (Abatemarco et al. 2004).

The Community Anti-Drug Coalitions of America (CADCA) is a nonprofit organization that provides technical assistance and training to community-based coalitions. The organization was established in 1992 by Jim Burke and Alvah Chapman and currently serves more than 5,000 anti-drug coalitions. The program provides community groups with lobbying handbooks, alerts on drug-related legislation, funding information, and coalition training on various drug abuse topics. One CADCA affiliate is Families in Action (FIA) of Wilson, North Carolina, incorporated in 1982. Formed by local leaders, agencies, and organizations, FIA attempts to address the growing problem of drug abuse in the community. FIA operates five local programs, including "I'm Special," a science-based program for third and fourth graders, and the Prom "Think Card" campaign, targeting community merchants to discourage high school students from drinking alcohol at their prom.

Main Points

- There seems to be no argument about the seriousness of the drug problem in the United States and throughout the world. In 2005, 19.7 million Americans age 12 and older reported they were current illicit drug users.

- Functionalists argue that society provides us with norms or guidelines on alcohol and drug use. A set of social norms identifies the appropriate use of drugs and alcohol.

- Conflict theorists argue that powerful political and business interest groups have made intentional decisions about which drugs are illegal.

- Feminists argue that theorists and practitioners in the field of alcohol and drug abuse have ignored experiences unique to women and other marginalized groups. However, there is increasing recognition of the importance of gender-sensitive treatment models.

- The interactionist perspective argues that drug abuse is learned from others; it addresses how individuals or groups are labeled abusers and how society responds to them.

- Alcohol is the most abused drug in the United States. Other abused drugs include nicotine, marijuana, methamphetamine, and cocaine.

Alcohol problems can be both a cause and an excuse for intimate partner violence and child abuse. Alcohol abuse is more often suspected in crimes than abuse of any other drug.

- Employers have always been concerned about the impact of substance abuse. About 8 percent of full-time workers between the ages of 18 to 64 years used an illicit drug in the past month, costing businesses billions of dollars a year.

- Binge drinking among college students has been called a major public health problem. Some research shows that students who engaged in high-risk behaviors were more likely to be binge drinkers, whereas students who were involved in community service, the arts, or studying were less likely to be binge drinkers. Thousands of college students are injured or die each year from alcohol-related driving or injuries.

- In the early 1990s, the primary sources of meth-amphetamines were super laboratories. More recently, there has been an increase in the number of small-scale labs operated by independent "cooks." The waste and residue remaining from meth cooking can contaminate the surrounding area, and cleanup is very dangerous and costly.

- Ecstasy, Rohypnol, GHB, and Ketamine are drugs commonly used in club and rave scenes. Ecstasy has become a mainstream drug and can cause physical problems, irreparable damage to the brain and memory, psychological difficulties long after use, and death.

- Three offices—the NIDA, the ONDCP, and the NIAAA—are federally funded agencies that research and educate about drug and alcohol abuse. Extensive use of illegal drugs continues despite the efforts of these three lead agencies and the War on Drugs.

- Some have explored alternatives to the War on Drugs, including **legalization** (removing drugs from the control of the law). Proponents argue that current laws fail to eradicate the problem; incarceration does not alleviate the drug problem, they say, and drug crimes are victimless. Proponents argue that legalization would reduce drug-related crimes and violence, make drugs cleaner, and return to citizens a basic civil liberty. Legalization is generally opposed by the medical and public health community because research shows that drugs are harmful and cause risky behaviors and criminal activity.

- As conflict and symbolic interaction theories suggest, drug laws are not enforced equally, with certain minority groups (particularly Blacks) and the lower class being singled out. Recently, however, key legislative changes suggest a shift in thinking to treatment rather than incarceration.

- Drug addiction is considered a "treatable disorder." Treatment may be either behavioral or medical and either short or long term. Research in the past 25 years indicates that treatment works to reduce drug intake and drug-related crimes.

- Certain employers are required by law to have drug free workplace programs. A drug-free workplace program likely decreases administrative work losses, hidden losses, legal losses, and health care services.

- Drug-testing programs have been subject to lawsuits in the past decade over constitutional issues. There have also been challenges to the accuracy of drug tests. Consistent with conflict theories, some have argued that drug testing promotes various political agendas and reflects the manipulation of interest groups. Yet, many U.S. companies consider drug-testing programs part of an effective policy against substance abuse.

- Employee assistance programs are the most common intervention used in the workplace to prevent and treat alcohol and other drug abuse among employees. There is some evidence of the effectiveness of these programs.

- The U.S. Supreme Court ruled that drug testing in schools is legal and an important deterrent to drug use for student athletes (1993) and for students in other extracurricular activities (2002). However, a national study of 76,000 high school students indicated that education, not testing, may be the most effective weapon against abuse.

- In general, prevention models that discourage or deglamorize alcohol and drug use are associated with better outcomes than are those that merely ban or restrict substance use.

- Recent acts and initiatives to reduce substance abuse include the Drug Free Communities Act and the Community Anti-Drug Coalitions of America.

On Your Own

Log on to the Web-based student study site at www.pineforge.com/leonguerrero2study for interactive quizzes, e-flashcards, journal articles, Community and Policy Guides, a Service Learning Guide, the end-of-chapter Web exercises, and additional Web resources.

Internet and Community Exercises

1. Several advocacy groups are committed to promoting alternative solutions to the drug problem in the United States. Two groups are Students for a Sensible Drug Policy and Stop the Drug War (DCRNet). Log on to *Study Site Chapter 12* for links to their Web site. Examine how both organizations define the drug problem. Are there any differences in their definitions? What solutions does each group support?

2. According to the U.S. Drug Enforcement Administration, the illegal drug market in the United States is one of the most profitable in the world. The trafficking of illegal drugs has benefited from globalization, linking producers, dealers, and users more easily. The DEA posts state fact sheets on its Web site, identifying the drug trafficking situation in each state, along with a list and description of the illicit drugs that are smuggled in the state. For information about drug trafficking in your state, go to *Study Site Chapter 12*. To what extent does drug trafficking occur in your state?

3. According to Maria Alaniz, "Alcohol outlet density is an important determinant of the amount of alcohol advertising in a community. Merchants use storefronts and the interiors of alcohol outlets to advertise alcohol products. Therefore, areas with a high density of outlets have a greater number of advertisements" (1998:286). Alaniz cites a study showing that a student walking home from school in a predominately Latino neighborhood in northern California may be exposed to between 10 and 60 storefront alcohol advertisements. The same study found that there are five times more alcohol advertisements in Latino neighborhoods than in predominately White neighborhoods. Count the number of alcohol outlets around your college-university, along with billboard advertising within a five-mile radius. Do these ads target college students? Do you think exposure to alcohol advertising increases alcohol consumption? Why or why not?

4. The Campaign for Tobacco-Free Kids is a national campaign effort to protect children from tobacco addiction and exposure to secondhand smoke. The campaign's Web site includes information on state initiatives, as well as statistics on tobacco use. Log on to *Study Site Chapter 12*.

5. The World Health Organization (WHO) supports a global alcohol database. It reports trends in alcohol use and related mortality for selected countries since 1961, allowing you to compare international data. Log on to *Study Site Chapter 12*.

CHAPTER 13

Crime and Criminal Justice

Think of a crime, any crime. Picture the first "crime" that comes into your mind. What do you see? The odds are you are not imagining a mining company executive sitting at his desk, calculating the costs of proper safety precautions and deciding not to invest in them. Probably what you see with your mind's eye is one person physically attacking another or robbing something from another via the threat of a physical attack.

—Reiman (1998:57)

W hen we think of crime, we imagine violent or life-threatening acts, not white-collar crimes committed by men and women using accounting ledgers and calculators as deadly weapons. Yet, an act does not have to be violent or bloody to be considered criminal. For our discussion, a **crime** is any behavior that violates criminal law and is punishable by fine, jail, or other negative sanctions. Crime is divided into two legal categories. **Felonies** are serious offenses, including murder, rape, robbery, and aggravated assault; these crimes are punishable by more than a year's imprisonment or death. **Misdemeanors** are minor offenses, such as traffic violations, that are punishable by a fine or less than a year in jail. In this chapter, we will examine crimes as a social problem. We will consider a full range of crimes, not just violent events that make the headlines on your evening news. Before we review the specifics about crimes, let's review sociological explanations of why people commit crime.

Sociological Perspectives on Crime

Biological explanations of crime tend to address how criminals are "born that way." Early explanations were intended to classify criminal types by appearance and genetic factors, such as Cesare Lombroso's nineteenth-century theory of "born criminal" types. Lombroso argued that criminals could be easily identified by distinct physical features: a huge forehead, large jaw, and a longer arm span. Contemporary biological explanations focus on biochemical (diet and hormones) and neurophysical (brain lesions, brain dysfunctions) characteristics related to violence and criminality. Like biological theories, psychological perspectives also focus on inherent criminal characteristics. Researchers link personality development, moral development, or mental disorders to criminal behaviors. Both biological and psychological theories address how crime is determined by individual characteristics or predispositions to crime, but they fail to explain why crime rates vary between urban and rural areas, different neighborhoods, or social or economic groups (Adler, Mueller, and Laufer 1991). Sociological theories attempt to address the reasons for these differences, highlighting how larger social forces contribute to crime.

Functionalist Perspective

Functionalists offer several explanations for criminal behavior. For the first explanation, we return to one of the first sociologists, Émile Durkheim. Criminal behavior, according to Durkheim, is normal and inevitable. Criminal behavior is functional because it separates acceptable from non-acceptable behavior in society. Though not a criminologist, Durkheim provided the field with one of its most enduring concepts—**anomie** (Walsh and Ellis 2007). Recall from our discussion in Chapter 1 that Durkheim argued that society and its rules are what make man human; without any social regulation, man is able to pursue his own desires (even criminal ones). He defined anomie as a state of normlessless, a structural condition where there was no or little regulation of behavior, which leads to deviant or criminal behavior.

Robert K. Merton applied Durkheim's theory of anomie to develop the **strain theory** of criminal behavior. He argued that we are socialized to attain traditional material and social goals: a good job, a nice home, or a great-looking car. We assume that society is set up in such a way that everyone has the same opportunity or resources to attain these goals. Merton explains that society isn't that fair; some experience blocked opportunities or resources because discrimination, social position, or talent. People feel strained when they are exposed to these goals but do not have the access or resources to achieve them. This disjunction between cultural goals and the structural impediments is anomic and this is where crime is bred (Walsh and Ellis 2007).

This anomie creates an opportunity to establish new norms or break the old ones to attain these goals. Merton's strain theory explains how people adapt to life in this anomic situation. Merton presents five ways in which people adapt to society's goals and means, as presented in Table 13.1. Most individuals fit under the first category, conformity. Conformers accept the traditional goals and have the traditional means to achieve them. Attending college is part of the traditional means to attain a job, an income, and a home. Criminal behavior comes under the innovation category. Innovators accept society's goals, but they don't have the legitimate means to achieve them. Individuals in this category innovate by stealing from their boss, cheating on their taxes, or robbing a local store to achieve these goals.

Working from Merton's assumptions, scholars argue that criminal activity would decline if economic conditions improved. Solutions to crime would target strained groups, providing access to traditional methods and resources to attain goals. Actually, several studies have confirmed that when anomie is reduced among the poor, lower crime rates may result. Factors usually associated with anomie—the prevalence of female-headed families, the percentage of the population that is African American, and family poverty rates—are more weakly related to crime in areas with higher levels of welfare support (Hannon and Defronzo 1998).

The second functionalist explanation links social control (or the lack of it) to criminal behavior. Whereas Merton's theory asks why someone does commit a crime, social control theorists ask why someone *doesn't* commit crime. Society functions best when everyone behaves. Durkheim identified how well society provides us with a set of norms and laws to regulate our behavior. According to sociologist Travis Hirschi (1969), society controls our behavior through four elements: attachments, our personal relationships with others; commitment, our acceptance of conventional goals and means; involvement, our participation in conventional activities; and beliefs, our acceptance of conventional values and norms. Delbert Elliot, Suzanne Ageton, and Rachelle Canter (1979) redefined Hirschi's elements as integration (involvement with and emotional ties to external bonds) and commitment (expectations linked with conventional activities and beliefs). They believe that when all these elements are strong, criminal behavior is unlikely to occur.

Conflict Perspective

Sociologist Austin Turk (1969) explains that criminality is not a biological, psychological, or behavioral phenomenon; rather, it is a way to define a person's social status according to how that person is perceived and treated by law enforcement. An act is not inherently criminal; society defines it that way. Theorists from this perspective argue that criminal laws do not exist for our good; rather, they exist to preserve the interests and power of specific groups.

In this view, criminal justice decisions are discriminatory and designed to sanction offenders based on their minority or subordinate group membership (race, class, age, or gender) (Akers 1997). Turk (1969, 1976) states that criminal status is defined by "authorities" or members of the dominant class. Criminal status is imposed on "subjects," members of the subordinate class, regardless of whether a crime has actually been committed. Laws serve as a means for those in power to promote their ideas and interests against others. Law enforcement agents protect the interests and power of the dominant class at the expense of subjects.

From this perspective, problems emerge when particular groups are disadvantaged by the criminal justice system. Although the powerful are able to resist criminal labels, the

Photo 13.1

More than 4,000 Enron employees lost their jobs and pension savings when irregular accounting and business practices were revealed in 2001. While Enron was losing money, Andrew Fastow (Enron's CFO pictured here) and other chief executives continued to personally profit by millions of dollars. Fastow was sentenced to six years in prison for his role in the company's collapse.

Table 13.1

Merton's Strain Theory

Mode of Adaptation	Agrees With Cultural Goals	Follows Institutional Means	Method of Adaptation
Conformity	Yes	No	Nondeviant
			Accepts cultural goals and institutional means to achieve them
Innovation	Yes	No	Criminal
			Accepts cultural goals, adopts nontraditional means to achieve them
Ritualism	No	No	Deviant, noncriminal
			Rejects cultural goals but continues to conform to approved means
Retreatism	No	No	Deviant, could be criminal
			Rejects cultural goals and the approved means to achieve them
Rebellion	No	No	Deviant
			Challenges cultural goals and the approved means to achieve them

labels seem to stick to minority power groups—the poor, youth, and ethnic minorities. Minority power groups' interests are on the margins of mainstream society, so that most of their activities can be criminalized by dominant authorities (Walsh and Ellis 2007). Conflict theorists argue that the criminal justice system is intentionally unequal and serves as the vehicle for conflicts between opposing groups. We will discuss this further in the section "The Inequalities of Crime."

Feminist Perspective

For a long time, criminology ignored the experiences of women, choosing to apply theories and models of male criminality to women. Feminist researchers have been credited with making female offenders visible (Naffine 1996) and with documenting the experiences of women as the victims or survivors of violent men and as victims of the criminal justice system (Chesney-Lind and Pasko 2004). Feminist scholarship has attempted to understand how women's criminal experiences are different from those of men and how experiences of women differ from each other based on race, ethnicity, class, age, and sexual orientation (Chesney-Lind and Pasko 2004; Flavin 2001).

Freda Adler (1975) was one of the first to explain that women were "liberated" to commit crime when they were no longer restrained by traditional ideals of feminine behavior and could take on more masculine traits including criminal behavior. Called the "liberation approach," the logic of the argument is that as gender equality increases, women are more likely to commit crime. Although the approach was met with wide public acceptance, it has been discredited because of lack of empirical evidence (Chesney-Lind and Pasko 2004).

Recently, gender inequality theories have been presented as explanations of female crime. According to D. Steffensmeier and E. Allan (1996), patriarchal power relations shape gender differences in crime, pushing women into criminal behavior through role entrapment, economic marginalization, and victimization or as a survival response. The authors point out, "Nowhere is the gender ratio more skewed than in the great disparity of males as offenders and females as victims of sexual and domestic abuse" (p. 470). The logic of the inequality argument is that female crime increases as gender inequality increases.

Women are more likely to kill intimate partners, family members, or acquaintances than men are, "so the connection between women's overall homicide offending rates and gender inequality lies largely in the connection between gender stratification and women's domestic lives," explains Vicki Jensen (2001:8). Jensen describes gender inequality as being composed of economic, political-legal, and social inequalities experienced by women. Lower gender equality can negatively affect women's freedom and opportunities; these situations can push women into situations in which lethal violence seems to be the only way out. In the case of women killing abusive intimate partners, low levels of economic security limit women's opportunities to escape abusive situations. Low levels of gender equality can increase the emphasis on traditional gender norms, placing the responsibility on women to please men and requiring that women must be submissive and accept whatever their partner does (including violence). Jensen explains that most women who kill an intimate partner do so in response to abuse situations, in imminent self-defense, or when all other strategies have failed.

Feminists have also challenged the masculine bias of criminal justice programs. For example, feminist criminologists have criticized boot camp programs, claiming that these programs embody "a distorted image of masculinity, one that emphasizes aggressiveness, unquestioned authority, and insensitivity to others' pain while deemphasizing 'feminine' characteristics such as group cooperation and empathy" (Flavin 2001:281). Feminist scholarship has also been credited with encouraging the reconsideration of what is known about men's experiences and has led to more studies of masculinity and crime (Flavin 2001).

Since 1995, the annual rate of growth in the number of U.S. female prisoners averaged 5 percent, higher than the 3 percent for male prisoners. (Refer to Table 13.2.) During 2005, 107,518 females were in prison, accounting for 7 percent of all U.S. prisoners (Harrison and Beck 2006). The International Centre for Prison Studies analyzed data on the number of incarcerated women worldwide at the end of April 2006. Based on data from 187 prison systems in countries and dependent territories, the Centre reported that more than half a million women and girls were currently being held in penal institutions as pre-trial detainees or having been convicted and sentenced. The highest number of female detainees or prisoners was in the United States (183,400), China (71,280), the Russian Federation (55,400), and Thailand (28,450) (Walmsley 2006).

Table 13.2

Number of Prisoners Under the Jurisdiction of State or Federal Correctional Authorities and Incarceration Rate by Gender, 1995, 2004, and 2005

	Males	Females
All inmates—2005	1,418,406	107,518
All inmates—2004	1,392,278	104,822
All inmates—1995	1,057,406	68,468
Incarceration Rate*—2005	929	65
Incarceration Rate—2004	920	64
Incarceration Rate—1995	781	47

Source: Harrison and Beck 2006.

*The number of persons with a sentence of more than 1 year per 100,000 U.S. residents.

Research indicates that the surge in women's incarceration has little to do with any major change in women's criminal behavior. Actually, the number of U.S. women incarcerated since 1999 has increased by 75 percent despite decreases in most violent crime offenses committed by women. There is no more differential treatment of women in sentencing and imprisonment. The public's "get tough with crime" approach, along with a legal system that encourages treating women equally as men, has resulted in the greater use of imprisonment as punishment for female criminal behavior (Chesney-Lind and Pasko 2004). The increase in the number of women inmates has caused prison officials to reconsider custody procedures for women, especially specific programs and services available to women. Parenting has become a focus of many programs. In 2000, 80,000 women in prisons and jails were estimated to have about 200,000 children under the age of 18 (Amnesty International 2000).

Photo 13.2

During 2005, 107,518 females were in prison, accounting for 7 percent of all U.S. prisoners (Harrison and Beck 2006). The female prison population has increased faster than the male prison population since 1995. Pictured here are inmates of a female chain gang at Estrella Jail in Phoenix, Arizona.

Interactionist Perspective

Interactionists examine the process that defines certain individuals and acts as criminal. The theory is called **labeling theory**, highlighting that it isn't the criminal or his or her act that's important but, rather, the audience that labels the person or his/her act as criminal. As Kai Erickson explains, "Deviance is not a property inherent in certain forms of behavior; it is a property conferred upon these forms by audiences which directly or indirectly witness them" (1964:11).

The basic elements of labeling theory were presented by sociologist Edwin Lemert (1967), who believed that everyone is involved in behavior that could be labeled delinquent or criminal, yet only a few are actually labeled. He explained that deviance is a process, beginning with primary deviation, which arises from a variety of social, cultural, psychological, and physiological factors. Although most primary acts of deviance go unnoticed, they may lead to a social response in the form of an arrest, punishment, or stigmatization. Secondary deviation includes more serious deviant acts, which follow the social response to the primary deviance. Once a criminal label is attached to a person, a criminal career is set in motion.

John Braithwaite (1989) observed that nations with low crime rates are those where **shaming** has great social power. Braithwaite defined shaming as all processes of "expressing disapproval which have the intention or effect of invoking remorse in the person being shamed and/or condemnation by others who become aware of the shaming" (p. 9). Braithwaite agreed with Lemert that shaming could be "stigmatizing" (increasing the distance between the offender and society) and lead to additional criminal acts. However, he proposed that shaming should be "reintegrative" (restoring the link between the offender to society) and lead to less crime.

Some groups may be more resistant to the criminal label. Criminologist William Chambliss in his classic 1973 (reprinted in 1995) study, "The Saints and the Roughnecks," noted the differential treatment given two groups of boys. The Saints, a group of eight boys from White upper-class families, were always engaged in truancy, drinking, theft, and vandalism. Yet, throughout his study, a Saint was never arrested. The local police saw the Saints "as good boys who were among the leaders of the youth in the community" (p. 258). On the other hand, a group of six lower-class White boys, the Roughnecks, were "constantly in trouble with police and community even though their rate of delinquency was about equal with that of the Saints" (p. 254). The community and the police viewed the Roughnecks, who were "not so well dressed, not so well mannered, not so rich boys," as a group of delinquents. According to Chambliss, "From the community's viewpoint, the real indication that these kids were trouble was that they were constantly involved with the police" (p. 259).

Henry Brownstein (2001) argues that race and class matter in our perception of crime. It matters in how we conceptualize victims and offenders, even whether we believe that a person could have been a perpetrator or a victim of violence. In 1995, when Susan Smith told police that a Black man carjacked her car with her two young sons still buckled in their car seats, no one questioned her story. There was no doubt that this young White mother of two was the victim and that an unknown Black man was the offender. Police in Union, South Carolina, released a composite sketch based on Smith's description of the carjacker. For days, police and community members searched for the man and Smith's two sons. The Fox television program, *America's Most Wanted,* came into town to film a segment about Smith and her two sons. But the segment never aired. Nine days after the alleged carjacking, Smith confessed to rolling her car into a local lake, murdering her two sons.

What Does It Mean to Me?

What are your perceptions of a typical victim or perpetrator? What roles do the media—news, television, and movies—play in creating these perceptions? From what other sources are these perceptions learned?

Interactionists also attempt to explain how deviant or criminal behavior is learned through association with others. Edwin Sutherland's 1949 theory of differential association states that individuals are likely to commit deviant acts if they associate with others who are deviants. This criticism is often raised about our jail and prison systems; instead of rehabilitation, prisoners are able to learn more criminal activity and behavior while serving their sentences. Sutherland's theory does not address how the first criminal learned criminal behavior, but he does highlight how criminal behavior emerges from interaction, association, and socialization.

For a summary of sociological perspectives, see Table 13.3.

Sources of Crime Statistics

We rely on three sources of data to estimate the nature and extent of crime in the United States. The primary source is annual data collected by the Federal Bureau of Investigation (FBI). Since 1930, the FBI has published the Uniform Crime Report (UCR), data supplied by 17,000 federal, state, and local law enforcement agencies. The UCR reports two categories

Table 13.3

Summary of Sociological Perspectives: Crime and Criminal Justice

	Functional	Conflict/Feminist	Interactionist
Explanation of crime and criminal justice	Crime emerges from the social order. People experience "strain" when they are exposed to cultural goals but do not have the access or resources to achieve them. Individuals are likely to make some new rules (or break the old ones) to attain these goals. Society also controls criminal behavior through four elements: attachments, commitment, involvement, and beliefs.	Criminal justice decisions are discriminatory and designed to sanction offenders based on their minority or subordinate group membership (race, class, age, or gender). Problems emerge when particular groups are disadvantaged more than others are by the criminal justice system.	Interactionists examine the process that defines certain individuals and acts as criminal. Interactionists also examine how criminal or deviant behavior is learned through association with others.
Questions asked about crime and criminal behavior	How/why are individuals denied access to resources to achieve their goals? What social controls are in place to reduce criminal behavior?	How do our criminal justice policies reflect political, economic, and social interests? Why/how are particular groups targeted as "criminals"? How do women's experiences as crime victims and offenders differ from men's?	Is criminal behavior the result of being labeled a "criminal"? Is criminal behavior learned? How are our perceptions of criminals and victims socially created?

of crimes: **index crimes** and **nonindex crimes**. Index crimes include murder, rape, robbery, assault, burglary, motor vehicle theft, arson, and larceny (theft of property worth $50 or more). All other crimes except traffic violations are categorized as nonindex crimes. The UCR provides law enforcement officers and agencies with useful data about serious rates across states, counties, and cities, as well as trend and longitudinal data since its inception. The second source of crime data emerged in 1982, when the FBI began to use the National Incident-Based Reporting System (NIBRS), which adds detailed offender and victim information to the UCR data. Currently, only 18 states provide NIBRS information.

However, the often-cited problem with the UCR and the NIBRS is that the data only reflect reported crimes. The FBI cannot collect information on crimes that have not been reported, but it is estimated that only 3 to 4 percent of crimes are actually discovered by police (Kappeler, Blumberg, and Potter 2000). Also, being reported doesn't mean that a crime has actually occurred. The FBI does not require that a suspect has been arrested or that a crime is investigated and found to have actually occurred; it only needs to be reported (Kappeler et al. 2000).

In addition to the UCR, the FBI releases the Crime Clock, a graphic display of how often specific offenses are committed. Although it may make for good newspaper copy or give law

Table 13.4

Number of Crime Victims, Personal and Property Crimes, 2005

Type of Crime	Number of Incidents
All crimes	23,440,720
All personal crimes	5,400,790
Crimes of violence	5,173,720
Rape/sexual assault	191,680
Robbery	624,850
Assault	4,357,190
Purse snatching/pocket picking	227,070
All property crimes	18,039,930
Household burglary	3,456,220
Motor vehicle theft	978,120
Theft	13,605,590

Source: Catalano 2006a.

enforcement and political officials clout (Chambliss 1988), the Crime Clock has been accused of exaggerating the amount of crime, leaving the public with the impression that they are in imminent danger of being victims of violence (Kappeler et al. 2000).

The third data source about crime is the National Crime Victimization Survey (NCVS or NCS), which has been published by the Bureau of Justice Statistics since 1972. The survey is based on victimization surveys first conducted in Denmark (in 1720) and Norway (in the late 1940s). Twice a year, the U.S. Census Bureau interviews members of about 77,200 households regarding their experience of crime. The NCVS identifies crime victims whether or not the crime was reported. The survey includes information about victims and crimes but covers only six offenses (compared with the eight index crimes reported by the UCR). Also included is information on the experiences of victims with the criminal justice system, self-protective measures used by victims, and possible substance abuse by offenders. NCVS crime victim data for 2005 is presented in Table 13.4.

The results of the NCVS are often compared with the UCR to indicate that the number of crimes committed is actually higher than the number of crimes reported, suggesting that the UCR may not be an adequate measure of violent crime. However, "a more thoughtful interpretation of the inconsistency between these statistical reports concludes that while neither the UCR nor the NCVS is by itself an adequate measure of violence, each is an estimate of the scope and nature of violent crime" (Brownstein 2001:8–9).

Types of Crime

Violent Crime

Violent crime is defined as actions that involve force or the threat of force against others and includes aggravated assault, murder, rape, and robbery. The victimization rate for crimes of violence in 2005 was 21.2 victimizations per 1,000 people age 12 or older, declining slightly from 21.4 in 2004 (Catalano 2006a). The number of victimizations declined as well from 5.18 million in 2004 to 5.17 million in 2005.

Except for rape/sexual assault, males had higher victimization rates than females. The victimization rate was 25.2 victimizations per 1,000 men and 17.6 victimizations per 1,000 women (Catalano 2006a). Male homicide victimization rates are higher than females worldwide. Comparing homicide victimization rates from 1950 to 2001, the countries with the highest male homicide victimization rate were Mexico, 35.13 per 100,000, followed by Puerto Rico, 24.64. For female homicide, the highest victimization rate was in Puerto Rico, 3.53 per 100,000; Mexico is second at 3.40. For the same period, rates for the United States were reported at 11.88 for males and 3.37 for females (Lafree and Hunnicutt 2006).

Males are more likely to be victimized by a stranger, whereas women are more likely to be violently victimized by a friend, an acquaintance, or an intimate partner (Bureau of Justice Statistics 2003a). In 2005, women reported being raped or sexually assaulted by a friend or acquaintance in 38 percent of cases (Catalano 2006a).

Intimate violence (violence at the hands of someone known to the victim) is primarily committed against women in both the developed and developing world. The World Health Organization (WHO) (2005), in its study of women from 10 different countries, revealed that violence by an intimate partner is "common, widespread, and far-reaching in its impact" (WHO 2005:viii). WHO reports that the proportion of ever-partnered women who had experienced physical or sexual violence or both by an intimate partner ranged from 29 to 62 percent in most studied countries. Women in Japan were the least likely to experience either type of violence and the greatest amount of violence was reported by women living in provincial rural areas of Bangladesh, Ethiopia, Peru, and the United Republic of Tanzania (WHO 2005).

In 2005, U.S. women experienced 389,100 rape, sexual assault, robbery, aggravated assault, and simple assault victimizations at the hands of an intimate partner; for men, 78,180 were victims of violent crimes by an intimate partner (Catalano 2006a). The number of men and women killed by an intimate partner has declined since 1976. In 1976, 1,600 women and 1,357 men were killed by an intimate partner; in 2004, 1,159 women and 385 men were murdered by an intimate partner (Catalano 2006b).

Since 1973, Blacks have had the highest violent crime victimization rates. In 2005, 27 of 1,000 Black people experienced a violent crime versus 20 out of 1,000 Whites (Catalano 2006a). Among Hispanics, the rate of violent crime fell from a high of 70 crimes per 1,000 in 1993 to 25 violent crimes per 1,000 in 2005. Research by William Julius Wilson (1996) and Robert Sampson and Wilson (1995) reveals that structural disadvantages, rather than race, contribute to higher levels of crime and victimization in Black communities (Ackerman 1998). The structural factors include neighborhood poverty, unemployment, social isolation, and economic disadvantage.

The younger the person, the more likely that person was to experience a violent crime. In addition, the young (those 12 years of age or older) had higher rates of injury from crime than older people (Simon, Mercy, and Perkins 2001). People who never married were more likely to be victims of violent crime than were those married, widowed, or divorced (Catalano 2006a). In general, the number of violent crimes was higher among households with an annual income under $7,500. Victimization rates were lowest among households earning $75,000 or more. Violent crime rates were highest in the West and in urban areas (Catalano 2006a). Refer to Table 13.5 for a summary of violent crime rates by selected social characteristics.

Table 13.5

Number of Violent Crimes per 1,000 People Age 12 or Older by Selected Social Characteristics, 2005

	Number of Violent Crimes per 1,000 Persons Age 12 or Older
Income	
Less than $7,500	37.7
$7,500–$14,999	26.5
$15,000–$24,999	30.1
$25,000–$34,999	26.1
$35,000–$49,999	22.4
$50,000–$74,999	21.1
$75,000 or more	16.4
Region	
Northeast	19.3
Midwest	22.8
South	18.5
West	25.2
Urban	29.8
Suburban	18.6
Rural	16.4

Source: Catalano 2006a.

What Does It Mean to Me?

Crime rates have dropped significantly in most large cities since the 1990s. Take, for example, the homicide rate. The rate peaked in 1980 at 10.2 homicides per 100,000 people (Blumstein and Rosenfeld 1998). Since 2000, the homicide rate has been stable between 5.5 and 5.7 per 100,000 persons, a little more than half of the 1980 rate (Fox and Zawitz 2006). In 1994, about 25.1 million households or household members experienced one or more violent or property crimes (Klaus 2004), but by 2005, the number had dropped to 23.4 million households (Catalano 2006a). There is less violent crime, but do we still live in fear? Why or why not?

Property Crime

Property crime consists of taking money or property from another without force or the threat of force against the victims. Burglary, larceny, theft, motor vehicle theft, and arson are examples of property crimes. Property crimes make up about three-fourths of all crime in the United States (Catalano 2006a). In 2005, there were an estimated 18 million property crimes, including 3.4 million household burglaries and 978,000 motor vehicle thefts.

Urban households, households earning less than $7,500, and households located in the Western United States had higher rates of property victimization (burglary, motor vehicle theft, and theft). Households that rented their homes were more likely to be victims of property crime than homeowners (Catalano 2006a).

Juvenile Delinquency

The term **juvenile delinquent** often refers to a youth who is in trouble with the law. Technically, a **juvenile status offender** is a juvenile who has violated a law that only applies to minors 7 to 17 years old, such as cutting school or buying and consuming alcohol (Sanders 1981). In certain cases, minors can also be tried as adults. Crimes committed by juveniles are more likely to be cleared by law enforcement than crimes committed by adults.

The Office of Juvenile Justice and Delinquency Prevention monitors data on rates of **juvenile crime**. For 2003, the total number of juvenile arrests was 2,220,300. Almost half of all juvenile arrests involved larceny-theft, simple assault, drug abuse violation, disorderly conduct, or liquor law violation. Only 15 percent of all violent crimes involved offenders younger than age 18 (Synder and Sickmund 2006). Unlike the United States, many countries do not collect systematic data on delinquency and among the countries that do, their data are described as "incomplete because of faulty record keeping" (Stafford 2004:486).

Though the majority of juvenile offenders are male (71 percent in 2003), juvenile justice researchers have noted the increase in the number and percent of female juvenile offenders. In 1980, 20 percent of all juvenile arrests involved female offenders, increasing to 29 percent in 2003. Although male juvenile arrests have declined in several crime index categories, arrests for female juvenile offenders have increased in nearly all categories since 1980. Refer to Table 13.6 for a comparison between the percent change in male and female juvenile arrest rates.

For 2003, the racial composition of juvenile offenders was 71 percent White (Hispanics are also classified as White), 21 percent Black, 2 percent Asian/Pacific Islander, and 1 percent American Indian. Black youth were overrepresented in juvenile arrests given their proportion of the juvenile offender population; Black youth composed 16 percent of the juvenile population compared with White youth, who composed 78 percent. Female and male Black juveniles had higher rates of personal offenses but lower rates of property offenses than White female and male White juveniles did (Snyder and Sickmund 2006).

Delinquency is often explained by the absence of strong bonds to society or the lack of social controls. In studies of serious adolescent crime, research indicates that the economic isolation of inner-city neighborhoods, along with the concentration of poverty and unemployment, leads to an erosion of the formal and informal controls that inhibit delinquent behavior (Laub 1983). Juveniles without any or much social control are likely to engage in illegal behavior when they live in any environment that offers opportunities for illegal activities. Youth who are strongly bonded to conventional role models and institutions (parents, teachers, school, community leaders, and law-abiding peers) are least likely to engage in delinquent behaviors.

White-Collar Crime

The term **white-collar crime** was first used by Sutherland in 1939. He used the term to refer to "a crime committed by a person of respectability and high social status in the course of his occupation" (Sutherland 1949:9). Since then, the term has come to include three categories of crimes: those committed by an offender as described by Sutherland (someone of high social

Table 13.6

Percentage Change in Arrest Rates for Juvenile Males and Females, 1980 to 2003

Offense or Charge	Males	Females
All offenses	−20	22
Violent Crime Index	14	135
Aggravated assault	75	186
Property Crime Index	−57	−28
Burglary	−69	−49
Larceny-theft	−54	−26
Simple assault	174	284
Stolen property	−51	21
Vandalism	−42	26
Weapons violation	119	522
Sex offense	116	186
Drug abuse violation	95	143
Disorderly conduct	89	244
Curfew	101	228
Runaway	−51	−36

Source: Synder and Sickmund 2006.

status and respectability), crimes committed for financial or economic gain, and crimes taking place in a particular organization or business (Barnett n.d.).

The FBI has defined white-collar crime by the type of crime:

[White-collar crime includes] illegal acts which are characterized by deceit, concealment, or violation of trust and which are not dependent upon the application or threat of physical force or violence. Individuals or organizations commit these acts to obtain money, property or services; to avoid payment or loss of money or services; or to secure personal and business advantage. (FBI 1989:3).

Such acts include credit card fraud, insurance fraud, mail fraud, tax evasion, money laundering, embezzlement, or theft of trade secrets. Corporate crime may also include illegal acts committed by corporate employees on behalf of the corporation and with its support.

From 1997 through 1999, white-collar crime accounted for 3.8 percent of the incidents reported to the FBI. Most of the offenses involved fraud, counterfeiting, or forgery (Barnett n.d.). Check fraud and counterfeiting are among the fastest-growing problems affecting our financial system, producing estimated annual losses of about $10 billion (National White Collar Crime Center 2002a). As many as three-quarters of all employees steal from their employers at least once, and some may regularly engage in theft at work. Losses resulting from employee theft can range from $20 billion to $90 billion annually (National White Collar Crime Center 2002b). White-collar crimes cost taxpayers more than all other types of crime.

One of the most widespread forms of white-collar crime is Internet fraud and abuse, also known as **cybercrime**. In 2006, the Internet Fraud Complaint Center referred more than 86,000 cases to the law enforcement officials (National White Collar Crime Center and the Federal

Bureau of Investigation 2006). The crimes include identity theft, online credit card fraud schemes, theft of trade secrets, sales of counterfeit software, and computer intrusions (a hacker breaking into a system). Years ago, when computer systems were relatively self-contained, there was no concern about cybercrime. Committing cybercrime is easier with the growth of the Internet, increasing computer connectivity, and the availability of break-in programs and information. As computer-controlled infrastructure and networks have expanded, many systems—power grids, airports, rail systems, hospitals—have become vulnerable (Wolf 2000).

In response to Internet crimes, the FBI and the Department of Justice established computer crime teams or offices. Some states, such as Massachusetts and New York, have created high-technology crime units. Despite the increasing attention to cybercrime, "law enforcement at all levels is losing the battle" (Wolf 2000). Law enforcement does not have enough resources or technical support to detect and prosecute a significant number of cyberthieves. The government catches only 10 percent of those who break into government-controlled computers and fewer who break into computers of private companies (Wolf 2000).

What Does It Mean to Me?

Have you been a victim of any one of these types of crimes? Do you know someone who may have been a victim? Of the crimes that we have just reviewed, which type do you think is more problematic than others?

The Inequalities of Crime—Offenders and Victims

Offenders

Reports consistently reveal that African American males are overrepresented in incarceration statistics. Most jail or prison inmates are male and African American (refer to Table 13.7). As reported by the Bureau of Justice Statistics (2003b), the lifetime chances of a person going to state or federal prison are higher for men (9 percent) than women (1.1 percent) and higher for Blacks (16.2 percent) and Hispanics (9.4 percent) than Whites (2.5 percent). An estimated 28 percent of Black males will enter prison during their lifetime, compared with 16 percent of Hispanic males and 4.4 percent of White males (Bureau of Justice Statistics 2003b). That a category of people is overrepresented among violent offenders does not necessarily mean that this group is responsible for more violent acts (Brownstein 2001). Keep in mind that these statistics are based only on those who were caught by the criminal justice system.

A number of studies confirm that regardless of the seriousness of the crime, racial and ethnic minorities, particularly African Americans and Hispanics, are more likely to be arrested or incarcerated than are their White counterparts. This is also true for minority juvenile delinquents. Minority youth are overrepresented at every stage in the juvenile justice system; they are arrested more often, detained more often, overrepresented in referrals to juvenile court, and institutionalized at a disproportionate rate compared with White youth (Joseph 2000).

An early criminological explanation was offered by Marvin Wolfgang and Franco Ferracuti (1967), who argued that Blacks have adopted violent subcultural values, creating a "subculture of violence." Although this is an often-cited theory, there is insufficient empirical evidence to

Table 13.7

Percentage of Prisoners Under State or Federal Jurisdiction by Race, Based on Inmates With Sentences of More Than One Year

	1995	2005
White	33.5	34.6
Black	45.7	39.5
Hispanic	17.6	20.2
Other	3.2	2.7
Two or more races	—	3.0

Source: Harrison and Beck 2006.

support the idea that Blacks are more likely to embrace a violent value system. Actually, studies have indicated that White males are more likely to express violent beliefs or attitudes than Black males are (Cao, Adams, and Jensen 2000).

Family structure, specifically the presence of female-headed households in African American communities, has also been identified as a potential source for racial crime disparities. Yet, criminological research has not articulated how family structure or family processes are related to crime (Morenoff 2005). Does the structure of the family itself increase the likelihood of crime? Is the incidence of crime related to the amount of parental supervision, the level of parental effectiveness, or the nature of the parent-child relationship? These questions remain the focus of researchers.

Criminologists and sociologists have examined patterns of racial bias or discrimination in the law enforcement and criminal justice system. Research has identified patterns of discrimination by law enforcement officers designed to sanction offenders based on extralegal variables (e.g., race, age, low socioeconomic status, and unemployment) (Cureton 2001). In its analysis of contacts between police and the public for 2005, the U.S. Bureau of Justice reported that Black (9.5 percent) and Hispanic (8.8 percent) motorists were more likely to be searched during a traffic stop than White drivers (3.6 percent) (Durose, Smith, and Langan 2007). Blacks (3.5 percent) and Hispanics (2.5 percent) were more likely than Whites (1.1 percent) were to report being involved in an incident where police force was used. The data reveal that though Blacks accounted for 1 of 10 contacts with police, they were involved with 1 of 4 contracts where force was used (Durose et al. 2007).

Victims

Early studies on crime victims tend to perpetuate the image that the victim was simply at the "wrong place at the wrong time" (Davis, Taylor, and Titus 1997). Following that reasoning, not being a victim of crime could be explained simply by good luck. However, research indicates that some individuals, by virtue of their social group or social behavior, are more prone than are others to become victims (Davis et al. 1997). What people do, where they go, and whom they associate with affect their likelihood of victimization (Laub 1997).

Victimization is distributed across key demographic dimensions (Laub 1997). We reviewed some of the characteristics of crime victims in the section on "Types of Crime." Victimization rates are substantially higher for the poor, the young, males, Blacks, single people, renters, and central city residents (Davis et al. 1997). The likelihood of being injured because of a violent crime is higher among the young, the poor, urban dwellers, Blacks, Hispanics, and American Indians (Simon et al. 2001). Injury rates are lower for the elderly, people with higher income or higher educational attainment, and people who are married or widowed (Simon et al. 2001).

Black males have the highest rate of violent victimization, and White females have the lowest (Laub 1997). Blacks also have the highest rate of overall household victimization. Although rates of violent crimes declined during the 1990s, mortality from homicide among minority groups is still high. Homicide is the leading cause of death among Black males between the ages of 15 and 24, and it is the second leading cause of death for Latino males in the same age group (Rich and Ro 2002). People who have been victims once are at an elevated risk of becoming victims again. Repeat victimization is likely to occur in poor, predominately Black areas (Davis et al. 1997).

What Does It Mean to Me?

Have you been a victim of nonviolent or violent crime? Were you victimized because of your social behavior or your membership in a particular social group? Or were you at the wrong place at the wrong time?

Our Current Response to Crime

The Police

As of June 2004, federal agencies employed 106,000 full-time personnel authorized to make arrests and carry firearms. Among all federal officers, 33 percent were racial or ethnic minorities and 16 percent were women (Bureau of Justice Statistics 2006). Local police departments had 580,749 full-time employees, including 451,737 sworn personnel. Among all local police officers, 24 percent were racial or ethnic minorities (Bureau of Justice Statistics 2007).

We rely on the police force to serve as the first line of defense against crime, and some officers lose their lives in the line of duty. In 2005, 55 law enforcement officers were feloniously killed in 24 states and Puerto Rico and 67 were killed in accidents. From 1996 through 2005, 26 percent of officers who were killed were involved in arrest situations at the time, 18 percent were involved in ambush situations or traffic stops, and 16 percent were responding to disturbance calls (FBI 2006).

American policing has gone through substantial changes during the past several decades (MacDonald 2002). Traditional models of policing emphasized high visibility and the use of force and arrests as deterrents to crime. Policing under these models relies on three tactics: police patrols, rapid response to service calls, and retrospective investigations (Moore 1999). These models reinforce an "us" versus "them" division, sometimes pitting the police against the public it was sworn to protect. High-profile incidents of police brutality and violence, such as the cases that involved Rodney King in Los Angeles and Amadou Diallo and Sean Bell in New York City, increased public distrust and tainted the image of policing.

Studies indicate that Whites trust police more and have more positive interactions with them than do Blacks and that Hispanics fall between these two groups (Norris et al. 1992). Black youth tend to have the most negative or hostile feelings toward police (Norris et al. 1992). Residents from poor or disadvantaged areas have a much lower regard for the police than the general public does. However, research has also indicated that when citizens believe that they are treated fairly, they tend to grant police more legitimacy and are more likely to comply with police (Stoutland 2001).

Police departments are now incorporating new methods of policing based on the community and problem-solving approaches (Goldstein 1990, MacDonald 2002). The community policing approach refers to efforts to increase the interaction between officers and citizens, including the use of foot patrols, community substations, and neighborhood watches (Beckett and Sasson 2000). By 2000, two-thirds of all local police departments and 62 percent of sheriff's offices had full-time sworn personnel engaged in community policing activities (Bureau of Justice Statistics 2003c). We will learn more about community policing in the section, "Community, Policy, and Social Action."

Photo 13.3

For most individuals, their interaction with police officers is limited to traffic violation stops. Police departments are incorporating new methods of policing based on community and problem-solving approaches, hoping to increase the interaction between officers and citizens.

Prisons

Despite the decline in crime rates, prison populations are increasing (Anderson 2003). At the end of 2005, the federal and state inmate population was more than 2.2 million (see U.S. Data Map 13.1). Mandatory sentencing, especially for nonviolent drug offenders, is a key reason why inmate populations have increased for 30 years. Drug offenders now make up more than half of all federal prisoners (Anderson 2003).

Along with the increase in the number of prisoners comes an increase in prison budgets. In 2001, the average state inmate cost about $22,650 per year; the average federal inmate about $22,632 (Stephan 2004). For each U.S. resident, the state prison costs were about $104 for 2001. Average operating costs per inmate varied by state, indicating differences in costs of living, wage rates, and other related factors. The five states with the highest annual operating costs per inmate were Maine ($44,379), Rhode Island ($38,503), Massachusetts ($37,718), Minnesota ($36,836), and New York ($36,835). States with the lowest annual operating costs per inmate included Alabama ($8,128), Mississippi ($12,795), Missouri ($12,867), Louisiana ($12,951), and Texas ($13,808). Expenditures for state prison activities increased from $11.7 billion in 1986 to $29.5 billion in 2001, including construction, staffing, and maintenance of prisons. State correctional spending increased an average of 6.4 percent per year, more than the annual increase for health care (5.8 percent), education (4.2 percent), and natural resources (3.3 percent) (Stephan 2004).

Taking a World View

Policing in Brazil

Policing in Brazil has a dark and ugly history. The death squads of Brazil began in 1958, when Army General Amaury Kruel, chief of the police forces in Rio de Janeiro, handpicked a group of special policemen to combat rising theft and robberies in the city. These "bandit hunters" were given permission to hunt and kill these criminals. In other states and cities, teams of police hunters were formed to pursue *pistoleiros* (armed criminals), undesirables, and gangsters. Each new death squad, whether targeting economic or political criminals, became more distant from the formal criminal justice system (Huggins 1997). Through several political regimes, Brazilian police have never abandoned their practices of violent enforcement and vigilantism.

After 21 years of military dictatorship (1964–1985), the civilian government (inaugurated in 1985) set out to reform Brazil's authoritarian practices. In 1988, armed with a new constitution, the democratic leadership lifted the barriers to political participation and attempted to restore the legal premises of universal citizenship rights (Mitchell and Wood 1999). In 1996, President Fernando Henrique Cardoso released the National Human Rights Plan, a comprehensive set of measures to address human rights violations in Brazil, including cases of police abuses (Human Rights Watch 1997).

Yet, police violence and human rights violations all increased dramatically under democratic rule (Caldeira and Holston 1999). Police are some of the primary agents of violence in Brazil. According to organizations such as Amnesty International and the Human Rights Watch, many citizens continue to suffer systematic abuse and violations at the hands of their own police force. Data from 1986 to 1990 reveal that police committed 10 percent of the killings in Sao Paulo; in 1991, it was 15.9 percent, and in 1992, it was 27.4 percent. By comparison, police only accounted for 1.2 percent of killings in New York City during the 1990s and 2.1 percent of killings in Los Angeles. In 1991, Sao Paulo police killed 1,171 civilians, compared with only 27 people killed by New York police and 23 killed by Los Angeles police (Caldeira and Holston 1999). In 2006, 1,603 people were killed in alleged confrontations with police in the state of Rio, according to Brazil's Institute of Public Security. Within the first months of 2007, 449 deaths were officially registered by the institute, an increase of 36 percent compared to the same period last year (Human Rights Watch 2007).

Although Brazilian law endorses due process, criminal proceedings and police methods subvert this principle (Mitchell and Wood 1999), supporting extralegal conduct in the majority of cases (Huggins 1997; Mitchell and Wood 1999). Government leaders also offer their support of extralegal activities. Three days after state civil police officers killed 13 suspected drug traffickers, Marcello Alencar, governor of Rio de Janeiro, was quoted as saying, "These violent criminals have become animals.... They are animals. They can't be understood any other way.... These people don't have to be treated in a civilized way. They have to be treated like animals" (Human Rights Watch 1997).

According to Martha Huggins (1997), police violence in Brazil comes in two forms: on-duty police violence and death squads. Highly organized, elite police units carry out extralegal killings while on duty. The police violence is usually deliberately planned and conducted during routine street sweeps and dragnets, actions justified by the state's war on drugs and crime. There is also death squad violence conducted by a group of murderers, usually off-duty police, who are paid by local businesses or politicians for their services. Huggins calls them privatized security guards serving commercial or political interests.

In 1985, Brazil implemented women's police stations, *Delegacias de Policia dos Dieritos da Mulher* (DPM or *delegacias*). The delegacias were created in response to pressure from feminist groups demanding that violence against women be addressed. The first groups were formed in Rio de Janeiro and San Paulo, where the first DPM was established in 1985

(Hautzinger 1997). The female officers at each station are conventionally trained police, with no specialized training or qualifications for serving at the delegacias except for the fact that they are women. There are more than 250 delegacias in Brazil today (Hautzinger 2002). Although domestic violence complaints were the original mission of the DPMs, these account for only 80 percent of the caseloads (Hautzinger 2002). The Human Rights Watch reports that despite the presence of DPMs, "many rural and urban women have found police to be unresponsive to their claims and have encountered open hostility … when they attempted to report domestic violence" (Human Rights Watch 1995).

U.S. DATA MAP 13.1

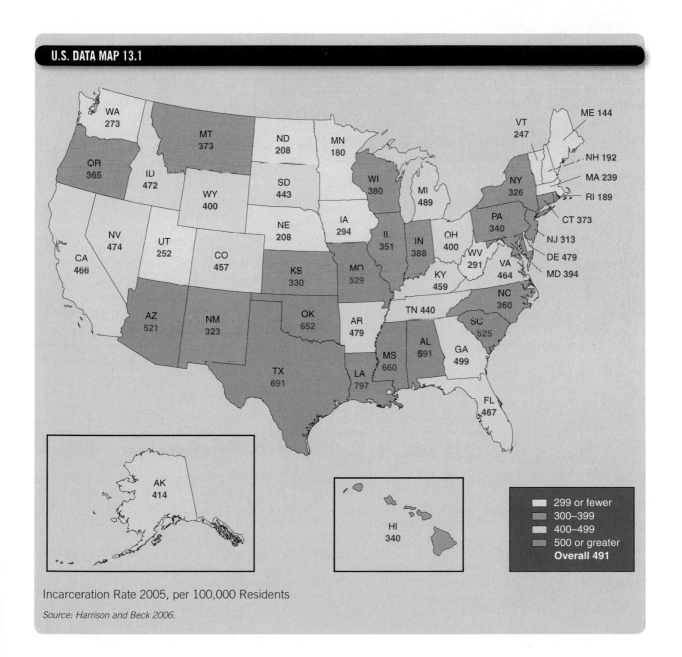

Incarceration Rate 2005, per 100,000 Residents

Source: Harrison and Beck 2006.

Photo 13.4

In 2007, Paris Hilton was ordered to spend 45 days in jail for violating the terms of her probation for alcohol-related reckless driving. Hilton served only 23 days, her sentence reduced with credits for good behavior. She described being locked in a cell as a "traumatic experience."

What is the purpose of the prison system? Some argue that the system is intended to rehabilitate offenders, to prevent them from committing crime again. However, beginning the mid-1970s, U.S. rehabilitation programs and community-based programs began to lose funding. Probation and parole offices redefined their core missions from treatment to control and surveillance (Tonry 2004). Other Western governments, such as Germany, the Netherlands, and Scandinavia, continue to make large investments in treatment and educational programs in comparison with the United States (Albrecht 2001), with evidence revealing that some programs are able to reduce future reoffending (Gaes et al. 1999). For example, Belgium, Denmark, England, France, the Netherlands, and Sweden have established crime prevention agencies that develop programs for developmental, community, and situational crime prevention (Tonry and Farrington 1995); there are no comparable agencies in the United States (Tonry 2004).

The U.S. record of recidivism (or repeat offenses) indicates how badly our prison system is working. Data from the Bureau of Justice Statistics revealed that among 300,000 prisoners released in 1993, 67.5 percent were rearrested within three years (Langan and Levin 2002). Among those rearrested, 46.9 percent were reconvicted for a new crime, and 51.8 percent were back in prison for a new sentence or for a technical violation of their release (failing a drug test, missing an appointment with their parole officer). Men, Blacks, non-Hispanics, and prisoners with longer prior records were more likely to be rearrested. Younger prisoners were also more likely to reoffend than were older prisoners (Langan and Levin 2002).

What Does It Mean to Me?

The latest in American corrections is home detention. Electronic surveillance equipment allows law enforcement to monitor large numbers of offenders through monitors strapped to their legs. Paris Hilton was briefly released by the Los Angeles County Sheriff during her 2007 incarceration, shifting her sentence to home confinement. (Her home confinement lasted a day before she was returned to jail.) Nonviolent offenders can leave for work by 7 a.m. and must be home by 6 p.m., continue to follow scheduled appointments and drug tests, and are continuously monitored. Keeping offenders imprisoned but not behind a prison wall, the system is said to be less expensive than prison and more flexible. What do you think about these surveillance systems? Is this an effective system of punishment or rehabilitation?

The Death Penalty

Since 1979, 1,075 men and women have been executed in the United States. The majority of executions have occurred in the South, with Texas and Virginia having 490 executions since 1979 (Death Penalty Information Center 2007). As of May 2007, 45 percent of death row inmates are White, 42 percent are Black, and about 2 percent are women. The number of death sentences has declined since 1998, when 298 men and women were sentenced to death. In 2005, 128 were sentenced.

The death penalty was instituted as a deterrent to serious crime. The penalty applies only to capital murder cases, where aggravating circumstances are present. However, research indicates that capital punishment has no deterrent effect on committing murder. In fact, states with the death penalty have murder rates significantly higher than do states without the death penalty. Victor Kappeler, Mark Blumberg, and Gary Potter (2000) explain that only a small proportion of people charged with murder can be sentenced to death. For example, between 1980 and 1989, 206,710 murders were reported to the police, but during the same period, only 117 executions were carried out. This is about 1 execution for every 1,767 murders committed during the same period (Kappeler et al. 2000). Worldwide, 89 countries have abolished the use of the death penalty (Amnesty International 2007).

Opponents of the death penalty point out the racial disparities in its application. The most significant studies of racial disparities point to the race of the victims as the critical factor in sentencing. Those convicted of committing a crime against a White person are more often sentenced to death. Data reported by the Death Penalty Information Center (2007) reveal that among persons executed for interracial murders, 15 White defendants were executed for killing a Black victim, but 216 Black defendants were executed for killing a White victim. In their analysis of 1990 to 1999 California death penalty cases, Michael Radalet and Glenn Pierce (2005) reported that those who killed a non-Hispanic African American were 56 percent less likely to be sentenced to death than were those who kill non-Hispanic Whites. The difference increases to 67 percent when comparing those sentenced to death for killing Whites versus killing Hispanics.

In federal cases, racial minorities are being prosecuted beyond their proportion in the general population or in the population of criminal offenders. In an analysis of prosecutions under the federal death penalty provisions of the Anti-Drug Abuse Act of 1988, 89 percent of defendants were either African American or Mexican American (Subcommittee on Civil and Constitutional Rights 1994). The number of prisoners on death row is presented in Table 13.8.

Table 13.8

Race of Death Row Inmates, as of May 15, 2007

Race	Number	Percentage of Death Row Inmates	Percentage of U.S. Population (2000)
White, not Hispanic	1517	45	69.1
Black, not Hispanic	1397	42	12.1
Hispanic	359	11	12.5
Other (Native American, Asian, Iraqi)	77	2	6.3

Source: Death Penalty Information Center 2007.

Community, Policy, and Social Action

U.S. Department of Justice

We may perceive our criminal justice system as a single system when, actually, we have 51 different criminal justice systems, 1 federal and 50 state systems. The federal system is led by the U.S. Department of Justice. Headed by the Attorney General of the United States, the Department of Justice comprises 39 separate component organizations, including the FBI; Drug Enforcement Administration (DEA); the Bureau of Alcohol, Tobacco, Firearms and Explosives; U.S. Citizenship and Immigration Services, which controls the border and provides services to lawful immigrants; the Antitrust Division, which promotes and protects the competitive process in business and industry; and the Bureau of Prisons, which oversees correctional operations and programs (U.S. Department of Justice 2002).

Funding for the U.S. Department of Justice and its organizations comes from federal legislation. The 1994 Violent Crime Control and Law Enforcement Act was the largest crime bill in history, providing funds for 100,000 new officers, $9.7 billion for prisons, and $6.1 billion for prevention programs. Under the Violence Against Women Acts of 1994 and 2000, the Office on Violence Against Women, administered by the Department of Justice, has awarded more than $1 billion in grant funds to U.S. states and territories. These grants have helped state, tribal, and local governments and community-based agencies to train personnel, establish domestic violence and sexual assault units, assist victims of violence, and hold perpetrators accountable (Office on Violence Against Women 2003).

Juvenile Justice and Delinquency Prevention Programs

The U.S. Department of Justice also supports the Office of Juvenile Justice and Prevention Programs. As its mission, the office attempts to provide national leadership, coordination, and resources to prevent and respond to juvenile delinquency and victimization. The office is guided by the Juvenile Justice and Delinquency Prevention Act of 1974. As stated in the 2002 reauthorization bill, "Although the juvenile violent crime arrest rate in 1999 was the lowest in the decade, there remains a consensus that the number of crimes and the rate of offending by juveniles nationwide is still too high" (U.S. Congress 2002).

The office sponsors more than 15 programs targeting juveniles and their communities. Through the Program of Research on the Causes and Correlates of Delinquency, the office sponsors three longitudinal programs in Denver, Colorado; Pittsburgh, Pennsylvania; and Rochester, New York, to examine how delinquency, violence, and drug use occur among juveniles. The Tribal Youth Program is part of the Indian Country Law Enforcement Initiative to support tribal efforts to prevent and control juvenile delinquency. The Child Protection Division administers programs related to crimes against children, such as the National Center for Missing and Exploited Children and the Internet Crimes Against Children Task Force. In one of its ongoing programs, the Juvenile Mentoring Program or JUMP, the office supports one-to-one mentoring programs for youth at risk of failing in school, dropping out of school, or becoming involved in delinquent behavior. JUMP matches adults 21 years or older with youths ages 5 through 17 years. Congress has appropriated more than $56 million to support one-to-one mentoring programs in schools. Since 1994, 204 JUMP sites in 47 states and two territories have been funded, serving 9,200 students.

Preliminary evaluation data suggest significant improvement in peer relationships, aggressive behavior, and delinquency risks in youth who have participated in JUMP projects (JUMP 2003). Both youth and their mentors were extremely positive when rating aspects of their mentoring experiences, and both had similar perceptions of how the program benefited the youth. Among the most beneficial aspects of the program were helping youths to get better grades, avoid friends who start trouble, and stay out of fights (JUMP 1998).

Faith-based and community organizations have also been involved in delinquency prevention programs. One such program is the Blue Nile Passage rites-of-passage program based in Harlem's Abyssinian Baptist Church (established before President George W. Bush's initiative). Since 1994, the program has relied on a community-based mentoring component to address the spiritual, cultural, and moral character development of African American youth (Harlem Live 1999). In general, the program was designed to have "a positive impact on the family relationships, self-esteem, sexual behaviors, drug use, peer and sibling relationships, and the religious and social values of the youth" (Irwin 2002:30). Darrell Irwin (2002) says that faith-based initiatives contain much of what has been found to be valuable in traditional mentoring programs, with the church as a backdrop. In the Harlem program, students were assigned to a mentor, either a church minister or a congregation member. During weekly meetings, mentors worked with their partnered youth, emphasizing problem-solving and decision-making skills. In addition, the mentors introduce the youth to different educational and cultural activities as examples of potential community interactions. The program also includes a media literacy component where students learn how to film, edit, direct, and produce their own videos (Harlem Live 1999).

The New American Prison

In an effort to reduce the costs of incarceration to state and federal correctional agencies, the idea of private correctional institutions has gained momentum. The perception that public prisons are deteriorating and overcrowded helped to encourage the growth of private prisons in the 1980s (Pratt and Maahs 1999). Private prisons incarcerate about 5 percent of the sentenced adult population (Camp and Gaes 2002), usually housing low-risk offenders. It has been argued that private correctional institutions save taxpayers' money, providing more services with fewer resources; studies reveal a slight advantage to private prisons and demonstrate a reduction in per-inmate cost over time (Larason Schneider 1999). State correction officials are also responding to the problem of chronic prison overcrowding. For example, one-third of Hawaii's 6,000 state inmates are jailed in private prisons in Arizona, Kentucky, Mississippi, and Oklahoma because Hawaii does not have enough room in its own state facility (Moore 2007). Studies have indicated that the savings from private prisons are likely to come from lower wages and benefits, fewer staff, more efficient uses of staff, or a combination of these factors (Camp and Gaes 2002).

S. Camp and Gerald Gaes (2002) found private prisons did not perform better than public ones and in some cases did worse. According to the researchers, one of the most reliable indicators of prison operations is the rate at which inmates test positive for use of drugs or alcohol. If substance abuse is high, it indicates a pattern of poor security practices. For the private institutions, tests showed no drug use in 34 percent of the facilities and low to high drug use in 66 percent of the facilities. Nearly 62 percent of public prisons showed no evidence of drug use. In their analysis of another important performance measure, William Bales et al. (2005) found no significant difference in recidivism rates between public and private prisoner offenders. There were no significant differences among three groups in their study—adult males, adult females, or youth offender males.

Passed in 1996, the Prison Litigation Reform Act requires prisoners to exhaust all internal and administrative remedies before they can file federal lawsuits to challenge the conditions of their confinement or to report civil rights violations. The act was intended to prevent frivolous or unfounded lawsuits, but it has made it impossible for prisoners with valid complaints to be heard. Prisoners cannot file lawsuits for mental or emotional injury unless they can also show that physical injury occurred. Prisoners are required to pay for their own court filing fees; monthly installments can be taken out of their prison commissary account (American Civil

In Focus

A Public Health Approach to Gun Control

On April 16, 2007, Virginia Tech University student Cho Seung-Hui shot and killed 32 people, including himself, in the nation's deadliest shooting. In 2005, a Virginia court declared Cho mentally ill and a danger to himself and sent him for psychiatric treatment. A month before the shooting, he purchased two semi-automatic pistols and 50 rounds of ammunition. Under federal law, Cho should have been prohibited from purchasing a gun after the Virginia court declaration. Federal law prohibits anyone who has been "adjudicated as a mental defective," as well as those who have been involuntarily committed to a mental health facility, from purchasing a gun. This tragedy reignited the debate over the regulated sale of firearms.

According to David Hemenway (2004:85),

> The United States has more firearms in civilian hands than any other high income nation. About 25 percent of adults in the United States personally own a firearm. Many gun owners have more than one firearm; some 10 percent of adults own over 75 percent of all firearms in the country. The percentage of households with a firearm has declined in the past two decades; about one in three households now contains a firearm.

He notes that public health approaches have reduced the burden of infectious disease, tobacco related illness, and motor vehicle injuries, suggesting that a similar approach could also be successfully applied to reducing gun violence, focusing on the impact of gun violence on one's community and encouraging collaboration, research, and comprehensive policies. Hemenway explains what would be necessary for a public health approach to gun control:

> Like the approach to reducing motor vehicle injuries, a public health approach to curtailing problems caused by firearms suggests pursuing a wide variety of policies while maintaining the ability of law-abiding Americans to use guns responsibly. This approach emphasizes the importance of obtaining accurate, detailed and comparable information each year on the extent and nature of the problem. For each motor vehicle death in the United States, the Fatality Analysis Reporting System collects data on more than one hundred variables … This information suggests interventions and permits evaluation of which policies are effective and which are not.
>
> A major problem is that the detailed national information about firearm injuries does not exist. For example, whether most unintentional firearm injuries occur at home or away from home, or with long guns or firearms, is unknown … Many groups have backed the creation of a national violent death reporting system to provide detailed information on all homicides, suicides, and unintentional firearm deaths …
>
> Although firearms are among the most lethal consumer products, killing tens of thousands of civilians each year, firearm manufacturing is one of the least-regulated industries in the United States. No federal regulatory body has specific authority over firearm manufacturing, which is exempt from regulation by the Consumer Product Safety Commission. The industry has also escaped any comprehensive examination by Congress. Instead, Congress is considering giving the industry immunity from tort liability for negligence.
>
> A public health approach would create incentives for firearm manufacturers to make products that reduce rather than increase the burden on law enforcement. Rather than producing and promoting firearms that appear primarily designed for criminal use, such as those who do not retain fingerprints, manufacturers could produce guns with unique, tamper resistant serial numbers. They could also make guns that "fingerprint" each bullet to permit authorities to match bullet and firearm with a high degree of accuracy.

Policies common in other developed countries—registration of handguns, licensing of owners, and background checks for all gun transfers—could reduce the U.S. homicide rate substantially by making it harder for adolescents and criminals to obtain handguns ...

The public health approach to reducing gun violence emphasizes the need for prevention as well as punishment, recognizes that alternations in the product and the environment are more likely to be effective than attempts to change individual behavior, and urges the pursuit of multiple strategies to tackle the problem. The public health community understands the importance of involving the entire community and sees roles for many groups, including educational institutions, religious organizations, medical associations and the media. (Hemenway 2004:91–93, 95)

Liberties Union [ACLU] 1999). The act also prohibits prison officials from settling lawsuits by agreeing to make changes in unconstitutional prison conditions. The ACLU and its National Prison Project have challenged the constitutionality of the Prison Litigation Reform Act, claiming that it "slams the courthouse door on society's most vulnerable members" (ACLU 1999). Despite its questionable constitutionality, many government leaders and courts have applauded the act.

Voices in the Community

Max Kenner

"The brainchild of Max Kenner, the Bard Prison Initiative (BPI), was created in 1999 to address the educational needs of prisoners and to provide them with the opportunity and the means to attain higher education while remaining within the correctional system.

"To understand the logic behind such a program as BPI, one must revisit the 1970s, a time when the federal government looked favorably upon college in prison programs. Since then, numerous studies have shown that college in prison programs reduce the rate of recidivism, lower the number of violent incidents that occur within prisons, reestablish broken relationships between incarcerated parents and their children, and create a general sense of hope among inmates. Despite these beneficial consequences, in 1994, President Clinton signed the Violent Crime Control and Law Enforcement Act into law, essentially abrogating federal support and funding for existing programs. As a result, of the 350 programs that had arisen, only three remained.

"'The prison system is so large,' Kenner muses, 'because it locks people up at a young age, and when they return home, they are less equipped to work, to attend school, and to function as social beings.' These deficiencies result in an increased chance that released prisoners will commit another crime of a greater magnitude, thereby paving the road back to prison, but this time for a much longer sentence.

"As an undergraduate of Bard College, Kenner immersed himself in the prevailing culture of social justice advocacy on campus. In 1999, he and a group of like-minded individuals made the unsettling discovery that of the 72,000 men and women in the New York State prison system, four out of every five inmates were from New York City. Armed with this finding, and an increasing frustration with governmental divestment from education in social services, the group set out to tackle the issue of educating prison inmates. 'We felt that if we were really going to commit ourselves to some kind of effort to improve social justice it should be broad-based, and it should be based on public institutions,' explains Kenner.

"With that in mind, Kenner embarked on a mission to make Bard College an institutional home that would allow either faculty or students to gain access to prisons by lending its transcript services and by offering credit-bearing courses and degrees to prison inmates.

"After the national collapse of the college in prison programs, however, there was an incredible distrust among people in corrections who wanted to see the colleges come back and people in higher education who wanted colleges in the prison. According to Kenner, colleges only wanted to offer courses if they could make a profit or if they could do so under ideal circumstances. Some colleges were simply not interested.

"It took Kenner one and a half years to begin working with prisons. He was able to organize student volunteer programs that allowed students to conduct writing, GED, literacy, and theology workshops within the prison. 'By the spring of my senior year, we had some 40 students volunteering at the prison on a weekly basis. Many of them said that it was the single most profound and influential thing that they had done at their time at Bard,' says Kenner.

"Upon graduation, Kenner made a proposal to Bard College President Leon Botstein, requesting that the college provide him with an office and grant him access to its transcripts so that they could begin offering college credit to prison inmates. The only stipulation was that Kenner would have to find a way to raise money to support the program.

"Following graduation, Kenner was given a salaried position by Episcopal Social Services (EPS). 'The Bard Prison Initiative officially started as a partnership between EPS and Bard College,' says Kenner, 'and five months later, in 2001, we began offering credit bearing courses to 17 students.'

"Since then, the program has continued to expand. In the fall, two more prisons, one of which is a women's prison, will be joining BPI and is expected to have about 125 enrolled students. BPI employs a blind admissions procedure and tuition for the program is completely waived by the college. Through grants, BPI acquires enough funding to enroll 15 students per facility in any given year.

"Currently, BPI offers two educational programs to inmates. Anyone with a GED can apply for the pre-college program and those with a higher level of education can apply for the associate's degree program. In the fall, BPI will begin offering a bachelor's program that is consistent with the degree conferred to Bard College students. Those who have successfully completed the associate's degree program in two or three years can then reapply for admission into the bachelor's degree program.

"Kenner hopes that the programs that have been implemented thus far will remain active and prove to be self-sustaining. He remains a passionate advocate for the return of college in prison programs and will continue to play an integral role in enhancing their opportunities."

Source: Quoted material is from Malik 2005.

Community Approaches to Law

Enforcement and Crime Prevention—COPS

Community-Oriented Policing Services or COPS was created in response to the Violent Crime Control and Law Enforcement Act of 1994. The goal of the program is to shift from traditional law enforcement to community-oriented policing services, a change that includes putting law enforcement officers within a community and emphasizing crime prevention rather than law enforcement (COPS 2003). As stated on the COPS Web site, "By earning the trust of the members of their communities and making those individuals stakeholders in their own safety, community policing makes law enforcement easier and more efficient, and makes America safer" (COPS 2003). Researchers found that "police administrators have hailed community oriented policing as the preferred strategy for the delivery of services" (Novak, Alarid, and Lucas 2003:57).

One of the central premises of community policing is the relationships among the police, citizens, and other agencies. Since 1981, the National Night Out program has worked to strengthen police-community partnerships in anti-crime efforts. Usually scheduled in August, National Night Out activities first involved just turning the front lights of houses on but now include block parties, cookouts, parades, and neighborhood walks involving community members and police officers (National Night Out 2003). The community plays an important role in ensuring its own safety. In community policing, "Problem solving requires that police and the community work together in identifying neighborhood problems, and that the community assumes greater 'guardianship' of the neighborhood" (Greene and Pelfrey 1997:395).

COPS sponsors grants and initiatives in selected communities, offering specialized training and programs to police professionals, such as technological innovations (mobile computing, computer-aided dispatch, automated fingerprint identification systems) or policing methods. In addition, the program supports innovative strategies linking police with their communities. For example, COPS supported community organizations in New York City and Los Angeles as they developed police magnet schools. In New York, the East Brooklyn Congregations/East New York High School of Public Safety and Law was established in 1999 to serve largely Hispanic and African American neighborhoods in East Brooklyn. The college-track high school program, developed in partnership with the John Jay College of Criminal Justice, exposes students to public safety, security, and law courses. COPS reports that the program has been successful in increasing attendance and retention rates, increasing student grades, and helping students exceed the benchmarks set for scores on the state's Regents tests. Graduating students are encouraged to pursue careers in public safety, law, forensics, and corrections (Pressman, Chapman, and Rosen 2002).

It is difficult to assess the effectiveness of community policing methods because many of these community approaches have not been sufficiently implemented (MacDonald 2002). J. Zhao, M. Scheider, and Q. Thurman (2002) report that an additional problem is that much of the research designed to assess COPS programs is limited to the individual programs or cities. However, based on their analysis of COPS programs in 6,100 cities, the researchers discovered that COPS hiring and innovative programs resulted in significant reductions in local violent and nonviolent crime rates in cities with populations greater than 10,000. In addition, an increase of $1 in grant funding per resident for hiring contributed to a decline of 5.26 violent crimes and 21.63 property crimes per 100,000 residents. However, COPS grants had no significant impact on violent and property crime rates in cities with less than 10,000 residents.

Prison Advocacy and Death Penalty Reform

Several national, state, and local organizations are committed to reforming our prison system and advocating for prisoner rights. Most of their work comes in the form of advocacy, educational campaigns, and litigation.

Not With Our Money is a network of student and community activists working to end the use of prisons for profit. The group successfully mobilized against Sodexho Marriott, a food service provider that owned more than 10 percent of Corrections Corporation of America (CCA), one of the largest owners and operators of U.S. private prisons. Efforts to pressure university administrators to end their food service contracts with Sodexho Marriott were successful at several colleges and universities. Since protests against Sodexho Marriott began, its parent company Sodexho Alliance divested all its interest in CCA (Bigda 2001).

State and local grassroots organizations, such as Coloradans Against the Death Penalty, Mississippians for Alternatives to the Death Penalty, and the Texas Coalition to Abolish the Death Penalty, support prisoners' rights and legislation to abolish the death penalty in their states. Citizens United for Alternatives to the Death Penalty (CUADP) attempts to "end the death penalty through aggressive campaigns of public education and the promotion of tactical grassroots activism" (CUADP 2003a). CUADP is using public education, funding, direct action, and professional media campaigns to send its message to the public. CUADP (2003b) believes that "we as a people, and the media in particular, have a responsibility to the public to expose wrongful convictions." On its Web site, the organization includes a list of cases that the organization believes may merit the attention of investigative reporters.

One organization that has accepted the mission of correcting wrongful convictions is the Innocence Project. Established in 1992 by attorneys Barry Scheck and Peter Neufeld, the Innocence Project is a nonprofit legal clinic at the Benjamin N. Cardozo School of Law at Yeshiva University in New York. The clinic is dedicated to "exonerating the innocent through postconviction DNA testing" (Innocence Project 2007a). By July 2007, the project had exonerated 205 individuals. These cases highlight the problems of misidentification, corrupt scientists and police, overzealous prosecutors, inept defense attorneys, and the influence of poverty and race in the criminal justice system. The Innocence Project is currently working to establish the Innocence Network, a group of law and journalism schools and public defender offices that assist inmates trying to prove their innocence, even if their cases do not involve biological or DNA evidence. Several states have established their own innocence or justice projects.

The Innocence Project and similar organizations consistently draw the public's attention when an innocent inmate is released. Byron Halsey was convicted in 1988 of the brutal sexual assault and murders of two of his girlfriend's children, Tyrone and Tina Urquhart. He was exonerated in 2007 after the Innocence Project showed that DNA evidence used to convict Halsey actually belonged to Cliff Hall, Halsey's next door neighbor during the time of the children's murders. Already in prison for other sex crimes committed in the early 1990s, Hall has been charged with the Urquhart murders. In police custody and prison for 22 years, Halsey was released in May 2007 and returned home (Innocence Project 2007b).

Photo 13.5

Byron Halsey hugs a supporter as he leaves the Union County Courthouse following his hearing. Halsey served 22 years before being released in 2007 through efforts by the Innocence Project.

Main Points

- A **crime** is any behavior that violates criminal law and is punishable. Crime is divided into two legal categories: **felonies** (serious offenses) and **misdemeanors** (minor offenses).

- Biological and psychological theories of crime address how crime is determined by individual characteristics or predispositions but do not explain why crime rates vary in certain areas. Sociological theories attempt to address these reasons.

- Functionalists argue that society sets goals and expectations, but people feel strain when they do not have the access or resources to achieve these goals. They then experience anomie and create new norms or rules. Functionalists argue that solutions to crime should target strained groups, providing access to resources to attain goals.

- A second functionalist explanation links social control to criminal behavior. Social control theorists ask why someone *doesn't* commit crime. When societal elements are strong, criminal behavior is unlikely to occur.

- Conflict theorists believe that criminality is a way to define a person's social status according to how that person is perceived and treated by law enforcement. An act is not inherently criminal; society defines it that way. Theorists argue that criminal laws exist to preserve the interests and power of specific groups. Problems emerge when particular groups are disadvantaged by the criminal justice system, which conflict theorists argue is intentionally unequal.

- In feminist scholarship, gender inequality theories have been presented as explanations of female crime. In this view, patriarchal power relations shape gender differences in crime, pushing women into criminal behavior through role entrapment, economic marginalization, and victimization or as a survival response.

- Interactionists examine the process that defines certain individuals and acts as criminal. The theory is called **labeling theory**, highlighting that it isn't the criminal or act that's important but, rather, the audience that labels the person or act as criminal. Once a criminal label is attached, a criminal career is set in motion.

- There are different types of crime, including **violent crime** and **property crime** (three fourths of all crime in the United States). **Juvenile crime** refers to youths in trouble with the law. **White-collar crime** includes crimes committed by someone of high social status, for financial gain, or in a particular organization.

- Within the prison system, African American males are overrepresented. A number of studies confirm that regardless of the seriousness of the crime, racial and ethnic minorities are more likely to be arrested or incarcerated than are their White counterparts. Studies also show that police in high-crime areas are more likely to be harsh on suspected criminals, to use coercive authority, and to make arrests than they are in lower crime areas.

- No consistent pattern can be identified in sentencing practices in state and federal courts. The exception is that Blacks are more likely to receive a death sentence than are Whites, especially in cases involving a White victim.

- Research indicates that some individuals are more prone than others to become victims. Victimization rates are substantially higher for the poor, the young, males, Blacks, single people, renters, and central city residents.

- American policing has gone through substantial changes during the past several decades. Traditional models emphasized high visibility and the use of force and arrests as deterrents. These models reinforce an "us" versus "them" division. Police departments are now incorporating new methods based on the community and problem-solving approaches.

- Despite the decline in crime rates, prison populations are increasing. Mandatory sentencing, especially for nonviolent drug offenders, is a key reason why inmate populations have increased for 30 years.

■ Some argue that the prison system is intended to rehabilitate offenders. But rates of recidivism, or repeat offenses, indicate how badly the system is working.

■ The death penalty was instituted as a deterrent to serious crime. However, research indicates that capital punishment has no deterrent effect on committing murder. In fact, states with the death penalty have murder rates significantly higher than do states without the death penalty.

■ Several national, state, and local organizations are committed to reforming our prison system and advocating for prisoner rights. Most of their work comes in the form of advocacy, educational campaigns, and litigation.

On Your Own

Log on to the Web-based student study site at www.pineforge.com/leonguerrero2study for interactive quizzes, e-flashcards, journal articles, Community and Policy Guides, a Service Learning Guide, the end-of-chapter Web exercises, and additional Web resources.

Internet and Community Exercises

1. Investigate the two official sources of U.S. crime data: the U.S. Department of Justice's Bureau of Justice Statistics and the FBI's Uniform Crime Reports. On both sites, search and compare information for one type of crime (e.g., homicide, property crime, rape). Log on to *Study Site Chapter 13* for the Web links.

2. The debate about gun control has two vocal sides: those advocating for gun control and those supporting the right to own guns. Investigate local or national organizations on both sides of this issue. You can start by visiting the following sites: Women Against Gun Control, The Brady Campaign to Prevent Gun Violence, and The National Rifle Association (log on to *Study Site Chapter 13* for links). How do these organizations define the problem of handgun violence? How does each side identify the pros and cons of gun control? What solutions does each side offer?

3. Do your local police support the National Night Out program? Contact the local police station to learn more about the program and other community policing efforts in your city. What community programs are implemented? What residents or social groups are police attempting to reach? To learn more about the program, visit the National Association of Town Watch, the sponsor of the program. The Web site includes links to selected city and state programs. Log on to *Study Site Chapter 13*.

4. The International Crime Victims Survey (ICVS) is administered approximately every 4 years. The ICVS uses the same questions, definitions of crime, and data collection method for each country. The next survey will include data from more than 90 countries. For more information about the ICVS, go to *Study Site Chapter 13*.

CHAPTER 14

Cities and Suburbs

W elcome to the Urban Millennium. During the past century, the percentage of the world's population living in urban areas grew from 13 percent in 1900 to 49 percent in 2005. The United Nations Population Fund (2007) predicts that by 2008, more than half the world's population or about 3.3 billion men, women, and children will be living in cities or urban areas. (Some researchers already declared that the world had reached its urban tipping point in 2007 [North Carolina State University 2007]). China is experiencing the "greatest migration in human history" (French 2007); in 1978, 18 percent of the Chinese population lived in cities and urban areas, but by 2010, more than 50 percent likely will. The urban growth is inevitable; as sociologist Tang Jun observes, "To ask whether China wants urbanization is like asking whether a person needs to eat" (French 2007:A13).

Cities have always maintained an allure of better living and opportunities. But an examination of our cities and their surrounding areas reveals a "profound duality" (Stanback 1991). Although our urban areas are shining examples of economic and social progress, they also harbor significant social problems such as poverty, crime, crowding, pollution, and collapsing infrastructures. Moreover, opportunities and resources are unevenly distributed in cities: Some neighborhoods have safer streets and better services and may offer a better quality of life than others do (Massey 2001). "Cities in different countries with different socioeconomic and political systems often face quite similar problems, although their scales, trends, or causes differ from place to place" (Kim and Gottdiener 2004:172).

Before we begin our analysis of urban problems, we will first review two sociological fields of study. Both remind us that cities don't just happen overnight; rather, social and demographic factors help shape our urban areas and their problems. Our urban areas are produced by economic, political, and cultural forces operating at international, national, and local levels (Kim and Gottdiener 2004).

Urban Sociology and Demography

In the 1920s, sociologists from the University of Chicago examined their city and the impact of city life and its problems on its residents. Their research provided the basis for urban study and for understanding the determinants of urbanization. **Urban sociology** examines the social, political, and economic structures and their impact within an urban setting. Rural sociology is the study of the same structures within a rural setting.

The first studies on urbanization or urban sociology adopted a functionalist approach, comparing a city to a biological organism. The growth of a city was likened to the development of a social organism, with each part of the city serving a specific and necessary function. A city's core, for example, served as its business or industrial center; and areas outside of a city were reserved for residential or commuting activity. Out of the Chicago School of Sociology came two dominant traditions in urban study, one focusing on **human ecology** (the study of the relationship between individuals and their physical environment) and population dynamics and the second focusing on community studies and ethnographies (Feagin 1998a).

Although this chapter's primary focus is on urban problems, an essential part of urbanization is the number of residents in an area, its population. The second sociological field we will rely on is **demography**: the study of the size, composition, and distribution of human populations. Demography isn't just about counting people. Demographers analyze the changes and trends in the population. Their work begins with two fundamental facts: We are born and then we die. Recall how in Chapter 10, "Health and Medicine," we reviewed two demographic elements, fertility and mortality, identifying how each is determined by biological and social factors.

An additional demographic element is **migration**, the movement of individuals from one area to another. Migration is distinguished by the type of movement: Immigration is the movement of people into a geographic area; emigration is the movement of people out of a geographic area. **Domestic migration** (the movement of people within a country) plays a large role in the population redistribution in the United States (Perry 2006). In the United States, about 40 million people moved from 2002 through 2003 (Schachter 2004). People migrate to pursue employment opportunities, to be closer to family, to find a more temperate climate, and to seek the opportunity of a better life. Most movers have housing-related reasons: They move to a new, better, or more affordable home or apartment (Schachter 2004).

The Processes of Urbanization and Suburbanization

Urbanization, the process by which a population shifts from rural to urban, took off in the later half of the nineteenth century (Williams 2000). Urbanization in the United States, as in other developed countries, was closely linked with economic development and industrialization. The U.S. economy in the middle of the nineteenth century was divided: The northern economy was characterized by a mixture of family-based agriculture, commerce, finance, and an increasing industrial base, whereas the southern economy remained dependent on agriculture (Gordon 2001). But as the industrial economy began to grow in the North and extended into the Midwest, thousands of people were attracted to these emerging urban centers, drawn by the promise of work in factories and mills (Williams 2000). Also contributing to early urban growth was the immigration of Europeans and the migration of rural Blacks and Whites from the South to northern and midwestern urban areas (Dreier 1996).

Table 14.1

Urban Population Distribution of the World by Development Groups, 1975, 2000, 2005, and 2015 (population in millions)

	1975	2000	2005	2015
World—Total	1,516	2,845	3,150	3,819
More developed regions	701	874	898	945
Less developed regions	815	1,971	2,252	2,874

Source: UN Population Fund 2007.

The process of global urbanization is described in waves. The first wave occurred in North America and Europe from 1750 to 1950 (United Nations Population Fund 2007), closely linked with industrialization and economic development (Kim and Gottdiener 2004). This wave involved a few hundred million people and produced urban industrial societies that now dominate the world (United Nations Population Fund 2007).

The second wave of urbanization took place during the past half century in developing countries. Some have referred to the shift as **overurbanization**, the process where an excess population is concentrated in an urban area that lacks the capacity to provide basic services and shelter. Overurbanization is characterized by a lack of employment, housing, and education or health infrastructures for its residents (Kim and Gottdiener 2004). This second urbanization wave is problematic because it involves large populations of poor people—instead of hundreds of millions as in the first wave, the second wave involves billions residing mostly in Africa and Asia (United Nations Population Fund 2007). Refer to Table 14.1 for a comparison of urban populations between developed and less developed regions.

After World War II, the United States experienced another significant population shift: **suburbanization**. Although suburbanization has come to represent the outward expansion of central cities into suburban areas (Smith 1986), it has also been linked with two additional population shifts: from the Snow Belt (industrial regions of the North and Midwest) to the Sun Belt (South and Southwest) and from rural to metropolitan areas (Dreier 1996). Though many factors contributed to suburbanization, the key players were government leaders and their policies. The U.S. Congress passed the Housing Act of 1949, which encouraged construction outside city boundaries and made home purchasing easier through the Federal Housing Administration (FHA) and Veterans Association home mortgage loan programs. The 1956 Federal Aid Highway Act, which established the modern interstate highway system, made rural areas more accessible. President Dwight Eisenhower, a chief proponent of the act, believed in the importance of the interstate highway system. Eisenhower declared,

> More than any single action by the government since the end of the war, this one would change the face of America.... Its impact on the American economy—the jobs it would produce in manufacturing and construction, the rural areas it would open up—was beyond calculation. (1963:548–49)

The U.S. Census Bureau defines an **urban population** as an area with 2,500 or more individuals. An **urbanized area** is a densely populated area with 50,000 or more residents, and

Photo 14.1

For 19 consecutive years, 1986 to 2005, Nevada was the fastest growing state in the nation. After Arizona was named the fastest growing state during 2005 to 2006, Nevada regained the title in 2006 to 2007. Much of Nevada's population growth has occurred in Las Vegas/Clark County (pictured here).

a **metropolitan statistical area** is a densely populated area with 100,000 or more people. U.S. Census data indicate a rapid increase in the urban population, especially after World War II (Table 14.2). For 2005, the five largest metropolitan statistical areas were New York City, Los Angeles, Chicago, Philadelphia, and Dallas (U.S. Census Bureau 2007). The largest area was New York City–northern New Jersey–Long Island with 19,747,000 people. A complete list of the 10 largest metropolitan statistical areas is presented in Table 14.3. The New York City–Newark area remains one of the world's largest mega-cities (10 million residents or more). Refer to Table 14.4 for the complete list of mega-cities.

The Population Composition

Changes in the fertility, mortality, and migration rates affect the population composition, the biological and social characteristics of a population. For example, **age distribution**, the distribution of individuals by age, is particularly important because it provides a community with some direction in its social and economic planning, assessing its education, health, housing, and employment needs. As discussed in Chapter 6, "Age and Aging," demographers have long predicted the graying of the U.S. population, warning of the growing number of elderly and the increasing demands they will make on health care and social security programs.

Table 14.2

U.S. Population, Percentage of Urban Residents

Year	Percentage Urban
1900	40.0
1910	45.8
1920	51.4
1930	56.2
1940	56.5
1950	64.0
1960	69.9
1970	73.5
1980	73.7
1990	75.2

Source: U.S. Census Bureau 1995.

Table 14.3

Ten Largest U.S. Metropolitan Statistical Areas, 2005

	Population
New York, NY	19,747,000
Los Angeles, CA	12,924,000
Chicago, IL	9,443,000
Philadelphia, PA	5,823,000
Dallas, TX	5,819,000
Miami, FL	5,422,000
Houston, TX	5,290,000
Washington, DC	5,215,000
Atlanta, GA	4,918,000
Detroit, MI	4,488,000

Source: U.S. Census Bureau 2007.

Table 14.4

Largest Mega-Cities in the World—1950, 2000, 2005, and 2015 (population in millions)

1950	Pop.	2000	Pop.	2005	Pop.	2015 predicted	Pop.
New York–Newark, USA	12.3	Tokyo, Japan	34.4	Tokyo, Japan	35.2	Tokyo, Japan	35.5
Tokyo, Japan	11.3	Ciudad de Mexico (Mexico City), Mexico	18.1	Ciudad de Mexico, Mexico	19.4	Mumbai, India	21.9
		New York–Newark, USA	17.8	New York–Newark, USA	18.7	Ciudad de Mexico, Mexico	21.6
		Sao Paulo, Brazil	17.1	Sao Paulo, Brazil	18.3	Sao Paulo, Brazil	20.5
		Mumbai, India	16.1	Mumbai, India	18.2	New York–Newark, USA	19.9
		Shanghai, China	13.2	Delhi, India	15.0	Delhi, India	18.6
		Calcutta, India	13.1	Shanghai, China	14.5	Shanghai, China	17.2
		Delhi, India	12.4	Calcutta, India	14.3	Calcutta, India	17.0
		Buenos Aires, Argentina	11.8	Jakarta, India	13.2	Dhaka, Bangladesh	16.8
		Los Angeles–Long Beach–Santa Ana, USA	11.8	Buenos Aires, Argentina	12.6	Jakarta, India	16.8

Source: United Nations Population Division 2006.
Note: In 1950, there were only two mega-cities in the world. By 2005, their number had increased to 20 (though only the top ten are listed here). By 2015, there will be 22 mega-cities in the world, 17 located in developing countries.

In addition, demographers have noted that the **ethnic composition** (the composition of ethnic groups within a population) affects social and human services. In 2007, the U.S. Census Bureau announced that one in three Americans were ethnic minorities, about 101 million. Roberto Suro and Audrey Singer (2002) explain how the Latino population has spread out further and faster across the nation than any previous wave of immigrants. Hispanics are the fastest-growing U.S. minority group—their numbers increased 58 percent during the 1990s, from 22.4 million in 1990 to 35.3 million in 1999. In 2003, the U.S. Census Bureau announced that Hispanics are the largest minority group in the United States, numbering 37 million in 2001 (Ramirez and de la Cruz 2003).

Housing, education, health, and public transportation demands are affected by the rate of Latino population growth. For example, cities and areas with large established Latino communities (e.g., New York, Chicago, Miami, Southern California) should prepare for a growing Latino population characterized by low-wage workers, large families, and significant numbers of adults with little English proficiency (Suro and Singer 2002). In California, researchers from the UCLA Medical School recommended shifting health care services because Hispanic Californians tend to live longer than non-Hispanics and are less likely to die from heart disease or cancer (Murphy 2003).

The increasing number of ethnic Americans and their age distribution has caught demographers' attention. When data for the 2007 U.S. Census were announced, researchers noted that younger Americans are more ethnically diverse than are older generations—the average age of non-Hispanic Whites was 40.5 years, whereas among Hispanics, the average age was 27.4 years. This age gap may lead to competing political and social agendas. Communities may be divided between older Americans advocating for social security, lower taxes, and better health care and younger ethnically diverse Americans demanding better education, jobs, and social services (Roberts 2007).

What Does It Mean to Me?

What is the age and ethnic composition of your state? Would you characterize your state as a "young" state or as ethnically diverse? Log on to *Study Site Chapter 14* for U.S. Census Bureau links.

Sociological Perspectives on Urbanization

Functionalist Perspective

Early functionalists were critical about the transition from simple to complex social communities. Émile Durkheim described this transition as a movement from **mechanical solidarity** to **organic solidarity**. Under mechanical solidarity, individuals in small simple societies are united through a set of common values, beliefs, and customs and a simple division of labor. In contrast, Durkheim argued that organic solidarity was the result of increasing industrialization and the growth of large complex societies, where individuals are linked through a complex division of labor. Under organic solidarity, individuals begin to share the responsibility for the production of goods and services. New relationships are created according to what people can do or provide for each other. Durkheim believed that as a result of industrialization, the social bonds that unite us will eventually weaken, leading to social problems.

Taking a World View

Global Urbanization and Population Growth

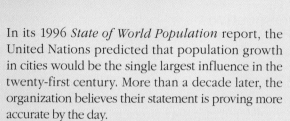

In its 1996 *State of World Population* report, the United Nations predicted that population growth in cities would be the single largest influence in the twenty-first century. More than a decade later, the organization believes their statement is proving more accurate by the day.

In its 2007 report, the United Nations predicted, "At the global level, all future population growth will [thus] be in towns and cities" (United Nations Population Fund 2007:6). More than half of the world's population will be living in urban areas by 2008, approximately 3.3 billion individuals. In contrast, the world's rural population will decline by about 28 million between 2005 and 2030. Urban population growth is expected to occur in developing nations such as Africa and Asia, with slower expansion expected in Latin America and the Caribbean. Most of the growth is attributed to natural increases (more births than deaths) rather than migration. In the developed world, the urban population is expected to grow very little, from 870 million to 1 billion.

There are many negative consequences of urban population growth for individuals, nations, and the world: increased demand for social and human services, increased economic and political burdens particularly for poorer developing countries, and global environmental degradation (Desai 2004). Some segments of the population are more vulnerable than others are. By 2030, 60 percent of all urban dwellers will be younger than age 18. Cities need to ensure that appropriate levels of basic services, education, housing, and medical care are available for these youth; if not, life on these urban streets will threaten the quality of youths' health, education, safety, and future (United Nations Population Fund 2007).

Despite its dire message, the United Nations concludes, "Urban and national governments, together with civil society, and supported by international organizations, can take steps now that will make a huge difference for the social, economic and environmental living conditions of a majority of the world's population" (2007:3). Suggesting the need for more proactive and creative approaches, the organization recommends strategies to improve the social conditions of the poor, promote gender equality, and ensure environmental sustainability. A major component of the recommendations is to empower women and increase the level of reproductive health services available to families, believing that these interventions will influence individuals' fertility preferences (the number of children and the timing of births) and their ability to meet them. Such strategies "empowers the exercise of human rights and gives people greater control over their lives" (United Nations Population Fund 2007:70).

Although industrialization and urbanization have been functional, creating a more efficient, interdependent, and productive society, they have also been problematic. Because of the weakening of social bonds and an absence of norms, society begins to lose its ability to function effectively. As our social bonds with each other have loosened, our sense of obligation or duty to one another has declined.

Urbanization can lead to social problems such as crime, poverty, violence, and deviant behavior. Functional solutions to these problems may encourage reinforcing or re-creating social bonds through such existing institutions as churches, families, and schools or instituting societal changes through political or economic initiatives. For example, under mechanical solidarity, the strong social bonds linking an individual to society (through one's family and friends) deterred criminal behavior. A person would not think of committing a criminal act because it was inherently wrong or would harm the individual's relationship with other members of society. Under organic solidarity, criminal laws, police, and prison systems serve as formal structures to deter criminal activity.

Conflict and Feminist Perspectives

Since the late 1960s, a new perspective on urban study has emerged. Referred to as the **critical political-economy** or **socio-spatial perspective**, this approach uses a conflict perspective to focus on how cities are formed on the basis of racial, gender, or class inequalities. From this perspective, cities are shaped by powerful social and political actors from the private and public sectors, working within the modern capitalistic structure (Feagin 1998b). Social problems are natural to the system, rising from the unequal distribution of power between politicians and taxpayers, the rich and the poor, or the home owner and the renter.

Within this tradition, scholars examine the role of capitalism and capitalists in shaping cities (Feagin 1998b). Land use decisions are made by politicians and businesspeople (Gottdiener 1977), real estate developers and financiers (Molotch 1976), or coalitions between public officials and private citizens (Rast 2001). Joe Feagin (1998b) presented a theory of urban ecology that accented the role of class structure and powerful land-oriented capitalist actors in shaping the location, development, and decline of American cities. Land speculators shape the internal structure of cities by identifying and packaging particular parcels of land for business or residential use. As Feagin describes it,

> Powerful land-interested capitalists have contributed substantially to the internal physical structure and patterning of cities themselves. The central areas of cities such as San Francisco have been intentionally remade, in the name of private profit, by combinations of speculators and other capitalists, such as developers. (1998b:154)

Although women play a pivotal role in urban life, theories about urbanization have taken a gender-blind approach (Women's International Network News 1999). Urban studies have not systematically considered cities as sites of institutionalized patriarchy (Garber and Turner 1995) and have not legitimately considered the role of women in urban development. Feminist urbanists have argued for the development of a comprehensive field of theory and research that acknowledges the role of women in urban structures (Masson 1984).

By incorporating feminist theory in patriarchy and urban studies, we can understand an additional dimension of urban life, namely the complex ways in which cities reproduce and challenge patriarchy (Appleton 1995) and the problems these create. Judith Garber and Robyne Turner explain,

> Urban environments are constructed around the delivery of public services and the development of policies. These shape women's ability to cope with complex urban locations, largely through the responsiveness of public and private organizations to the needs of diverse groups of women and children. (1995:xxiii)

Turner (1995) argues that the living conditions of lower-income, inner-city women have been affected by the economic restructuring of cities and the patterns of downtown development. Woman-headed households increasingly make up the majority of inner-city households. Turner explains that although low-income women may find inner-city housing less costly and more accessible than housing in the suburbs is, urban living also presents a unique set of challenges in transportation, housing, employment, services, and safety. Inner-city women have less control over their living situations than suburban women do. City development decisions are made by those in power, often men, whereas the recipients of these decisions are female, the young, or the elderly. Turner concludes, "It is important to recognize the implications for women, as the heads of households, in the debate on economic restructuring, land based economies, and the portrayal of political power" (1995:287–88).

Interactionist Perspective

Georg Simmel (1903/1997) was the first sociologist to explain how city life is also a state of mind. In his 1903 essay, "The Metropolis and Mental Life," Simmel described how life in a small town is self-contained; interactions with others are routine and rather ordinary. But a city's economic, personal, and intellectual relationships cannot be defined or confined by its physical space; rather, they are as extensive as the number of interactions between its residents. City dwellers must interact with a variety of people for goods and services but also for personal and professional relationships. City living stimulates the intellect and individuality of its residents (Karp, Stone, and Yoels 1991). Within this complex web of city life, Simmel argued that man would struggle to define his own individuality "in order to preserve his most personal core" (1903/1997:184). A city represents an opportunity for individuals to find self-expression while being connected with fellow city dwellers.

Photo 14.2

These businessmen may commute daily on a Japanese commuter train, but how likely are they to interact with one another? Does urban living unite or divide its residents?

But how well are city dwellers connected with their neighbors? The answer is that they may not be as connected as Simmel predicted. The way a city is constructed might actually interfere with social interaction. Our dependency on automobiles compartmentalizes neighborhood relations (we only know neighbors on our block or street) and interferes with street life (no space for ball games, block parties, bike riding, and joggers) (Gottdiener 1977). Home and residential designs limit our face-to-face contact with our neighbors. Without porches, there is no place to sit out front and visit with one's neighbors; without sidewalks or local parks, families find it less appealing to take walks around their neighborhood and less easy to meet with neighbors.

Tridib Banerjee and William Baer (1984) discovered that how we define our cities is linked to what we value or use within them. What goods and services do people use in their community? Is it the coffee shop, the local dry cleaner, or the neighborhood grocery store? Or is it the local park, the bicycle lanes, or the athletic center down the street? The researchers asked residents of several Southern California cities to draw maps of their residential areas and discovered that illustrations by middle- and upper-income individuals contained more details and area than illustrations by lower-income people did. Upper-income groups included amenities such as tree-lined streets, wooded areas, and golf courses, whereas middle- and low-income groups included commercial and retail locations, such as gas stations, discount stores, or drug stores. Corporate symbols were commonly used to define landmarks in middle- and low-income illustrations. Banerjee and Baer concluded that income was the single most important variable in explaining the quality of residential experiences and residents' judgments about what constitutes a "good place" to live.

Based on his examination of the transformation of European cities from industrial centers to recreational destinations, Mathis Stock (2006) discovered that a city may be defined by conflicting constituent groups (residents versus tourists) and vastly different experiences of the city they share. The transformation of cities such as Paris, Venice, and Florence into tourist destinations is often intentional, with city leaders and businesses embracing the symbols of tourism—the language, images, practices, and customs—in their cities. To entice visitors to their city, they promote their city's annual festivals and package them along with well-known cultural and historical sites as part of a cultural heritage experience. This transformation has its detractors, who though acknowledging the revenue benefit of tourism, also question whether this process has stripped these historical European cities of their authentic identities and subordinated native residents as a result.

For a summary of the different sociological perspectives, see Table 14.5.

What Does It Mean to Me?

Can't a person find some personal space? Symbolic interactionists have noted that urban dwellers are able to create a "public privacy" while living in a demanding urban world (Karp et al. 1991). Using props like the newspaper or an iPod, individuals send messages that they aren't interested in talking with others. You may bump into people while walking on a busy street but never stop to say "excuse me." The proportion of unlisted phone numbers is greater in the city than in small towns or suburban areas (Karp et al. 1991). Your coffee barista may have memorized your morning coffee order, but does your barista know your name or any other personal information? How do you create and maintain your public privacy?

Table 14.5

Summary of Sociological Perspectives: Cities and Suburbs

	Functional	Conflict/Feminist	Interactionist
Explanation of urbanization and its social problems	This perspective focuses on the weakening of social bonds and the functions and dysfunctions of urbanization.	Both perspectives focus on how cities are formed on the basis of racial, gender, or class inequalities.	City living is a state of mind. Urban living and its related social problems are socially defined.
Questions asked about urbanization and its social problems	How does urbanization enhance or destroy our social bonds? In what ways does urbanization affect existing institutions such as churches, families, and schools? How can we strengthen our social bonds?	Does one's race, gender, and social class determine the quality of urban living? How can we address the needs of all urban dwellers? How does one urban group gain power over the others?	How is urban living defined? How can we establish common ideas on urban life and its social problems?

The Consequences of Urbanization and Suburbanization

Along with suburbanization came the decentralization, some may even say the demise, of American cities. Inner cities became repositories for low-income individuals and families, as the suburbs enjoyed higher tax bases and fewer social programs (Massey and Eggers 1993). Researchers have suggested that the poor economic outcomes of racial minorities, particularly African Americans, are partly the result of patterns of housing prejudice and discrimination that have prevented minority groups from moving at the same pace as the suburbanization of employment (Massey 2001; Pastor 2001). According to Douglas Massey and Mitchell Eggers,

> The simultaneous proliferation of poverty and affluence created a situation in which social problems among those at the bottom of the income hierarchy multiplied rapidly at a time when more and more people had the means to escape these maladies. (1993:313)

Many of the social problems we discuss in this text seem to be magnified in urban areas. In this next section, we will review specific social problems plaguing urban areas: affordable housing, crowding, quality housing, homelessness, gentrification, and transportation.

Affordable Housing

Today, 68 percent of Americans own their own homes. Not until the passage of the 1949 Housing Act did the majority of Americans become home owners. In 1940, only 43.6 percent of Americans were home owners, but by 1950, more than half, 55 percent, were home owners (Bunce et al. 1996). Although most Americans still aspire to own a home, for many poor and working Americans, home ownership is just a dream (Freeman 2002). For 2002, Whites had the highest homeownership rates (72 percent), followed by Asians/Pacific Islanders (54 percent) and Blacks (47 percent) (U.S. Census Bureau 2006). Central city residents of all income levels are less likely to own a home than are suburban residents with the same income (U.S. Department of Housing and Urban Development [HUD] 1999). Refer to U.S. Data Map 14.1 for the rate of homeownership by state.

As rates of home ownership have increased, the affordability of homes has declined (Savage 1999). The generally accepted definition of affordability is for a family to pay no more than 28 percent of its annual income on housing (30 percent for a rental unit). Twelve million renter and home owner households pay more than 50 percent of their annual incomes for housing (HUD 2004), especially among the poorest households (Pugh 2007). Lance Freeman

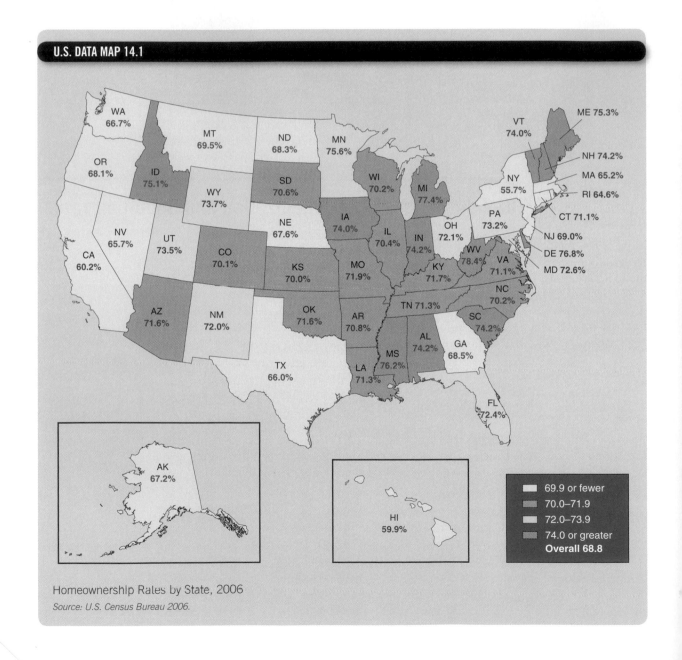

U.S. DATA MAP 14.1

Legend:
- 69.9 or fewer
- 70.0–71.9
- 72.0–73.9
- 74.0 or greater
- **Overall 68.8**

Homeownership Rates by State, 2006

Source: U.S. Census Bureau 2006.

explains that because housing is the single largest expenditure for most households, "housing affordability has the potential to affect all domains of life that are subject to cost constraints, including health" (2002:710). Most families pay their rent first, buying basic needs such as food, clothing, and health care with what they have left. The lack of public assistance, increasing prices, slow wage growth, and a limited inventory of affordable apartments and houses make it nearly impossible for some to find adequate housing (Pugh 2007).

The National Housing Conference, a nonprofit coalition of industry experts, advocates, and academics, reports that the average janitor earns enough to rent a one-room apartment and pay for other necessities in only 6 of the nation's 60 largest cities (Loven 2001). For instance, in Las Vegas, Nevada, the monthly rent for a one-bedroom apartment was $582, and the median hourly wage for a janitor was $8.92. The wage is below the $11.19-per-hour minimum recommended by the National Housing Conference to comfortably afford an apartment and additional expenses.

Renters are limited by more than just the increasing cost of housing; there is also a shortage of rental properties. The gap between the demand and the supply of affordable housing continues to grow. Higher land prices have made it less profitable to build new low- and moderate-priced housing. The National Low Income Housing Coalition (2007) reports that the nation is 2.8 million apartments short of meeting the need among people with the lowest incomes. For every new affordable housing unit constructed, two are demolished, abandoned, or converted into condominiums (Pugh 2007).

Consider the housing situation of Alice Greenwood and her six-year-old son, Makalii (Magin 2006). When the home she rented for 30 years for $300 a month was sold, she and her son joined some 1,000 people living in tents along the 13-mile stretch of beaches on the Waianae Coast of Oahu, Hawaii. In Waianae, homes that rented for $200 to 300 per month a couple of years ago now average more than $1,000 per month. For 2007, the median price of a home on Oahu was $665,000. Most of Greenwood's tent neighbors are employed in service and construction sectors. The state's homeless shelter is not a viable option for Greenwood and other beach residents—it would take them away from their communities (it is about 10 miles from the beach park) and would make commuting difficult (it is located several miles from the nearest bus route).

Although economic factors account for much of the disparity in home ownership, discrimination and prejudice also play a role. African Americans and Latinos are more likely than Whites are to be turned down for a home loan, even if they have similar financial, employment, and neighborhood backgrounds (Oliver 2003). HUD estimates that more than two million instances of housing discrimination occur each year. The highest number of complaints comes from African Americans, people with disabilities, and families with children. Hispanic renters experience higher rates of discrimination than African American renters do (Turner et al. 2002) and are discriminated against as much as 70 percent of the time in their housing search (Fair Housing Law 2004).

Household Crowding

Interior residential density refers to the number of individuals per room in a dwelling. The criterion for **crowding** is more than one person per room in the household. In the United States, household crowding is more likely to be found among poor, immigrant, or urban families. Research indicates that children who live in more crowded homes have greater behavioral problems in the classroom. Crowding also leads to greater conflict between parents and children. In crowded homes, parents have been found to be more critical of and less responsive to their children (Evans, Saegert, and Harris 2001).

W. Clark, M. Deurloo, and F. Dieleman (2000) argue that household crowding is linked to inequalities in housing consumption. The researchers explain that the rising income of a large segment of U.S. society has led to increases in the overall quality of housing in the United States. But at the same time, growing income inequalities create affordability and crowding problems for very-low-income households. Affluent households demand better quality and larger housing, increasing their consumption of livable space, pushing housing outside of inner-city boundaries. Clark et al. explain that as middle-class families move to the suburbs, they leave behind inner cities plagued with increased density and housing shortages.

Cities with large numbers of immigrants—such as those in such states as California, Texas, Arizona, and Florida—are especially subject to crowding. A study based in Southern California reveals that households of Hispanics who immigrated in the 1970s are two and a half times more likely to be overcrowded than are those of earlier Hispanic immigrants or White immigrants (Myers and Lee 1996). Clark et al. (2000) also report that in metropolitan areas with high levels of recent immigration, overcrowding is higher than in nonimmigrant areas. Poor immigrant households experience the most crowding and have the most difficulty in transitioning to better, more affordable housing. Studies suggest that competition for housing in cities with many immigrants may increase the cost of housing and can lead to a housing squeeze. Cultural norms may also play a role in encouraging crowding among Hispanic immigrant homes.

Substandard Housing

Many aspects of urban life—quality of air and drinking water, sanitation and fire services, and the availability and affordability of health care—have well-established connections to the health of urban dwellers (Cohen and Northridge 2000). One area that is often overlooked is the quality of housing. Substandard housing is a major public health issue (Krieger and Higgins 2002), particularly among urban dwellers. About 7 percent of all housing and 15 percent of all low-income rental housing has severe or moderate structural problems (malfunctioning plumbing, heating, or electrical systems, inadequate maintenance) (Freeman 2002). Housing quality has been associated with morbidity from infectious diseases, chronic illnesses, injuries, poor nutrition, and mental disorders (Krieger and Higgins 2002).

People of color and people with low income are disproportionately exposed to substandard housing (Krieger and Higgins 2002). For example, data from the American Housing Survey (U.S. Census Bureau 2002) showed that of those residing in an estimated 13.3 million housing units occupied by Blacks, about 5 percent reported having no working toilets in the previous three months, 10 percent reported being uncomfortably cold for 24 hours or more the previous winter, about 15 percent said they were subjected to bothersome street noise or traffic, and 27 percent reported the presence of neighborhood crime. About 10 percent of Black respondents reported having unsatisfactory police protection.

Homelessness

By its very nature, homelessness is impossible to measure with 100 percent accuracy (National Coalition for the Homeless 2002). Most estimates are based on head counts in shelters, on the streets, or at soup kitchens. These estimates do not include those who live in temporary or unstable housing (e.g., those who move in with friends or relatives). There are several national estimates of homelessness. The National Law Center on Homelessness and Poverty (1999) estimates that as many as 2 million people experience homelessness during one year, and more than 700,000 people are homeless on any given night. Most public and private sources agree that the number of homeless people is at least in the hundreds of thousands, not counting those who live with relatives or friends (Choi and Snyder 1999).

Globally, the highest concentrations of homeless people tend to be located in urban settings and segregated in some of the traditionally poorest areas (Toro 2007). In contrast to the United States, European countries experience lower rates of homelessness because their social welfare system guarantees some level of income, health care, and housing for all citizens. Japan is facing a rapidly growing problem of homelessness because its welfare system is even more underdeveloped than is that of the United States (Toro 2007).

In its 2006 *Status Report on Hunger and Homelessness* in the United States, the U.S. Conference of Mayors reported that single men make up 51 percent of the homeless population, families with children 30 percent, single women 17 percent, and unaccompanied minors 2 percent. Forty-two percent of the homeless population was estimated to be African American, 39 percent White, 13 percent Hispanic, 4 percent Native American, and 2 percent Asian. People were homeless for an average of eight months, an increase from the previous year, among a third of the cities surveyed.

Peter Rossi (1989) maintains that the homeless are likely to be more visible, female, a member of an ethnic minority, somewhere in their twenties or thirties, and unemployed, with no or low monthly incomes. Other developed countries have also seen an increase in the numbers of homeless families and youth (Toro 2007). Before the mid-1970s, the majority of the homeless were older, single males with substance abuse or physical or mental problems (Choi and Synder 1999). Their individual disabilities or personal pathologies were likely to have caused their homelessness. Since the mid-1970s, however, the increasing number of homeless men, women, and families indicates that more than individual disabilities or personal characteristics are causing homelessness (Choi and Snyder 1999). Studies indicate that family homelessness in the 1980s and 1990s was primarily attributable not to individual deficits but to the increased number of the poor, especially minority, single, female-headed households, and to the lack of affordable low-income housing units (Choi and Snyder 1999).

The U.S. Conference of Mayors (2006) identified several interrelated causes of homelessness: mental illness and the lack of needed services, lack of affordable housing, substance abuse and the lack of needed services, low-paying jobs, domestic violence, prisoner reentry, unemployment, and poverty. For nearly 70 percent of the surveyed cities, the lack of affordable housing was identified as the primary cause of homelessness. Requests for housing by low-income families and individuals increased in 86 percent of the surveyed cities.

What Does It Mean to Me?

What is the extent of homelessness in your community? Contact a housing agency or homeless shelter to determine the estimated number of homeless. How are the experiences of the homeless invisible to mainstream society? to students on your campus?

Gentrification

In their report, "Dealing with Neighborhood Change," Maureen Kennedy and Paul Leonard reveal that "**gentrification**, the process of neighborhood change which results in the replacement of lower-income residents with higher-income ones, has changed the character of hundreds of urban neighborhoods in America over the last 50 years" (2001:1). Gentrification has occurred in waves: the urban renewal efforts in the 1950s and 1960s and the "back-to-the-city" movement of the late 1970s and 1980s. Gentrification is a global experience, with renewal efforts documented in Tokyo, London, Mexico City, Cape Town, Paris, Shanghai, and Sydney, as well as in many other countries and cities (Smith 2002).

Kennedy and Leonard explain that gentrification is reemerging in the United States for three basic reasons. First, the nation's strong economy creates a greater demand for labor and housing, making housing in central cities and inner suburbs attractive to higher-income newcomers. Second, federal, state, and local governments, along with nonprofit organizations, have increased their level of motivation, funding, and policy initiatives to revitalize central cities. Under some circumstances, these revitalization efforts may lead to gentrification. Finally, in an effort to reduce the concentration of poverty in inner cities, public officials have attempted to attract higher-income families to the area.

The researchers describe gentrification as a "double edged sword." Officials and developers point to the increasing real estate values, tax revenues, and commercial activity that take place in revitalized communities. But is there a price for these capital and economic improvements? The most contentious by-product of gentrification is the involuntary displacement of a neighborhood's low-income residents. Gentrification is most often associated with the disproportionate pressure it puts on marginalized poor, elderly, or minorities, particularly renters. No consistent data exist on the number of individuals who have been displaced through gentrification; yet the evidence suggests that where housing markets are tight (or limited), the amount of displacement is likely to be greater and the impacts on those displaced more serious.

Urban Sprawl and Commuting

As urban areas spread out, they create a phenomenon referred to as **urban sprawl**. Urban sprawl began with land development after World War II. Sprawl is defined as the process in which the spread of development across the landscape outpaces population growth (Ewing, Pendall, and Chen 2002). Sprawl creates four conditions for an urban area: a population that is widely dispersed in low-density developments; rigidly separated homes, shops, and workplaces; a network of roads marked by huge blocks and poor access; and a lack of well-defined activity centers, such as downtowns or town centers (Ewing et al. 2002). Sprawl increases stress on urban livability as goods and services, along with economic and educational opportunities, become less accessible to inner city residents (Johnson 2007; Powell 2007).

As sprawl increases, so do the number of miles traveled, number of vehicles owned per household, traffic fatality rates, air pollution (Corvin 2001; Ewing et al. 2002), and eventually, our risk of asthma, obesity, and poor health. On average, an American spends 443 hours per year behind the wheel (Crenson 2003). In areas with high urban sprawl, only 2.3 percent of workers take public transportation to work, whereas in areas that have less urban sprawl, an average of 5.1 percent of workers take public transportation (Ewing et al. 2002). New suburban residential developments don't include sidewalks, and automobiles are needed to get from place to place.

The Centers for Disease Control reports that urban sprawl increases our time on the road and decreases our time spent exercising, including walking, jogging, or riding a bike (Corvin

2001). Residents who live in spread-out areas spend fewer minutes each month walking and weigh about six pounds more on average than do those who live in densely populated areas (Stein 2003).

Americans spend more than 100 hours per year commuting to work (U.S. Census Bureau 2005). On average, an American worker's daily commute is about 24 minutes (one-way commute time). Residents from larger cities tended to have longer commutes—New York (38 minutes); Chicago, Illinois (33 minutes); Newark, New Jersey (31 minutes); and Riverside, California (31 minutes). Long commutes are also a global experience, particularly for poor disadvantaged workers. For example, commuting to Brazil's largest industrial city, São Paulo, can take as much as four hours each way. São Paulo commuters are mostly minimum wage laborers, commuting daily from their working-class poor suburbs to the city's factories.

What Does It Mean to Me?

Examine the public transportation system for your college city. First, is one available? Second, assess its effectiveness. Does the system serve all areas of your community? How much does it cost to use the system? How does the system serve disadvantaged populations—elderly, poor, or disabled residents? Does the system provide discounted fares for students?

Photo 14.3

The average American spends 443 hours per year behind the wheel. In areas with high urban sprawl, only 2.3 percent of workers take public transportation to work.

In Focus

Living Car Free

North Americans are known for their love affair with, maybe even addiction to, their automobiles (Boddy 2000). According to the U.S. Department of Transportation (2001), there are more cars than people per household: The average U.S. household has 1.80 drivers and 1.90 personal vehicles. About 88 percent of persons 15 years of age or older are licensed drivers; in contrast, only 17 percent of the same age group used public transportation in the past two months. Between 2000 and 2005, 77 percent of commuters were driving alone to work. Despite rising fuel prices (reaching a high of $3.186 per gallon in 2007), U.S. housing and work patterns make it hard for suburban commuters to change their driving habits (Ohlemacher 2007).

However, many European countries and cities have found innovative and community friendly ways to deal with gas prices consistently higher than $6.00 per gallon. For example, most of the 4,700 residents of Vauban, Germany, live car free. The rate of Vauban car ownership is 150 per 1,000 inhabitants compared with 640 per 1,000 residents in the United States. How does Vauban do it? Extensive city planning and innovative public policy. The city was built with an extensive system of bike paths and little parking spots. Parking spots for vehicles are available in a garage at the edge of the community for €17,500 (more than $20,000 per year). Vauban city planners also encouraged residents to use public transportation such as tramways and buses. Many of the city's streets were designed to be too narrow for cars (de Pommereau 2006).

Photo 14.4

In 2007, the city of Paris introduced Velib', a self-service bicycle transit system. From more than 700 locations throughout the city, individuals can rent bikes by the hour. There are 230 miles of cycling lanes in Paris.

Car use is discouraged in many different ways throughout the world. In Germany, there is a yearly car tax based on the automobile engine size—the bigger the engine, the higher the tax. Several European cities maintain fleets of bikes for public use to encourage bike riding as a transportation alternative. European Union countries, such as France, Italy, and Germany, have closed off parts of their city centers to cars for a day or permanently. In 1983, Bogotá, Columbia, initiated a program called *ciclovia* (bike path). Designated streets are closed to cars every Sunday, encouraging residents to jog, walk, or ride their bikes on the streets and revitalizing their neighborhoods as a result. One and a half million people are said to turn out for the Sunday ciclovia (Wood 2007).

Car-free zones have also been embraced in several American cities. New York, Atlanta, Philadelphia, Cleveland, Chicago, and El Paso have begun promoting car-free days in public parks, designated neighborhoods, and green spaces. Every Saturday from May 26 through September 30, a mile of San Francisco's John F. Kennedy Drive is designated car-free, open to pedestrians, joggers, and bicyclists. Advocates argue that these car-free practices promote family activities, active lifestyles, and closer communities (Wood 2007).

Community, Policy, and Social Action

Department of Housing and Urban Development (HUD)

The federal agency responsible for addressing the nation's housing needs and improving and developing the nation's communities is HUD. Created in 1965 as part of President Lyndon Johnson's War on Poverty, HUD was given the authority to enforce fair housing laws and to administer a variety of federal programs to provide a decent, safe, and sanitary living environment for every American (HUD 2003b; Martinez 2000).

HUD's history extends back to the National Housing Act of 1934 and to the 1937 amendment that created the U.S. Housing Authority for low-rent housing. HUD's efforts to encourage home ownership are rooted in the Housing Act of 1949, a declaration that all Americans have the right to become home owners. Despite its expressed goal of creating "well planned and integrated residential neighborhoods," the Housing Act did not improve housing conditions for nonminority households (Martinez 2000). The goals of the Housing Act of 1949 were reaffirmed in the Fair Housing Act of 1968, authorizing the FHA to make sure that home ownership was affordable and accessible for every American family, including minorities and the poor (Martinez 2000). HUD continues its housing mission, expanding services to elderly residents, and oversees health care facilities and lead hazard control.

In addition, HUD has been a major player in influencing land use decisions in urban areas (Williams 2000), spurring economic growth and development in distressed communities (HUD 2003b). HUD's major urban initiatives have included the Housing and Development Act of 1970, which established a national growth policy that emphasized new community and inner-city development; the Housing and Community Development Act of 1974, which established community development block grants; and the Omnibus Budget Reconciliation Act of 1993, which created the first enterprise zones to stimulate economic development in distressed areas. Enacted in 2000, the Community Renewal and New Markets Initiative reinforced HUD's focus on fostering economic opportunity, enhancing the quality of life, and building a stronger sense of community in impoverished inner-city neighborhoods (Williams 2000). Renewal communities, urban empowerment zones, and urban enterprise communities were part of this initiative, and each will be discussed in the following section.

Renewal Communities, Empowerment Zones, and Enterprise Communities

With oversight provided by HUD, the renewal communities, empowerment zones, and enterprise communities program takes an innovative approach to revitalization that targets inner cities and rural areas. The program attempts to bring communities together through public and private partnerships to attract the social and economic investment necessary for sustainable economic and community development (HUD 2003c). Each program integrates four principles: a strategic vision for change, community-based partnerships, economic opportunity, and sustainable community development. The program begins with the assumption that local communities can best identify and develop local solutions to the problems they face (HUD 2003c).

Enterprise communities and empowerment zones rely on local and regional partnerships. Proposed by President Bill Clinton's administration, the Federal Omnibus Budget Reconciliation Act of 1993 provided federal support to distressed cities and communities in the form of funding, increased spending on social programs, tax credits, or deductions to encourage business growth, and affordable housing production. These empowerment zones have been characterized as urban laboratories where a variety of economic development strategies are tested and where lessons can be applied to future community policies (Butler 1991). Since the program's initiation, 32 empowerment zones have been created. The selected communities have census tracts with poverty rates of at least 20 percent of median family income or areas with pervasive poverty, unemployment, and general distress. The only difference between enterprise communities and empowerment zones is the level of support and tax credits. Communities designated as empowerment zones receive larger direct grants, as much as $100 million, whereas enterprise communities receive as much as $3 million in grant and loan guarantees. Rural empowerment zones may receive as much as $40 million. Businesses located in empowerment zones also receive tax credits not available in enterprise communities.

The renewal communities program was created under the Community Renewal Tax Relief Act of 2000, which authorized as many as 40 renewal communities. Rather than providing federal funds or grants to support business development and economic growth, the renewal communities program offers federal tax breaks and incentives for local businesses and communities. For example, the program offers a Work Opportunity Tax Credit to businesses, federal tax credits of as much as $2,400 for each new hire from groups that have high unemployment rates or other special employment needs, including youth ages 18 to 24 years, and summer hires of teens 16 to 17 years old who live in the renewal community area. To be eligible, each renewal community must meet criteria related to population, unemployment, poverty, and general distress.

Voices in the Community

Magic Johnson

Since his retirement from the Los Angeles Lakers, Earvin "Magic" Johnson has become a commercial developer opening state of the art multiplex theaters including restaurants, retail, personal service, and Starbucks locations. Johnson's company, Johnson Development Corporation, specifically targets business opportunities in minority inner-city and suburban neighborhoods.

At the opening of the 12-screen multiplex, the Magic Johnson Theaters, in South Central Los Angeles, Johnson was confident that such a business would succeed in inner cities because African Americans make up about 13 percent of the movie-going audience (Dretzka 1995). Johnson said,

> We're the No. 1 movie goers of any (minority) group but you can't find any theaters in your neighborhood. That's why our theaters are doing so much business. We have great numbers, and for everybody in the neighborhood, it means more to them than just a theater. It's a pride situation, bringing the community together. (Dretzka 1995:1)

The Magic Johnson Theaters cost an estimated $11 million and feature an art deco lobby with a large concession stand and a two-level parking garage. The Johnson Development Corporation has also opened theaters in Atlanta, Houston, Cleveland, and Harlem. The five theater complexes grossed $30 million in revenue in 2002 (Johnson 2003).

Johnson brought his understanding of the inner-city community to the business. In his movie concessions, knowing that inner-city children grew up drinking Kool-Aid, Johnson sells flavored sodas. "Used to be we couldn't afford to go to dinner and the movie afterward. I told Loew's [his theater partner], 'Black people are going to eat dinner at the movies,'" says Johnson (Wilborn 2002). As a result, in addition to hot dogs and popcorn, the concessions sell chicken wings and buffalo shrimp.

At his theaters, no gang colors or hanging out in large groups is allowed. Before each movie, a clip of Johnson is played, reminding his audiences: "So we got a few policies that apply to everyone. They are not meant to disrespect. They're there so we can all have a good time. So if you have a problem, leave it in the street" (Wilborn 2002).

Johnson has joined Howard Schultz, chief executive officer of Starbucks, in a franchise deal. Their partnership opened 37 locations by the end of 2002, with plans to expand to 125 locations within the next five to seven years. Their first location was in Ladera Center, a few miles away from the Los Angeles International Airport. The location is one of the biggest-grossing in the Starbucks chain. Schultz explains that through the partnership, "We could create unique opportunities for the community—employment opportunities, opportunities for vendors—and also some hope and aspiration about a leading consumer brand doing business in underserved communities, and perhaps other companies would follow us in" (Johnson 2003:77).

Johnson's business philosophy is simple, "All of my businesses deal with people, customer service, and entertainment because that's what I'm good at. Everything flows together from that, and all the companies help each other" (Smith 1999:80).

In 2002, Johnson established the Magic Johnson Technology Initiative Program, currently in 14 states, including the District of Columbia. Attempting to address the digital divide (refer to Chapter 8, "Education") among poor and minority children, the program establishes Community Empowerment Centers providing youth and community members access to computer technology, training, and experience.

Photo 14.5

Magic Johnson greets children at the Branch Brook School in Newark, New Jersey. Along with corporate sponsors Samsung and Microsoft, the Magic Johnson Foundation awarded the school $200,000 in software and technology products.

Urban Revitalization Demonstration Program: HOPE VI

The HOPE VI program was established by Congress in 1992. Originally called the Urban Revitalization Demonstration Program, HOPE VI spent nearly $4.8 billion to tear down 115,000 public housing facilities and revitalize some 85,000 others into larger modern town-homes and detached homes with the goal of creating mixed income communities in inner cities (Armas 2003; HUD 2003a). Funding was awarded to 146 housing authorities in 37 states, the District of Columbia, Puerto Rico, and the Virgin Islands (HUD 2003a).

The HOPE VI program was created based on the recommendations from the National Commission on Severely Distressed Public Housing (HUD 2003a). The commission recommended revitalization in three areas: physical improvements, management improvements, and social and community services to address residents' needs. Program grants pay for demolition of distressed public housing and rehabilitation or new construction. The program has been criticized for worsening the local housing situation because not all demolished units are replaced and program data reveal that not all residents return to the redeveloped HOPE VI sites. As of September 2002, the HOPE VI program planned to demolish more than 78,000 public housing units but to replace only 34,000 units (Center for Community Change 2003).

Although the program's primary focus is on the quality of housing units, HUD officials report that the HOPE VI program has made an impact on its residents through community and supportive programs for residents. When the program was honored in 2000 by the Institute for Government Innovation, HOPE VI officials reported that nearly 3,500 public housing residents left welfare and more than 6,500 residents found new jobs as a result of the program (Institute for Government Innovation 2000).

In 1995, the Seattle Housing Authority was one of the first public housing authorities to receive HOPE VI funding (Naparstek et al. 2000). With $48 million, the authority tore down and replaced Holly Park, one of the most distressed public housing projects in Seattle, Washington. Holly Park, with 893 units, has been replaced with NewHolly, a mixed-income community with 400 public housing units, 400 tax credit rental units for low-income households, and 400 affordable home ownership units. Of the original 893 residents, 393 moved into NewHolly. Most residents who did not return moved into other public housing or into Section 8 rental housing. The renovation of Holly Park also involved a collection of supportive service programs and educational initiatives called the Campus of Learners, each program promoting self-sufficiency. The Campus of Learners includes a career development center, the Atlantic Street Family Center, a branch campus of South Seattle Community College, child care resources, and an onsite public library, among other community partners. Through the program, "The housing authority moved to a new kind of partnership—a collaborative, community-resource, community building model" (Naparstek et al. 2000:16). The HOPE VI program continues, with funding provided to smaller communities to rejuvenate their older downtown business districts.

Creating Sustainable Communities

Tyler Norris (2001) chronicled the emergence and importance of the sustainable community movement. Since the early 1960s, thousands of public-private partnerships have been formed to work for economic development, educational improvement, environmental protection, health care, social issues, and other issues critical to communities. An array of private and public community groups form these partnerships. These alliances have been identified by several names and terms: healthy communities, sustainable communities, livable communities, safe communities, whole communities, or smart growth. Much of the improvement in public health, community revitalization, and quality of life can be attributed to these alliances.

Table 14.6

Best Practices for Sustainable Communities

1. Define community broadly, using not only physical space but also community of interest (e.g., youth assets).

2. Make the community's vision reflect the core values of all its members.

3. Define health as the optimum state of well-being—physical, mental, emotional, and spiritual.

4. Address the quality of life as experienced by all residents.

5. Invite diverse participation and promote widespread community ownership.

6. Focus on system change to address how people live and work together.

7. Use local assets and resources to build capacity.

8. Measure and report your progress and outcomes to keep citizens informed and to keep partners accountable.

Source: Adapted from Norris 2001.

The best partnerships, according to Norris, bring together traditional leaders and community members often not included in the decision-making process. A summary of best practices from successful sustainable communities is presented in Table 14.6.

Examples of sustainable communities include the following:

- Urban Resources Initiative, Detroit, Michigan. The initiative is a program of the Michigan State Department of Forestry, a community forestry program that uses a "bottom up" approach to address community needs (Sustainable Communities 2002). Since 1990, Detroit residents have been reclaiming vacant city land and using it for innovative forestry projects that offer social, economic, and environmental benefits. When a community is ready to start its own project, it contacts the Urban Resources Initiative. The work begins with a community assessment. The initiative builds on the skills and materials already available in the community, for example, identifying the resident "gardener" in the community and someone who has a spare lawnmower that could be used. The Appoline Street project may exemplify the best of these efforts. Appoline community members have worked on a community garden for several years. Block Club President Alice Dye teaches neighborhood children about plants, bugs, and ecology while they work together in the garden. The children sell their produce at a vegetable stand for a nominal fee and use the profits for other community projects. Funding for the program expired in 1996; the organization's primary objective was to provide resources for community sustainability, rather than rely on the Urban Resources Initiative.

- Highlander Research and Education Center, New Market, Tennessee. The center was established in 1932, working primarily on social change and education in the areas of labor, civil rights, and Appalachian issues. In the 1990s, the center expanded its program focus to be a multi-issue, multicultural, and intergenerational movement for social and economic change. Current programs include an environmental economic education program, a community environmental health program, a cultural and diversity initiative, a global education project, a residential education program, and the Southern Appalachian Leadership Training program. Each program serves as an invaluable resource to community groups in Appalachia and the South.

What Does It Mean to Me?

Go to *Study Site Chapter 14* to identify the sustainable communities and initiatives in your area. What principles and values does your local group embrace? Is there evidence that their efforts have worked in the community? Why or why not?

Housing and Homelessness Programs

The one community response to homelessness that most of us are familiar with is the homeless shelter. These shelters have been referred to as "band aid" solutions, helping but not really fixing the problem. However, these emergency programs can provide immediate and necessary assistance and, in particular, security for families with children. If these shelters are to be truly effective, more humane and supportive shelter environments should be promoted to assist families and to better prepare them for independent living (Choi and Snyder 1999). For example, existing social and human services programs will be more effective if the homeless are able to obtain the benefits that they are already eligible for (Rossi 1989), such as social security, disability payments, and food stamps.

In 1987, Congress passed the Stuart B. McKinney Homeless Assistance Act, which established assistance programs for homeless individuals and families. Under this act, 20 programs were authorized to provide emergency food and shelter, transitional and permanent housing, education, mental health care, primary health care, and veteran's assistance services. In an effort to create more affordable housing, under Title II of the 1998 Cranston-Gonzalez National Affordable Housing Act, the HOME program provides grants to state and local governments to build, buy, or rehabilitate affordable housing for rent or home ownership. Working with community groups, the HOME program targets low-income families.

Community efforts are important for the homeless. Project Homeless Connect began in San Francisco in 2004 when Mayor Gavin Newsome had the idea of bringing city hall staff and programs to the homeless community. The project, replicated in more than 125 cities, attempts to reach out to a city's homeless population by delivering an array of services—social, medical, mental health, housing—all under one roof at a local venue. Quality of life services are also offered for the day including haircuts, wheel chair repair, eye glasses, and dental services. The project connects homeless men and women with program representatives and members of the community. The event is staffed by volunteers and is supported through donations from local businesses.

The best-known community based housing program is Habitat for Humanity International. Habitat serves primarily low- and very-low-income families with support from local volunteers, churches, and businesses, as well as the sweat equity of future home owners. At the time of a 1998 HUD review of the program, 84 percent of Habitat's participants were families with an average family income of $24,251 (Applied Real Estate Analysis 1998). The average sales price for a Habitat home was $37,782. Habitat home buyers reported that the greatest benefit of home ownership was the pride and security they felt. Habitat's U.S. housing production volume puts it in the ranks of the nation's top 20 home builders (Applied Real Estate Analysis 1998). Other community-based initiatives include Trinity Housing Corporation (Columbia, South Carolina), established by the Trinity Episcopal Church in 1989. It relies primarily on volunteers to provide safe, affordable, transition housing for homeless families (Grass-Roots.org 2003b). Another is the Light Street Housing Corporation (Baltimore, Maryland), established in 1985 by a group of local churches. The program purchases and renovates (one at a time) low-cost housing for the homeless (Grass-Roots.org 2003a).

Although supportive services are necessary for the homeless, homelessness cannot be prevented or eliminated without enough housing for the poor. Homelessness cannot be prevented or eliminated without a livable wage, employment opportunities for inner-city residents, more efficient management of public housing projects, emergency rent assistance programs, and the expansion of low-income housing subsidies (Choi and Snyder 1999).

Main Points

- Although urban areas provide examples of economic and social progress, they also have many social problems, with opportunities and resources distributed unevenly. **Urban sociology** examines social, political, and economic structures and their impact within an urban setting.

- The first studies in urban sociology adopted a functionalist approach, comparing a city to a biological organism. The Chicago School developed two dominant traditions in urban study, one focusing on **human ecology** (the study of the relationship between individuals and their physical environment) and population dynamics and the second focusing on community studies and ethnographies.

- **Demography**, an essential part of urban studies, is the study of the size, composition, and distribution of human populations, as well as changes and trends in those areas. An additional demographic element is **migration**, the movement of individuals from one area to another.

- **Urbanization**, the process by which a population shifts from rural to urban locales, expanded in the latter half of the nineteenth century. **Overurbanization**, the process where an excess population is concentrated in an urban area that lacks the capacity to provide basic services and shelter, began during the late twentieth century in developing countries like Africa and Asia. After World War II, the United States experienced another significant population shift: **suburbanization**, or the outward expansion of central cities into suburban areas.

- Changes in fertility, mortality, and migration rates affect the population composition. **Age distribution**, or the distribution of individuals by age, is particularly important as it provides a community with some direction in its social and economic planning. Demographers have predicted the graying of the U.S. population and accompanying problems. Ethnic and age compositions will also affect community services and priorities.

- Functional solutions to urban social problems encourage reinforcing social bonds through existing institutions or instituting societal changes.

- Since the late 1960s, a new approach called the **critical political-economy** or **socio-spatial perspective** uses a conflict perspective to focus on how cities are formed on the basis of racial, gender, or class inequalities stemming from capitalism. In this view, social problems are natural to the system, arising from the unequal distribution of power among various groups.

- Feminist theorists have argued for the development of a comprehensive field of theory and research that acknowledges the role and experiences of women in urban environments.

- Urban communities continue to be segregated by income, race/ethnicity, or immigrant status. This contributes to isolation, not just physically but also through the meanings attached to different neighborhoods, a focus of the interactionist perspective. This social isolation leads us to believe that certain social problems are unique to someone else's neighborhood.

- Along with suburbanization came the decentralization of American cities. Inner cities became repositories for low-income individuals and families.

- Substandard housing and **crowding** are major public health issues, particularly among urban dwellers. Housing quality has been associated with morbidity from infectious diseases, chronic illnesses, injuries, poor nutrition, and mental disorders. People of color and those with low income are disproportionately exposed to substandard housing.

- The increasing number of homeless people in the 1980s and 1990s was primarily attributable to the increased number of the poor, especially minority, single, female-headed households, and to the lack of affordable low-income housing units. Substance abuse, mental illness, domestic violence, poverty, and low-paying jobs also have contributed.

- Another problem is **gentrification**, or the process of neighborhood change that results in the replacement of lower-income residents with higher-income ones, a process most often associated with the disproportionate pressure it puts on marginalized groups.

- As urban areas spread out, they create **urban sprawl**, in which development across the landscape outpaces population growth. As sprawl increases, so do the number of miles traveled, number of vehicles owned, traffic fatality rates, air pollution, and eventually, the risk of poor health.

- The federal agency responsible for addressing the nation's housing needs and improving and developing the nation's communities is the U.S. Department of Housing and Urban Development (HUD). With HUD oversight, renewal communities, empowerment zones, and enterprise communities have attempted to bring communities together through public and private partnerships to attract necessary social and economic investments.

- Since the early 1960s, a sustainable community movement composed of thousands of public-private partnerships has worked for economic development, educational improvement, environmental protection, health care, social issues, and other issues critical to communities. Much of the improvements in public health, community revitalization, and quality of life can be attributed to these alliances.

On Your Own

Log on to the Web-based student study site at www.pineforge.com/leonguerrero2study for interactive quizzes, e-flashcards, journal articles, Community and Policy Guides, a Service Learning Guide, the end-of-chapter Web exercises, and additional Web resources.

Internet and Community Exercises

1. Use U.S. Census (log on to *Study Site Chapter 14*) or local data to determine the population growth or decline for your community. What areas of your community are thriving? What areas are declining? Overall, how would you characterize the state of your community?

2. Investigate the Center for Neighborhood Technology, the Civic Practices Network, and U.S. Conference of Mayors Web sites. These organizations attempt to address various urban living, quality of life, and urbanization issues. What solutions, policies, and programs are promoted by each? How does this affect your community? Log on to *Study Site Chapter 14.*

3. If you can't live without a car, perhaps you might consider sharing one. Car-sharing programs began in the 1980s in Switzerland and Germany, with Canadian and U.S. programs established in the 1990s and early 2000s. By 2007, there were 18 car-sharing programs in the United States. FlexCar, a national car-sharing service, has provided access to a vehicle for a monthly user fee since 2004. Subscribers pre-pay for hours per month and are able to pick up and drop off a vehicle in a designated location. FlexCar also has a campus-based program. Investigate whether a car-sharing program exists in your city, state, or campus. What advantages and disadvantages can you identify with sharing a car?

4. Community interest developments are private subdivisions where residents own their homes and pay an annual or monthly home association fee. For their nominal or sometimes large fee, residents ensure that their "neighborhood" is strictly managed by a residential board backed by a series of covenants, codes, and restrictions dictating the color of one's home, how many cars can be parked in the driveway, or where fences or a garden shed can be built. These communities have also been referred to as "gated communities," reinforcing the notion that their residents are separating themselves physically and psychically from the general population in their city or town. Although the exact number of community interest developments is not known, industry estimates suggest that by 2000, some 48 million Americans were living under such agreements (Drew 1998). Determine if there are any community interest developments in your community. If available, interview a member or officer of a development's board. What does the development's codes statement cover? How does the board member describe the quality of life in the development? Would you want to live in this development? Why or why not?

5. Log on the United Nations Human Development Report Web site and review data for the total population and the percent (rate) of urban population. Identify the five countries with the highest rates and the five countries with the lowest for 2004 and 2015 (predicted). How would you characterize the five countries with the highest rates? What do these countries have in common? Log on to *Study Site Chapter 14* for more information.

CHAPTER 15

The Environment

A zerbaijan, China, India, Peru, Russia, Ukraine, and Zambia are among the world's most polluted places. These countries (refer to Table 15.1) are exposed to organic and industrial pollutants caused by mining, manufacturing, and transportation. Although much of these pollutants can be attributed to the countries' substandard infrastructures or the absence of regulatory controls, even if both were brought up to industrial country standards, "the legacy of old contamination from the past would continue to poison the local population" (Blacksmith Institute 2007:3). This pollution is a major cause of illness, death, and long-term environmental damage.

The Shanxi Province is at the center of China's coal mining industry. The country's industrial and urban growth has increased the demand for coal, one of the dirtiest sources of energy (Kahn and Yardley 2007). Coal provides China with two-thirds of its energy supply. During recent years, Shanxi residents have witnessed the development of hundreds of unregulated coal mines, steel factories, and refineries. In Linfen, a Shanxi city referred to as the most polluted city in China, there has been an increase in the number of bronchitis, pneumonia, and lung cancer cases linked with air pollution. In addition, lead poisoning and arsenicosis (a disease caused by drinking water with high levels of arsenic) are found in higher levels among Linfen residents than among residents in other parts of China (Blacksmith Institute 2007).

China's pollution problem is a world problem. Sulfur dioxide and nitrogen oxides from its coal fired power plants turn into acid rain falling on Seoul, South Korea, and Tokyo, Japan (Casey 2007; Kahn and Yardley 2007). It takes 5 to 10 days for China's coal pollution to reach

Table 15.1

Top Ten Worst Polluted Places (alphabetically by country)

Site Name and Location	Major Pollutants and Sources
Sumgayit, Azerbaijan	Organic chemicals and mercury, from petrochemical and industrial complexes
Linfen, China	Particulates and gases from industry and traffic
Tianying, China	Heavy metals and particulates from industry
Sukinda, India	Hexavalent chromium from chromium mines
Vapi, India	Wide variety of industry effluents from industrial estates
La Oroya, Peru	Lead and other heavy metals from mining and metal processing
Dzerzhinsk, Russia	Chemicals and toxic byproducts, lead from chemical weapons and industrial manufacturing
Norilsk, Russia	Heavy metals, particulates from mining and smelting
Chernobyl, Ukraine	Radioactive materials from nuclear reactor explosion
Kabwe, Zambia	Lead from mining and smelting

Source: Blacksmith Institute 2007.

the United States. Scientists have identified China as the source of mercury (a by-product of coal-mining) in fish living in Oregon's Willamette River (Casey 2007) and particulate pollution over Los Angeles, California (Kahn and Yardley 2007).

Environmental Problems Are Human Problems

Humans create environmental problems through intentional efforts to exploit or manage nature. Rivers that are dammed, straightened, or treated as sewers may create unintended downstream environmental problems (Caldwell 1997). The removal of rainforests to harvest wood or to create farmland decreases the number of plants and trees that absorb carbon dioxide, leading to higher amounts of greenhouse gases in the air. But environmental problems don't exist just because of our actions. As exemplified by the coal mining industry in the Shanxi Province, our pursuit of economic development, growth, and jobs has also led to the degradation of the environment (Caldwell 1997). The state of the environment is also influenced by our cultural values and attitudes toward the environment, our social class, our technology, and our relationship with others (Cable and Cable 1995).

The field of **environmental sociology** considers the interactions between our physical and natural environment, on the one hand, and our social organization and social behavior on the other (Dunlap and Catton 1994). Human beings are an integral part of the ecosystem (Irwin 2001). When we use a sociological perspective to understand environmental problems, we acknowledge that "human activities are causing the deterioration in the quality of the environment and that environmental deterioration in turn has negative impacts on people" (Dunlap 1997:27).

Photo 15.1

This nickel refinery in Norilsk, Russia, releases 2.8 million metric tons of sulfur dioxide into the atmosphere, six times the emissions of the entire U.S. nonferrous metals industry. Norilsk is one of the Blacksmith Institute's most polluted places in the world (refer to Table 15.1).

Environmentalist Paul Hawken (1997) refers to the Biosphere II experiment to demonstrate just how vital and fragile our ecosystem is. The Biosphere II was a three-acre glass-enclosed ecosystem intended to sustain eight people for a two-year experiment, from September 1991 through September 1993. The Biosphere II's $200 million budget was not enough to create a viable ecosystem for eight people. By the time the experiment ended, the Biosphere's air and drinking water were polluted, crops and trees had been killed by other vegetation, and 19 of the 25 small animal species they brought with them had died. The scientists who lived in the biosphere showed signs of oxygen starvation from living at the equivalent of an altitude of 17,500 feet. Even with scientific knowledge and planning, there are no human-made substitutions for essential natural resources.[1] Hawken explains,

> We have not come up with an economical way to manufacture watershed, gene pools, topsoil, wetlands, river systems, pollinators, or fisheries. Technological fixes can't solve problems with soil fertility or guarantee clean air, biological diversity, pure water, and climatic stability; nor can they increase the capacity of the environment to absorb 25 billion tons of waste created annually by America alone. (1997:41)

Sociological Perspectives on Environmental Problems

Functionalist Perspective

Whether they are looking at a social system or an ecosystem, functionalists examine the entire system and its components. Where are environmental problems likely to arise? Functionalists would answer that problems develop from the system itself. Agricultural and industrial modes of production are destabilizing forces in our ecosystem. Agriculture replaces complex natural systems with simpler artificial ones to sustain select highly productive crops. These crops require constant attention in the form of cultivation, fertilizers, and pesticides, all foreign elements to the natural environment (Ehrlich, Ehrlich, and Holdren 1973). When it first began, industrialization entered a society that had fewer people, less material well-being, and abundant natural resources. But modern industrialization uses "more resources to make few people more productive" and as a result, "more people are chasing fewer natural resources" (Hawken 1997:40). As much as agriculture, industrialization, and related technologies have improved the quality of our lives, we must also deal with the negative consequences of waste, pollution, and the destruction of our natural resources. Human activities have become a dominant influence on the Earth's climate and ecosystems (Kanter 2007).

Biologists Paul Ehrlich and Anne Ehrlich (1990) contend that the impact of any human group on the environment is the product of three different factors. First is the population, second is the average person's consumption of resources or level of affluence, and third is the amount of damage caused by technology. The Ehrlichs present a final formula: Environmental Damage = (Population Growth) × (Level of Affluence) × (Technological Damage). A high rate of population growth or consumption can lead to a "hasty application" of new technologies in an attempt to meet new and increasing demands. "The larger the absolute size of the population and its level of consumption, the larger the scale of the technology must be, and, hence, the more serious are the mistakes that are made" (Ehrlich et al. 1973:15).

There is no simple way to stop the escalation of environmental problems. Halting population growth would be a good start but by itself could not solve the problem. Reducing technology's impact on the environment might be useful, but not if our population and affluence were allowed to grow. According to Ehrlich et al., the only way to address environmental problems is to simultaneously attack all components.

What Does It Mean to Me?

Whether we look at individuals, cities, or nations, everyone and everything has an impact on the Earth because we consume the finite products and services of nature. As a result, we each leave an **ecological footprint**, some environmental impact on the amount of natural resources we use and waste output we create. Footprints are calculated for countries by measuring the amount of resources (e.g., fossil fuel, acreage and land, housing, and transportation) consumed in a given year. Individual footprints can also be estimated. To estimate your personal footprint, visit *Study Site Chapter 15.*

Photo 15.2

A woman stands atop a mountain of rubbish in a China landfill. China generates an estimated 150 million tons of rubbish per year, as much as 400 million tons by 2020.

Conflict Perspective

Public discourse on environmental problems is often framed in terms of costs and interests. Do you save the spotted owl habitat or hundreds of logging jobs? Should you close a factory or save the river where its waste is being dumped? From this sociological perspective, environmental problems are created by humans competing for power, income, and their own interests.

Our capitalist economic system has been identified as a primary source of the conflict over polluting (or conserving) our natural world. Competing political and economic interests ensure that this conflict will continue. J. Clarence Davies (1970) argues that the capitalist system encourages pollution, simply because air and water are treated as infinite and free resources. Polluters don't really consider who or what is being affected by environmental problems. If a paper mill is polluting the river, it doesn't affect the paper mill itself but, rather, the users of the water or the residents downstream. If a power plant is polluting the air, the plant doesn't pay for the cost of using the air, only the cost of cleaning up a polluted area (Davies 1970).

Environmental problems occasionally make life unpleasant and inconvenient, but most Americans will tolerate this in exchange for the benefits and comforts associated with a developed industrial economy (Tobin 2000). A higher standard of living has been confused with consumption: More is better. Politicians encourage lower taxes so that we have more money to spend. Television and print media overwhelm us with products and services and tell us that we cannot live without them. But increased consumption requires increased production and energy, which in turn leads to environmental damage. David Korten explains,

> About 70 percent of this productivity growth has been in … economic activity accounted for by the petroleum, petrochemical, and metal industries; chemical intensive agriculture; public utilities; road building; transportation; and mining … the industries that are most rapidly drawing down natural capital, generating the bulk of our toxic waste, and consuming a substantial portion of our renewable energy. (1995:37–38)

What Does It Mean to Me?

Hawken (1993) contends that we should be able to put an economic value on our renewable (forests, fisheries) and nonrenewable (coal, oil) natural resources. He advocates for **natural capitalism**, the awareness of the value of nature as a system, no different than assessing the value of human or financial capital. We are able to attach a dollar amount to a tree once it is cut down for its timber, but what is its value as a living part of our ecosystem? Is it possible to put a price tag on nature?

Polluters target those with the least amount of power. Robert Bullard defines **environmental racism** as "any environmental policy, practice or directive that differentially affects or disadvantages individuals, groups, or communities based on race or color" (1994:98). Research consistently indicates that low-income people and people of color are exposed to greater environmental risks than are those who live in White or affluent communities. Evidence also indicates that low-income people and people of color suffer higher levels of environmentally generated diseases and death as a result of their elevated risk (Ringquist 2000). Environmental racism has been expanded to include members of other disadvantaged communities, examining heightened environmental risk based on race, class, gender, education, and political power.

Feminist Perspective

The feminist perspective argues that a masculine worldview is responsible for the domination of nature, the domination of women, and the domination of minorities (Scarce 1990). Ecofeminism may be the dominant feminist perspective for explaining the relationship between humans and the environment (Littig 2001). Ecofeminism was introduced in 1974 in an effort to bring attention to the power of women to bring about an ecological revolution. Ecofeminists argue, "Men driven by rationalism, domination, competitiveness, individualism, and a need to control, are most often the culprits in the exploitation of animals and the environment" (Scarce 1990:40). According to ecofeminists, "Respect for nature generally promotes human welfare and genuine respect for all human beings tends to protect nature" (Wenz 2001:190). Other feminist approaches include the feminist critique of natural science, feminist analyses of specific environmental issues (work, garbage, consumption), and feminist contributions to sustainable development (Littig 2001).

Cynthia Hamilton (1994) argues that environmental conflicts mirror social injustice struggles in other areas—for women, for people of color, for the poor. In environmental movements, Hamilton explains, what motivates activist women is the need to protect home and children. As the home is defined as the woman's domain, her position places her closest to the dangers of hazardous waste, providing her with an opportunity to monitor illnesses and possible environmental causes within her family and among her neighbors. As Hamilton sees it, these women are not responding to "'nature' in the abstract but to their homes and the health of their children" (p. 210).

The modern **environmental justice** movement emerged out of citizen protests at Love Canal, near Niagara Falls, New York (Newman 2001). The movement is based on the principle that "all peoples and communities are entitled to equal protection of environmental and public health laws and regulations" (Bullard 1994). For many years, the movement has effectively brought racial and economic discrimination in waste disposal, polluting industries, access to services, and the impacts of transportation and city planning to the public's attention (Morland and Wing 2007).

At the center of Love Canal's citizens' protest movement was a group of local women who called themselves "housewives turned activists." Lois Gibbs and Debbie Cerillo formed the Love Canal Homeowners Association in 1978. Concerned about the number of miscarriages, birth defects, illnesses, and rare forms of cancer among their families and neighbors, the women worked with Beverly Paigen, a research scientist, to document the health problems in their community (Breton 1998). The information they collected became known as "housewife data" (Newman 2001). The women held demonstrations, wrote press releases, distributed petitions, and provided testimony before state and federal officials (Newman 2001). In 1978, Love Canal was declared a disaster area, some 800 residents were evacuated and relocated, and the site was cleaned up. Gibbs went on to form the Center for Health, Environment, and Justice and continues to work on behalf of communities fighting toxic waste problems. More information about Love Canal is presented in the section, "Hazardous Waste Sites and Brownfields."

Interactionist Perspective

Theorists working within the interactionist perspective address how environmental problems are created and defined. Riley Dunlap and William Catton explain, "Environmental sociologists have a long tradition of highlighting the development of societal recognition and definition of environmental conditions as 'problems'" (1994:20). Environmental problems do not materialize by themselves (Irwin 2001). As John Hannigan (1995:55) describes, the successful construction of an environmental problem requires six factors: the scientific authority for and validation of claims; the existence of "popularisers" (activists, scientists) who can frame and package the "problem" to journalists, political leaders, and other opinion makers; media attention that frames the problem as novel and important (such as the problems of rainforest destruction or ozone depletion); the dramatization of the problem in symbolic or visual terms; visible economic incentives for taking positive action; and the emergence of an institutional sponsor who can ensure legitimacy and continuity of the problem.

Social constructionists do not deny that real environmental problems exist. Rather their interest is in "the process through which environmental claims-makers influence those who hold the reins of power to recognize definitions of environmental problems, to implement them and to accept responsibility for their solution" (Hannigan 1995:185). This perspective helps us understand how environmental concerns vary over time and how some problems are given higher priority than others.

Table 15.2

Summary of Sociological Perspectives: The Environment

	Functional	Conflict/Feminist	Interactionist
Explanation of environmental problems	Environmental problems are dysfunctions of modern living, the result of agricultural and industrial modes of production.	Problems are created by humans competing for power, income, and their own interests. Our capitalist economic system has been identified as a primary source of the conflict over polluting (or conserving) our physical and natural worlds. According to the feminist perspective, a masculine worldview is responsible for the domination of nature, the domination of women, and the domination of other minorities.	Theorists from this perspective address how environmental problems are created and defined.
Questions asked about the environment and environmental problems	How are environmental problems related to our modes of production? To our patterns of consumption? Are environmental problems inevitable consequences of modern living?	How do environmental problems emerge from our capitalist economic system? from a patriarchal society? Which particular groups are at risk for experiencing environmental problems or their impacts?	How are environmental problems created? What factors are included in the process? How is a problem legitimized? What individuals or groups play a role in the process?

What Does It Mean to Me?

Which sociological perspective offers the best explanation of environmental problems? Based on your answer, what solution(s) would be appropriate? For a summary of these perspectives, see Table 15.2.

Social Problems and the Environment

Global Warming

To sustain life on Earth, a certain amount of surface heat is required. Heat becomes trapped through the buildup of greenhouse gases—water vapor, carbon dioxide, and other gases—making the Earth's average temperature a comfortable and sustainable 60 degrees Fahrenheit (EPA 2003b). The term **global warming** refers, literally, to the warming of the Earth's surface.

Photo 15.3

Global warming is causing glaciers to melt or recede throughout the world. This photo offers an aerial view of a receding glacier in Los Glaciares National Park, Argentina. The loss of glaciers affects human populations and ecosystems.

It refers to perceptible climate trends on a time scale of decades or more rather than weather conditions or atmospheric events in a specific area at a particular time. According to the National Academy of Sciences, the surface temperature has risen by about 1 degree Fahrenheit in the past century, with warming accelerating during the last two decades. The problem is the accumulation of specific greenhouse gases—carbon dioxide, methane, and nitrous oxide—primarily attributable to human activity during the past 50 years (EPA 2003b).

Since the Industrial Revolution, concentrations of carbon dioxide have increased nearly 30 percent, methane concentrations have increased by 145 percent, and nitrous oxide concentrations have increased 15 percent (Ehrlich and Ehrlich 1996). Fossil fuels used to run cars and trucks, heat homes and businesses, and power factories are responsible for about 98 percent of carbon dioxide emissions, 24 percent of methane emissions, and 18 percent of nitrous oxide emissions. Agriculture, deforestation, landfills, and mining also add to the amount of emissions.

Although they are unable to predict specifically what will happen, where it will happen, and when it will happen, scientists have identified how our health, agriculture, resources, forests, and wildlife are vulnerable to the changes brought about by global warming. Soil moisture may decline in many regions; rainstorms may become more frequent. Changing regional climates could alter forests, crop yields, and water supplies. Sea levels could rise two feet along most of the U.S. coast. The Intergovernmental Panel on Climate Change (IPCC) projects that global warming should increase by 2.2 to 10 degrees Fahrenheit by 2100 (EPA 2003b). In its 2007 report, IPCC (Kanter and Revkin 2007) predicted that the most severe effects would be felt in poor countries and areas facing existing dangers from climate and coastal hazards. The report also confirmed that human activity has been the main case of warming since 1950 (Kanter and Revkin 2007).

Americans believe that global warming is a real problem and that human activity is a contributing factor. A 2007 national poll revealed that if asked to choose between stimulating the economy and protecting the environment, 52 percent of Americans chose protecting the environment. There was strong agreement to several strategies to reduce global warming—more than 90 percent of those surveyed favored requiring car manufacturers to produce more energy efficient cars and 75 percent were willing to pay more for electricity generated by renewable energy resources such as wind or solar power. Americans indicated that the United States should be a global leader in addressing environmental problems and in the development of alternative energy resources (Broder and Connelly 2007).

During his presidency, President George W. Bush downplayed the problem of global warming, referring to the lack of scientific evidence confirming the causes and consequences of global warming. In March 2001, President Bush announced that the United States would not support the Kyoto Treaty, which was drawn up in 1997 to implement the United Nations Framework Convention for Climate Change. The treaty limits the emissions of greenhouse gases by an average of 5.2 percent below 1990 levels. According to Bush, the Kyoto regulations would have become too burdensome for U.S. industry at a time when businesses were struggling with a slowing economy. In addition, the president noted that leading world polluters, India and China, were exempted from the treaty. Responding to the president's announcement, European Union Environment Commissioner Margot Wallstrom told reporters, "We don't see that it's such a good idea to sort of let the Americans off the hook, those who are among the biggest emitters of greenhouse gases." The United States produces 25 percent of the world's greenhouse gases but includes only 4 percent of its population (NewsMax.com 2001).

In his second term, Bush began to reverse his position on global warming. In 2005, he agreed for the first time that human action was responsible for global warming, and in 2007, the president called for a long-term plan for cutting greenhouse gas emissions in the United States and throughout the world. Acknowledging how science has "deepened our understanding of climate change," the president called for other countries to set national targets to reduce greenhouse gas emissions in 10 and 20 years (Stolberg 2007). As of 2007, the United States had not ratified the Kyoto Treaty. Days after former Vice President Al Gore and the United Nations' Intergovernmental Panel on Climate Change were awarded the Nobel Peace Prize for their efforts to increase climate change awareness, President Bush referred to the Kyoto Treaty as "bad policy" and expressed interest in playing a significant role in negotiating its successor.

What Does It Mean to Me?

Do you believe that global warming is a social problem? How has global warming affected your life? What has influenced your opinion about global warming?

Air Quality

The quality of the air we breathe is subject to pollution from two sources: particulate matter and smog. Research has linked air pollution to acute and chronic illnesses (e.g., burning eyes and nose, asthma), as well as death. Air pollution also leads to environmental and property damage.

Particulate or particle pollution is caused by the combustion of fossil fuels—the burning of coal, diesel, gasoline, and wood. Particulate matter includes road dust, diesel soot, ash, wood smoke, and sulfate aerosols that are suspended in the air (NRDC 2007). The U.S. Environmental

Protection Agency (EPA) is concerned about small particulate matter, 10 micrometers in diameter or smaller because these smaller particles are able to pass through the throat and nose and enter the lungs causing respiratory problems (EPA 2007a).

Particulate pollution caused by traffic and diesel engines is a growing problem in many European Union cities. The World Health Organization (WHO) set the acceptable air quality standard at 10 micrograms of particles per cubic meter. However, nowhere in Europe is this standard being met—at the lower end of the spectrum are Paris and London (16 micrograms per cubic meter) and at the highest are cities such as Warsaw (34), Turin (41), and Milan (38) (Rosenthal 2007). The standard in the United States is 15 micrograms per cubic meter.

Smog or ground-level ozone has been referred to as a public health crisis, affecting people in nearly every U.S. state (Clean Air Network 2003). Smog is formed when nitrogen oxides emitted from electric power plants and automobiles react with organic compounds in sunlight and heat. Our reliance on automobiles has been blamed for much of the increase in smog levels.

The EPA monitors smog levels throughout the nation. The EPA sets federal eight-hour smog standards and collects data on the number of days that exceed the standard. A day is considered unhealthy if smog levels exceed the eight-hour standard. The EPA reported that 2002 was the worst recorded smog season; the eight-hour health standard was exceeded 8,818 times nationwide. The states with the highest number of unhealthy ozone days were California, Texas, and Tennessee (Clean Air Network 2003). In 2007, the EPA announced its proposal to strengthen quality air standards and reducing ground-level ozone, the first time standards have been tightened since 1997.

Photo 15.4

Cars travel on a Beijing highway under smog-covered skies. City officials attempted to reduce smog pollution and traffic congestion in time for the 2008 Summer Olympic Games.

Scientists report that one of every three people in the United States is at a higher risk of experiencing ozone-related health effects. Those most vulnerable to the health effects of smoggy air are children, people who work or exercise regularly outdoors, the elderly, and people with respiratory diseases. Short-term effects of smog mostly attack the lung and lung functioning: irritating the lungs, reducing lung function, aggravating asthma, and inflaming and damaging the lining of the lungs (EPA 1999a).

During the past two decades the prevalence of asthma has increased globally, suggesting growing problems associated with indoor and outdoor air quality. Among children, who tend to be outdoors more than adults, asthma is the most common chronic disorder worldwide (WHO 2006), the leading cause of missing school, and the leading cause of hospitalization (Eisele 2003). In the same way that the ozone damages human health, it also affects the health of other animals and vegetation and damages buildings (Palmer 1997).

Hazardous Waste Sites and Brownfields

The story of Love Canal awakened the world to chemical dumping hazards (Breton 1998). During the 1940s and 1950s, the Hooker Electrochemical Company dumped 20,000 tons of chemicals into Love Canal, New York (Center for Health, Environment, and Justice 2001). In 1953, after filling the canal and covering it with dirt, the company sold the land to the Board of Education for a dollar. Homes and an elementary school were built next to the canal. By the late 1970s, dioxin and benzene chemicals began seeping through backyards and basements. Because of the efforts of the Love Canal Neighborhood Association, state and federal agencies responded by cleaning the area and relocating many residents. In 1995, the Occidental Chemical Corporation (which bought out the Hooker Electrochemical Company) agreed to pay the government $129 million to cover the costs of the incident.

As a result of the Love Canal incident, the EPA created the Superfund program to clean hazardous waste sites. Hazardous materials may come from chemical manufacturers, electroplating companies, petroleum refineries, and common businesses such as dry cleaners, auto repair shops, hospitals, and photo processing centers (EPA 2003a). Sites may be placed on the national priority list (NPL) by their state if the site meets specific hazard and cleanup criteria. To date, 1,569 NPL sites have been identified; of these, 320 sites have removed from the list after cleanup efforts were completed (EPA 2007b).

Toxic sites continue to be identified today. The Child Proofing Our Communities Campaign is a collaboration of groups concerned about children's environmental health. The campaign focuses on where children spend most of their time: in school. In their 2001 report, the campaign identified more than 1,100 public schools within a half-mile radius of known contaminated sites in California, Massachusetts, Michigan, New Jersey, and New York. The campaign estimates that more than 600,000 students attend classes in schools near contaminated land (Center for Health, Environment, and Justice 2001).

In 2002, President Bush signed the Brownfields Revitalization Act, which authorized up to $250 million annually for the cleanup of brownfields, into law. **Brownfields** are abandoned or underused industrial or commercial properties where expansion or redevelopment is complicated by the presence or potential presence of hazardous substances, pollutants, or contaminants. There are more than 450,000 sites throughout the United States. Redevelopment efforts have included restoring waterfront parks and converting landfills to golf courses, as well as commercial or business expansion (EPA 2003a).

Water Quality

With the passage of the Water Pollution Control Act or the Clean Water Act in 1972, the federal government declared its commitment to cleaning the nation's waterways, which had become badly polluted from industrial contaminants and untreated sewage (Ehrlich and Ehrlich 1996). Since the passage of the act, the overall quality of water in U.S. lakes and streams has improved or at least has not significantly deteriorated. Yet, a variety of harmful substances have found their ways into bodies of water, including chemical compounds unknown to nature. A recent study conducted by the Pew Oceans Commission revealed a "crisis" in U.S. waters caused by pollution and fishing practices (Weiss 2003). The commission expressed concern about runoff from agricultural fields, lawns, and roads. Oil from gas stations and nutrients from agricultural fields disrupt the balance of river and ocean ecosystems. A "dead zone" in the Gulf of Mexico near the mouth of the Mississippi River has been attributed to contaminated runoff (Revkin 2003).

Toxic substances are turning up in greater frequency in groundwater, the source of drinking water for one of every two Americans (Ehrlich and Ehrlich 1996). The EPA (1999b)

Taking a World View

Home Sweet Landfill

China and the United States produce more waste than any other countries in the world. Traditional waste disposal includes incineration and landfills. Most U.S. waste, about 222 million tons per year, is sent to landfills. Landfills have been identified as one of the largest emitters of greenhouse gases, primarily methane (an odorless, colorless glass caused by the decomposition of animal and plant matter).

Yet, in many countries, many individuals and families call landfills their homes. In Mexico, landfill residents are called *pepenadore*. To make money, they scavenge through garbage piles, finding items (appliances, metals, or clothing) that can be reused or sold. Meals are harvested from discarded waste, and they build their homes using scrap material. Matthew Power explains, "Household and industrial trash has become for the world's poor a more viable source of sustenance than agriculture and husbandry . . ." (2006:62).

In his article, "The Magic Mountain," Power focuses on life on Payatas, a 50-acre landfill in Quezon City, Philippines. Here the scavengers are called *mangangalahigs,* which translated means "chicken scratcher," describing the way they pick through piles of trash. Payatas has been called the "Second Smokey Mountain." The first Smokey Mountain was a landfill site in Manila that supported about 30,000 men, women, and children who lived on the landfill. In 1995, the Philippine government closed the site, moving residents to temporary housing. However, over time some Smokey Mountain residents moved to Payatas and resumed their lives as mangangalahigs.

Power describes that so much garbage is piled at the Payatas dump, it would take 3,000 trucks a day for 11 years to move it all to another landfill. He writes,

As trucks dump each new load with a shriek of gears and a sickening glorp of wet garbage, the scavengers surge forward, tearing open plastic bags, spearing cans and plastic bottles with choreographed efficiency.... The ability to discern value at a glimpse, to sift the useful out of the rejected with as little expenditure of energy as possible, is the great talent of the scavenger. (2006:62)

Scavengers can make as much as 150 pesos a day, about three dollars for their work on the dump.

Scientists have collected global evidence identifying the unhealthy consequences of landfills on human health. Increased risk in adverse health effects (e.g., low birth weight, respiratory illnesses, birth defects, and certain types of cancers) have been found near individual landfill sites and in several multisite and multicountry studies (Vrijheid 2000).

reports, "While tap water that meets federal and state standards generally is safe to drink, threats to water quality and quantity are increasing." Groundwater contaminant problems are being reported throughout the country (Kraft and Vig 2000). Our drinking water is monitored in more than 55,000 community water systems for more than 80 known contaminants, including arsenic, nitrate, human and animal fecal waste, or legionella (the cause of Legionnaire's Disease). Nearly 1,000 deaths per year and at least 400,000 cases of waterborne illness may be attributed to contaminated water (Kraft and Vig 2000).

Frank and Ernestene Roberson from DeBerry, Texas, first complained about their water quality in 1996. Mrs. Roberson attributed her discolored bath tub and stomach problems to using the water from her well (Blumenthal 2006). According to the EPA (2007c), in 2003 State of Texas officials determined that DeBerry's ground water was contaminated with arsenic, benzene, lead, and mercury. Residents in this historically Black Texas enclave were informed that they could not use their groundwater and began to rely on bottled water for drinking, cooking, and bathing. In 2005, the EPA began delivering bottled water to residents. The EPA determined that groundwater contamination was caused by oil and gas waste deposited by an oilfield services company. In 2007, cleanup of the area and construction of a new water line began (EPA 2007c).

Our Water Supply

Only 1 percent of the world's water can be used for drinking. Nearly 97 percent of the water is salty or undrinkable; the other 2 percent is in ice caps and glaciers (EPA 2003b). Fresh water comes from surface water sources (lakes, rivers, and streams) and groundwater sources (wells and underground aquifers). About 66 percent of people get their drinking water from surface water sources. Large metropolitan areas rely on surface water whereas small communities and rural areas depend on groundwater sources. In recent years, there has been growing concern about the availability of fresh water sources. Because of pollution, increasing urbanization, and sprawling development, we may be running out of fresh water.

Almost 30 million residents in seven western states (Colorado, Utah, Wyoming, New Mexico, Nevada, Arizona, and California) rely on the Colorado River for their drinking water. The river basin covers 240,000 miles in the United States and a portion of northwestern Mexico. So much of the river is diverted for drinking and agricultural use that by the time it reaches the Sea of Cortez, it isn't much more than a trickle. A 2007 report by the National Academies on the Colorado River concluded that the combination of limited water supplies, increasing population demands, warmer temperatures, and the prospect of future droughts are likely to cause conflict among existing and future water users. The report predicted that this would "inevitably lead to increasingly costly, controversial, and unavoidable trade-offs among water managers, policy makers, and their constituents" (National Academies 2007).

Humans have attempted to harness water and its power through the construction of dams. About 75,000 dams in the United States provide water for irrigation, drinking, water control, and hydroelectricity. Once lauded as engineering miracles, dams have been blamed for serious environmental impacts affecting surrounding forests, watershed, beaches, habitat, and life. Former Secretary of Interior Bruce Babbitt (1998) explains,

> The public is now learning that we have paid a steadily accumulating price for these projects in the form of: fish spawning runs destroyed, downstream rivers altered by changes in temperature, unnatural nutrient load and seasonal flows, wedges of sediment piling up behind structures, and delta wetlands degraded by lack of fresh water and saltwater intrusion. Rivers are always on the move and their inhabitants know no boundaries; salmon and shad do not read maps, only streams.

More than 400 dams have been decommissioned nationwide (American Rivers 2002).

Drought and development exacerbate the pressure on water supply. An intense battle continues to be waged over Oregon's Klamath River. As described by Bruce Barcott (2003), the Klamath was born "a cripple." The river begins in Klamath Falls, Oregon, at the southern outlet of Klamath Lake, in "a body of water so shallow that a tall man could nearly cross it without wetting his hat" (Barcott 2003:46). First, water is diverted to irrigate more than 1,400 farms and ranches. Later, some of the water drains back into local lakes and refuges and is eventually pumped back into the Klamath River itself. The area used to be the picture

Photo 15.5

Various interest groups and stakeholders have attempted to establish their claim on the dwindling water supply from the Klamath River. The river runs from southern Oregon to Northern California. The Klamath is used for recreation, wildlife habitat, irrigation, fishing, and power.

of growth and prosperity, flourishing with housing development, farming, ranching, and expanding recreational activities.

But in recent years, as the area has suffered severe droughts and with the Klamath water supply dwindling, contentious battle lines have been drawn between the farmers, ranchers, fishermen, native tribes, recreational outfitters, and environmental groups over the benefits of removing the river's dams. Add to the list PacifiCorp, a regional power company. PacifiCorp owns four hydroelectric dams on the Klamath that provide power to about 70,000 homes. The company asserts that keeping the dams in place would maintain a cheap and clean power source (hydroelectricity versus coal or natural gas). Those supporting the removal of the dams say that new sources of energy, such as wind and solar, can replace the lost hydroelectric power (Yardley 2007).

What Does It Mean Me?

Where does your drinking water come from? Is there a concern about limits to the water supply? Contact your local water company or public works department for information.

Land Conservation and Wilderness Protection

Efforts in land conservation and wilderness protection seem to be successful largely because of federal protection policies. Since adopting the 1964 Wilderness Act, Congress has designated more than 106 million acres as "wilderness areas" through the National Wilderness Preservation System (2004). Under the act, timber cutting, mechanized vehicles, mining, and grazing activities are restricted. Human activity is limited to primitive recreation activities. The wilderness lands are protected for their ecological, historical, scientific, and experiential resources. The areas range in size from the smallest, Pelican Island, Florida (five acres), to the largest, Wrangell–St. Elias, Alaska (almost 10 million acres of land).

The Endangered Species Act of 1973 attempts to preserve species of fish, wildlife, and plants that are of "aesthetic, ecological, educational, historical, recreational and scientific value to the Nation and its people." The act has been controversial because it preserves the interests of the species above economic and human interests. For example, if endangered species are present, the act will restrict what landowners can do on their land (Palmer 1997). Currently, 1,921 U.S. and foreign animal and plant species have been listed as endangered or threatened, with recovery plans approved or implemented for 1,119 species (U.S. Fish and Wildlife Service 2007). Although federal funding for the Endangered Species Act expired in October 1992, Congress has appropriated funds in each fiscal year to support the program.

The National Park System includes 387 areas covering more than 83 million acres. Unlike the National Wilderness Preservation System, the National Park System allows and supports recreational activities. However, human activity in the form of motorized access, road and highway developments, logging, and pollution threaten the health of several national parks. For 2003, the National Parks Conservation Association listed 10 parks on its most endangered list: Big Thicket National Preserve (Texas), Denali National Park and Preserve (Alaska), Everglades National Park (Florida), Glacier National Park (Montana), Great Smoky Mountains National Park (North Carolina and Tennessee), Joshua Tree National Park (California), Ocmulgee National Monument (Georgia), Shenandoah National Park (Virginia), Virgin Islands National Park, and Yellowstone National Park (Montana).

The Great Smoky Mountains National Park is the most visited national park, with about 10 million visitors each year (National Parks Conservation Association 2003). The park features an ecosystem of rare plants and wildlife along with historical structures representing southern Appalachian culture, all of which are endangered according to the National Parks Conservation Association. The park has been listed as "endangered" for several years primarily because of chronic air pollution problems. The pollution has been attributed to coal-fired power plants and other industrial sources. Local developers are allowed to build right up to the park's boundaries.

Community, Policy, and Social Action

Federal Responses

The government's first response to the environment was directed at cleaning the nation's polluted water, air, and land. In 1969, Congress adopted the National Environmental Policy Act (NEPA), a comprehensive policy statement on our environment. For the first time in our nation's history, the government was committed to maintaining and preserving the environment (Caldwell 1970). The EPA, established in 1970, is charged with providing leadership in the nation's environmental science, research, education, and assessment efforts (EPA 2004). As the chief environmental agency, the EPA sets national standards and delegates to states and tribes the responsibility for issuing permits and monitoring and enforcing compliance. Beginning in the 1980s, the agency shifted its policies from cleanup to pollution management or prevention through market-based and collaborative mechanisms with business and industry and environmental strategic planning (Mazmanian and Kraft 1999). Additional environmental legislation, some of which we have already reviewed, includes the following:

- The Land and Water Conservation Act and the Wilderness Act of 1964. In our discussion on land conservation, we already reviewed the Wilderness Act. The Land and Water Conservation Act provides the necessary funds and assistance to states in planning, acquiring, and developing recreational lands and natural areas. The act also regulates admission and special user fees at national recreational areas. These two acts have been referred to as the "initial building blocks of environmental action" (Caulfield 1989:31).
- The Clean Air Act of 1970 regulates air emissions from area, stationary, and mobile sources. The act helped establish maximum pollutant standards. In 1990, the Clean Air Act was amended to address acid rain, ground-level ozone, ozone depletion, and air toxins. In 2003, seven state attorney generals filed a lawsuit against the EPA, accusing the agency of neglecting to update air pollution standards. The suit seeks regulations on carbon dioxide emissions, which are not listed under the Clean Air Act (Lee 2003).
- The Clean Water Act followed in 1977. This act established standards and regulations regarding the discharge of pollutants into the waters of the United States. The EPA was authorized to implement pollution control programs, setting wastewater standards and water quality standards for all contaminants in surface waters.
- The Endangered Species Act of 1973 created a program for the conservation of threatened and endangered plants and animals and their habitats.
- Under the Toxic Substances Control Act of 1976, the EPA has the authority to track 75,000 industrial chemicals being produced or imported into the United States.
- The most recent law is the Food Quality Protection Act of 1996. The law modified earlier statutes and created a single health-based standard for all pesticides in all foods.

In its strategic plan for 2003 to 2008, the EPA identified five long-term goals: achieve clean air, ensure clean and safe water, preserve and restore the land, build healthy communities and ecosystems, and develop a compliance and environmental stewardship.

State and Local Responses

The EPA highlights work done by states in "developing and implementing a range of programs and strategies that are cost-effectively reducing greenhouse gases, improving air quality, enhancing economic development and increasing the nation's energy security" (EPA 2007d). Calling state action "a key component of the U.S. response to climate change," the agency notes that 29 states and Puerto Rico have completed or implemented action plans for reducing greenhouse gas emissions or enhancing greenhouse gas capture (refer to U.S. State Map 15.1 for a complete list of states with action plans in place).

Recognizing that there are no real borders for carbon emissions and becoming increasingly frustrated with the slow progress of federal legislation on the matter (Broder 2007), many state leaders have formed regional coalitions to combat climate change. For example, the Regional Greenhouse Gas Initiative (RGGI) is a cooperative effort by Northeastern and

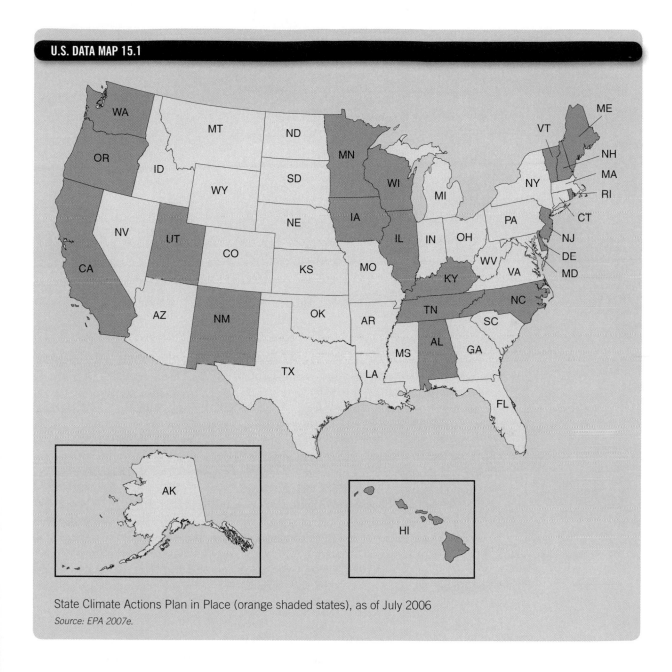

U.S. DATA MAP 15.1

State Climate Actions Plan in Place (orange shaded states), as of July 2006

Source: EPA 2007e.

Mid-Atlantic states to determine a regional strategy for controlling greenhouse gas emissions. In 2006, the Clean and Diversified Energy Initiative was signed by governors of 19 Western states, along with those from American Samoa, Guam, and Northern Marianas islands. The goal of their initiative is to identify and produce affordable, sustainable, and environmentally responsible energy for the western states and islands.

Cities throughout the nation have embraced sustainability goals, along with environmentally responsible planning and consumption. For Earth Day 2006, the Green Guide praised 10 U.S. cities for their green achievements in improving and maintaining air and water quality, recycling programs, green spaces, and environmental policies (refer to Table 15.3).

Table 15.3

Top Ten Green U.S. Cities, 2006

1. Eugene, Oregon
2. Austin, Texas
3. Portland, Oregon
4. St. Paul, Minnesota
5. Santa Rosa, California
6. Oakland, California
7. Berkeley, California
8. Honolulu, Hawaii
9. Huntsville, Alabama
10. Denver, Colorado

Source: McRandle and Smiley Smith 2006.

Recently, two major U.S. cities made news for their environmental vision and leadership. Declaring New York City as the "first environmentally sustainable twenty-first-century city," Mayor Michael R. Bloomberg proposed a master plan to reduce energy consumption in six areas: land, transportation, water, energy, air quality, and climate change. One of the most contentious parts of his 2007 plan involves the city's transportation burdens. Called "congestion pricing," Bloomberg proposed charging a fee for vehicles entering the city from 6 a.m. to 6 p.m. on weekdays—$8 for cars and $21 for trucks. The plan would reduce traffic congestion and improve air quality, as it has in parts of London and Singapore. The plan includes improvements in public transportation options such as bus services and subways (Lueck 2007).

On the West Coast, San Francisco was the first city in the nation to ban the use of petroleum-based shopping bags. These plastic shopping bags take many years to degrade, meanwhile contaminating our land and water and endangering animal and marine life. Nearly 90 percent of all U.S. shopping bags are plastic. Beginning 2007, plastic bags would be replaced with biodegradable plastic bags or recyclable bags in San Francisco grocery and drug stores. Several countries, including Butan, Bangladesh, and China, have banned the use of non-biodegradable plastic bags. Other countries, such as Germany and Ireland, charge a nominal fee for plastic shopping bags.

Environmental Interest Groups

Organizations concerned with the protection of the environment have played an important role in American politics since the foundation of the Sierra Club in 1892. The first wave of environmental interest groups included the National Audubon Society (1905), the National Parks Conservation Association (1919), and the National Wildlife Federation (1935). These groups were concerned with land conservation and the protection of specific sites and wildlife species. These first-wave groups depended on member support and involvement. All these organizations remain among the most influential groups in the environmental movement (Ingram and Mann 1989).

As public attention shifted to the problems of environmental pollution, a second wave of environmental groups emerged during the 1960s and 1970s. These new organizations focused their efforts on fighting pollution. In general, the second wave of environmental groups adopted an ecological approach to our natural environment, recognizing the inter-relationship between all living things and using science as a tool for understanding and protecting the environment. The Environmental Defense Fund (1967) and the Natural Resources Defense Council (1970) were started with funding support from the Ford Foundation. Both organizations relied on litigation as their instrument of reform. Other second-wave groups include Friends of the Earth, the Environmental Policy Institute, and Environmental Action. After the 1970s, environmental groups began to direct their appeals to policymakers rather than to the general public (Ingram and Mann 1989).

Although each group is committed to the environment, each has adopted its own cause, from broad environmental themes to such specific problems as toxic pollution or land conservation. The groups also have different strategies and tactics. Environmental groups may attempt to influence political policy, litigate environmental disputes, form coalitions with other environmental or interest groups, or endorse specific political candidates (Ingram and Mann 1989). Some, such as the Sea Shepherd Society, which rams and sinks whaling vessels throughout the word's oceans, adopt "in your face" tactics.

Now more than 70 years old, the National Wildlife Federation has 9 field offices and 46 state affiliates. The affiliates operate at the grassroots level by working to educate, encourage, and facilitate conservation efforts at the state level. One such organization is the Arkansas Wildlife Federation, established in 1936 by a group of sportsmen. The federation's goal has been to serve as a leader in educating people about conservation issues and in encouraging responsible stewardship of the state's natural resources. The federation, which represents a variety of constituents—hunters, sportsmen, hikers, anglers, and campers (Arkansas Wildlife Federation 2003a)—sponsors a variety of educational and community activities and projects. Seminars are offered to the public covering issues such as forest management, wetlands and water management, hunting, and fishing. The federation sponsors annual conservation achievement awards to honor citizens and organizations dedicated to natural resource endeavors. The organization also sponsors political candidate forums where citizens can ask candidates about their positions on various conservation and environmental issues (Arkansas Wildlife Federation 2003b).

A new environmental interest group is the Earth Island Institute, founded in 1982 by David Brower, who was the first executive director of the Sierra Club and cofounder of Friends of the Earth. The institute supports more than 30 projects worldwide, pledging "to provide activists with the freedom to develop program ideas, supported by services to help them pursue those ideas, with a minimum of bureaucracy" (Earth Island Institute 2003). The organization also honors youth for their environmental community work. Among those honored in 2007 was Erica Fernandez, a 16-year-old from Oxnard, California. Fernandez became an environmental activist when she learned that a liquefied gas facility was proposed for Oxnard and Malibu. A 36-inch pipeline would have been routed through the area's low-income neighborhoods, outraging Fernandez and many community leaders. Fernandez helped organize participation among high school youth and Latinos in local marches and rallies. Her testimony at the California State Lands Commission was credited with convincing commission members to deny the project's approval. The Commission voted 12–0 against the project (Earth Island Institute 2007).

Grassroots Movements

Sherry and Charles Cable (1995) argue that the grassroots environmental movement has improved the lives of many individuals and has spread environmental awareness among the

In Focus

The 100-Mile Diet

According to the Worldwatch Institute, food in the average American meal travels about 1,500 miles from farm to home. Food is brought into most communities via trucks, trains, or airplanes in refrigerated or storage containers. The 1,500-mile meal is said to consume as much as four times the energy and produces four times as much greenhouse gas emissions as a locally grown meal (Ketcham 2007). In addition to energy expended for transport, concern has also been expressed over the amount of fossil fuel energy used in producing and manufacturing food products to our table.

In response, environmentalists and foodies (people who really love food) throughout the United States and elsewhere in the world have adopted what is referred to as the 100-Mile Diet. For a week, a month, a year, or for life, the rules of the diet are simple: You may only consume food grown and produced within a 100-mile radius of your home. Called *locavores,* these 100-mile eaters advocate the value in being able to know the source of your food, supporting the local economy, and eating fresh and healthy food in season. They argue, "The distance from which our food comes represents our separation from the knowledge of how and by whom what we consume is produced, processed and transported" (Locavores 2007). This is not a diet for fast-food junkies.

Canadians Alisa Smith and J. B. McKinnon first coined the term "the 100-Mile Diet," when they began a year-long local food experiment in Vancouver, British Columbia. The couple discovered several food items that they had to live without for most or part of their year, such as chocolate, coffee, and wheat. (Most locavores make exceptions to these items, including salt and spices.) Unable to purchase local bread and wheat for baking, for one lunch meal, McKinnon, the family chef, fashioned bread out of turnip slices. But the couple found success (eventually finding a local supplier of wheat) and satisfaction in their diet. Their experience helped fuel broader interest in the benefits of a 100-mile diet.

One way to support local agriculture is through community sustainable agriculture (CSA), a partnership between community members and an independent local farm. The CSA movement began in Japan almost 30 years ago with a group of women who were concerned about pesticides, the increase in processed foods, and their country's shrinking rural population. Community members purchase seasonal shares, for about $300 to $400, which entitles them to weekly food allowances throughout the growing season. There are more than 1,200 CSA groups in the United States, some serving more than 1,000 families (Roosevelt 2003). According to FoodRoutes.org, independent local farms encourage biodiversity by diversifying the local landscape and natural environment. The CSA arrangement is beneficial to the farmer and to his or her customers. Customers receive fresh produce and have the satisfaction of supporting a local business. Customers can help at the farm and provide input and suggestions to their farmer. Instead of spending time marketing produce, farmers can focus their efforts on growing quality produce and working with their community members.

public. In contrast with national environmental organizations, grassroots organizations usually consist of working-class participants, people of color, and women. Although experienced organizers or community activists lead some groups, many grassroots groups are led by inexperienced but passionate leaders. In the fight against environmental racism, these grassroots environmental groups have given a voice to communities of color (Epstein 1995).

Cable and Cable (1995) suggest that the motivating factor for most grassroots organizations is a desire to protect the health and safety of families against some immediate environmental threat. Grassroots organizations emphasize environmental justice, acting in the belief that some injustice has been committed by a corporation, business, or industry and that appropriate action should be taken to correct, improve, or remove the injustice(s). These organizations have tackled toxic waste issues and have sought justice for housing, transportation, air quality, and economic development issues (Bullard 1994).

Sociologists Riley Dunlap and Angela Mertig note that the environmental movement is among the few movements that "significantly changed society" (1992:xi). Nicolas Freudenberg and Carol Steinsapir (1992:33–35) identify seven achievements of grassroots organizations:

1. A number of environmentally hazardous facilities have been controlled by cleaning up contaminated sites, blocking the construction of new facilities, and upgrading corporate pollution control equipment.

2. Grassroots organizations have forced businesses to consider the environmental consequences of their actions.

3. These groups encourage preventative approaches to environmental problems such as reducing or limiting the use of environmental contamination.

4. The grassroots movement has expanded citizens' rights to participate in environmental decision making.

5. Grassroots organizations have served as psychological and social support networks for victims and their families.

6. The movement has brought environmental concerns and action to working-class and minority Americans.

7. The grassroots movement has influenced how the general public thinks about the environment and public health.

Radical Environmentalists

Rik Scarce (1990) explains that radical environmentalists want to preserve our biological diversity. This isn't just a question of preserving the living things in our ecosystem, such as plants and animals; nonliving entities such as mountains, rivers, and oceans must also be protected. Radical groups confront problems through direct action, such as picketing an office building, breaking the law, or performing acts of civil disobedience. Other radical environmentalists may destroy machinery, property, or equipment used to build roads, kill animals, or harvest trees. Most "eco-warriors" act on their own and without the leadership of an organizational hierarchy.

Scarce (1990) says radical environmentalists adopt lifestyles that have minimal impact on the environment: not owning cars, adopting vegetarian diets, and avoiding occupations that involve the destruction of the environment. Although they are committed to their issues, radical environmentalists recognize that on their own, they will never be able to end the practices they protest. Their actions usually attract media attention, creating a groundswell of public support for their particular issues. Or their actions are done in concert with mainstream efforts; for example, radicals might stage tree sittings in Oregon to delay the cutting of timber until courts can hear a more mainstream group's request for injunction.

Tree sitting is a form of protest targeting timber companies at the point of production, slowing or even stopping tree cutting. Scarce (1990) explains how tree sitters make their protest in areas with active cutting, finding and choosing trees that will make the right statement.

The trees they choose are the tallest, most impressive, and clearly visible from a road, or those that overlook a recently cut area. Although we may think of tree sitting as a lonely activity, tree sitters require a support group for assistance, food, and clothing, including the hauling of waste (including human) from the site. The group will carry about 250 pounds of gear and provisions—food, water, clothing, and platform materials—as far as 10 miles to their intended site. Suspended about 80 to 150 feet off the ground, the 2½ by 6-foot wooden platform becomes the tree sitter's home for days, weeks, or months. Julia "Butterfly" Hill sat in a 600-year-old California redwood tree for 738 days, withstanding eviction threats and legal action from the Pacific Lumber Company, which owned the tree. Tree sitting has been effective in bringing public and media attention to the practices of timber and logging companies and in rallying support for the tree sitter's message. In several cases, agreements have been made with lumber companies to divert logging to other areas or to pursue viable logging programs.

College and University Sustainability

"Universities are huge institutions with huge carbon footprints, but they are also laboratories for concepts of sustainability," says Michael Crow, president of Arizona State University (Deutsch 2007:A21). Colleges and universities are taking the environmental lead by constructing green buildings, purchasing alternative energy, and investing in efforts to make campuses carbon neutral. Their efforts include (Deutsch 2007; Lipka 2006) the following:

- Bowdoin and Evergreen State Colleges purchase 100 percent of their energy from renewable sources or pay for energy offsets from solar or wind power.
- Arizona State University distributes free bus passes to every student, employee, and faculty member.
- Dickinson College students operate an organic garden, using some of the produce in their campus dining hall.
- Students at Central Oregon Community College and the University of Kentucky voted to pay additional fees to cover their institutions' clean-energy purchases.

Crow was among the first university presidents who signed the American College and University Presidents Climate Commitment. The presidents acknowledge their role in leading environmental responsibility in their campuses and broader communities. The commitment, signed by more than 250 college and university presidents, includes a pledge to develop and implement a plan to achieve carbon neutrality on their campuses.

Writer Sara Lipka (2006) portrays college students as the "watchdogs for sustainability." She explains,

> Armed with Internet research, they are investigating institutional operations like energy use, food purchasing, investments, transportation, and waste disposal. They are pushing administrators to approve new projects and set higher goals for sustainability. National networks are helping students share strategies with one another and organize sophisticated, often successful proposals for campus innovations and reforms. (2006:11)

Similar environmental programs have been established in elementary and secondary schools across the country.

To learn more about what you can do to protect the environment, refer to Table 15.4.

> **Table 15.4**

Living Green—Tips for College and University Students

1. Practice the three R's: Reduce, Reuse, and Recycle.

2. Printer and paper use. If possible, print on both sides of the page. Think twice about whether you need a hard copy of a Web page or document—could you bookmark a page or save a file on your computer?

3. Limit your use of disposable products. This includes cups, plates, and paper napkins. About those napkins, the next time you grab a handful at your dining hall or in a restaurant, ask yourself if you really need than many. One might be just enough.

4. Use compact florescent light bulbs in your dorm room or apartment. They may cost more, but will last longer and save you money.

5. Walk, bike, and limit the use of your car.

6. Carry a refillable water bottle. No more bottled water.

7. Buy recycled products, this includes paper for printing.

8. Use refillable binders instead of notebooks. Or go electronic and take all your notes on your laptop.

9. Buy used clothing and furniture. It is great way to save money, and it is a great thing to do for the environment.

10. Share your message—tell others how you are living green on your campus.

Source: Adapted from Rockler-Gladen 2007.

Voices in the Community

Chad Pregracke

Chad Pregracke started his career as an environmental activist in his own backyard. A native of Hampton, Illinois, he spent most of his life on the shores of the Mississippi River as a resident, shell diver, and commercial fisherman. Pregracke often saw garbage floating on the river or caught within its banks.

> I woke up like that many times—loving the natural world and hating the despicable mess. I saw the garbage on the river, and it didn't look right. It may have been commonplace, but I didn't like it and I didn't accept it. And the more garbage sites I noticed, the more it made me want to do something about it. (Pregracke and Barrow 2007:12–13)

He admits that his attempts to organize and fund his first river clean up were less than perfect. He recalls how he did everything wrong—he didn't have a clear message, didn't have a plan, didn't talk to the right people at local agencies. But he persevered and received funding from his first corporate sponsor, the Alcoa Corporation. Pregracke started cleaning up the river in 1997, collecting 45,000 pounds of debris from its shores. In 1998, Pregracke established his nonprofit organization, Living Lands and Waters, dedicated to cleaning up the Mississippi River.

Since its establishment, Living Lands and Waters has removed 4 million pounds of trash from the Mississippi and its tributary rivers Missouri and Ohio. The cleanup efforts are staffed primarily by community volunteers. Their annual haul of trash includes items such as refrigerators, tires, bowling balls, baby dolls, and Styrofoam. In 2003, Living Lands and Waters established the Riverbottom Restoration Project. Through the project, volunteers remove invasive plant species from the river banks, planting native trees in their place. In August 2005, Pregracke and his crew provided assistance with clean up and rebuilding efforts in New Orleans after Hurricane Katrina struck the area.

Pregracke is modest in describing his life's mission,

I've been on mission to clean up the rivers. It was a simple concept then and it's simple concept now, but doing it has been far from simple or easy. I realize that we're not solving all the problems or necessarily saving America's rivers—we're simply doing our part, just as I hope you're doing yours. (2007:282–283)

We all make a difference, even if we don't intend to, and it's either negative or positive. My question to you is, how big a difference do you want to make? I can't say how big a difference I've made or will continue to make, but I know this—I will plant a lot of trees and leave a cleaner river. (p. 288)

For more information about Living Lands and Waters, visit their Web site, www.livinglandsandwaters.org.

Main Points

- Daily human activity, economic development, cultural values, social class, and technology all affect the health of the environment. **Environmental sociology** considers the interactions between our physical and natural environments and our social organization and behavior.

- Functionalists believe that environmental problems develop from the system itself. Thus, although agriculture, industrialization, and related technologies have improved quality of life, they have also led to waste, pollution, and the destruction of natural resources. Functionalists believe that the only way to address environmental problems is to simultaneously halt population growth and reduce technology's impact.

- Conflict theorists believe that environmental problems are created by humans competing for power, income, and their own interests in a capitalist system. Those with the least amount of power are targeted in **environmental racism**, in which marginalized groups are disadvantaged. Research indicates that low-income people and people of color are exposed to greater environmental risks than White or affluent people.

- Feminists argue that a masculine worldview is responsible for the domination of nature, women, and minorities. One strand of thought, ecofeminism, believes that men are the primary culprits in the exploitation of animals and the environment. Other feminist approaches include feminist critiques of natural science, analyses of specific environmental issues, and contributions to sustainable development.

- Interaction theorists believe that environmental problems do not materialize by themselves.

Social constructionists focus on the political process affecting the environment rather than on the problems themselves.

- Environmental problems include global warming, air pollution, hazardous waste sites, brownfields (abandoned or underused commercial properties with hazardous substances), and polluted and toxic water.

- Because of pollution, increasing urbanization, and sprawling development, we may be running out of water. Human efforts to conserve water, such as building dams, are now recognized as causes of serious and negative environmental problems.

- Efforts in land conservation and wilderness protection seem to be successful largely because of federal protection policies and legislation. Examples, among many, are the Wilderness Act, the Endangered Species Act, and the formation of the Environmental Protection Agency (EPA).

- Established in 1970 as the chief environmental agency, the EPA provides leadership in the nation's environmental science, research, education, and assessment efforts. It sets national standards and delegates certain functions of enforcement to states. The EPA's five long term goals through 2008 are to achieve clean air, ensure clean and safe water, preserve and restore the land, build healthy communities and ecosystems, and develop a compliance and environmental stewardship.

- Environmental organizations have played an important role in American politics since the foundation of the Sierra Club in 1892. Groups include the National Audubon Society and the National Wildlife Federation. These first-wave groups are concerned with land conservation and the protection of specific sites and species, and they rely on member support and involvement.

- A second wave of environmental groups, such as the Environmental Defense Fund and the Natural Resources Defense Council, focus on fighting pollution and emerged during the 1960s and 1970s. They adopt an ecological approach, recognizing the interrelationship between all living things and using science as a tool for understanding and protecting the environment. After the 1970s, environmental groups began to direct their appeals to policymakers rather than the general public.

- Each environmental group has its own cause and strategies. Groups may attempt to influence political policy, litigate disputes, form coalitions with other environmental or interest groups, endorse specific political candidates, or even adopt more radical tactics.

- Grassroots organizations usually consist of inexperienced but passionate working-class participants, people of color, and women. These groups are motivated by a desire to protect the health and safety of families against some immediate environmental threat. Grassroots organizations emphasize environmental justice.

- Radical groups confront problems through direct action, such as breaking the law, destroying property or equipment, or performing acts of civil disobedience. Most "eco-warriors" act on their own and without the leadership of an organizational hierarchy. They recognize that they cannot effect change on their own, but their actions usually draw media attention and thus engender public support. They also work with mainstream groups.

- The environmental movement has had a significant impact in controlling hazardous facilities, urging businesses to consider environmental impact, encouraging preventative approaches to problems, expanding citizens' rights to participate in decision making, serving as psychological and social support networks for victims and their families, bringing concerns and action to working class and minority Americans, and influencing how the general public thinks about the environment and public health.

- Colleges and universities are demonstrating environmental leadership by constructing green buildings, purchasing alternative energy, and investing in efforts to make campuses carbon neutral and sustainable.

On Your Own

Log on to the Web-based student study site at www.pineforge.com/leonguerrero2study for interactive quizzes, e-flashcards, journal articles, Community and Policy Guides, a Service Learning Guide, the end-of-chapter Web exercises, and additional Web resources.

Internet and Community Exercises

1. Access "Envirofacts," a one-stop source for environmental information about your community sponsored by the EPA. Go to *Study Site Chapter 15*. You can use the "Quick Start" on the page and type your area's zip code, city, county, or state. Or you can select a topic—water, waste, toxics, air, or radiation—to find out if the EPA is monitoring any local emissions, sites, violations, or companies. Based on the information provided, how would you rate the environmental quality of your community? Of your state?

2. The Edible Schoolyard, located at the Martin Luther King Middle School in Berkeley, California, was established in 1995. Founded by Chef Alice Waters, the nonprofit organization embraces a "Seed to Table" philosophy, allowing students to experience first hand the cycle of food production from garden to table via gardening and cooking classes. The organization promotes the establishment of garden-kitchens in schools, with a listing of school gardens throughout the United States. Explore whether a school garden exists in your university or in an elementary or secondary school in your area.

3. Invite a faculty member from the Biology or Natural Sciences Department to talk about the ecosystem of your college campus. How does your campus affect its environmental habitat? How much waste is produced on campus? Does your campus have a recycling program? What environmentally friendly practices are supported on campus?

4. What sustainability programs does your university have in place? For faculty, students, and the campus? Is your school part of the American College and University Presidents Climate Commitment?

Note

1. After a second experiment during 1993–1994, no other human experiments were conducted. In 1996, Columbia University took over the management of the Biosphere and was using the facility as a research and education center. In 2003, Columbia University announced that it would end its relationship with Biosphere II.

CHAPTER 16

War and Terrorism

The secure life many Americans took for granted changed on the morning of September 11, 2001 (Hoge and Rose 2001). For the first time since the attack on Pearl Harbor in 1941, U.S. citizens were under attack on our nation's soil. (Although some historians and social commentators refer further back to the War of 1812, noting that the Pearl Harbor attack should not count because Hawaii was not a state at the time.) In New York City, 2,752 were killed at the World Trade Center; at the Pentagon, 189 were killed; and in Pennsylvania, 44 died when their hijacked plane crashed into an open field.

In his address before the nation that evening, President George W. Bush (2001) stated, "Today, our fellow citizens, our way of life, our very freedom came under attack in a series of deliberate and deadly terrorist acts." His administration vowed to bring those responsible to justice, yet the events of September 11th brought home the need for the cooperation and support of U.S. allies (Booth and Dunne 2002; Hoge and Rose 2001). Within several months, the United States launched a war on terror, invading Afghanistan and Iraq, where U.S. military personnel remain.

As a nation, we continue to debate whether we are safer as a country since 2001. The Bush administration was accused of playing the terrorist card, cultivating a culture of fear among Americans, while highlighting its progress to ensure a safer democratic world. Furthermore, the administration was criticized for ignoring the true social, political, and economic consequences of the military campaign in the Middle East, globally and at home.

Just before the sixth anniversary of September 11th, former co-chairs of the 9-11 Commission Thomas H. Kean and Lee H. Hamilton stated that the United States "has not stemmed the rising tide of extremism in the Muslim world" (Kean and Hamilton 2007). The beginning of the twenty-first century is an era of globalized terrorism, marked by the emergence of a new international terrorist environment, where terrorists are capable of using and willing to use weapons of mass destruction to "inflict unprecedented casualties and destruction on enemy targets" (Martin 2006:3). Since 2001, the number of terrorists and terrorist groups has increased; some believe this is largely because of the U.S. invasion and occupation of Iraq and Afghanistan. The National Intelligence Council (2007) concluded that despite increased worldwide antiterrorism efforts, the United States was in a "heightened threat environment," at risk of attacks by al-Qaeda and other non-Muslim terrorist groups.

Defining Terrorism and War

Terrorism

According to the U.S. federal code, **terrorism** is "the unlawful use of force or violence against persons or property to intimidate or coerce a government, the civilian population, or any segment thereof, in furtherance of political or social objectives" (28 C.F.R. Section 0.85). Gus Martin (2006) offers what he calls an instinctive definition of terrorism as politically motivated violence, usually directed against "soft targets" (civilians and government targets) and with the intention to affect a target audience.

Domestic terrorism is defined as terrorism supported or coordinated by groups or individuals based in the United States. **Foreign or international terrorism** is defined as terrorism supported or coordinated by foreign groups threatening the security of U.S. nationals or the national security of the United States. International terrorism can occur outside of the United States but is directed at U.S. targets. Acts of terrorism have occurred on every continent, and perpetrators come from diverse religious and ethnic groups; however, Islamic governments and networks have committed the most extreme acts of terror (Booth and Dunne 2002). Al-Qaeda has come to symbolize terrorism in the twenty-first century.

Paul Pillar (2001) explains that there are five elements of terrorism: First, terrorism is a premeditated act. It requires intent and prior decision to commit an act of terrorism. Terrorism doesn't happen by accident; rather, it is the result of an individual's or group's policy or decision. Second, terrorism is purposeful; it is political in its motive to change or challenge the status quo. Religiously oriented or national terrorists are driven by social forces or shaped by circumstances specific to their particular religious or nationalistic experiences (Reich 1998). Third, terrorism is not like a war, in which both sides can shoot at one another. Terrorism targets noncombatants, such as civilians who cannot defend themselves against the violence. The direct targets of terrorist activity are not the main targets. Fourth, terrorism is usually carried out by subnational groups or clandestine agents. If uniformed military soldiers attack a group, it is considered an act of war; an attack conducted by nongovernmental perpetrators is considered terrorism. Individuals acting alone may also commit terrorism. Finally, terrorism includes the threat of violence. It does not involve only terrorist acts that may have occurred; it also involves the potential for future attacks.

Terrorist activity has changed little over the years. Six basic tactics account for 95 percent of all incidents: bombings, assassinations, armed assaults, kidnappings, hijackings, and other kinds of hostage seizures. As Brian Jenkins states, "Terrorists blow up things, kill people, or seize hostages. Every terrorist attack is merely a variation on these three activities" (1988:257).

What Does It Mean to Me?

Are you concerned about a terrorist attack on U.S. soil? Do you believe we are more at risk of an attack from domestic or foreign terrorists? Explain the reason for your answer.

A Brief History of U.S. Conflicts

War is a violent political instrument (Walter 1964). It is a legitimate violent activity between armed combatants, one side hoping to impose its will on the other. "The United States was born of violence and revolution," writes Ken Cunningham, "violence against the native population and among and against the various European imperial powers" (2004:556).

Our nation's birth was marked by a war—the American Revolution of 1775 to 1783. In 1776, the Declaration of Independence was adopted, a public statement of a new nation's independence from Great Britain and its rights to "life, liberty, and the pursuit of happiness." The British were defeated in 1783. More than 4,000 lives were lost in the revolution against the British.

The American Revolution was followed by the War of 1812 (1812–1815) and the Mexican War (1846–1848), both part of the economic and continental expansion of the United States. The effort was justified under the principle of manifest destiny. U.S. leaders felt it was their mission to extend freedom and democracy to others. At the end of the war, the United States acquired the northern part of Mexico, later dividing the area into Arizona, California, Nevada, New Mexico, and Utah.

The Civil War (1861–1865) between northern and southern states has been referred to as the bloodiest battle on U.S. soil. Although we tend to think that the war was waged solely over the issue of slavery, it was also based on deep economic, political, and social differences between the two groups of states. During his second inaugural address in 1865, President Abraham Lincoln said, "One of them would make war rather than let the nation survive, and the other would accept war rather than let it perish, and the war came." More than 600,000 died.

Beginning with the Spanish-American War (1898), U.S. troops and soldiers began to wage war in other countries, mostly responding to tyranny, oppression, and communism. The Spanish American War was fought to liberate Cuba from Spain and to protect U.S. interests in Cuban sugar, tobacco, and iron industries. U.S. participation during World War I (1917–1918) and World War II (1940–1946) helped establish its dominance as a worldwide military force. That accomplishment, however, came with great cost: The United States had heavy losses, 53,000 deaths in World War I and 400,000 deaths in World War II. In the latter war, the United States and its allies claimed victory against Germany, Japan, and Italy. After the North Korean People's Army invaded the Republic of Korea, the United States joined United Nations forces in the Korean War (1950–1953). No winner of the war has ever been declared. An armistice was agreed upon in 1953, forever separating North and South Korea. Fifty years after the invasion, Communist and UN soldiers still guard their sides of the demilitarized zone in Panmunjom. The United States fought against North Vietnamese communists in the jungles of South Vietnam from 1964 to 1973. Although U.S. troops had begun withdrawing from Vietnam in 1969, the formal ceasefire was declared in January 1973.

U.S. engagement in Middle East wars began with the Persian Gulf War of 1990–1991. The United States was joined by UN forces to liberate Kuwait from Iraqi forces. Operations Desert Shield and Desert Storm were the first display of high tech warfare: cruise missiles, stealth fighters, and precision guided munitions. Iraq accepted the terms for the cease fire in April 1991. After September 11, 2001, U.S. forces attacked Afghanistan in an effort to destroy al-Qaeda forces and to locate their leader, Osama bin Laden. The first war of the twenty-first

Photo 16.1

The U.S. invasion of Iraq began in 2003. By March 2008, 4,000 U.S. military personnel had been killed.

Table 16.1

America's Wars and Casualties (dates indicate U.S. troop involvement)

War	Participants	Deaths in Service
American Revolution 1775–1783	Unknown	4,435
War of 1812 1812–1815	286,730	2,260
Mexican War 1846–1848	78,718	13,283
Civil War, Union Only	2,213,363	364,511
Spanish-American War 1898–1902	306,760	2,446
World War I 1917–1918	4,734,991	116,516
World War II 1941–1945	16,112,566	405,399
Korean War 1950–1953	5,720,000	36,576
Vietnam War 1964–1973	8,744,000	58,209
Persian Gulf War 1990–1991	2,322,000	382

Source: U.S. Department of Veterans Affairs 2007.

century was the U.S. war against Iraq (2003–) or Operation Iraqi Freedom. On May 1, 2003, President Bush declared that the period of major combat was over. At this writing, U.S. troops still occupy Iraq. By March 24, 2008, 4,000 military casualties were reported in Operation Iraqi Freedom. A listing of U.S. wars and casualties is presented in Table 16.1.

Sociological Perspectives on War and Terrorism

Functionalist Perspective

Functionalists examine how war and terrorism help maintain the social order, creating and reinforcing social, religious, or national boundaries. War creates social stability by letting everyone know what side they are on: the good guys versus the bad guys or us versus them. There are norms and boundaries in war: Individuals will know what their roles are, what they should believe in, how they should respond in case of an attack, and how they should interact with members of the other side. But unlike war, the social boundaries in terrorist activities are less certain. In some cases, the identity of terrorists and their goals may never be known.

War and terrorism provide a "safety valve" function, giving marginalized or oppressed groups a means to express their discontent or anger. Acts of terrorism, according to Martha Crenshaw (1998), are selected as a course of action from a range of alternatives. Groups may choose terrorism because the other methods may be expected not to work or may be too time consuming for the group. Radical groups choose terrorism when they want immediate action and when they want their message to be heard. As they act, they also spread their group's message and recruit others for their cause.

Finally, warfare establishes power and domination. The victor is able to acquire the "spoils of war"—a country's land, people, and resources. Beyond those tangible fruits of victory, the winning side also gets to make new rules or impose its rules about appropriate political, social, and economic structures.

Conflict Perspective

War and terrorism are forms of conflict. Conflict may be based on disputes over resources or land. Modern conflict theorists have focused on how war is used to promote economic and political interests. In 1961, President Dwight D. Eisenhower cautioned the nation about the **military-industrial complex**, the growing collaboration of the government, the military, and the armament industry. Years earlier, Eisenhower (1953) warned,

> Every gun that is made, every warship launched, every rocket fired, signifies in the final sense a theft of those who hunger and are not fed, those who are cold and are not clothed. The world in arms is not spending money alone. It is spending the sweat of its laborers, the genius of its scientists, the hopes of its children.

He explained that the complex had a corrupting influence—economic, political, and spiritual—in every city, state, and federal office of government. A decision to go to war could be motivated not by ideals to preserve or promote freedom, but to ensure the economic well-being of the defense contractor. For example, during the Iraq war, the role of oil services firm Halliburton and its subsidiary company, KBR, has been closely scrutinized. Critics identify that Halliburton has an unfair advantage with no-bid military contracting because of its relationship with Vice President Dick Cheney. Cheney served as Halliburton's CEO from 1996 to 1998 and briefly in 2000. In 2007, federal investigators alleged that Halliburton was responsible for $2.7 billion in contractor waste and overcharging in Iraq.

Social scientists have noted the far-reaching and devastating impact of U.S. militarism for Americans and the rest of world. Cunningham (2004) identifies a "vast, entrenched, bureaucratic national security apparatus" that reaches into all areas of American life and politics, including businesses, universities, primary and secondary schools, the media, and popular culture. U.S. militarism also reaches into foreign countries, through its direct military intervention and war and covert operations (conducted by the Central Intelligence Agency [CIA], the National Security Agency [NSA], and other agencies).

Feminist Perspective

From this perspective, war is considered a primarily male activity that enhances the position of males in society. "War is a patriarchal tool always used by men to create new structures of dominance and to subjugate a large mass of people" ("The Events" 2002:96). In our military system, decision making and economic power are held primarily by men; as a result, international relations and politics are played out on women's bodies (Cuomo 1996). According to Cynthia Enloe (1990), local and global sexual politics shape and are shaped through the presence of national and international U.S. military bases—through the symbolism of the

Photo 16.2

U.S. Army PFC Lynndie England was convicted on charges stemming from her involvement in the Iraq Abu Ghraib prison scandal. England was sentenced to three years in prison. Six other soldiers were convicted and sentenced.

U.S. soldier, the reproduction of family structures on bases, and systems of prostitution that coexist alongside bases. She explains, "Bases are artificial societies created out of unequal relations between men and women of different races and classes" (p. 2).

Men reserve the right to make war themselves and claim that they fight wars to protect vulnerable people, such as women and children, who are viewed as not being able to protect themselves (Tickner 2002). During wars, women are charged with caring for their husbands, sons, and the victims of war. The ideal of the "caretaking woman" helps exclude women from public and political institutions by reminding women that their first responsibility is to the family. According to Laura Kaplan, this ideal "helps co-opt women's resistance to the war by convincing women that their immediate responsibility to ameliorate the effects of war takes precedence over organized public action against war" (1994:131).

Feminist theorists focus on the gender rhetoric used in war. After September 11th, Judith Lorber observed, "The gender rhetoric heard over and over portrayed our men as heroes and theirs as terrorists, our women as sad widows and theirs as equally sad oppressed wives" (2002:379). Most of the images represented men "acting" while women were "reacting" to the events of September 11th. The one exception was First Lady Laura Bush, who after maintaining a low profile before September 11th, became highly visible in the archetypal role of nurturer for the nation (Jansen 2002). Instead of focusing on the female soldiers deployed against Afghanistan, the media chose to focus on faceless, helpless Afghan women in blue burqas (Tickner 2002). As mostly men perpetuate the violence, respond with additional violence, and present media stories about it, world media portrayals of war are dominated by patriarchal images and logic (Lemish 2005).

As more women are serving as active duty personnel, however, their profile in military service has increased—negatively in some cases. Mary Ann Tétreault (2006) examined the portrayal of women involved in the torture of male prisoners at Abu Ghraib prison. In 2004, a *60 Minutes II* story and a series of articles in the *New Yorker* magazine first exposed the crimes being committed by U.S. military personnel. At the center of the story was a set of sensational photographs of naked and tortured prisoners taken by U.S. military men and women. Tétreault explains that although both male and female soldiers were involved in the torture, the popular face of the scandal and the official scapegoat were both women—Private Lynndie England and General Janis Karpinski. The focus on these women detracted public attention and ultimately deflected responsibility from the male soldiers responsible for the prison abuse. Tétreault writes,

> Those who chose to focus on the sex of General Karpinski, like those who chose to focus on that of Lynndie England, could reframe the meaning of the torture and humiliation taking place at Abu Ghraib as nothing more than what might have been expected from putting a woman in charge of a group of impressionable young men. (2006:43)

Interactionist Perspective

Interactionists focus on the social messages and meaning of war and terrorism. Terrorism is a word with intrinsically negative connotations. According to political scientist Crenshaw, the word "projects images, communicates messages, and creates myths that transcend historical circumstances and motivate future generations" (1995:12). In addition, the concept serves as an "an organizing concept that both describes the phenomenon as it exists and offers a moral judgment" (p. 9). Jenkins explains,

> What is called terrorism thus seems to be dependent on one's point of view. Use of the term implies a moral judgment; and if one party can successfully attach the label terrorist to its opponent, then it has indirectly persuaded others to adopt its moral viewpoint. (1980:10)

Crenshaw cautions that once political concepts such as terrorism are constructed, "They take on a certain autonomy, especially when they are adopted by news media, disseminated to the public, and integrated into a general context of norms and values" (1995:9). Use of the word terrorism promotes condemnation of the actors and may reflect an ideological or political bias (Gibbs 1989).

Terrorism is particularly useful for agenda-setting (Crenshaw 1998) by the terrorist group and its target. If the reasons behind the violence are articulated clearly, terrorists can put their issues on the public agenda. Instantly, the public is aware of the group and its cause. The act may make some sympathetic to the group or could elicit anger and calls for retaliation. But the target—such as the U.S. government—can also use terrorism to set its own agenda. According to Crenshaw, "conceptions of terrorism affect the ways in which governments define their interests, and also determine reliance on labels or their abandonment when politically convenient" (1998:10). When the problem is labeled terrorism or a group terrorists, a set of predetermined preferred solutions begins to emerge. When these terms depict the group as "fanatical and irrational," making attempts at diplomacy or compromise seems impossible, the inference is that the U.S. government has nothing left to do but retaliate with force. But the labeling doesn't stop: Such defensive actions are often "appropriate" and "legitimate" expressions of "self-defense," but not "terrorism."

For a summary of sociological perspectives, see Table 16.2.

Table 16.2

Summary of Sociological Perspectives: War and Terrorism

	Functional	Conflict/Feminist	Interactionist
Explanation of war and terrorism	Functionalists examine how war and terrorism help maintain the social order.	War and terrorism may be based on conflict over resources, territory, and power. From a feminist perspective, war is considered a primarily male activity that enhances the position of males in society.	An interactionist focuses on the social messages and meaning of war and terrorism.
Questions asked about war and terrorism	What functions do war and terrorism serve? For terrorist groups? For the targets of terrorism? For groups in conflict?	What groups are in conflict and why?	How do our conceptions of war and terrorism shape political and diplomatic responses? How do they shape our own behavior?

What Does It Mean to Me?

The rhetoric of September 11th also highlighted social differences based on social class and ethnicity/race. Media stories focused on the heroism of White firefighters and financial brokers, rather than on the victims and heroes who were busboys and maintenance or mailroom workers who were poor, non-White, working-class, immigrant men (Lorber 2002). Why do you think the public and media ignored their stories?

The Problems of War and Terrorism

Domestic Terrorism

Although the United States has shifted its focus to the threat of international terrorism, domestic terrorist groups continue to pose a threat. Between 1980 and 1999, the FBI recorded 327 incidents or suspected incidents of terrorism in the United States. Eighty-eight were attributed to international terrorists, and the remaining 239 were attributed to domestic terrorists. These acts resulted in the deaths of 205 people and the injury of 2,037 (FBI Counterterrorism Unit 1999).

On April 19, 1995, the worst domestic terrorism attack occurred around 9:00 a.m. in Oklahoma City, Oklahoma. A rental truck loaded with a mixture of fertilizer and fuel oil exploded in front of the Alfred P. Murrah Federal Building. The blast blew off the front side of the nine-story building, killing 169 and injuring hundreds more. The attack was conducted by Timothy McVeigh and Terry Nichols, motivated by antigovernment sentiment over the failed 1993 federal raid on the Branch Davidian compound in Texas. The bombing occurred on the second anniversary of the Branch Davidian incident. At the time, it was called the worst terrorist attack on U.S. soil. McVeigh was executed for his crime in 2001; Nichols is serving a life sentence.

Louis J. Freeh (2001), former FBI director, identified three types of domestic groups operating in the United States: right-wing extremists, left-wing and Puerto Rican extremists, and special interest extremists. Right-wing groups, such as the World Church of the Creator, Aryan Nations, and the Southeastern States Alliance, advocate for the principles of racial supremacy and tend to embrace antigovernment or antiregulatory beliefs. They have also been characterized as hate groups. The Southern Poverty Law Center (2007) counted 803 active hate groups in the United States in 2005. According to the Southern Poverty Law Center, "All hate groups have beliefs or practices that attack or malign an entire class of people, typically for their immutable characteristics." (See U.S. Data Map 16.1 for the distribution of known hate groups by state.)

Left-wing groups want to bring about revolutionary change, adopting a socialist doctrine, and see themselves as protectors of the people. Groups in this category include terrorist or separatist groups seeking Puerto Rico's independence from the United States and anarchists and extremist socialist groups such as the Workers' World Party, Reclaim the Streets, and Carnival against Capitalism. Many of these anarchist groups were blamed for the damage caused at the 1999 World Trade Organization meeting in Seattle, Washington. A comparison of left-wing and right-wing groups is presented in Table 16.3.

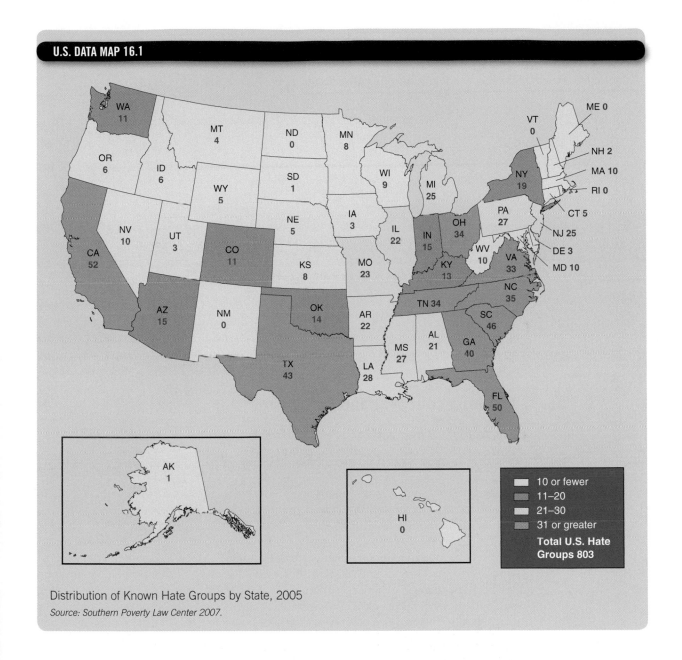

U.S. DATA MAP 16.1

Distribution of Known Hate Groups by State, 2005

Source: Southern Poverty Law Center 2007.

Special interest groups want to resolve specific issues, rather than effect political change. These groups are at the fringes of animal rights, pro-life, environmental, antinuclear, and other political and social movements. Animal rights and environmental groups, such as the Animal Liberation Front (ALF) and the Earth Liberation Front (ELF), have recently increased their activities. Activists from these groups usually use arson and incendiary devices equipped with timers to target government and company facilities. James Jarboe (2002), Domestic Terrorism Section Chief of the FBI, reported that ALF/ELF committed more than 600 criminal acts in the United States since 1996, causing more than $45 million in damages, and characterized the group as one of the most "active extremist elements" in the United States.

Table 16.3

Characteristics of Left-Wing and Right-Wing Terrorist Groups in the United States

Characteristics	Left Wing	Right Wing
Ideology	Political focus, primarily Marxism	Religious focus, ties to Christian Identity Movement (belief that Aryans are God's chosen people)
Economic views	Pro-communist/socialist; belief in Marxist maxim "receive according to one's needs"	Strongly anti-communist; belief in Protestant work ethic, distributive justice
Base of operations	Urban areas	Rural areas
Tactical approach	Cellular structure; use of safe houses	National networking; camps and compounds
Targets for funding	Armored trucks	Armored trucks
Targets for terrorism	Seats of capitalism/government buildings	Federal law enforcement agencies, opposing racial or religious groups
Average age at indictment	35; 18 percent were older than age 40	39; 36 percent older than age 40
Gender	73 percent male	93 percent male
Race	29 percent White 71 percent minority	97 percent White 3 percent American Indian

Source: Smith 1994.

The Impact of War and Terrorism

Psychological Impact

According to Crenshaw (1983), small incremental societal changes in trust, social cohesion, and integration occur as a result of terrorism. Terrorism has a particular impact in small or homogenous societies. Research on the long-term impact of terrorism is limited, although some case studies have been conducted in Northern Ireland, where residents have lived with domestic terrorism since the late 1960s. (See this chapter's Taking a World View discussion for more about the IRA.)

War takes a psychological toll, particularly for soldiers involved in battle. Mental health experts have identified posttraumatic stress disorder (PTSD) as a common aftereffect of battle. Those suffering from PTSD feel depressed and detached and have nightmares and flashbacks of their war experience. This anxiety disorder may also occur with related issues, such as depression, substance abuse, cognition problems, and other problems of physical or mental health (National Center for PTSD 2003a). About 8 percent of the U.S. population will experience PTSD symptoms in their lifetimes.

Vietnam veterans were particularly hard-hit because of extensive war-related trauma: serving hazardous duty, witnessing death and harm to self or others, and being on frequent or prolonged combat missions. The estimated lifetime prevalence of PTSD among Vietnam veterans is 30.9 percent for men and 26.9 percent for women. An additional 22.5 percent of men and 21.2 percent of women have had partial PTSD sometime in their lives (National Center for PTSD 2003b). In a study of post-service mortality comparing more than 9,000 U.S. Army veterans who served in Vietnam with more than 8,000 veterans who served in Korea, Germany, or the United States during the same period, the total mortality of those who served in Vietnam was reported to be 17 percent higher than that of other veteran groups. The excess mortality occurred primarily in the first five years after discharge, as a result of motor vehicle accidents, suicide, homicide, and accidental poisonings. Drug-related deaths among Vietnam vets were also higher than among other veteran groups (Centers for Disease Control 2003a).

Photo 16.3

The 2004 11-M Spain train bombing was the worst worldwide terrorist attack since September 11, 2001. More than 190 people were killed and 1,800 wounded when a series of bombs were set off during peak commuting hours in Madrid. Twenty-one men were found guilty for their role in the bombing or for their association with a terrorist organization.

For information on the war's impact on Iraq war veterans, turn to this chapter's In Focus feature.

Economic Impact

President Bush pledged that the United States would win the fight against terrorism, "whatever it takes, whatever it costs." Although the United States now has a smaller army and is using technologies that are less human-intensive, military expenditures have been increasing, and with those increases comes the greater likelihood of waste, inefficiency, and "good old fashioned pork," according to opponents (Knickerbocker 2002).

In its 2008 budget request, the U.S. Department of Defense explained that $481.4 billion would sustain President Bush's commitment to ensure a high state of military readiness and ground force strength, enhance the combat capabilities of U.S. Armed Forces, continue the development of capabilities to maintain U.S. superiority against potential threats, and continue the department's support of service members and their families. Though some believe that these additional defense expenses are necessary to win the global war on terrorism (GWOT), others have criticized the escalation in military expenditures. The final cost of the Iraq war is estimated to be more than $2 trillion dollars (and unlike past U.S. wars, the Iraq war is being paid almost entirely with debt). A comparison of U.S. war expenditures with other countries is presented in Table 16.4 (see page 412).

Do you recall President Eisenhower's warning? Billions of dollars spent on the war means that other programs are not receiving any funding. The National Priorities Project, a nonprofit, nonpartisan organization, analyzes the impact of federal spending policies on city and state funding and budgets. Since the beginning of the Iraq War, the project has tracked the expanding war budget along with increasing budget cuts in non-security (nonwar) discretionary spending. Based on its 2008 budget, the Bush administration proposed cutting $13 billion from social service and education programming—for example, a 35 percent budget cut in

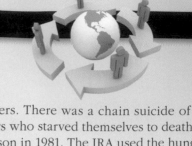

Taking a World View

Ireland and the Decommissioning of the IRA

"The trouble with the Irish is the English."

—A line from a
popular Irish song
(quoted in Whittaker 2002)

In August 2007, the British Army ended Operation Banner, a 38-year security operation in Northern Ireland. Violence and bloodshed first began in 1969, the result of confrontations between Nationalist and Unionist (or Loyalist) forces. The Nationalist or Catholic groups were led by the Irish Republican Army (IRA) and its splinter groups. In 1970, the Provisional IRA was created (*provisional* to honor the provisional government declared by the leaders of the 1916 Easter Rising in Dublin). The Provisional IRA formed to defend the Catholic community and to throw out the British army and police (Wilkinson 1993). The Unionist or Protestant forces represented the Ulster Defence Association and its splinter groups: the Ulster Volunteer Force, the Red Hand Commandos, and the Ulster Freedom Fighters. Ulster Protestants saw themselves as Britons, loyal to the Protestant crown. These groups represent "Orange Extremism" as described by Paul Wilkinson (1993). Orange extremism was also been blamed for provoking the open conflict in the late 1960s and for creating the conditions in which the Provisional IRA could grow. The division between the groups emphasized their polarization in religion, politics, and economics (Whittaker 2002).

Between them, both sides amassed an impressive armory of rifles, homemade machine guns, grenade throwers, anti-tank weapons, and explosives. Bombing seemed to be a favorite tactic of both sides. In more than 30 years of conflict, 3,500 civilians have been killed, some 30,000 have been injured, and there has been a loss of property amounting to millions of pounds (Whittaker 2002). IRA terrorists also have taken their own lives without attempting to harm others. There was a chain suicide of 11 IRA members who starved themselves to death in a Belfast prison in 1981. The IRA used the hunger strikes and deaths to their organizational advantage, "to reap emotive propaganda, to restore the flow of cash and weapons from the previously dwindling U.S. sources, and to regroup and rearm" (Wilkinson 1993). Many Irish Americans embraced the IRA as freedom fighters and supported their cause politically and financially.

In 1998, both sides agreed to the Good Friday Agreement, a 65-page document that sought to define relationships within Northern Ireland, between Northern Ireland and the Republic, and between Ireland, England, Scotland, and Wales. The agreement acknowledges that the people of Northern and Southern Ireland are the only ones with the power to bring about a united Ireland. Through the agreement, new government institutions were set in place. A cabinet-style Executive of Ministers, with members in proportion to party support, allows the participation of all groups (Ahern 2003). The new government composed of unionists and nationalists was established with the first elections for the new Northern Ireland Assembly in June 1998.

A number of political obstacles slowed down the implementation of the agreement, including difficulties over the total disarmament of the Provisional IRA and the dismantling of British military installations in Nationalist areas (Ahern 2003). After experiencing sectarian squabbling, the assembly failed to hold a session for five years (2001–2006); however, elections were held in March 2007 to reestablish the Northern Ireland Assembly and determine how many members from unionist (Protestant) and nationalist (Catholic) parties would be represented (Lyall and Quinn 2007). Self-rule was restored in Belfast in May 2007.

Community Development Block grants, 6 percent cut in the Head Start program, 40 percent cut in the Low Income Energy Assistance program, and 6 percent in Special Education programming. At the same time, the administration requested an additional $100 billion for war related spending (National Priorities Project 2007).

In Focus

The Hidden Costs of War

Of the 1.4 million U.S. forces deployed in Iraq and Afghanistan, almost 200,000 have sought care and assistance from the U.S. Department of Veterans Affairs. The number is predicted to increase to at least 700,000 by the end of the war (Donn and Hefling 2007). Government, health, and veterans service administrators have observed how different the needs of Iraq war veterans are from veterans of past wars and conflicts.

Because of the nature of the conflict itself, forces deployed in Iraq have had more contact with the enemy and more exposure to terrorist attacks than did troops in the earlier Iraq war (O'Conner 2004). Estimates range from one in six to one in three soldiers and Marines seeking help for mental health problems or PTSD as a result of their Iraq experiences. Between 80 and 85 percent of military personnel have witnessed or been part of traumatic events (Vedantam 2006).

In contrast with past wars, more wounded soldiers are returning home alive from Iraq. In the past, many would not have survived traumatic brain injuries (TBI) as a result of improvised explosive device (IED), bomb, or rocket attacks. This shockwave injury is caused when the brain is literally shaken in the soldier's skull, damaging brain tissue. Warfare and medical technology have increased survival rates, however, by making it possible to treat and mobilize wounded soldiers more quickly and closer to combat zones. TBI has been referred to as the signature injury of the Iraq war, much like Agent Orange was for the Vietnam War. Doctors at the Army's Walter Reed Hospital screened all incoming patients between January 2003 and January 2005 and discovered that 60 percent of all wounded soldiers had some form of TBI (Zoroya 2005). TBI symptoms include severe headaches, impaired memory, and sensitivity to light and sound. These symptoms may be temporary or permanent.

Policy analysts blame the lack of planning by the U.S. Department of Veterans Affairs, noting that the administration did not expect the conflict to last as long as it has and did not predict the nature and extent of mental and physical injuries for its active duty and retired personnel (Donn and Hefling 2007). In 2002, Veterans Affairs guaranteed two years of free care to returning combat veterans for any combat-related medical condition. The cost of these benefits, as well as that of providing long-term disability support for soldiers and veterans and their families, is straining the department's budget and government resources. Some predict that in the end, the cost of caring for returning veterans may cost as much as the war itself.

What Does It Mean to Me?

What is the impact of the war's budget on your state? The National Priorities Project provides state level fact sheets on the impact of funding the Iraq War. Go to *Study Site Chapter 16* and review your state's fact sheet.

Table 16.4

Top Ten Military Spenders (2006)

Country	2006 Expenditures (in billions)	Share of World Arms Expenditures (%)
United States	$528.7	46
Britain	$59.2	5
France	$53.1	5
China	$49.5	4
Japan	$43.7	4
Germany	$37.0	3
Russia	$34.7	3
Italy	$29.9	3
Saudi Arabia	$29.0	3
India	$23.9	2

Source: Stockholm International Peace Research Institute 2007.

Political Impact

Terrorism can effect change in two political areas: the overall distribution of political power and government policies (Crenshaw 1983). Terrorism may result in radical changes in the power relationships within a state, involving shifts in who governs and under what rules. As a target of terrorism, the U.S. government has also experienced a redistribution of power as federal and state agencies sought to improve intelligence gathering and security procedures. Institutional changes have occurred, with each major intelligence agency improving its anti-terrorism activities, culminating in the creation of the Department of Homeland Security. In extreme cases, terrorism may lead to the replacement of one government by another.

Government policies usually have two goals: to destroy the terrorist group and to protect potential targets from attack. Policies may include foreign policy efforts seeking the cooperation and support of international allies. After September 11th, the U.S. government attempted to destroy terrorist groups and to ensure U.S. security through a series of executive orders, regulations, and laws. In 2001, Congress passed the Patriot Act, which established a separate counterterrorism fund, expanded government authority to gather and share evidence with wire and electronic communications, allowed agencies to detain suspected foreign terrorists, and provided for victims of terrorism.

Constitutional and civil rights attorneys have been critical of the Patriot Act, alarmed that it would erode individual liberties and increase law enforcement abuses. Often cited is the act's disregard for the principles of political freedom, due process, and the protections of privacy—all principles at the core of a democratic society (Cole and Dempsey 2002). More than 330 cities, towns, and counties, as well as four states, have passed resolutions critical of the federal antiterrorism law (Egan 2004). Two provisions of the act have caused particular concern among citizens. One is the provision that empowers authorities to search people's homes without notification. The provision may have been used by federal agents to search the home of Brandon Mayfield, a lawyer from Oregon who was suspected, but later cleared, in connection with the 2004 bombings in Madrid, Spain. A second clause allows government officials the right to review a person's library, business, and medical records.

As the Iraq conflict continues, the reputation of our country and presidential administration has been tarnished. Though acknowledging that terrorism is everyone's problem, several world leaders have described the war effort as a failure, calling for immediate troop withdrawals and increased diplomatic and humanitarian efforts. Coalition forces from Norway, Japan, Italy, Poland, Great Britain, and Denmark began to withdraw their troops in Iraq as support for the war began waning at home. Here in the United States, the unpopularity of the war is credited with helping Democrats retake both houses of Congress in 2006. Strong support for Democratic candidates was seen as a vote of no confidence in the Bush administration's foreign policy (Block 2007).

How Did This Happen? The Role of Intelligence and Security

In the aftermath of September 11th, investigations were launched to discover how teams of hijackers could coordinate and carry out the hijacking of four domestic aircraft. After all, no U.S. plane had been bombed in 13 years, and none had been successfully hijacked since 1987 (Easterbrook 2001). What was airport security doing? What did our intelligence sources indicate? Except for the events of September 11th, U.S. intelligence and services have generally done very well in protecting the country; there is no evidence that spending more money would have prevented the September 11th attacks (Betts 2001).

Before September 11th, the airline industry was designed to maximize passenger miles and minimize costs (Easterbrook 2001). The industry's imperatives were to attract more passengers, to reduce prices, and to get more flights in the air. Airport security cost money and created hassles that passengers and the industry did not want to deal with. Sewing scissors, fingernail clippers, and tweezers may be everyday items, but they have been considered potential weapons since September 11th. Airport security is tighter and more thorough—and also slower. In addition, airplane cockpit doors have been reinforced, sky marshals have been deployed, and there has been some discussion of arming airline pilots. After a bombing plan was thwarted in London in 2006, containers holding more than 3 milliliters of liquids, gels, or aerosols have been banned from all carryon items.

Did no one imagine that terrorist attacks were possible? Convened in 1998, the bipartisan U.S. Commission on National Security in the 21st Century reported, "Mass-casualty terrorism directed against the U.S. homeland [is] of serious and growing concern" (2001:vi), predicting that "a direct attack against American citizens on American soil is likely over the next quarter century" (p. viii). The commission suggested that America's first priority should be to "defend the United States and ensure that it is safe from the dangers of a new era" (p. 5).

Immediately after September 11, 2001, weaknesses in U.S. intelligence and security were revealed. Land and sea borders were relatively unprotected. On September 10, 2001, about 300 border agents and one single analyst were responsible for patrolling the 4,000-mile land and water border we share with Canada. The U.S. Coast Guard, which is charged with maintaining port security and patrolling 95,000 miles of shoreline, had been forced to reduce its ranks to the lowest level since 1964 (Flynn 2001). One highly publicized intelligence failure was FBI Agent Coleen Rowley's request to investigate Zacarias Moussaoui under the Foreign Intelligence Surveillance Act. Moussaoui had been detained in Minneapolis, Minnesota, for a violation of his immigration status. FBI headquarters denied Rowley's request, saying that there was insufficient evidence that Moussaoui was connected to a foreign terrorist organization. He was later charged as a co-conspirator in the September 11th suicide hijackings.

The National Commission on the Terrorist Attacks upon the United States, also known as the 9-11 Commission, was an independent bipartisan commission created by congressional legislation. The 9-11 Commission was charged with documenting and preparing a full account of the September 11th terrorist attacks. Throughout 2003 and 2004, hearings were held investigating Osama bin Laden's network, the performance of the intelligence community, emergency preparedness and response, and national policy coordination. The commission concluded that U.S. intelligence gathering by the FBI and CIA was inadequate, fragmented, and poorly coordinated.

In 2002, President Bush established the Department of Homeland Security. (Before September 11th, the U.S. Commission on National Security in the 21st Century [2001] had recommended the creation of a National Homeland Security Agency, responsible for planning, coordinating, and integrating all U.S. agencies responsible for security.) The primary mission of the department is to prevent terrorist attacks and reduce the vulnerability of the United States to terrorism through coordination with component agencies: U.S. Secret Service, U.S. Coast Guard, U.S. Citizenship and Immigration Services, U.S. Immigration Customs Enforcement, U.S. Customs and Border Protection, the Federal Emergency Management Agency, and the Transportation Security Administration. The Department of Homeland Security is also responsible for the Homeland Security Advisory System, which informs the public of the current level of terrorist threat.

Where Will the Next Attack Come From?

Bioterrorism

Chemical, biological, and radiological weapons have the potential to kill and injure large numbers of people. Use of these weapons or threat of their use has steadily increased since 1995. Recent bioterrorism cases included the intentional release of ricin (a deadly toxin derived from the caster beans) and anthrax in the early 2000s.

Just 24 days after September 11th, the Florida Department of Health and the Centers for Disease Control confirmed the first case of inhalation anthrax in more than 25 years (Perkins, Popovic, and Yeskey 2002). Anthrax is a disease caused by a bacterium that forms spores. There are three types of anthrax: skin (cutaneous), lung (inhalation), and digestive (gastrointestinal). The Centers for Disease Control classifies anthrax as a Category A agent: It can spread across a large area, poses the greatest possible threat to public health, and requires a great deal of planning to protect the public. In 2001, anthrax was deliberately spread through the U.S. postal system by someone sending letters with powder containing anthrax (Centers for Disease Control 2003b). Twenty-two cases of anthrax infection were reported in seven East Coast states: Connecticut (1), Florida (2), Maryland (3), New Jersey (5), New York (8), Pennsylvania (1), and Virginia (2). Five people died as a result of their exposure. No one has been charged with these crimes.

Nuclear Weapons

The nuclear weapons age began on July 16, 1945, when the United States exploded the first nuclear bomb in Alamogordo, New Mexico. Three weeks later, an atomic bomb was used on the city of Hiroshima, Japan, killing 100,000 residents. Three days later, an atomic bomb was used on the city of Nagasaki, Japan, killing about 74,000 and injuring 75,000. During the 1950s and 1960s, the United States was engaged in a Cold War with Russia and other nuclear countries, locked in a stalemate over who would be the first to launch a nuclear attack. In 1963, the countries agreed to sign a partial test ban treaty, banning nuclear tests in the atmosphere, under water, and in space, and a nonproliferation treaty was signed in 1968, prohibiting nonnuclear countries from possessing or developing nuclear weapons. In 1996, President Bill Clinton was the first world leader to sign the Comprehensive Nuclear Test Ban Treaty, which prohibits all nuclear test explosions in all environments. However, the treaty has yet to be ratified by the U.S. Senate. All North Atlantic Treaty Organization (NATO) members, except for the United States, have ratified the treaty.

Nuclear weapons are still held by more than eight nations in the world. Suspected weapons have been identified in China (400), France (35), India (60), Pakistan (24 to 48), Russia (about 10,000), the United Kingdom (185), and the United States (10,656) (Center for Defense Information 2003). In 2003, North Korean officials informed the Bush administration that they had enough plutonium to create six nuclear devices (Sanger 2003). In 2007, Iran's President Mahmoud Ahmadinejad announced that his country was moving closer to nuclear-generated electricity. Later that year, Iran refused to comply with the UN Security Council's attempts to curtail Iran's nuclear program. Nuclear watchdog groups have reported that based on their current assessment, Iran does not have the capacity to build nuclear weapons.

Community, Policy, and Social Action

Political terrorism expert Grant Wardlaw (1988) states, "Terrorism is a phenomenon that is increasingly coming to dominate our lives." The effects of terrorism are widespread:

> It influences the way governments conduct their foreign policy and corporations transact their business. It causes changes to the structure and role of our security forces and necessitates huge expenditures on measures to protect public figures, vital installations, citizens and perhaps in the final analysis, our system of government. It affects the way we travel, the places we visit and the manner in which we live our daily lives. (p. 206)

Although terrorism is a worldwide growth industry, modes of counterterrorism have been characterized as piecemeal and ineffective (Whittaker 2002). The current counterterrorist policy includes several tenets: Make no concessions and deals with terrorists, bring terrorists to justice for their crimes, isolate and apply pressure on states that sponsor terrorism, and improve the counterterrorist capabilities of countries that work with the United States (Pillar 2001). As a nation, the United States has used several approaches to combat terrorism and to reduce the risk of warfare: diplomacy, sanctions, and military force. Usually, war is justified as being the last resort in circumstances where there are severe domestic rights violations or international aggression by an offending state (Garfield 2002).

Political Diplomacy

According to Christopher Harmon, "Political will, more than new laws or new direction[s] in international politics, is the most important component of an enhanced effort against foreign supported terrorism" (2000:236). Political diplomacy includes articulating policy to foreign leaders, persuading them, and reaching agreements with them (Pillar 2001). Wars reshape diplomacy; victory becomes the goal of foreign policy, and diplomatic relationships are adjusted to achieve it (Mandelbaum 2003).

Relationship building and persuasion are at the heart of U.S. diplomatic efforts. As the lead foreign affairs agency, the Department of State attempts to formulate, represent, and implement the president's foreign policy (U.S. Department of State 2003). The secretary of state is the president's principal adviser on foreign policy and represents the United States abroad in foreign affairs. Primarily, the department manages diplomatic relations with other countries and international institutions (such as the United Nations, the NATO, the World Bank, and the International Monetary Fund [IMF]). The Department of State conducts negotiations and concludes agreements and treaties with other countries on issues ranging from trade to nuclear weapons. The United States maintains diplomatic relations with more than 180 countries (U.S. Department of State 2003). Diplomacy is conducted by the secretary of state and by foreign service officers, immigration officers, FBI special agents, intelligence officers, transportation specialists, defense attachés, and other officials (Pillar 2001).

In its final report, the 9-11 Commission argued for "putting foreign policy at the center of our counterterrorism efforts." In 2007, commission leaders Thomas Kean and Lee Hamilton concluded that the United States had actually lost ground on its foreign policy efforts.

The Use of Economic Sanctions

For many years, the United States and the United Nations have used nonviolent approaches in the form of economic sanctions to punish or pressure countries that have violated U.S. laws or values. Sanctions are considered an alternative to diplomacy or military force. By July 1999, the United States had imposed sanctions on no fewer than 28 countries. From World War I to 1990, the U.S. imposed sanctions 77 times. The United Nations has imposed economic sanctions 13 times (Marks 1999).

Economic sanctions, in the form of trade embargoes and the termination of development assistance, are the most commonly applied form of sanctions, and they have the most significant public health consequences. U.S. sanctions were used against Iraq in 1990 to force its withdrawal from Kuwait and against Yugoslavia (1991–1996) and Serbia and Montenegro (1992–1996) during the Serbian war (Marks 1999). According to Richard Haas, "Economic sanctions are popular because they offer what appears to be a proportional response to challenges in which the interests at stake are less than vital" (1997:75). Sanctions can also serve as a signal of official displeasure with a country's behavior or action.

Many attempts have focused on cutting aid to countries sponsoring or supporting terrorism. The problem is that most terrorist states do not receive significant aid from the United States, and based on past experience, sanctions have only made target countries and nations more angry and impassioned against the United States (Flores 1981). Shaheen Ayubi et al. (1982) concluded that economic sanctions in U.S. foreign policy may not be effective in their objectives and have often failed to achieve their political purposes. For example, sanctions applied against Cuba since 1960 have failed to destabilize the Castro regime.

Sanctions may not hurt dictators and terrorists, but they may increase suffering and death among civilians (Garfield 2002). A report issued in 1993 by the Harvard Center for Population and Development Studies maintained that sanctions exacerbated malnutrition in Haiti and increased child deaths caused by misgovernment (Neier 1993). One approach has been to freeze only economic assets while continuing to provide food and medical aid through private agencies (Neier 1993). However, although only economic sanctions were directed against Iraq in the early 1990s, the Harvard Study Team (1991) reported that essential goods—food, medicine, infrastructure support—were not reaching those in need. All health facilities surveyed reported major drug shortages. Children were most vulnerable, dying of preventable diseases and starvation. The chance of dying before age five in Iraq more than doubled, from 54 per 1,000 between 1984 and 1989 to 131 per 1,000 between 1995 and 1999 (Ali and Shah 2000).

Military Response

Military action is based on the idea that the most effective way to defeat an enemy is by inflicting destruction of the enemy's armies, equipment, transport systems, industrial centers, and cities. Brian Jenkins (1988) identifies four types of military response to terrorism: First, preemptive operations range from evacuation of U.S. citizens or interests to invasion of a location. Most preemptive operations are based on credible and accurate intelligence. Second, search and recovery operations recover stolen weapons or nuclear material that might have fallen into terrorist hands. Third, rescue operations deploy specially trained units to extract hostages taken by terrorists. Finally, the military may use retaliatory or punitive raids to attack terrorist bases or targets.

Retaliation has been the most important counterterrorist use of U.S. military force. The United States first used it against Libya in 1986, responding to the April 4th bombing of a nightclub in Berlin where two Americans were killed and 71 were wounded. One hundred military aircraft were used to attack military targets in and around Tripoli and Benghazi in Libya (Pillar 2001). After September 11, 2001, U.S. troops were deployed to destroy Taliban operations based in Afghanistan.

According to Pillar (2001), evidence suggests that military retaliation does not serve as an effective deterrent to terrorism. First, terrorists who threaten the United States present few suitable military targets. It is tough to attack an enemy that can't be located. Many terrorist groups lack any high-value targets, whose destruction would be costly to their organization. Second, a military attack against a terrorist group may serve political and organizational goals of the terrorist leaders. Such attacks may increase recruiting, sympathy, and resources for terrorist groups. And finally, there is no evidence that terrorists will respond peacefully after a retaliatory attack. Terrorists may also respond by fighting back.

Michael Mandelbaum says that the war on terrorism will not end so neatly:

> Success in this conflict will be measured not, as in other wars, by what American military forces do, but rather by what terrorists do not do. In this new war, a day when nothing happens will be a good day for the United States. (2003:267)

Voices in the Community

Greg Mortenson

If we try to resolve terrorism with military might and nothing else, then we will be no safer than we were before 9/11. In the long term, we have to help feed and clothe people where terrorists are recruiting volunteers. And we have to educate them—especially the girls. We have to prove to them that the world can be a better place. If we truly want a legacy of peace for our children, we need to understand that this is a war that will ultimately be won with books, not with bombs.

—Greg Mortenson,
cofounder and executive director,
Central Asia Institute (quoted in Fedarko 2003:5)

In 1993, Greg Mortenson climbed K2, the world's second-highest mountain. Stricken by illness on the mountain, he was forced to return to a base camp. Two porters took him to their homes in Korphe, an isolated northern village in Pakistan. While recovering in Korphe, Mortenson was impressed by the sight of 84 village children sitting outside and scratching their school lessons in the dirt with sticks. The students were working alone; the village was unable to afford the $1-a-day salary to hire a teacher (Fedarko 2003).

When he returned to his home in Montana, Mortenson decided to pay back the villagers for their kindness by building them a school and providing money for a teacher's salary. He was able to raise $12,000 and kept his promise (Fedarko 2003). Mortenson met Jean Hoerni, who helped established the Central Asia Institute (CAI) in 1996 with a donation of $1 million. After Hoerni died in 1997, Mortenson continued his work through the institute. As of 2007, the CAI completed 58 primary schools, more than 20 water projects, and

14 women's vocational centers, providing more than 24,000 students (including 14,300 girls) with an education (CAI 2007).

Mortenson's program focuses on girls' education and women's empowerment. CAI requires that each village must agree to increase the enrollment of girls by 10 percent each year. He explains,

> You can hand out condoms, build roads, put in electricity, but nothing will change until the girls are educated. They are the ones who remain at home. They are the ones who instill values. Educating the girls is a long-term solution to the war on poverty, and will have a big impact on the war on terrorism. (Fedarko 2003:5)

According to a 2003 CAI press release, students in CAI schools scored higher than the national average on the Pakistan national exams in 2002, 74 percent versus the national average of 46 percent (CAI 2003).

Jahan, a 17-year-old girl from Korphe, is the first girl from her village who has learned to read and write. Mortenson provided her with the tuition to attend a maternal health-care program.

> Jahan is special not just because she was in the first graduating class of my school but also because her mother died while giving birth to her. When Jahan finishes her course in maternal health care, she will have broken the cycle of ignorance. She'll also have an immense impact on the future of the place. What could be more incredible than that? (Mortenson, quoted in Fedarko 2003:6)

Photo 16.4

Greg Mortenson meets students at one of the community schools built by his Central Asia Institute. As of January 2008, the institute has supported or built 58 schools and 14 women's vocational centers.

Antiwar and Peace Movements

Antiwar movements have been characterized as reactive, occurring only in response to specific wars or the threat of war. Every twentieth-century war conducted by the United States elicited organized protest and opposition (Chatfield 1992). On the other hand, peace movements represent organized coalitions that are "fundamentally concerned with the problems of war, militarism, conscription, and mass violence, and the ideals of internationalism, globalism and non-violent relations between people" (Young 1999:228). According to Nigel Young (1999), there are different peace traditions: groups that provide ideas and initiatives for the entire peace movement. For example, the tradition of liberalism and internationalism attempts to prevent war through reformed behavior of states: peace plans, treaties, international law, and arbitration between all groups. Another tradition, anticonscriptionism, linked the peace movement with individual civil rights.

Women as Peacemakers

Another peace tradition is feminist antimilitarism. Peace movements within this tradition are united by the ideal of a distinctive role for women on the issue of peace and female unity across national boundaries (Young 1999). Feminist antimilitary groups first began in the early 1900s. In 1914, Jane Addams, founder of Hull House, led a women's peace parade in New York to protest World War I. Addams, along with Carrie Chapman Catt, the main strategist and leader for the women's suffrage movement, and other women activists formed the Women's Peace Party in 1915. Later that year, the party was renamed the Women's International League for Peace and Freedom. The organization exists today, with chapters in Africa, Asia, South Asia, the Middle East, Europe, and the Americas. Its current global mission includes building a "culture of peace" rather than a "culture of war."

College Activists

Paul Knott (1971) explains that college students before World War II were mainly upper-middle-class students who treated their education as a privilege. Except for a few campuses, most undergraduates showed little social consciousness and were unwilling to challenge or question the status quo. But increasing diversity on college campuses—in students' ages, gender, and ethnic backgrounds—helped increase social awareness and infused students with a greater sense of empowerment. Student activism was at its peak in the 1960s and 1970s, supporting the civil rights movement and later protesting the Vietnam War. During the late 1970s, campuses institutionalized many of the gains made in the previous decade: establishing women's centers, Black student unions, and gay and lesbian organizations and ensuring that student government had a greater role in university operations (Vellela 1988).

The college student population is less politicized as a whole than it was 40 years ago. According to the University of California at Los Angeles (UCLA)'s Annual Freshmen Survey, more than 60 percent of students viewed "keeping up with politics" as "very important" or "essential" in 1966 compared with only 32.9 percent in 2002 (Hamilton 2003). Some argue, however, that progressive student activism didn't stop after the Vietnam War era (Vellela 1988). Student activism is on the rise, but it doesn't reproduce the civil rights and war protests of the 1960s and 1970s. Students are engaged in a broad range of issues: women's rights, discrimination, homophobia, the homeless, labor unions, and political action groups.

What Does It Mean to Me?

How important is keeping up with politics to you? If it is important, in what social or political action have you participated in the past year? What additional opportunities might exist on your campus or in your community?

The new wave of peace activism builds on existing networks established by the student anticorporate movement, which focused on economic justice related to sweatshop labor and unionization on campuses (refer to Chapter 9, "Work and the Economy"). The current wave of peace activism includes a diverse set of schools: rural southern schools (Appalachian State University, North Carolina, University of Southern Mississippi), historically Black colleges (Morehouse College, University of Georgia), community colleges from Hawaii to Massachusetts, and urban public universities (City University of New York and University of Illinois, Chicago), as well as high schools and middle schools. Student groups have held teach-ins, vigils, and fasts to call attention to a variety of issues (Featherstone 2003).

Although most recent peace activism has protested against the war in Iraq, this sentiment has not been universal. Student peace groups have been sensitive to the message that they send (Featherstone 2003). Peace groups are linking their opposition to war to the campaign for social justice, dealing with racism, economic inequality, and sexism at home. Student protesters seem to have learned from the protests of the 1960s, wanting to prevent the kind of alienation experienced by Vietnam War veterans (Rhoads 1991).

Photo 16.5

After her son Casey was killed in Iraq in 2004, Cindy Sheehan became an antiwar activist, taking her message to President Bush at his Crawford, Texas, ranch and at the White House. Sheehan and other protestors were arrested in front of the White House (pictured here). Sheehan stepped away from her activism in 2007, saying that she wanted to return home and be a mother to her surviving children.

Main Points

- September 11, 2001, brought terrorism directly to American soil. **Terrorism** is the unlawful use of force to intimidate or coerce compliance with a particular set of beliefs and can be either **domestic** (based in the United States) or **foreign** (supported by foreign groups threatening the security of U.S. nationals or U.S. national security).

- Terrorism is a premeditated, purposeful, political act targeting noncombatants to force the main target to change course, and it is usually carried out by subnational groups or clandestine agents.

- Terrorist activity has changed little over the years. Six basic tactics account for 95 percent of all incidents: bombings, assassinations, armed assaults, kidnappings, hijackings, and other hostage seizures.

- In contrast to terrorism, war is a violent but legitimate political instrument between armed combatants. From the American Revolution to the twenty-first-century war in Iraq, American history is replete with war.

- Functionalists examine how war and terrorism help maintain the social order, creating and reinforcing boundaries. War creates social stability by letting everyone know what side they are on; there are norms and boundaries in war. But unlike war, the social boundaries in terrorist activities are less certain.

- War and terrorism are forms of conflict that may be based on disputes over resources or land. Modern conflict theorists have focused on how war is used to promote economic and political interests, such as replacing social program funding with military expenditures.

- From the feminist perspective, war is considered a primarily male activity that enhances the position of males in society. Women are expected either to be caretakers or to take over certain jobs vacated by men, as occurred in World War II. More recently, women combatants have redefined notions of courage. Feminist theorists focus on the gender rhetoric used in war.

- Interactionists focus on the social messages and meaning of war and terrorism, believing that *terrorism* is a word with intrinsically negative connotations. Use of the word promotes condemnation of the actors and may reflect an ideological or political bias.

- Although the United States has shifted its focus to the threat of international terrorism, domestic terrorist groups continue to pose a threat. The worst domestic terrorist act was the 1995 Oklahoma City bombing.

- Three types of domestic groups operate in the United States. Right-wing ("hate") groups are antigovernment and advocate for the principles of racial supremacy. Left-wing groups want to bring about socialist revolutionary change. Special interest groups focus on resolving specific issues rather than bringing about political change.

- Terrorism can effect change in two political areas: the overall distribution of political power and government policies. In extreme cases, terrorism may lead to the replacement of one government by another.

- Government policies usually have two goals: to destroy the terrorist group and to protect potential targets from attack. One recent policy is the 2001 Patriot Act, which has brought criticism from constitutional and civil liberty groups.

- After September 11th, a significant question was how it could have happened. The 9-11 Commission is an independent bipartisan commission created to document and prepare a full account of the attacks. It concluded that U.S. intelligence gathering by the FBI and CIA was inadequate, fragmented, and poorly coordinated.

- In 2002, President George W. Bush established the Department of Homeland Security to prevent terrorist attacks and reduce vulnerability to terrorism.

- Modes of counterterrorism appear piecemeal and ineffective. The current counterterrorist policy includes several tenets: make no deals with terrorists, bring terrorists to justice, isolate and apply pressure to states sponsoring terrorism, and improve the counterterrorist capabilities of U.S. allies.

- Political diplomacy includes articulating policy to foreign leaders, persuading them, and reaching agreements with them. Wars reshape diplomacy; victory becomes the goal of foreign policy. As the lead foreign affairs agency, the Department of State attempts to formulate, represent, and implement the president's foreign policy.

- For many years, the United States and the United Nations have used nonviolent approaches, in the form of economic sanctions, to punish or pressure countries that have violated U.S. laws or values. Sanctions are considered an alternative to diplomacy or military force. Economic sanctions such as trade embargoes, which are the most commonly applied form, can significantly affect the health of the targeted country's citizens.

- Many counterterrorism attempts have focused on cutting off aid to countries sponsoring or supporting terrorism. However, most such states do not receive significant aid from the United States, and sanctions may have only made targets angrier with the United States.

- There are four types of military response to terrorism: preemptive operations, search and recovery operations, rescue operations, and retaliatory or punitive raids. Retaliation has been the most important counterterrorist use of U.S. military force, but evidence suggests that military retaliation does not serve as an effective deterrent to terrorism because terrorists present few suitable military targets, military attacks may actually help the goals of terrorist leaders, and terrorists may respond violently.

- Every twentieth-century war conducted by the United States elicited organized antiwar movements. At the college level, student activism was at its peak in the 1960s and 1970s. Now, the student population is less politicized than it was 40 years ago. Activism is on the rise, but not as much as in the 1960s and 1970s. Students now are engaged in a broad range of issues: women's rights, discrimination, homophobia, the homeless, labor unions, and political action groups.

On Your Own

Log on to the Web-based student study site at www.pineforge.com/leonguerrero2study for interactive quizzes, e-flashcards, journal articles, Community and Policy Guides, a Service Learning Guide, the end-of-chapter Web exercises, and additional Web resources.

Internet and Community Exercises

1. Investigate the history of women's peace movements through the Swarthmore College Peace Collection, established in 1930 by Jane Addams of Hull House, Chicago, Illinois. Current women's peace organizations include Women Waging Peace and the Women's International League for Peace and Freedom. The site for the Women's International League for Peace and Freedom also lists state branches. Log on to *Study Site Chapter 16* for more information.

2. Investigate peace and social justice organizations on your campus. (Student Peace Action Network [SPAN] is an example of a campus based organization. SPAN has chapters throughout the country and in Europe.) What is the mission of the organization? How many student and faculty members? What activities does the organization sponsor? If your campus does not have such an organization, can you determine why this is the case?

3. Select a country (other than the United States) and research whether peace or social justice organizations exist on university campuses. Examine one organization's mission and activities. What similarities and differences can you identify between this organization and the one based on your own campus?

PART V

Individual Action
and Social Change

Thanks to Student Activists . . .

Student activists have had considerable impact on social movements, past and present. Consider their role in some of the past century's expressive movements, which aim to change individuals and individual behavior in matters both small and large.

▲ Thanks to these 1946 high school students and many others like them, young women are no longer required to wear skirts and hosiery in school and other public places.

▶ Largely because of the voices of students after the terrorist attacks of September 11, 2001, many Americans understood that most Muslims and people who look Middle Eastern were not to blame.

▲ Thanks in part to the environmental concerns expressed by students in the 1970s, most Americans now support environmental causes and try to recycle and conserve the earth's resources.

Students have also been major players in instrumental movements meant to change the structure of society, institutions, and organizations. Although some of these movements have not yet achieved all their goals, they have made a noticeable difference in people's lives.

◀ Thanks to the dedication of students like these NAACP members working on a 1942 campaign for the equal treatment of African American teachers in Norfolk, Virginia, the movement for civil rights gathered strength and stayed on the path that led to far-reaching legislation during Lyndon Johnson's presidency.

◀ Amid massive antiwar protests in the late 1960s and early 1970s, students learned how powerful their voices could be, extended the civil rights movement, and helped to end the Vietnam War.

▲ Thanks to students joining protests against the 1999 World Trade Organization meetings in Seattle, we have all become more sensitive to the inequities of globalization.

What other student movements—local or global in scope—do you believe have made a difference? Have you participated in a social movement?

CHAPTER 17

Social Problems and Social Action

I n our first chapter, I introduced you to three connections that would be made throughout this text. The first was the connection between sociology and social problems. We began with two concepts offered by C. Wright Mills: personal troubles and public issues. Mills explained that personal troubles transform into public issues when we recognize that troubles exist not because of individual characteristics or traits but because of social forces. This book has not focused on "nuts, sluts, and perverts" (Liazos 1972) as the source of our social problems. Rather, using our sociological imagination, we've examined how social forces shape social problems in U.S. society. We have relied on four sociological perspectives—functionalist, conflict, feminist, and interactionist—to guide us through each chapter. Each perspective provides a unique look at social problems and, as a consequence, offers insights about how we may solve them.

According to a functionalist, all society's needs are met by its social institutions (family, education, politics, religion, and economics). Working interdependently, these institutions ensure social order. When society experiences significant social change (e.g., Industrial Revolution, war), the social order is particularly susceptible to social problems (e.g., crime, poverty, or violence). Social problems do not emerge from individuals; rather, problems emerge when the order is disrupted or tested. Functionalist solutions focus on restoring the social order, repairing the broken institutions, and avoiding dramatic social change.

Like the functionalists, conflict and feminist theorists examine social problems at the macro or societal level. For conflict theorists, social problems are the result of social, economic, or political inequalities inherent in our society. Feminist perspectives consider how gender inequalities lead to social problems. Whereas functionalists assume that order is normal for society, the conflict and feminist theorists believe that conflict over resources and power is the status quo. From this perspective, how does one eliminate social problems? The existing social order needs to be replaced with a more equitable society. Midrange solutions attempt to redefine opportunity and power structures to include the participation of marginalized individuals or groups.

The interactionist perspective focuses on social problems at the micro or individual level. According to this perspective, we create our reality through social interaction. Social problems are created by the labels we attach to individuals and their situations (e.g., the "welfare mom" or a "crack addict"). In addition, problematic behavior is also learned from others; for example, interactionists believe that criminal behavior is learned from other criminals. Social problems are not objective realities. Rather, they are subjectively constructed by religious, political, and social leaders who influence our opinions and conceptions of what is a social problem. This perspective leads us to many different solutions: changing the labeling process (being careful of who is being labeled and what the label is), resocialization for deviant or inappropriate behavior (if the behavior was learned, it can be unlearned), and recognizing the social construction of social problems (acknowledging that it is a subjective process).

What Does It Mean to Me?

Throughout this text, you have been introduced to four sociological perspectives. Which sociological perspective do you agree with most? Which perspective best explains the reasons and solutions for social problems?

Social Problems and Their Solutions

Since its inception, sociology has been considered a means to understand and improve what is wrong with the world. Early sociological thinking emerged from the late eighteenth and early nineteenth centuries during periods of dramatic social, economic, and political change, such as the Industrial Revolution, the French Revolution, and the Enlightenment period. The first sociological thinkers, Karl Marx, Émile Durkheim, and Max Weber, were preoccupied with these social changes and the problems they created for society. They spent their lives studying these problems and attempted to develop programs that would help solve them (Ritzer 2000).

Sociology provides us with the means to examine the social structure or "machinery" that runs our lives. In his book, *Invitation to Sociology,* sociologist Peter Berger likens our human experience to that of puppets on a stage:

> We located ourselves in society and thus recognize our own position as we hang from subtle strings. For a moment we see ourselves as puppets.... Unlike the puppets, we have the possibility of stopping in our movements, looking up and perceiving the machinery by which we have been moved. In this act lies the first step towards freedom. (1963:176)

Freedom comes first in identifying the social "machinery" that controls us and, second, in recognizing that the way society controls individuals is fundamentally different from the strings that control puppets. We have the power to transform or alter that machinery; we have the power to create social change.

The past 16 chapters reveal how we continue to experience many social problems—and, bear in mind that not every social problem could be addressed in the pages of this text. Yet, during the past decade, strong evidence suggests that problems such as crime, drug abuse, and poverty have been minimized because of effective social policies and solutions. We tend to think of the government as the only effective agent of social change because the politicians pass new laws and policies. But the government is not the only agent of social change. In each chapter, I introduced you to individuals, groups, and communities that have attempted to address a particular social problem. Each in their own way is making the second connection, the one between social problems and their solutions.

Our nation's history is filled with groups of people who attempted to promote or prevent change from taking place (Harper and Leicht 2002). **Social movements** are defined as conscious, collective, organized attempts to bring about or resist large-scale change in the social order (Wilson 1973). In today's society, almost every critical public issue leads to a social movement supporting change (and an opposing countermovement to discourage it) (Meyer and Staggenbord 1996). Social movements are the most potent forces of social change in our society (Sztompka 1994). Social movements lead the way for social reform and policies by first identifying and calling attention to social problems.

Understanding Social Movements

Social movements are classified by two factors. First, how much change is intended by the social movement: Is it limited or radical change? And second, what is the scope of the intended change: Is it a group of people or an entire society? Sociologists Charles Harper and Kevin Leicht (2002) distinguish between two dimensions of social movements in their book, *Exploring Social Change: America and the World.* The first dimension identifies how much change is intended by the movement, distinguishing between reform and revolutionary movements. According to Harper and Leicht, **reform movements** try to bring about limited social change by working within the existing system, usually targeting social structures such as education or medicine and directly targeting policymakers. Examples of reform movements are pro- or anti-abortion groups. On the other hand, **revolutionary social movements** seek fundamental changes of the system itself. These types of social movements, such as the U.S. civil rights movement or the anti-apartheid movement in South Africa, consider the political system the key to system change.

The second dimension of social movements identified by Harper and Leicht (2002) is instrumental versus expressive, addressing the scope of intended change. **Instrumental movements** seek to change the structure of society; examples are the civil rights movement and the environmental movement. **Expressive movements** attempt to change individuals and individual behavior. Based on these dimensions, John Wilson (1973) specified four types of social movements: reformative, transformative, alternative, and redemptive (see Table 17.1).

New social movements theory emphasizes the distinctive features of recent social movements. New social movements first appeared in cultural and radical feminist movements in the late 1960s, in some radical sections of the environmental movement in the 1970s, in parts of the peace movement of the late 1970s through the mid-1980s, and in radical sections of the gay rights movement since the 1980s (Plotke 1995).

Photo 17.1

Supporters of Pakistan People's Party shout slogans during a December 2007 protest rally against the assassination of former premier Benazir Bhutto. Despite death threats, Bhutto returned to Pakistan committed to what she referred to as a "movement to democracy." Political observers noted that after Bhutto's death, the country's progress toward democracy was jeopardized.

Table 17.1

Types of Social Movements

	Instrumental	Expressive
Reform	*Reformative:* Partial change within the social structure via policy reform Examples: labor movement, NAACP, antiabortion	*Alternative:* Partial change in individuals via individual reform Examples: Christian evangelism, temperance movement
Revolutionary	*Transformative:* Total change of the social structure Examples: Bolsheviks, Islamic fundamentalism	*Redemptive:* Total change in the individuals Examples: millenarian movements, cults, the People's Temple

John Hannigan (1991) and David Plotke (1995) distinguish between new social movements and early social movements. First, new movements have different ideologies than earlier movements did. Instead of fighting for human rights, such as voting or freedom of speech, new movements are framed around concerns about cultural and community rights, such as the right to be different, to choose one's lifestyle, and to be protected from particular risks like nuclear or environmental hazards (Hannigan 1991). Second, new social movements distrust formal organizations. Consequently, they tend to be small-scale, informal organizations. Finally, whereas previous movements were identified with the economic oppression of workers or minorities, new social movements are associated with a new middle class of "younger, social and cultural specialists" (Plotke 1995). Instead of acting on behalf of their own interests, this new middle class acts on behalf of groups who cannot act on their own.

How Do Social Movements Begin?

Social scientists offer several explanations of how social movements emerge. Individual explanations focus on the psychological dispositions or motivations of those drawn to social movements. Women and men are depicted as either frustrated or calculating actors in political or social movements. Empirical studies have not consistently supported these explanations, demonstrating that individual predispositions are insufficient to account for collective action in social movements. In addition, such theories tend to deflect attention from the real causes of discontent and injustice in our social and political structures (Wilson and Orum 1976).

Social movements do not generally arise from a stable social context; rather, they arise from a changing social order (Lauer 1976). Social movements arise from the structure itself, primarily the result of social and economic deprivation. People are not acting just because of their suffering. They are likely to act when they experience **relative deprivation**, a perceived gap between what people expect and what they actually get. James Davies (1974) argued that social movements were likely to occur when a long period of economic and social improvement is followed by a period of decline. Relative deprivation theory has been used to explain the development of urban protests among African Americans during the 1960s, which were initiated by middle-class African Americans who perceived social and economic gaps between Black and White Americans (Harper and Leicht 2002). But relative deprivation alone isn't enough to create a social movement.

Neil Smelser (1963) explains that six structural conditions are necessary for the development of collective behaviors and social movements. These conditions operate in an additive fashion. First, particular structures in society are more likely to generate certain kinds of social movements than others. For example, societies with racial divisions are more likely to develop racial movements. Second, people will become dissatisfied with the current structure only if the structure is perceived as oppressive or illegitimate. Third, there must be growth of a generalized belief system. People need to share an ideology, a set of ideas, which defines the sources of the structural problems or strains and the solutions necessary to alleviate them. The civil rights movement was based on the ideology that racism was the source of restricted opportunities for minorities (Harper and Leicht 2002). Fourth, dramatic events sharpen and concretize issues. These events may initiate or exaggerate people's dissatisfaction with the current structure or redefine their beliefs about the sources of the structural problems. Examples of dramatic or precipitating events include the 1968 Watts riots in relation to the Black Power phase of the civil rights movement and the 1979 Three Mile Island nuclear disaster in relation to the antinuclear power movement (Harper and Leicht 2002). Fifth, the movement gains momentum with the mobilization of leaders and members for the movement. At this time, the social movement also begins to take the shape of a formal organization. Finally, forces in society (the existing political structure or countermovements) respond to the social movement either by accepting or suppressing it. One of the important features of Smelser's theory is his emphasis on the relationship between the social movement and society itself, a powerful force in shaping the development, direction, and ultimately, the success of the movement (Harper and Leicht 2002).

Photo 17.2

According to **resource mobilization theory**, no social movement can succeed without resources. J. McCarthy and M. Zald (1977) argue that human and organizational resources must be mobilized to create a social movement. A social movement requires human skills in the form of leadership, talent, and knowledge, as well as an organizational infrastructure to support its work. On the other hand, the **political process model** emphasizes the relationship between a mobilized social movement and a favorable structure of political opportunities. Social movements are political phenomena, attempting to change social policy and political coalitions, in this view. Political structures enhance the likelihood of a successful social movement by being receptive to change or by being more or less vulnerable at different points in time. For example, Doug McAdam (1982) noted that the efforts of the civil rights movement were enhanced by the expansion of the Black vote and the shift of Black voters to the Democratic Party. Without favorable support from the political structure, the civil rights movement might not have succeeded.

Social movements gain strength when they develop symbols and a sense of community, which generates strong feelings and helps direct this energy into organized action. People will form a social movement when they develop "a shared understanding of the world and of themselves that legitimate and motivate collective action" (McAdam, McCarthy, and Zald 1996:6). McAdam (1982) explains that resource mobilization must also include **cognitive liberation**. Much like Karl Marx's concept of class consciousness, cognitive liberation begins when members of an aggrieved group begin to consider their situation as unjust. They must recognize the situation they are in. The second part of cognitive liberation is the group's sense that its situation can be changed. Finally, those who considered themselves powerless begin to believe that

More than 200,000 people participated in the March on Washington demonstrations in March 1963. This march, along with other nonviolent protests and marches, brought the need of basic civil rights for all Americans, regardless of race, to the nation's attention. The U.S. Congress passed landmark legislation in the 1960s: the Civil Rights Act of 1964, the Voting Rights Act of 1965, and the Civil Rights Act of 1968 (also known as the Fair Housing Act).

they can make a difference (Piven and Cloward 1979). Individuals must move through all three stages to become cognitively liberated. They must organize, act on political opportunities, and instigate change; "In the absence of these necessary attributions, oppressive conditions are likely, even in the face of increased resources, to go unchallenged" (McAdam 1982:34).

What Does It Mean to Me?

What will be our next great social movement? After the civil rights, women's rights, and gay rights movements, what social movement will affect you next? What will contribute to the development of this movement? What will contribute to its success?

How Have Reform Movements Made a Difference?

"The interest of many scholars in social movements stems from their belief that movements represent an important force for social change" (McAdam, McCarthy, and Zald 1988:727); yet, "the study of the consequences of social movements is one of the most neglected topics in literature" (Giugni 1999:xiv–xv). Early in human history, most social change was the result of chance or trial and error (Mannheim 1940), but in modern history, social movements have been the basic avenues by which social change takes place (Harper and Leicht 2002).

According to Harper and Leicht (2002), the most dramatic social, cultural, economic, and political transformations come from revolutions. Successful revolutions are rare and dramatic events, such as the early revolutions in France (1789), Russia (1917), and China (1949), and include the political transformations in South America, Eastern Europe, and the former Soviet Union during the 1980s.

Most social movements that we're familiar with are reform movements that focus on either broad or narrow social reforms. They produce significant change, but in gradual or piecemeal ways (Harper and Leicht 2002). The most important U.S. reform movements in the first half of the twentieth century focused on grievances related to social class, such as the labor movements of the early 1900s, which helped ensure safer working conditions, eliminated child labor, and provided substantial increases in wages and benefits. After World War II, a new type of reform movement, which included the civil rights movement, student movement, feminist movement, gay liberation movement, and ethnic/racial movements, addressed inequalities based on social status rather than social class (Harper and Leicht 2002). Successful reform movements generate change in three areas.

1. **Culture.** Reform movements educate people and change beliefs and behaviors. Change can occur in our culture, identity, and everyday life (Taylor and Whittier 1995). By changing the ways individuals live, movements may effect long-term changes in society (Meyer 2000). The women's movement has established a clear record of cultural change. The women's movement changed the way women viewed themselves and altered our language, our schools, the workplace, politics, the military, and the media.

2. **New organizations or institutions.** Movements lead to the creation of new organizations that continue to generate change. Through these new organizations, social movements may influence ongoing and future initiatives by altering the structure of political

support, limiting resources to challengers, and changing the values and symbols used by supporters and challengers. D. Meyer argues that by changing participants' lives, "movements alter the personnel available for subsequent challenges" (2000:51).

From the women's movement, the National Organization for Women (NOW) was created in 1966, along with the Women's Equity Action League (1968), National Women's Political Caucus (1971), National Women's Law Center (1972), and the Feminist Majority Foundation (1987). NOW is the largest organization of feminist activists in the United States, with more than 500,000 members and 550 chapters in all 50 states. The organization continues its advocacy and legislative efforts in guaranteeing equal rights for women, ensuring abortion rights and reproductive freedom, opposing racism, and ending violence against women.

3. **Social policy and legislation.** Successful social policies have been nurtured by partnerships between the government and social movements (Skocpol 2000). Movements generally organize and mobilize themselves around specific policy demands (Meyer 2000), attempting to minimize or eliminate social problems. Public policy can do many things: New laws can be enacted or old ones may be struck down, social service programs can be created or ended, taxes can be used to discourage bad behaviors (cigarette or alcohol taxes) or encourage other behaviors (tax breaks to build enterprise zones) (Loseke 2003).

For reform movements, the relationship between desired and actual change varies (Lauer 1976). So far, the women's movement has not achieved the passage of the Equal Rights Amendment, first proposed in 1923. By 2007, 35 of the necessary 38 states had ratified the ERA. The women's movement has made progress in revising laws pertaining to violence against women, creating family-friendly business practices, and enhancing women's roles in the military, clergy, sports, and politics.

Making the Last Connection

It was a Sunday evening, 31 January 1960, when four freshmen at North Carolina Agriculture and Technical College stayed up late talking about ending segregation in the South. They were extraordinarily poorly positioned to effect political or social change on campus, much less in the United States: young, Black, by no means affluent, and generally disconnected from the major centers of power in America. On Monday morning Ezell Blair, Jr., Franklin McCain, Joseph McNeill, and David Richmond dressed in their best clothes to visit the Woolworth's in downtown Greensboro. After buying some school supplies, they sat at the lunch counter and waited for service. They spent the rest of their day there.

The following day twenty-seven other Black students joined them and on Wednesday twice as many. By Thursday, a few sympathetic White students from nearby schools had enlisted and, with the lunch counter at Woolworth's filled, a few started a sit-in at another lunch counter down the street. By the end of the week, city officials offered to negotiate a settlement and, on Saturday night, 1,600 students rallied to celebrate this victory. News of the sit-in campaigns spread throughout the South and then elsewhere across the United States, spurring other activists to emulate their efforts. Sit-ins to desegregate lunch counters and restaurants, stores and libraries, and even buses swept the South. A new organization, the Student Non-Violent Coordinating Committee (SNCC) was formed in April 1960. SNCC would become a leading force in the civil rights movement, setting much of the agenda for liberal politics in the United States during the early 1960s, precipitating the passage of the Voting Rights Act of 1965 and politicizing student activists across the United States. (Meyer 2000:33)

Yes, solutions to social problems are complex and, as Mills advised, require attention to large social forces, such as those targeted by social movements. But social movements don't appear overnight. Social movements begin with individual efforts such as those taken by college students Blair, McCain, McNeill, and Richmond. Grassroots organizations with strong community and local leadership, such as those on the frontline of the modern environmental movement, have also proven effective in addressing social problems.

Some may believe that individual efforts don't amount to much, leading only to short-term solutions or effectively helping one person or one family at a time. But according to David Rayside (1998), the impact of any social movement should be measured over the long term. The isolated effort of thousands of individuals and groups "creates changes in social and political climates, which then enable particular groups to make more specific inroads into public policy and institutional practice" (p. 390).

College and university students have always played an important role in addressing social problems. According to longtime social activist Ralph Nader, it is up to students "to prod and to provoke, to research and to act" (1972:23). What can a student accomplish? Nader thinks the answer is quite a lot:

Take the corporate polluter. Sit-ins and marches will not clean up rivers and the air that he fouls. He is too powerful and there are too many like him. Yet, the student has unique access to resources that can be effective in confronting the polluter. University and college campuses have the means for detecting the precise nature of the industrial effluent, through chemical and biological research. Through research such as they perform every day in the classroom, students can show the effect of the effluent on an entire watershed, and thus alert the community to real and demonstrable dangers to public health—a far more powerful way to arouse public support for a clean environment than a sit-in. Using the expertise of the campus, students can also demonstrate the technological means available for abating the discharge, and thus meet the polluter's argument that he can do nothing to control his pollution. By drawing on the knowledge of economists, students can counter arguments that an industry will go bankrupt or close down if forced to install pollution controls. Law and political science students can investigate the local, state, or federal regulations that may apply to the case, and publicly challenge the responsible agencies to fulfill their legal duties. (1972:21)

Student action may lead to significant social change. For example, voting drives led by 17- and 18-year-olds produced the 26th Amendment of the U.S. Constitution, granting voting rights to those 18 years of age and older (Nader 1972). The driving force behind the 26th Amendment came from youth who raised questions about the legitimacy of a representative government that asked 18- to 20-year-olds to fight in the Vietnam War but denied them the right to vote on war-related issues (Close Up Foundation 2004). Norvald Fimreite, a graduate student at the University of Western Ontario, was the first to report unusual levels of mercury residues in fish caught in the Great Lakes (Nader 1972). Fimreite's data led to a worldwide alert about the problems of mercury and other chemicals in the fish we eat.

Photo 17.3

North Carolina Agriculture and Technical College students during the second day of protest at the Whites-only lunch counter at a Woolworth's in Greensboro, North Carolina.

The last connection presented in this text is the connection between social problems and your community. Throughout the country, college students have affirmed their commitment to community service. According to UCLA's annual survey of entering college freshmen, one of four freshmen surveyed believed that it was important to be involved in community programs. About 80 percent of freshmen had participated in community service during their senior year in high school, and 67 percent of them believed that they would continue volunteering in college (Engle 2006). High school students are also increasing their involvement in community service. School districts in every state except Wyoming, North Dakota, and South Dakota require community service.

If you think there is nothing that you can do to effect change, you've not been paying attention. The first step is to recognize that you can make a difference. You do not have to believe in quick fixes, universal solutions, or changing the entire world to solve social problems. You do not have to join a national organization. To begin, you can join other college and university students who have chosen to become personally involved in their community. Most efforts are small and practical, but as one college student says, "I can't do anything about the theft of nuclear grade weapons materials in Azerbaijan, but I can clean up the local pond, help tutor a troubled kid, or work at a homeless shelter" (Levine and Cureton 1998:36).

Photo 17.4

Volunteers rally around an issuc or an event. Pictured here are volunteers preparing meals for recovery workers at the World Trade Center after the September 11, 2001, attack. You can explore service opportunities on your own campus or community, finding one that fits your interests or skills.

What does it take to start making that connection with your community? The second step is to explore opportunities for service on your campus and in your community. Take the chapters in this text or the material presented by your instructor to consider what social problems you are passionate about. Determine what issues you'd like to address and determine what individuals or groups you would like to serve. Even though you may be in your college community for only four or five years, act as if you're there for life: Take an interest in what happens in your community (Hollender and Catling 1996). Whatever your interests are, you can be sure that there are people and programs in your community who share them. And if they don't exist, what would it take to create such a program?

The third step is to do what you enjoy doing. When you know what you like, when you know what you can contribute, you will find the right connection. Whatever your talent, your community program will appreciate your contribution. It could be that you are an excellent writer; if so, you could help with a program's monthly newsletter, develop an informational brochure, or design the program's Web site. Do you enjoy working with others? Volunteer to work with clients, to answer phones, or to help at a rally. In addition to providing invaluable service to the program, recognize the experience and skills that you will gain from your efforts.

And the final step? Go out and do it. It doesn't have to last an entire semester or school year; you could just volunteer for a weekend or a day. Change doesn't happen automatically, it begins with individual action. As Paul Rogat Loeb (1994) explains, the hard questions must come from us:

> We need to ask what we want in this nation and why; how should we run our economy, meet human needs, protect the Earth, achieve greater justice? … The questions have to come from us, as we reach out to listen and learn, engage fellow citizens who aren't currently involved, and spur debate in environments that are habitually silent.

Charles Lemert reminds us of the most valuable sociological lesson:

> Sociology … is different for all because each [must] find a way to live in a world that threatens even while it provides. Grace is never cheap. In the end, what remains is that we all have a stake in the world. Like it or not, life is always life together. Social living is the courage to accept what we cannot change in order to do what can be done about the rest. (1997:191)

Put your sociological imagination to work to see where change is possible. Do you have the courage?

What Does It Mean to Me?

Chapters 2 through 16 each included a "Voices in the Community" feature, highlighting an individual or individuals who saw a social problem and made a decision to do something about it. Several began their activism during college or in their youth—Max Kenner (Crime and Criminal Justice), Wendy Kopp (Education), and Chad Pregracke (Environment). Here is your chance to be a voice in your community—what social problem do you see and what can you do about it? Better yet, if you've already begun work in your community and have a story to tell, let me know. Your story may be featured in the next edition!

Main Points

- In this book, we have reviewed several social problems; the theoretical perspectives that examine them; and the policies, organizations, and individuals that affect them.
- To functionalists, all of society's needs are met by its social institutions, which ensure social order. When there is significant social change, the social order is particularly susceptible to social problems. In this view, social problems stem from the disruption or testing of social order, and solutions focus on restoring the social order, repairing the broken institutions, and avoiding dramatic social change.
- For conflict theorists, social problems are the result of social, economic, or political inequalities inherent in our society. Feminist

perspectives specifically consider how gender inequalities lead to social problems. In contrast to functionalists, conflict and feminist theorists argue that conflict over resources and power is the status quo and that the existing social order needs to be replaced with a more equitable society offering more opportunity to marginalized individuals or groups.

- Interactionists focus on social problems at the individual level, contending that we create our own reality through social interaction. Social problems are subjectively constructed by societal leaders who label people and influence our opinions of what is a social problem. Problematic behavior is also learned from others. This perspective leads us to many different solutions: changing the labeling process, resocializing people who exhibit deviant or inappropriate behavior, and recognizing the social construction of social problems.

- Sociology provides us with the means to examine the social structure or "machinery" that runs our lives. Evidence from the past decade suggests that some social problems have been minimized because of effective social policies and solutions. But the government is not the only source of change; individuals, groups, and communities also make the difference.

- Throughout American history, groups have acted to promote change. **Social movements** are conscious, collective, organized attempts to bring about or resist large-scale change in the social order. They are the most potent forces of social change in our society.

- Social movements are classified by two factors: the scope and the depth of change. **Instrumental movements** seek to change the structure of society itself, whereas **expressive movements** attempt to change individuals. And while **reform movements** try to bring about limited social change by working within the existing system, **revolutionary social movements** seek fundamental changes of the system itself.

- **New social movements theory** emphasizes the distinctive features of recent social movements. Instead of fighting for human rights, new movements focus on cultural and community rights. New social movements also distrust formal organizations and thus tend to be small-scale, informal organizations. Also, new social movements are associated with a new middle class that acts on behalf of groups that cannot act on their own.

- People are more likely to act when they experience **relative deprivation**, a perceived gap between what people expect and what they actually get. Relative deprivation, in conjunction with six necessary conditions surrounding social structure, works to produce social movements.

- In **resource mobilization theory**, no social movement can succeed without human and organizational resources. In contrast, the **political process** model presents social movements as political phenomena, attempting to change social policy and political coalitions.

- Social movements gain strength when they develop symbols and a sense of community, which helps direct energy into organized action. Resource mobilization must also include **cognitive liberation**, which begins only when members of an aggrieved group start to consider their situation unjust, to believe the situation can be changed, and to believe they can make a difference.

- Revolutions produce the most dramatic social, cultural, economic, and political transformations. However, most movements center on gradual but significant reforms. Successful reform movements generate change in three areas: culture, new organizations or institutions, and social policy and legislation.

- You also can make a difference by taking four steps. First, recognize that you can make a difference. Second, explore opportunities for service on your campus and in your community. Third, do what you enjoy doing. And finally, go out and do it. Change doesn't happen automatically; it begins with individual action. Put your sociological imagination to work to see where change is possible.

On Your Own

Log on to the Web-based student study site at www.pineforge.com/leonguerrero2study for interactive quizzes, e-flashcards, journal articles, Community and Policy Guides, a Service Learning Guide, the end-of-chapter Web exercises, and additional Web resources.

Internet and Community Exercises

1. The Corporation for National and Community Service is part of the USA Freedom Corps, a White House initiative to "foster a culture of citizenship, service, and responsibility." The organization includes three programs: AmeriCorp, Senior Corps, and Learn and Serve. The AmeriCorp program is a network of local, state, and national service programs. The Senior Corp includes volunteers older than age 55 serving local nonprofits and public agencies. Learn and Serve supports service-learning programs, kindergarten through 12th-grade students, combining community service with student learning. For more information on local initiatives and volunteer opportunities in your state, go to *Study Site Chapter 17.*

2. To learn more about the Woolworth's Greensboro sit-ins go to *Study Site Chapter 17,* which includes recordings with sit-in participants, along with related stories and photos and links to other civil rights sites.

3. How can one person serve as an agent of social change? Eduardo Arias, a Kuna Indian from one of the San Blas islands off Panama's Caribbean coast, did something in 2007 that affected business and trade on 6 continents and in 34 countries. What did he do? He read the ingredient label on a 59-cent tube of toothpaste and saw two words, diethylene glycol, the same poisonous ingredient used in antifreeze. Arias purchased a tube and brought it as evidence to the Health Ministry office. His report set off a global recall of the product, along with increasing examination and scrutiny of Chinese manufacturers and their products (the source of the tainted toothpaste). On your own, find a story of one person who made a difference, who was part of some social change. The story could be based in the United States or elsewhere, and though the person might not have worldwide influence as Arias did, what impact did that person have?

4. If you do nothing else, make sure to show up on Election Day to support candidates who support your views (Hollender and Catling 1996). Tracking presidential election years, research indicates that electoral participation of Americans younger than 25 years of age has declined since 1972. In 2000, 42 percent of 18- to 24-year-olds voted, compared with 70 percent of citizens 25 years or older (Levine and Lopez 2002). Rock the Vote is a nonprofit, nonpartisan organization that attempts to mobilize youth to create positive social and political change in their lives and communities (Rock the Vote 2004). The organization works year round, encouraging youth to become involved in a range of issues: education, economy, environment, violence, and national and personal debt. You may have seen a Rock the Vote Community Street Team register voters at a local concert or event. Go to *Study Site Chapter 17* for more information.

References

Chapter 1

Adams, Bert and R. A. Sydie. 2001. *Sociological Theory*. Thousand Oaks, CA: Pine Forge.

Advisory Committee on Student Financial Assistance. 2002. "Empty Promises: The Myth of College Access in America." Retrieved May 15, 2004 (www.ed.gov/about/bdscomm/list/acsfa/emptypromises.pdf).

Ballesteros, P., J. L. Estrada, G. Barriga, F. Molinar, M. C. Hernandez, L. Huerta, G. Cocho, and C. Villarreal. 2006. "Comparative Analysis of Gender Differences in HIV-1 Infection Dynamics." *AIP Conference Proceedings* 854(1):48–50.

Berger, Peter and Thomas Luckmann. 1966. *The Social Construction of Reality*. Garden City, NY: Anchor.

Bernard, Jessie. 1982. *The Future of Marriage*. 2nd ed. New Haven, CT: Yale University Press. (Original work published 1972)

Boehner, J. and H. McKeon. 2003. "The College Cost Crisis." Retrieved May 15, 2004 (http://www.bgsu.edu/downloads/finance/file17013.pdf).

Callan, Patrick and Joni Finney. 2002. "State Policies for Affordable Higher Education." Pp. 10–11, 13 in *Losing Ground: A National Status Report on the Affordability of American Higher Education*. San Jose, CA: National Center for Public Policy and Higher Education.

Centers for Disease Control. 2007a. "Basic Information." Retrieved April 15, 2007 (http://www.cdc.gov/hiv/topics/basic/index.htm).

———. 2007b. "A Glance at the HIV/AIDS Epidemic." Retrieved April 7, 2007 (http://www.cdc.gov/hiv/resources/factsheets/At-A-Glance.htm).

Chakrapani, Venkatesan, Peter Newman, Murali Shunmugam, Alan McLuckie, and Frederick Melwin. 2007. "Structural Violence against Kothi-Identified Men Who Have Sex with Men in Chennai, India: A Qualitative Investigation." *AIDS Education and Prevention* 19(4):346–64.

Chodorow, Nancy. 1978. *The Reproduction of Mothering: Psychoanalysis and the Sociology of Gender*. Berkeley: University of California Press.

CNN. 1996. "Anti Gay Marriage Act Clears Congress." Retrieved September 27, 2003 (www.cnn.com/US9609/10/gay.marriage/).

Collins, Patricia Hill. 1990. *Black Feminist Thought: Knowledge, Consciousness, and Empowerment*. Boston, MA: Unwin Hyman.

Coser, L. 1956. *The Functions of Social Conflict*. New York: Free Press.

Dahrendorf, Ralf. 1959. *Class and Class Conflict in Industrial Society*. Stanford, CA: Stanford University Press.

Durkheim, Émile. 1973. "The Dualism of Human Nature and Its Moral Condition." Pp. 149–63 in *Emile Durkheim on Morality and Society,* edited by R. Bellah. Chicago, IL: University of Chicago Press. (Original work published 1914)

Ellis, R. A. 2003. *Impacting Social Policy: A Practitioner's Guide to Analysis and Action.* Pacific Grove, CA: Thomson Brooks/Cole.

Fine, Gary Alan. 2006. "The Chaining of Social Problems: Solutions and Unintended Consequences in the Age of Betrayal." *Social Problems* 53(1):3–17.

Frankfort-Nachmias, Chava and Anna Leon-Guerrero. 2006. *Social Statistics for a Diverse Society.* Thousand Oaks, CA: Pine Forge.

Fresco, Robert. 2004. "A High Price to Pay." *Newsday,* March 21, 2004. Retrieved March 30, 2004 (http://www.newsday.com/business/ny-epcoll0321,0,4637857.story?page=1&coll=ny-business-big-pix).

Fumaz, C. R., J. A. Munoz-Moreno, A. L. Ballesteros, R. Paredes, M. J. Ferrer, A. Salas, D. Fuster, E. Masmitja, N. Perez-Alvarez, G. Gomez, C. Tural, and B. Clotet. 2007. *AIDS Care* 19(1):138–45.

Glater, Jonathan. 2007. "College Costs Outpace Inflation Rate." *New York Times,* October 23, p. A21.

Habitat for Humanity. 2004. "Millard Fuller." Retrieved June 9, 2004 (www.habitat.org/how/millard html).

Hytrek, Gary and Kristine Zentgraf. 2007. *America Transformed: Globalization, Inequality and Power.* New York: Oxford University Press.

Irwin, Alan. 2001. *Sociology and the Environment.* Cambridge, England: Polity Press.

Kaplan, Laura Duhan. 1994. "Women as Caretaker: An Archetype That Supports Patriarchal Militarism." *Hypatia* 9(2):123–33.

Katzer, Jeffrey, Kenneth Cook, and Wayne Crouch. 1998. *Evaluating Information: A Guide for Users of Social Science Research.* 4th ed. Boston, MA: McGraw-Hill.

Konradi, Amanda and Martha Schmidt. 2001. *Reading Between the Lines: Toward an Understanding of Current Social Problems.* Mountain View, CA: Mayfield.

Lemert, Charles. 1997. *Social Things: An Introduction to the Sociological Life.* Lanham, MD: Rowman & Littlefield.

Lichtenstein, Bronwen. 2004. "AIDS as a Social Problem: The Creation of Social Pariahs in the Management of an Epidemic." Pp. 316–34 in *Handbook of Social Problems: A Comparative International Perspective,* edited by George Ritzer. Thousand Oaks, CA: Sage Publications.

Loseke, Denise. 2003. *Thinking about Social Problems.* New York: Aldine de Gruyter.

Loseke, Denise and J. Best 2003. *Social Problems: Constructionist Readings.* New York: Aldine De Gruyter.

Madoo Lengermann, Patricia and Jill Niebrugge-Brantley. 2004. "Contemporary Feminist Theory." Pp. 436–80 in *Sociological Theory,* edited by G. Ritzer and D. Goodman. Boston, MA: McGraw-Hill.

Marx, Karl. 1972. "Theses on Feuerbach." In *The Marx Engels Reader,* edited by R. C. Tucker. New York: Norton.

Mead, George Herbert. 1962. *Mind, Self, and Society: From the Standpoint of a Social Behaviorist.* Chicago, IL: University of Chicago Press. (Original work published 1934)

Merton, Robert. 1957. *Social Theory and Social Structure.* Glencoe: IL: Free Press.

Mills, C. Wright. 2000. *The Sociological Imagination.* New York: Oxford University Press. (Original work published 1959)

National Center for Public Policy and Higher Education. 2002. *Losing Ground: A National Status Report on the Affordability of American Higher Education.* San Jose, CA: Author.

Ritzer, George. 2000. *Sociological Theory.* New York: McGraw-Hill.

Schneider, Anne and Helen Ingram. 1993. "Social Construction of Target Populations: Implications for Politics and Policy." *American Political Science Review* 87(2):334–47.

Schroedel, Jean R. and Daniel R. Jordan. 1998. "Senate Voting and Social Construction of Target Populations: A Study of AIDS Policy Making, 1987–1992." *Journal of Health Politics, Policy and Law* 23(1):107–31.

Schutt, Russell K. (2006). *Investigating the Social World: The Process and Practice of Research.* Thousand Oaks, CA: Pine Forge.

Simoni, Jane, Karina Walters, Kimberly Balsam, and Seth Meyers. 2006. "Victimization, Substance Abuse, and HIV Risk Behaviors Among Gay/Bisexual/Two-Spirit and Heterosexual American Indian Men in New York City." *American Journal of Public Health* 96(12):2240–45.

Smith, Dorothy. 1987. *The Everyday World as Problematic: A Feminist Sociology.* Boston, MA: Northeastern University Press.

Spector, Malcolm and John Kituse. 1987. *Constructing Social Problems*. New York: Aldine de Gruyter.

Tabi, Marian and Robert Vogel. 2006. "Nutritional Counseling: An Intervention for HIV Positive Patients." *Journal of Advance Nursing* 54(6):676–82.

Tong, Rosemarie. 1989. *Feminist Thought: A Comprehensive Introduction*. Boulder, CO: Westview Press.

Turner, Jonathan. 1998. *The Structure of Sociological Theory*. 6th ed. Belmont, CA: Wadsworth.

UNAIDS. 2006. *2006 Report on the Global AIDS Epidemic: Executive Summary*. Geneva, Switzerland: UNAIDS.

World Health Organization. 2007. "HIV Surveillance, Estimations and Monitoring and Evaluation." Retrieved April 15, 2007 (http://www.who.int/hiv/topics/me/en/index.html).

Chapter 2

Abramovitz, Mimi. 1996. *Regulating the Lives of Women: Social Welfare Policy from the Colonial Times to the Present*. Boston, MA: South End Press.

Acs, G. and P. Loprest. 2001. "Synthesis Report of the Findings from ASPE's 'Leavers' Grants." Washington, DC: U.S. Department of Health and Human Services.

Adams, Gina and Monica Rohace. 2002. "Child Care and Welfare Reform." Pp. 121–42 in *Welfare Reform: The Next Act,* edited by A. Weil and K. Feingold. Washington, DC: Urban Institute.

Alder, J. 1994. "The Name of the Game Is Shame." *Newsweek* 124(24):41.

Banfield, Edward. 1974. *The Unheavenly City Revisited*. Boston, MA: Little, Brown.

Besharov, D. 2002. "Post-Welfare-Reform Welfare." Retrieved August 30, 2003 (www.aei.org/include/news_print.asp? newsID=15142).

Bhargava, D. and J. Kuriansky. 2002. "Defining Who's Poor: Families Suffer as the Government Continues to Rely on the Outdated Measure of the Poverty Line." Retrieved August 30, 2003 (www.sixstrategies.org/files/Times%20Union%209-22-02.PDF).

Blum, Barbara and Jennifer Farnsworth Francis. 2002. *Welfare Research Perspectives: Past, Present, and Future*. New York: National Center for Children in Poverty.

Bradberry, Stephen. 2007. Testimony to the Financial Services Subcommittee on Housing and Community Opportunity, U.S. House of Representatives, February 22.

Briefel, R., Jonathan Jacobson, Nancy Clusen, Teresa Zavitsky, Miki Satake, Brittany Dawson, and Rhoda Cohen. 2003. *The Emergency Food Assistance System: Findings from the Client Survey: Executive Summary* (Food Assistance and Nutrition Research Report #32). Washington, DC: U.S. Department of Agriculture, Economic Research Service.

Cammisa, Ann Marie. 1998. *From Rhetoric to Reform? Welfare Policy in American Politics*. Boulder, CO: Westview Press.

Center for Women Policy Studies. 2002. *From Poverty to Self Sufficiency: The Role of Postsecondary Education in Welfare Reform*. Washington, DC: Center for Women Policy Studies.

Center on Budget and Policy Priorities. 2007. "Facts about the Earned Income Credit." Retrieved March 11, 2007 (http://www.cbpp.org/eic2007/).

Clemetson, Lynette. 2003. "Poor Workers Finding Modest Housing Unaffordable, Study Says." *New York Times,* September 9, p. A15.

Coalition on Human Needs. 2003. *Welfare: CHN Issue Brief*. Washington, DC: Author.

Conley, Dalton 1999. *Being Black, Living in the Red: Race, Wealth, and Social Policy in America*. Berkeley: University of California Press.

DeNavas-Walt, Carmen, Bernadette Proctor, and Cheryl H. Lee. 2006. *Income, Poverty and Health Insurance Coverage in the United States: 2005* (Current Population Reports, P60–231). Washington, DC: U.S. Government Printing Office.

DeNavas-Walt, Carmen, Bernadette Proctor, and Jessica Smith. 2007. *Income, Poverty and Health Insurance Coverage in the United States: 2006* (Current Population Reports, P60–233). Washington, DC: U.S. Government Printing Office.

Domhoff, G. William. 2002. *Who Rules America? Power and Politics*. Boston, MA: McGraw-Hill.

Dreier, Peter. 2006. "Katrina and Power in America." *Urban Affairs Review* 41(4):528–49.

European Commission. 1985. "On Specific Community Action to Combat Poverty (Council Decision of December 19, 1984)." 85/8/EEC. *Official Journal of the European Communities* 2(24).

Family Economic Self-Sufficiency Project. 2001. "The Self Sufficiency Standard." Retrieved February 27, 2008 (http://www.sixstrategies.org/sixstrategies/selfsufficiencystandard.cfm).

Fass, Sarah and Nancy Cauthen. 2006. "Who Are America's Poor Children?" Retrieved March 10, 2007 (http://www.nccp.org/pub_cpt06a.html).

Federal Interagency Forum on Child and Family Statistics. 2002. *America's Children 2002*. Washington, DC: U.S. Government Printing Office.

Federal Register, Vol. 73, No. 15, January 23, 2008, pp. 3971–72.

Fraser, Nancy. 1989. *Unruly Practices: Power, Discourse, and Gender in Contemporary Social Theory*. Minneapolis: University of Minnesota Press.

Freidman, P. 2000. *The Earned Income Tax Credit* (Issue Notes, Vol. 4, No. 4, April).Washington, DC: Welfare Information Network.

Funicello, Theresa. 1993. *Tyranny of Kindness: Dismantling the Welfare System to End Poverty in America*. New York: Atlantic Monthly Press.

Gans, Herbert. 1971. "The Uses of Poverty: The Poor Pay All." *Social Policy* 2:20–24.

Gennetian, L., A. Huston, D. Crosby, Y. Chang, E. Lowe, and T. Weisner. 2002. *Making Child Care Choices*. Washington, DC: Manpower Demonstration Research Corporation.

Gilbert, Dennis. 2003. *The American Class Structure in an Age of Growing Inequality*. Belmont, CA: Thomson Wadsworth.

Gilens, Martin. 1999. *Why Americans Hate Welfare: Race, Media, and the Politics of Antipoverty Policy*. Chicago, IL: University of Chicago Press.

Goffman, Erving. 1951. Symbols of Class Status. *British Journal of Sociology* 2:294–304.

Gordon, Linda. 1994. *Pitied But Not Entitled*. New York: Free Press.

Gornick, Janet and Marcia Meyers. 2002. "Support for Working Families: What We Can Learn from Europe about Family Policies." Pp. 90–107 in *Making Welfare Work*, edited by R. Kuttner. New York: New Press.

Guyer, Jocelyn. 2000. *Health Care after Welfare. An Update from the Findings from State-Level Leaver Studies*. Washington, DC: Center on Budget and Policy Priorities.

Haas, Richard. 2005. "How the Flood Compromises U.S. Foreign Policy." Retrieved April 15, 2007 (http://www.slate.com/id/2125994/nav/tap1).

Hachen, David. 2001. *Sociology in Action*. Thousand Oaks, CA: Pine Forge.

Handler, Joel and Yeheskel Hasenfeld. 1991. *The Moral Construction of Poverty: Welfare Reform in America*. Newbury Park, CA: Sage Publications.

Harrington, Michael. 1963. *The Other America: Poverty in the United States*. Baltimore, MD: Penguin Books.

Harris, K. M. 1993. "Work and Welfare among Single Mothers in Poverty." *American Journal of Sociology* 99:317–52.

Hennessy, Judith. 2005. "Welfare, Work, and Family Well-Being: A Comparative Analysis of Welfare and Employment Status for Single Female-Headed Families Post-TANF." *Sociological Perspectives* 48(1):77–104.

Iceland, John. 2003. *Poverty in America*. Berkeley: University of California Press.

Johnson, Lyndon. 1965. *Public Papers of the Presidents of the United States: Lyndon B. Johnson, 1965* (Vol. 2, entry 301, pp. 635–40). Washington, DC: U.S. Government Printing Office.

Jones, Arthur and Daniel Weinberg. 2000. *The Changing Shape of the Nation's Income Distribution 1947–1998* (Current Population Reports pp. 60–204). Washington, DC: Government Printing Office.

Kaufman, P., J. MacDonald, S. Lutz, and D. Smallwood. 1997. *Do the Poor Pay More for Food? Item Selection and Price Differences Affect Low-Income Household Food Costs* (Agricultural Economics Report #759). Washington, DC: U.S. Department of Agriculture, Economic Research Service.

Keister, Lisa and Stephanie Moller. 2000. "Wealth Inequality in the United States." *Annual Review of Sociology* 26:63–81.

Kuhn, Annette and AnnMarie Wolpe. 1978. *Feminism and Materialism: Women and Modes of Production*. London: Routledge and Kegan Paul.

Kulongoski, Theodore. 2007. Press Release, April 24, 2007. Retrieved July 4, 2007 (http://governor.oregon.gov/Gov/P2007/press_042407.shtml).

Lafer, Gordon. 2002. *Let Them Eat Training: The False Promise of Federal Employment Policy since 1980.* Ithaca, NY: Cornell University Press.

Lewis, Oscar. 1969. *On Understanding Poverty: Perspectives from the Social Sciences.* New York: Basic Books.

Lieberman, Robert C. 1998. *Shifting the Color Line: Race and the American Welfare State.* Cambridge, MA: Harvard University Press.

Llobrera, Joseph and Bob Zahradnik. 2004. *A Hand Up: How State Earned Income Tax Credits Help Working Families Escape Poverty in 2004.* Washington, DC: Center on Budget and Policy Priorities.

Loden, Marilyn and Judy Rosener. 1991. *Workforce America! Managing Employee Diversity as a Vital Resource.* New York: McGraw-Hill.

Logan, John. 2006. *The Impact of Katrina: Race and Class in Storm-Damaged Neighborhoods.* Retrieved May 5, 2007 (http://www.s4.brown.edu/Katrina/report.pdf).

Loprest, Pamela. 2001. *How Are Families That Left Welfare Doing? A Comparison of Early and Recent Welfare Leavers.* Washington, DC: Urban Institute.

———. 2002. *Families Who Left Welfare: Who Are They and How Are They Doing?* Washington, DC: Urban Institute.

Marger, Martin. 2002. *Social Inequality: Patterns and Processes.* Boston, MA: McGraw-Hill.

McCall, Leslie. 2002. *Complex Inequality: Gender, Class, and Race in the New Economy.* New York: Routledge.

Miga, Andrew. 2005. "New Orleans Activist Wins RFK Award." Retrieved April 20, 2007 (www.washingtonpost.com).

Mills, C. Wright. 2000. *The Sociological Imagination.* New York: Oxford University Press. (Original work published 1959)

National Campaign for Jobs and Income Support. 2001. *Leaving Welfare, Left Behind: Employment Status, Income, and Well-Being of Former TANF Recipients.* Washington, DC: Author.

National Center for Children in Poverty. 2001. "Child Poverty Fact Sheet: June 2001." Retrieved September 27, 2003 (http://cpmcent.columbia.edu/dept/nccp/ycpf).

Nelson, Kathyrn, Mark Treskon, and Danilo Pelletiere. 2004. *Losing Ground in the Best of Times: Low Income Renters in the 1990s.* Washington, DC: National Low Income Housing Coalition.

Neubeck, Kenneth and Noel Cazenave. 2001. *Welfare Racism: Playing the Race Card Against America's Poor.* New York: Routledge.

Nicholas-Casebolt, Ann and Patricia McGrath Morris. 2001. "Making Ends Meet: Private Food Assistance and the Working Poor." (Discussion Paper No. 1222-01). Madison, WI: Institute for Research on Poverty.

Norris, Donald and Lyke Thompson. 1995. "Introduction." Pp. 1–18 in *The Politics of Welfare Reform,* edited by Donald Norris and Lyke Thompson. Thousand Oaks, CA: Sage Publications.

O'Gorman, Angie. 2002. "Playing by the Rules and Still Losing Ground." *America,* July 29–August 5, pp. 12–15.

Oliver, Melvin L. and Thomas M. Shapiro 1995. *Black Wealth/White Wealth.* New York: Routledge.

Patterson, J. T. 1981. *America's Struggle against Poverty, 1900–1980.* Cambridge, MA: Harvard University Press.

Piven, Frances. 2002. "Globalization, American Politics, and Welfare Policy." Pp. 27–42 in *Lost Ground: Welfare Reform and Beyond,* edited by R. Albelda and A. Withorm. Cambridge, MA: South End Press.

Piven, Frances and Richard Cloward. 1993. *Regulating the Poor: The Functions of Public Welfare.* New York: Vintage Books.

Polk, D. 1997. "Check out the International Neighbors." Retrieved September 1, 2003 (http://www.4children.org/news/5-97intl.htm).

Quadagno, Jill. 1994. *The Color of Welfare Reform: How Racism Undermined the War on Poverty.* New York: Basic Books.

Rainwater, Lee and Timothy M. Smeeding. 2003. *Poor Kids in a Rich Country: America's Children in Comparative Perspective.* New York: Russell Sage Foundation.

Rice, Douglas and Barbara Sard. 2007. *Cuts in Federal Housing Assistance Are Undermining Community Plans to End Homelessness.* Washington, DC: Center on Budget and Policy Priorities.

Robert F. Kennedy Memorial. 2005. "Advocate for the Poor Receives the Robert F. Kennedy Human Rights Award for His Work in Louisiana." Retrieved April 20, 2007 (http://www.rfkmemorial.org/legacyinaction/2005_bradberry/).

Romney, L. 2003. "End of Welfare Leaves Rural Poor in a Bind." *Los Angeles Times,* April 6, pp. B1, B10.

Safran, D. G., P. Neuman, C. Schoen, M.S. Kitchman, I. B. Wilson, B. Cooper, A. Li, H. Chang, and W. H. Rogers. 2005. "Prescription Drug Coverage and Seniors: Findings from a 2003 National Survey." *Health Affairs,* April 19.

Sard, Barbara and Margy Waller. 2002. *Housing Strategies to Strengthen Welfare Policy and Support Working Families.* Washington, DC: Brookings Institution and the Center on Budget and Policy Priorities.

Seccombe, Karen and C. Amey. 1995. "Playing by the Rules and Losing: Health Insurance and the Working Poor." *Journal of Health and Social Behavior* 36:168–81.

Sernau, Scott. (2001). *Worlds Apart: Social Inequalities in a New Century.* Thousand Oaks, CA: Pine Forge.

Shapiro, Thomas M. 2001. "Introduction." Pp. 1–6 in *Great Divides: Readings in Social Inequality in the United States,* edited by Thomas M. Shapiro. Mountain View, CA: Mayfield.

Sherman, Arole and Aviva Aron-Dine. 2007. "New CBO Data Show Income Inequality Continues to Widen." Center on Budget and Policy Priorities, January 23, 2007.

Short, Kathleen. 2001. *Experimental Poverty Measures: 1999* (Current Population Reports Consumer Income, P60–216). Washington, DC: U.S. Census Bureau.

Skolverket/Swedish National Agency for Education. 2003. "Child Care in Sweden." Retrieved September 1, 2003 (www.skolverket.se/english/system/child.shtml).

Smeeding, Timothy, Katherine Ross, and Michael O'Conner. 1999. *The Economic Impact of the Earned Income Tax Credit (EITC): Consumption, Savings, and Debt.* Syracuse, NY: Syracuse University, Center for Policy Research.

Tillmon, Johnnie. 1927. "Welfare Is a Women's Issue." Retrieved September 28, 2003 (www.msmagazine.com/spring2002/tillmon.asp).

United Nations Children's Fund. 2005. *Child Poverty in Rich Countries 2005, Innocenti Report Card No. 6.* Florence, Italy: United Nations.

U.S. Census Bureau. 2006. "Poverty Status by State: 2005." Retrieved April 6, 2007 (http://pubdb3.census.gov/macro/032006/pov/new46_100125_09.htm).

———. 2007. "Poverty 2006." Retrieved April 26, 2008 (http://www.census.gov/hhes/www/poverty/poverty06/state.html).

———. 2008. "Poverty Thresholds 2007." Retrieved May 19, 2008 (http://www.census.gov/hhes/www/poverty/theshld/thresh07.html).

U.S. Conference of Mayors. 2006. *A Status Report on Hunger and Homelessness in American Cities, 2006.* Washington, DC: Author.

U.S. Department of Agriculture. 2006. "Food Security in the United States: Conditions and Trends." Retrieved March 10, 2007 (http://www.ers.usda.gov/Briefing/FoodSecurity/trends.htm).

U.S. Department of Health and Human Services. 2006. "Welfare Reform: Deficit Reduction Act of 2005." Retrieved March 11, 2007 (http://www.acf.hhs.gov/programs/ofa/drafact.htm).

Waldfogel, Jane. 2001. "International Policies toward Parental Leave and Child Care." *The Future of Children* 11(1):99–111.

Waxman, H. 2004. "Correspondence to Committee on Government Reform," June 14. Retrieved June 30, 2004 (www.house.gov/reform).

Weaver, R. Kent. 2000. *Ending Welfare as We Know It.* Washington, DC: Brookings Institution Press.

Webster, Bruce and Alemayehu Bishaw. 2007. *Income, Earnings, and Poverty Data from the 2006 American Community Survey.* Washington, DC: U.S. Government Printing Office.

Weil, Alan and Kenneth Feingold. 2002. "Introduction." Pp. xi–xxxi in *Welfare Reform: The Next Act,* edited by Alan Weil and Kenneth Feingold. Washington, DC: Urban Institute.

World Bank. 2007. "GNI per Capita 2005, Atlas Method and PPP." Retrieved April 15, 2007 (http://siteresources.worldbank.org/DATASTATISTICS/Resources/GNIPC.pdf).

———. 2008. "Understanding Poverty." Retrieved February 27, 2008 (http://www1.worldbank.org/prem/poverty/mission/up1.htm).

Wright, Beverly and Robert D. Bullard. 2007. "Washed Away by Hurricane Katrina: Rebuilding a 'New' New Orleans." Pp. 189–221 in *Growing Smarter: Achieving Livable Communities, Environmental Justice and Regional Equity,* edited by Robert D. Bullard. Cambridge, MA: MIT Press.

Yardley, William. 2007. "A Governor Truly Tightens His Belt." *New York Times,* May 1, p. A13.

Zedlewski, Shelia and Pamela Loprest. 2003. "Welfare Reform: One Size Doesn't Fit All." *Christian Science Monitor, Electronic Edition,* August 25. Retrieved August 27, 2003 (www.urban.org/url.cfm?ID=900648).

Zucchino, David. 1999. *Myth of the Welfare Queen.* New York: Touchstone/Simon & Schuster.

Chapter 3

Academy of Achievement. 2005. "Interview with Rosa Parks." Retrieved April 22, 2007 (http://www.achievement.org/autodoc/page/par0int-1).

Ackerman, Bruce. 2004. "The Emergency Constitution." *Yale Law Journal* 133(5):1029–91.

Allport, Gordon. 1954. *The Nature of Prejudice.* Garden City, NY: Doubleday.

American Association of Colleges and Universities. 2000. "More Colleges and Universities Require Students to Take Diversity Classes." Retrieved May 13, 2007 (http://www.aacu-edu.org/news_room/2000archive/irvinesurveypr.cfm).

American Council on Education and American Association of University Professors. 2000. *Does Diversity Make A Difference? Three Research Studies on Diversity in College Classrooms, Executive Summary.* Washington, DC: Authors.

Asahina, Robert. 2006. *Just Americans: How Japanese Americans Won the War at Home and Abroad.* New York: Gotham Books.

Bauman, Zygmunt. 2000. *Globalisation: The Human Consequences.* Cambridge, England: Polity Press.

Bendick, Mark Jr., Mary Lou Egan, and Suzanne M. Lofhjelm. 1998. *The Documentation and Evaluation of Antidiscrimination Training in the United States.* Washington, DC: Bendick and Egan Economic Consultants.

Bonacich, Edna. 1972. "A Theory of Ethnic Antagonism: The Split Labor Market." *American Sociological Review* 37:547–59.

Budrys, Grace. 2003. *Unequal Health: How Inequality Contributes to Health or Illness.* Lanham, MA: Rowman & Littlefield.

Byrd, W. Michael and Linda Clayton. 2002. *An American Health Dilemma: Race, Medicine, and Health Care in the United States, 1900–2000.* New York: Routledge.

Camarota, Steven. 2003. "Immigration in a Time of Recession: An Examination of Trends Since 2000." Retrieved April 22, 2007 (http://www.cis.org/articles/2003/back1603.html).

Carby, Hazel. 1985. "White Woman Listen! Black Feminism and the Boundaries of Sisterhood." Pp. 389–403 in *The Empire Strikes Back: Race and Racism in Seventies Britain,* edited by Paul Gilroy. London, England: Hutchinson.

Castles, Stephen and Mark Miller. 1998. *The Age of Migration.* New York: Guilford Press.

Collin, J. and K. Lee. 2003. *Globalization and Transborder Health Risk in the U.K.* London, England: Nuffield Trust.

DeNavas-Walt, Carmen, Bernadette Proctor, and Cheryl H. Lee. 2006. *Income, Poverty and Health Insurance Coverage in the United States: 2005* (Current Population Reports, P60–231). Washington, DC: U.S. Government Printing Office.

DeNavas-Walt, Carmen, Bernadette Proctor, and Jessica Smith. 2007. *Income, Poverty and Health Insurance Coverage in the United States: 2006* (Current Population Reports, P60–233). Washington, DC: U.S. Government Printing Office.

DeParle, Jason. 2007. "In a World on the Move, a Tiny Land Strains to Cope." *New York Times,* June 24, pp. 1, 8.

Du Bois, W. E. B. 1996. *Darkwater.* P. 532 in *The Oxford W. E. B. Du Bois Reader,* edited by Eric Sunquist. New York: Oxford University Press.

Düvell, Franck. 2005. "Globalization of Migration Control." Pp. 23–46 in *Crossing Over: Comparing Recent Migration in the United States and Europe,* edited by Holger Henke. Oxford, England: Lexington Books.

Farley, John. 2005. *Majority-Minority Relations.* Upper Saddle River, NJ: Pearson-Prentice Hall.

Feagin, Joe. 1999. "Excluding Blacks and Others from Housing: The Foundation of White Racism." *Cityscape* 4(3):79–91.

Feagin, Joe and Pinar Batur. 2004. "Racism in Comparative Perspective." Pp. 316–34 in *Handbook of Social Problems: A Comparative International Perspective,* edited by George Ritzer. Thousand Oaks, CA: Sage Publications.

Fry, Richard. 2004. *Latino Youth Finishing College: The Role of Selective Pathways.* Washington, DC: Pew Hispanic Center.

———. 2005. *The High Schools Hispanics Attend: Size and Other Key Characteristics.* Washington, DC: Pew Hispanic Center.

Gagné, Patricia and Richard Tewksbury. 2003. *The Dynamics of Inequality: Race, Class, and Sexuality in the United States.* Upper Saddle River, NJ: Prentice Hall.

Galagan, Patricia. 1993. "Navigating the Differences." *Training and Development* 47(4):28–33.

Global Commission on Immigration. 2005. *Migration in an Interconnected World: New Directions for Action.* Switzerland: SRO Kunding.

Godoy, J. 2002. "Rights-France: Anti-Immigrant and Racist Views on the Rise." *Inter Press Service.* Retrieved April 22, 2007, from Lexis-Nexis.

Gordon, Milton. 1964. *Assimilation in American Life.* New York: Oxford University Press.

Greenberg, Cheryl. 1995. "Black and Jewish Responses to Japanese Internment." *Journal of American Ethnic History* 1(2):3–38.

Greenhouse, Linda. 2003. "Justices Back Affirmative Action by 5 to 4, But Wider Vote Bans a Racial Point System." *New York Times,* June 24, pp. A1, A25.

———. 2007. "Justices, 5–4, Limit Use of Race for School Integration Plans." *New York Times,* June 29, pp. A1, A20.

Grinde, D. A. 2001. *ALANA/Ethnic Studies Assessment.* Retrieved May 13, 2007 (http://www.aacu-edud .org/KnowNet/general/alana.htm).

Heilman, M. E., C. J. Block, and P. Stahatos. 1997. "The Affirmative Action Stigma of Incompetence: Effects of Performance Information Ambiguity." *Academy of Management Journal* 40:603–25.

Herring, Cedric and Sharon Collins. 1995. "Retreat from Equal Opportunity? The Case of Affirmative Action." Pp. 163–81 in *The Bubbling Cauldron: Race, Ethnicity, and the Urban Crisis,* edited by Michael Smith and Joe Feagin. Minneapolis: University of Minnesota Press.

Human Rights Watch. 2001. *Caste Discrimination: A Global Concern.* New York: Author.

Hytrek, Gary and Kristine Zentgraf. 2007. *America Transformed: Globalization, Inequality and Power.* New York: Oxford University Press.

Idelson, Holly. (1995). "A 30-Year Experiment." *Congressional Quarterly Weekly Report* 53(22):1579.

International Organization for Migration. 2003. *World Migration Report.* Geneva, Switzerland: Author.

Jacobson, Matthew Frye. 1998. *Whiteness of a Different Color: European Immigrants and the Alchemy of Race.* Cambridge, MA: Harvard University.

Jaspin, Eliot. 2006. *Buried in the Bitter Waters: The Hidden History of Racial Cleansing in America.* New York: Basic Books.

Jeffreys, Kelly. 2006. *Refugees and Asylees: 2005.* Washington, DC: Office of Immigration Statistics, Office of Homeland Security.

Johnson, Lyndon. 1965. *Public Papers of the Presidents of the United States: Lyndon B. Johnson, 1965* (Vol. 2, entry 301, pp. 635–40). Washington, DC: U.S. Government Printing Office.

Kalev, Alexandra, Frank Dobbin, and Erin Kelly. 2006. "Best Practices or Best Guesses? Assessing the Efficacy of Corporate Affirmative Action and Diversity Policies." *American Sociological Review* 71:589–617.

Kaplan, W. and B. Lee. 1995. *The Law of Higher Education: A Comprehensive Guide to Legal Implications of Administrative Decision Making.* San Francisco, CA: Jossey-Bass.

Korgen, Kathleen, J. Mahon, and Gabe Wang. 2003. "Diversity of College Campuses Today: The Growing Need to Foster Campus Environments Capable of Countering a Possible Tipping Effect." *College Student Journal* 37:16–26.

Kotlowski, Dean. 1998. "Richard Nixon and the Origins of Affirmative Action." *Historian* 60(3):523–42.

Kramer, Andrew. 2007. "Markets Suffer after Russia Bans Immigrant Vendors." *New York Times,* April 14, p. A3.

Larsen, Luke. 2004. *The Foreign-Born Population in the United States: 2003* (P20-551). Washington, DC: U.S. Census Bureau.

Lee, Erika. 2006. "A Nation of Immigrants and A Gatekeeping Nation: American Immigration Law and Policy." Pp. 5–35 in *A Companion to American Immigration,* edited by Reed Ueda. Cornwall, England: Blackwell.

Marger, Martin. 2002. *Social Inequality: Patterns and Processes.* Boston, MA: McGraw-Hill.

McGary, Howard. 1999. *Race and Social Justice.* Oxford, England: Blackwell.

Myers, John P. 2005. *Minority Voices.* Boston, MA: Pearson Education.

Odell, Patricia, Kathleen Korgen, and Gabe Wang. 2005. "Cross-Racial Friendships and Social Distance Between Racial Groups on a College Campus." *Innovative Higher Education* 29(4):291–305.

Ohlemacher, Stephen. 2006. "Race Matters, and Minorities Still Suffer, Report Says." *News Tribune,* November 14, pp. A6–A7.

Oliver, Melvin and Thomas Shapiro 1995. *Black Wealth/White Wealth.* New York: Routledge.

Omi, Michael and Howard Winant. 1994. *Racial Formation in the United States: From the 1960s to the 1990s.* New York: Routledge.

Ore, Tracy E., ed. 2003. *The Social Construction of Difference and Inequality: Race, Class, Gender, and Sexual Orientation.* Boston, MA: McGraw-Hill.

Pew Hispanic Center. 2006. "Table 19. Educational Attainment by Race and Ethnicity, 2005." Retrieved May 5, 2007 (http://pewhispanic.org/files/other/middecade/Table-19.pdf).

Preston, Julia. 2007. "Immigration Is at Center of New Laws Around U.S." *New York Times,* August 6, p. A12.

Redfield, Robert. 1958. "Race as a Social Phenomenon." Pp. 66–71 in *Race: Individual and Collective Behavior,* edited by E. Thompson and E. C. Hughes. Glencoe, IL: Free Press.

Rosenbaum, Sara and Joel Teitelbaum. 2004. "Addressing Racial Inequality in Health Care." Pp. 135–50 in *Policy Challenges in Modern Health Care,* edited by D. Mechanic, L. Rogut, D. Colby, and J. Knickman. New Brunswick, NJ: Rutgers University Press.

Schemo, Diana Jean. 2001. "U. of Georgia Won't Contest Ruling on Admissions Policy." *New York Times,* November 10, p. A8.

Shuford, John. 2001. "Four DuBoisian Contributions to Critical Race Theory." *Transactions of the Charles S. Peirce Society* 37(3):301–37.

Silverman, M. 1992. *Deconstructing the Country: Immigration, Racism, and Citizenship in Modern France.* London, England: Routledge.

Springer, A. D. 2005. "Update on Affirmative Action in Higher Education: A Current Legal Overview, January 2005." Retrieved March 15, 2008 (www.aaup.org/AAUP/protect/legal/topics/aff-ac-update.htm).

St. John, Warren. 2007. "Outcasts United: On a Small Town's Soccer Fields, Refugees Find Hostility and Hope." *New York Times,* January 21, pp. 1, 18–19.

Swink, Dawn. 2003. "Back to Bakke: Affirmative Action Revisited in Educational Diversity." *BYU Education and Law Journal* 1:211–57.

Tsang, Chiu-Wai Rita and Tracy Dietz. 2001. "The Unrelenting Significance of Minority Statuses: Gender, Ethnicity, and Economic Attainment Since Affirmative Action." *Sociological Spectrum* 21:61–80.

U.S. Census Bureau. 2005. *Race and Hispanic Origin in 2005.* Retrieved July 5, 2007 (http://www.census.gov/population/pop-profile/dynamic/RACEHO.pdf).

Van Ausdale, Debra and Joe R. Feagin. 2001. *The First R: How Children Learn Race and Racism.* Lanham, MA: Rowman & Littlefield.

Wilgoren, J. 2000. "Affirmative Action Plan Is Upheld in Michigan." *New York Times,* December 14, p. A32.

Williams, Richard, Reynold Nesiba, and Eileen Diaz McConnell. 2005. "The Change Face of Inequality in Home Mortgage Lending." *Social Problems* 52(2):181–208.

Wittig, Michele and Sheila Grant-Thompson. 1998. "The Utility of Allport's Conditions of Intergroup Contact for Predicting Perceptions of Improved Racial Attitudes and Beliefs." *Journal of Social Issues* 54(4):795–812.

Woodhouse, Shawn. 1999. "Faculty Perceptions of the Impact of Affirmative Action on Employment Practices in the University of Missouri System." PhD dissertation, University of Missouri, Columbia, MO.

————. 2002. "The Historical Development of Affirmative Action: An Aggregated Analysis." *Western Journal of Black Studies* 26(3):155–59.

Yardley, Jim. 2002. "The Ten Percent Solution." *New York Times Magazine,* April 14, pp. 28–31.

Yee, Shirley. 2001. "The Past, Present, and Future of Affirmative Action: AHA Roundtable, January 1998—Introduction." *NWSA Journal* 10(3):135–41.

Zhou, Min. 2004. "Assimilation, the Asian Way." Pp. 139–53 in *Reinventing the Melting Pot: The New Immigrants and What It Means to Be American,* edited by Tamar Jacoby. New York: Basic Books.

Chapter 4

Associated Press. 2003. "Title IX Cornerstone Remains Intact, but Schools Will Be Educated." *News Tribune,* July 12, p. C10.

Beeghley, L. 2005. *The Structure of Social Stratification in the United States.* Boston, MA: Pearson.

Bem, Sandra Lipsitz. 1993. *The Lenses of Gender: Transforming the Debate on Sexual Inequality.* New Haven, CT: Yale University Press.

Bonvillian, Nancy. 2006. *Women and Men, Cultural Constructs of Gender.* New York: Prentice Hall.

Brady, Erik. 2003. "Poll: Most Adults Want Title IX Left Alone." Retrieved June 12, 2003 (http://www.usatoday.com/sports/college/other/2003-01-07-title-ix_x.htm).

Charles, Maria. 2003. "Deciphering Sex Segregation: Vertical and Horizontal Inequalities in Ten National Labor Markets." *Acta Sociologica* 46(4):267–87.

Charles, Maria and David Grusky. 2004. "The Four Puzzles of Sex Segregation." Pp. 3–37 in *Occupational Ghettos: The Worldwide Segregation of Women and Men,* edited by Maria Charles and David Grusky. Stanford, CA: Stanford University Press.

Collins, Gail. 1993. "Potty Politics: The Gender Gap." *Working Woman* 18(3):12.

Dey, Judy Goldberg and Catherine Hill. 2007. *Behind the Pay Gap.* Washington, DC: AAUW Educational Foundation.

Durkheim, Émile. 2007. "The Division of Labor and Marriage." Pp. 41–44 in *Before the Second Wave: Gender in the Sociological Tradition,* edited by Barbara Finlay. Upper Saddle River, NJ: Pearson Prentice Hall.

Elliot, Jane. 2005. "Comparing Occupational Segregation in Great Britain and the United States: The Benefits of Using a Multi-Group Measure of Segregation." *Work, Employment and Society* 19(1):153–74.

England, Paula. 2001. "Gender and Access to Money: What Do Trends in Earnings and Household Poverty Tell Us?" Pp. 131–53 in *Reconfigurations of Class and Gender,* edited by Janine Baxter and Mark Western. Stanford, CA: Stanford University Press.

Fackler, Martin. 2007. "Career Women in Japan Find a Blocked Path." *New York Times,* August 6, pp. A1, A6.

Farley, John. 2005. *Majority-Minority Relations.* Upper Saddle River, NJ: Pearson-Prentice Hall.

Fletcher, Michael and Greg Sandoval. 2003. "Proposals Would Relax Law Promoting Women's Sports." *News Tribune,* January 31, p. A4.

Garber, Greg. 2002. "Landmark Law Faces New Challenges Even Now." Retrieved June 12, 2003 (http://sports.espn.go.com/espn/print?id=1393193&type=feature).

Herbert, Bob. 2006. "Punished for Being Female." *New York Times,* November 2, p. A27.

Inter-Parliamentary Union. 2007. "Women in National Parliaments." Retrieved July 8, 2007 (http://www.ipu.org/wmn-e/classif.htm).

Kanter, Rosabeth Moss. 1977. *Men and Women of the Corporation.* New York: Basic Books.

King, Jacqueline. 2006. *Gender Equity in Higher Education: 2006.* Washington, DC: American Council on Education Center for Policy Analysis.

Marger, Martin. 2008. *Social Inequality: Patterns and Processes.* Boston, MA: McGraw-Hill.

Michel, Lawrence, Jared Bernstein, and Sylvia Allegretto. 2007. *The State of Working America: 2006/2007.* Ithaca, NY: ILR Press, an imprint of Cornell University Press.

National Committee on Pay Equity. 2007. "The Wage Gap Over Time: In Real Dollars, Women See a Continuing Gap." Retrieved July 8, 2007 (http://www.pay-equity.org/info-time.html).

National Women's Law Center. 2002a. *The Battle for Gender Equity in Athletics: Title IX at Thirty.* Washington, DC: National Women's Law Center.

———. 2002b. "Quick Facts on Women and Girls in Athletics." Retrieved June 12, 2003 (www.nwlc .org/pdf/quickfacts_June2002.pdf).

Pelosi, Nancy. 2007. "Press Release: Pelosi Celebrates Women's History Month, March 3, 2007." Retrieved July 10, 2007 (http://www.speaker.gov/newsroom/pressreleases?id=0084).

Picker, Lauren. 1993. "The Women's Room." *New York Times,* January 8, page unknown.

Ridgeway, C. and L. Smith-Lovin. 1999. "The Gender System and Interaction." *Annual Review of Sociology* 25:191–216.

Rose, Stephen J. and Heidi I. Harmann. 2004. *Still a Man's Labor Market: The Long-Term Earnings Gap.* Washington, DC: Institute for Women's Policy Research.

Sandler, Bernice. 1997. "'Too Strong for a Woman'—The Five Words That Created Title IX." Retrieved March 24, 2004 (http://bernicesandler.com/id44.htm).

"Some States' Constitutions Are Going Gender-Neutral." 2003. *New York Times,* May 22, p. A28.

Stolberg, Sheryl Gay. 2003. "Working Mothers Swaying Senate Debate, as Senators." *New York Times,* June 7, pp. A1, A11.

Talev, Margaret. 2006. "Bathrooms Could Reveal Women's Pull in D.C." *News Tribune,* December 25, p. AA8.

Tickner, J. A. 2002. "Gendering World Politics: Issues and Approaches in the Post-Cold War Era." *Political Science Quarterly* 117(2):336–37.

United Nations. 2006a. "Executive Summary: Ending Violence against Women: From Words to Action. Study of the Secretary-General." Retrieved April 28, 2007 (http://www.un.org/womenwatch/daw/ vaw/launch/english/v.a.w-exeE-use.pdf).

———. 2006b. "Violence against Women: Forms, Consequences and Costs: Ending Violence against Women: From Words to Action. Study of the Secretary-General." Retrieved April 28, 2007 (http:// www.un.org/womenwatch/daw/vaw/launch/english/v.a.w-consequenceE-use.pdf).

U.S. Census Bureau. 2002. *Current Population Survey, March 2002* (Table 1a). Washington, DC: Author.

U.S. Department of Education. 1997. *Title IX: 25 Years of Progress.* Washington, DC: Office of Educational Research and Improvement.

———. 2004. *Digest of Education Statistics, 2004.* Washington, DC: Author.

U.S. Department of Labor, Women's Bureau. 2006. "20 Leading Occupations of Employed Women: Full Time Wage and Salaried Workers, 2006." Retrieved July 8, 2007 (http://www.dol.gov/wb/factsheets/ 20lead2006.htm).

Weinberg, Daniel. 2004. *Evidence from Census 2000 About Earnings by Detailed Occupation for Men and Women, CENSR-15.* Washington, DC: U.S. Census Bureau.

Wharton, Amy. 2004. "Gender Inequality." Pp. 156–71 in *Handbook of Social Problems: A Comparative International Perspective,* edited by George Ritzer. Thousand Oaks, CA: Sage Publications.

Wildman, Stephanie and Adrienne Davis. 2000. "Language and Silence: Making Systems of Privilege Visible." Pp. 50–60 in *Readings for Diversity and Social Justice,* edited by Maurianne Adams, Warren Blumenfeld, Rosie Castañeda, Heather Hackman, Madeline Peters, and Ximena Zúñiga. New York: Routledge.

Woodhouse, Shawn. 2002. "The Historical Development of Affirmative Action: An Aggregated Analysis." *Western Journal of Black Studies* 26(3):155–59.

Chapter 5

Badgett, M. V. Lee. 1997. "Vulnerability in the Workplace: Evidence of Anti-Gay Discrimination." *Angles* 2(1):1–4.

Badgett, M. V. Lee and M. King. 1997. "Lesbian and Gay Men Occupational Strategies." Pp. 73–86 in *HomoEconomics: Capitalism, Community and Lesbian and Gay Life,* edited by A. Gluckman and B. Reed. New York: Routledge.

Bevacqua, Maria. 2004. "Feminist Theory and the Question of Lesbian and Gay Marriage." *Feminism and Psychology* 14(1):36–40.

Belkin, Aaron. 2003. "Don't Ask, Don't Tell: Is the Gay Ban Based on Military Necessity?" *Parameters* Summer:108–19.

Bumiller, Elizabeth. 2002. "The Most Unlikely Story Behind a Gay Rights Victory." *New York Times,* June 27, p. A19.

———. 2006. "The Egg Roll (Again!) Becomes a Stage for Controversy." *New York Times,* April 10.

Cass, V. C. 1979. "Homosexual Identity Formation: A Theoretical Model." *Journal of Homosexuality* 4:219–35.

Chou, W. S. 2001. "Homosexuality and the Cultural Practices of Tongzhi in Chinese Societies." *Journal of Homosexuality* 4:27–46.

D'Augelli, Anthony. 1998. "Developmental Implications of Victimization of Lesbian, Gay and Bisexual Youths." Pp. 187–210 in *Understanding Prejudice Against Lesbians, Gay Men and Bisexuals,* edited by G. M. Herek. Thousand Oaks, CA: Sage Publications.

D'Augelli, Anthony, Scott Hershberger, and Neal Pilkington. 1998. "Lesbian, Gay, and Bisexual Youths and Their Families: Disclosure of Sexual Orientation and Its Consequences." *American Journal of Orthopsychiatry,* 68:361–71.

Dubé, E. M. 2000. "The Role of Sexual Behavior in the Identification Process of Gay and Bisexual Men." *Journal of Sex Research* 48:139–62.

Feigenbaum, Erika Faith. 2007. "Heterosexual Privilege: The Political and the Personal." *Hypatia* 22(1):1–9.

Ferguson, Ann. 2007. "Gay Marriage: An American and Feminist Dilemma." *Hypatia* 22(2):39–57.

Greenhouse, Linda. 2003. "Texas Sodomy Law Held Unconstitutional—Scathing Dissent." *New York Times,* June 27, pp. A1, A17.

Gunther, Marc. 2006. "Corporate America Backs Gay Rights." Retrieved February 3, 2007 (http://money.cnn.com/2006/04/25/magazines/fortune/pluggedin_fortune/).

Hall, Randy. 2007. "Bush Expected to Veto Hate Crimes Bill." Retrieved June 20, 2007 (http://www.cnsnews.com/ViewPolitics.asp?Page=/Politics/archive/200705/POL20070503d.html).

Human Rights Campaign. 2006. "Adoption Laws State by State." Retrieved February 4, 2007 (http://www.hrc.org/issues/parenting/adoptions/2375.htm).

———. 2007a. "Profiles of Featured Speakers." Retrieved July 5, 2007 (http://www.hrc.org/legacyofservice/about_vets.asp).

———. 2007b. "Impact of Lifting the Ban: Other Agencies and Countries that Allow Open Service." Retrieved June 10, 2007 (http://www.hrc.org/issues/4882.htm).

Hunter, Nan. 1995. "Marriage, Law and Gender: a Feminist Inquiry." Pp. 107–22 in *Sex Wars: Sexual Dissent and Political Culture,* edited by L. Duggan and N. D. Hunter. New York: Routledge.

Katz, Jonathan Ned. 2003. "The Invention of Heterosexuality." Pp. 136–48 in *The Social Construction of Difference and Inequality: Race, Class, Gender, and Sexuality,* edited by Tracy E. Ore. Boston, MA: McGraw-Hill.

Katz, Marsha and Helen LaVan. 2004. "Legal Protection from Discrimination Based on Sexual Orientation: Findings from Litigation." *Employee Responsibilities and Rights* 16(4):195–209.

Lambda Legal. 2003a. "Family." Retrieved July 8, 2003 (www.lambdalegal.org/cgi-bin/iowa/issues/record?record=5).

———. 2003b. "Landmark Victory." Retrieved July 8, 2003 (http://cache.lamdalegal.org/cgi-bin/iowa/splash.html).

———. 2004. "California Supreme Court Invalidates Marriages from San Francisco, Without Resolving Whether Same-Sex Couples Have the Right to Marry." Retrieved August 27, 2004 (http://www.lambdalegal.org/news/pr/california-supreme-court-marriage-license.html).

Leff, Lisa. 2004. "Bay City Licensed 4,037 Couples." Retrieved March 24, 2004 (www.dailynews.com).

Lehne, Gregory K. 1995. "Homophobia among Men: Supporting and Defining the Male Role." Pp. 325–36 in *Men's Lives,* edited by M. Kimmel and M. Messner. Boston, MA: Allyn & Bacon.

Lind, Amy. 2004. "Legislating the Family: Heterosexist Bias in Social Welfare Policy Frameworks." *Journal of Sociology and Social Welfare* 31(4):21–35.

Marquis, C. 2003. "Gay Partners Too Are Separated by War, and by Their Need for Secrecy." *New York Times,* April 18, p. B9.

McLeod, A. and I. Crawford. 1998. "The Postmodern Family: An Examination of the Psychosocial and Legal Perspectives of Gay and Lesbian Parenting." Pp. 211–23 in *Stigma and Sexual Orientation: Understanding Prejudice Against Lesbians, Gay Men, and Bisexuals,* edited by G. Herek. Thousand Oaks, CA: Sage Publications.

Michael, Robert, John Gagnon, Edward Laumann, and Gina Kolata. 1994. *Sex in America: A Definitive Survey.* Boston, MA: Little, Brown.

Miga, Andrew. 2007. "House Approves Federal Ban on Gay Job Discrimination." *Honolulu Advertiser,* November 8, A10.

Patterson, Charlotte and Richard Redding. 1996. "Lesbian and Gay Families with Children: Implications of Social Science Research for Policy." *Journal of Social Issues* 52(3):29–50

Quittner, Jeremy. 2004. "Newsom to Bush: "Keep Your Hands Off the Constitution." Retrieved March 24, 2004 (www.advocate.com/html/stories/911/911_newsom.asp).

Radkowsky, M. and L. J. Siegel. 1997. "The Gay Adolescent: Stressors, Adaptations, and Psychological Interventions." *Clinical Psychology Review* 17:191–216.

Rivers, Ian and Daniel Carragher. 2003. "Social-Developmental Factors Affecting Lesbian and Gay Youth: A Review of Cross-National Research Findings." *Children and Society,* 17(5):374–85.

Rosenbloom, Stephanie. 2006. "Is This Campus Gay-Friendly?" *New York Times,* September 14, p. Style 1-2.

Saad, Lydia. 2007. "Tolerance for Gay-Rights at the High Water Mark." Retrieved November 17, 2007 (http://www.gallup.com/poll/27694/Tolerance-Gay-Rights-HighWater-Mark.aspx).

Schneider, Beth. 1986. "Coming Out at Work: Bridging the Private/Public Gap." *Work and Occupations* 13(4):463–87.

Servicemembers Legal Defense Network. 2007a. "Total 'Don't Ask Don't Tell' Discharges 1994–2006." Retrieved June 10, 2007 (http://www.sldn.org/binary-data/SLDN_ARTICLES/pdf_file/3864.pdf).

———. 2007b. "Fact Sheet Foreign Military Services Which Allow Open Service. Retrieved June 10, 2007 (http://www.sldn.org/binary-data/SLDN_ARTICLES/pdf_file/1904.pdf).

Shalikashvili, John. 2007. "Second Thoughts on Gays in the Military." *New York Times,* January 2, p. A19.

Smith, David and Gary Gates. 2001. *Gay and Lesbian Families in the United States: Same-Sex Unmarried Partnership Households: Preliminary Analysis of 2000 U.S. Census Data.* Washington, DC: Human Rights Campaign.

Stacey, Judith and Timothy Biblarz. 2001. "(How) Does the Sexual Orientation of Parents Matter?" *American Sociological Review* 66:159–83.

Surkis, Alisa and Colleen Gillespie. 2006. "The Great White House Easter Egg Roll Controversy." Retrieved June 10, 2007 (http://lesbianlife.about.com/od/families/a/EasterEggRoll.htm).

Vidal, Gore. 1998. "Someone to Laugh at the Squares With." In *At Home: Essays, 1982–1988,* edited by G. Vidal. New York: Random House.

Windmeyer, Shane. 2006. *The Advocate College Guide for LGBT Students.* New York: Allyson Books.

Chapter 6

Binstock, Robert. 2005. "Old-Age Policies, Politics and Ageism." *Generations* 29:73–78.

———. 2006–2007. "Older People and Political Engagement: From Avid Voters to 'Cooled-Out Marks.'" *Generations* 30:24–30.

Butler, Robert. 1969. "Ageism: Another Form of Bigotry." *Gerontologist* 9(3):243–46.

Bytheway, William. 1995. *Ageism.* Buckingham, England: Open University Press.

Calasanti, Toni and Kathleen Slevin. 2001. *Gender, Social Inequalities, and Aging.* Walnut Creek, CA: AltaMira Press.

Calasanti, Toni, Kathleen Slevin, and Neal King. 2006. "Ageism and Feminism: From 'Et Cetera' to Center." *NWSA Journal* 18(1):13–30.

Campbell, A. 2003. *How Policies Make Citizens: Senior Political Activism and the American Welfare State.* Princeton, NJ: Princeton University Press.

Centers for Medicare and Medicaid Services. 2006. "National Health Expenditure Data." Retrieved February 9, 2007 (www.cms.hhs.gov/NationalHealthExpendData/downloads/highlights.pdf).

Clark, Robert, Richard Burkhauser, Marilyn Moon, Joseph Quinn, and T. Smeeding. 2004. *Economics of an Aging Society*. Malden, MA: Blackwell.

Cowgill, Donald. 1974. "The Aging of Populations and Societies." *The Annals of the American Academy of Political and Social Science* 415:1–18.

Curl, Angela and M. C. Hokenstad. 2006. "Reshaping Retirement Policies in Post-Industrial Nations: The Need for Flexibility." *Journal of Sociology and Social Welfare* 33(2):85–106.

DeNavas-Walt, Carmen, Bernadette Proctor, and Jessica Smith. 2007. *Income, Poverty and Health Insurance Coverage in the United States: 2006* (Current Population Reports, P60–233). Washington, DC: U.S. Government Printing Office.

Duncan, Colin and Wendy Loretto. 2004. "Never the Right Age? Gender and Age-Based Discrimination in Employment." *Gender, Work and Organizations* 11(1):95–115.

Feder, Judith and Robert Friedland. 2005. "The Value of Social Security and Medicare to Families." *Generations* 29.78–85.

Fisher, Anne. 2004. "Older, Wiser, Job-Hunting." *Fortune* 149(3):46.

Fountain, John. 2002. "Age Counts in Hiring, the Older Jobless Find." *New York Times,* November 13, p. A16.

Freedman, Rita. 1986. *Beauty Bound*. Lexington, MA: D.C. Heath.

Frey, William. 2007. *Mapping the Growth of Older America: Seniors and Boomers in the Early 21st Century*. Washington, DC: Brookings Institution.

Goodman, John C. 1998. "Why Your Grandchildren May Pay a 55 Percent Payroll Tax." *Wall Street Journal,* October 7, p. A22.

Hagestad, Gunhild and Peter Uhlenberg. 2006. "Should We Be Concerned About Age Segregation? Some Theoretical and Empirical Explorations." *Research on Aging* 28(6):638–53.

He, Wan, Manisha Sengupta, Victoria Velkoff, and Kimberly DeBarros. 2005. *65+ in the United States: 2005* (P23–209). Washington, DC: U.S. Census Bureau.

Himes, C. L. 2001. "Elderly Americans." *Population Bulletin* 56(4):4.

Jayson, Sharon. 2007. "No Age Limit on Stages of Life." *USA Today,* June 13, pp. D1–D2.

Kincannon, Charles, Wan He, and Loraine West. 2005. "Demography of Aging in China and the United States and the Economic Well-being of their Older Populations." *Journal of Cross Cultural Gerontology* 20:243–55.

Landphair, Ted. 2007. "Aging Activist Still on Quest for Social Justice." Retrieved April 20, 2007 (http://www.commondreams.org/headlines07/0306-09.htm).

Levenson, A. J. 1981. "Ageism: A Major Deterrent to the Introduction of Curricula in Aging." *Gerontology and Geriatrics Education* 1:11–62.

Levin, Jack and William Levin. 1980. *Ageism: Prejudice and Discrimination against the Elderly*. Belmont, CA: Wadsworth.

Levy, B. and E. Langer. 1994. "Aging Free from Negative Stereotypes: Successful Memory in China and among the American Deaf." *Journal of Personality and Social Psychology* 66(6):989–97.

Mabry, J. Beth and Vern Bengston. 2005. "Disengagement Theory." Pp. 113–17 in *Encyclopedia of Ageism,* edited by E. Palmore, L. Branch, and D. Harris. New York: Haworth Press.

McConatha, Jasin, Frauke Schnell, Karin Volkwein, Lori Riley, and Elizabeth Leach. 2003. "Attitudes toward Aging: A Comparative Analysis of Young Adults from the United States and Germany." *International Journal of Aging and Human Development* 57(3):203–15.

Miller, Darryl, Teresita Leyell, and Julianna Mazachek. 2004. "Stereotypes of the Elderly in U.S. Television Commercials from the 1950s to the 1990s." *International Journal of Aging and Human Development* 58(4):315–40.

Miniño, Arialdi, Melonie Heron, Sherry L. Murphy, and Kenneth D. Kochanek. 2005. "Deaths: Final Data for 2004." Retrieved February 9, 2007 (www.cdc.gov/nchs/products/pubs/pubd/hestats/finaldeaths04/finaldeaths04.htm).

Moody, Harry. 2006. *Aging: Concepts and Controversies*. Thousand Oaks, CA: Pine Forge.

Moon, Marilyn. 1999. "Growth in Medicare Spending: What Will Beneficiaries Pay?" Retrieved March 16, 2003 (www.urban.org/health/medicare_growth).

National Center for Health Statistics. 2006. *Health, United States, 2006 with Chartbook on Trends in the Health of Americans*. Hyattsville, MD: Author.

Nelson, Todd. 2002. *Ageism: Stereotyping and Prejudice against Older Adults*. Cambridge, MA: MIT Press.

———. 2005. "Ageism: Prejudice against Our Feared Future Self." *Journal of Social Issues* 61(2):207–21.

OASDI Trustees. 2007. *The 2007 Annual Report of the Board of Trustees of the Federal Old-Age and Survivors Insurance and Federal Disability Insurance Trust Funds*. Washington, DC: U.S. Government Printing Office.

Ohlemacher, Stephen. 2007. "The First Baby Boomer Applies for Social Security." *Seattle Times*, October 16, pp. A1, A11.

Packer, Dominic and Alison Chasteen. 2006. "Looking to the Future: How Possible Aged Selves Influence Prejudice Toward Older Adults." *Social Cognition*, 24:218–47.

Pampel, Fred C. 1998. *Aging, Social Inequality, and Public Policy*. Thousand Oaks, CA: Pine Forge.

Robinson, Kristen. 2007. *Trends in Health Status and Health Care Use among Older Women, Aging Trends, No. 7*. Hyattsville, MD: National Center for Health Statistics.

Saucier, Maggi. 2004. "Midlife and Beyond: Issue for Aging Women." *Journal of Counseling and Development* 82:420–25.

Siegel, Michele, Charlotte Muller, and Marjorie Honig. 2000. *The Incidence of Job Loss: The Shift from Younger to Older Workers, 1981–1996*. New York: International Longevity Center–USA.

Social Security and Medicare Boards of Trustees. 2007. "The Summary of the 2007 Annual Reports." Retrieved June 17, 2007 (http://www.ssa.gov/OACT/TRSUM/trsummary.html).

Sontag, Susan. 1979. *The Double Standard of Aging*. Retrieved February 11, 2007 (http://www.mediawatch.com/wordpress/?p=33).

Springer, John. 2007. "Woman Defends Decision to Give Birth at 60." Retrieved June 17, 2007 (http://www.msnbc.msn.com/id/18841574/).

Street, Debra and Jeralynn Sittig Crossman. 2006. "Greatest Generation or Greedy Geezers? Social Spending Preference and the Elderly." *Social Problems* 53(1):75–96.

Swuwade, Philip. 1996. "U.S. Older Workers: Their Employment and Occupational Problems in the Labor Market." *Social Behavior and Personality* 24(3):235–38.

Thornton, James. 2002. "Myths of Aging or Ageist Stereotypes." *Educational Gerontology* 28:301–12.

Turner, B. 1996. *The Body and Society: Explorations in Social Theory*. Thousand Oaks, CA: Sage Publications.

Wilson, Duff. 2007. "Aging: Disease or Business Opportunity?" *New York Times*, April 15, pp. BU 1, 7–8.

Wood, Geoffrey, Mark Harcourt, and Sondra Harcourt. 2004. "The Effects of Age Discrimination Legislation on Workplace Practice: A New Zealand Case Study." *Industrial Relations* 35(4):359–71.

Yang, Frances and Sue Levkoff. 2005. "Ageism and Minority Populations: Strengths in the Face of Challenge." *Generations* 29:42–48.

Zaidi, Asghar. 2006. *Poverty of Elderly People in EU25*. Retrieved June 6, 2007 (http://www.euro.centre.org/data/1156245035_36346.pdf).

Chapter 7

AFL-CIO. 2002. "Fact Sheet: Bargaining for Family Leave." Retrieved August 28, 2002 (www.aflcio.org/women/f_fam).

Amato, E. R. and B. Keith. 1991. "Parental Divorce and the Well-Being of Children: A Meta-analysis." *Journal of Marriage and the Family* 53:895–915.

American Association of Retired Persons (AARP). 2002. "Facts about Grandparents Raising Children." Retrieved August 28, 2002 (http://www.aarp.org/congacts/grandparents/grandfacts).

Anderson, Kristin. 1997. "Gender, Status, and Domestic Violence: An Integration of Feminist and Family Violence Approaches." *Journal of Marriage and the Family* 59(3):655–79.

Annie E. Casey Foundation. 1998. *Kids Count Special Report: When Teens Have Sex: Issues and Trends*. Baltimore, MD: Author.

Armas, Genaro. 2002. "Census Shows More Grandparents Are Raising Their Grandchildren." *News Tribune,* July 8, pp. A1, A6.

Avellar, Sarah and Pamela Smock. 2005. "The Economic Consequences of the Dissolution of Cohabiting Unions." *Journal of Marriage and the Family* 67(2):315–27.

Bengston, V. L. 2001. "Beyond the Nuclear Family: The Increasing Importance of Multigenerational Bonds." *Journal of Marriage and the Family* 63(1):1–17.

Bent-Goodley, Tricia. 2005. "Culture and Domestic Violence: Transforming Knowledge Development." *Journal of Interpersonal Violence* 20(2):195–203.

Bethea, L. 1999. "Primary Prevention of Child Abuse." *American Family Physician* 59(6):1577–86.

Bianchi, Suzanne, John Robinson, and Melissa Milkie. 2006. *Changing Rhythms of American Family Life.* New York: Russell Sage Foundation.

Breines, W. and L. Gordon. 1983. "The New Scholarship on Family Violence." *Signs* 8:490–531.

Brooks Conway, Morgan, Teresa Christensen, and Barbara Herlihy. 2003. "Adult Children of Divorce and Intimate Relationships: Implications for Counseling." *Family Journal* 11(4):364–73.

Brownmiller, S. 1975. *Against Our Will: Men, Women, and Rape.* New York. Bantam Books.

Buckey, Cara. 2007. "Despite Alternatives, Many Newborns are Abandoned." *New York Times,* January 13, p. A14.

Bumpass, Larry. 1998. "The Changing Significance of Marriage in the United States." Pp. 63–82 in *The Changing Family in Comparative Perspective: Asia and the United States,* edited by K. Oppenheim Mason, N. Tsuya, and M. Choe. Honolulu, HI: East-West Center.

Bumpass, Larry and Hsien-Hen Lu. 2000. "Trends in Cohabitation and Implications for Children's Family Contexts in the United States." *Population Studies* 54(1):29–41.

Card, J. 1999. "Teen Pregnancy Prevention: Do Any Programs Work?" *Annual Review of Public Health* 20:257–85.

Casper, L. and K. Bryson. 1998. "Co-resident Grandparents and Their Grandchildren: Grandparent Maintained Families" (Population Division Working Paper No. 26). Washington, DC: U.S. Census Bureau.

Centers for Disease Control and Prevention. 2002. *Births, Marriages, Divorces, and Deaths: Provisional Data for October 2001* (National Vital Statistics Reports Vol. 50, No. 11, June 26, 2002). Atlanta, GA: Author.

————. 2006. *Births, Marriages, Divorces, and Deaths: Provisional Data for 2005* (National Vital Statistics Reports Vol. 54, No. 20, July 21, 2006). Atlanta, GA: Author.

Chase Goodman, Catherine and Merril Silverstein. 2006. "Grandmothers Raising Grandchildren: Ethnic and Racial Differences in Well-Being Among Custodial and Coparenting Families." *Journal of Family Issues* 27(11):1605–26.

Cherlin, A., K. Kiernan, and P. Chase-Lansdale. 1995. "Parental Divorce in Childhood and Demographic Outcomes in Young Adulthood." *Demography* 32:299–318.

Children's Bureau. 2006. *Child Maltreatment 2004.* Washington, DC: U.S. Department of Health and Human Services, Administration for Children and Families.

Child Welfare Information Gateway. 2006. *Long-term Consequences of Child Abuse and Neglect.* Washington, DC: U.S. Department of Health and Human Services, Administration for Children and Families.

Collier, L. 2001. "Havens for Abandoned Babies Occupy Tricky Terrain." *Chicago Tribune,* February 18, 2001, p. C1. Retrieved August 30, 2002 (www.adoptionnation.com/chicago_trib_02-18-01).

Copeland, A. P. and K. M. White. 1991. *Studying Families.* Newbury Park, CA: Sage Publications.

D'Agostino, J. 2000. "Infant Abandonment Has Become an Epidemic." *Human Events* 56(12):4.

DeFoe Whitehead, Barbara and David Popenoe. 2005. *The State of Our Unions: The Social Health of Marriage in America, 2005.* Retrieved May 6, 2007 (http://marriage.rutgers.edu/Publications/SOOU/TEXTSOOU2005.htm#Cohabitation).

Demo, D. H. (1992). "Parent-Child Relations: Assessing Recent Changes." *Journal of Marriage and the Family* 54:104–17.

Dodson, Lisa, Tiffany Manuel, and Ellen Bravo. 2002. *Keeping Jobs and Raising Families in Low-Income America: It Just Doesn't Work.* Cambridge, MA: Radcliffe Public Policy Center and 9 to 5 National Association of Working Women.

Engels, Fredrich. 1902. *The Origin of Family, Private Property, and the State*. New York: International Publishing.

EuroStat 2005a. "Divorces, per 1,000 Persons." Retrieved January 24, 2007 (http://epp.eurostat.ec.europa.eu/portal/page?_pageid=1996,39140985&_dad=portal&_schema=PORTAL&p_product_code=CAB10000).

———. 2005b. " Marriages, per 1,000 Persons." Retrieved January 24, 2007 (http://epp.eurostat.ec.europa.eu/portal/page?_pageid=1073,46870091&_dad=portal&_schema=PORTAL&p_product_code=CAB10000).

Fields, Jason. 2003. *America's Families and Living Arrangements: 2002* (Current Population Reports, P20–547). Washington, DC: U.S. Census Bureau.

———. 2004. *America's Families and Living Arrangements: 2003* (Current Population Reports, P20–553). Washington, DC: U.S. Census Bureau.

Furstenberg, F. F., J. Brooks-Gunn, and S. Morgan. 1987. *Adolescent Mothers in Later Life*. Cambridge, England: Cambridge University Press.

Garcia-Moreno, Claudia, Henrica Jansen, Mary Ellsberg, Lori Heise, and Charlotte Watts. 2006. "Prevalence of Intimate Partner Violence: Findings from the WHO Multi-country Study on Women's Health and Domestic Violence." *The Lancet* 368:1260–69.

Gelles, Richard and Peter Maynard. 1987. "A Structural Family Systems Approach to Intervention in Cases of Family Violence." *Family Relations* 33(2):270–76.

Hanson, Marci and Eleanor Lynch. 1992. "Family Diversity: Implications for Policy and Practice." *Topics in Early Childhood Special Education* 12(3):283–305.

He, Wan, Manisha Sengupta, Victoria Velkoff, and Kimberly De Barros. 2005. *65+ in the United States, 2005* (P23-209). Washington, DC: U.S. Dept. of Commerce, Economics and Statistics Administration, U.S. Census Bureau.

Heymann, Jody, Alison Earle, and Jeffrey Hayes. 2007. *The Work, Family, and Equity Index: How Does the United States Measure Up*. Montreal, Quebec: Project on Global Working Families.

Hoffman, Saul. 1998. "Teenage Childbearing Is Not So Bad After All . . . Or Is It? A Review of the New Literature." *Family Planning Perspectives* 30(5):236–39, 243.

Illinois Department of Children and Family Services. 2006. "Children and Families Celebrate Anniversary of Illinois' Safe Haven Law." Retrieved May 6, 2007 (http://www.state.il.us/DCFS/library/com_communications_pr_Nov142006.shtml).

Kalil, Ariel and Sandra Danzinger. 2000. "How Teen Mothers Are Faring Under Welfare Reform." *Journal of Social Issues* 56(4):775–98.

Kreider, Rose. 2005. *Number, Timing and Duration of Marriage and Divorces: 2001* (Current Population Reports, P70-97). Washington, DC: U.S. Census Bureau.

Kreider, Rose and J. Fields. 2002. *Number, Timing, and Duration of Marriages and Divorces: Fall 1996* (Current Population Reports, P70–80). Washington, DC: U.S. Census Bureau.

Lang, Susan. 1993. "Findings Refute Traditional Views on Elder Abuse." *Human Ecology* 21(3):30.

Lenski, G. and G. Lenski 1987. *Human Societies*. New York: McGraw-Hill.

Levendosky, A. and S. Graham-Bermann. 2000. "Behavioral Observations of Parenting in Battered Women." *Journal of Family Psychology* 14:80–94.

Levendosky, A., S. Lynch, and S. Graham-Bermann. 2000. "Mothers' Perceptions of the Impact of Woman Abuse on Their Parenting." *Violence against Women* 6(3):247–71.

Little, Kristin, Mary Malefyt, and Alexander Walker. 1998. "A Tool for Communities to Develop Coordinated Responses." In *Promising Practices Initiative of the STOP Violence Against Women Grants Technical Assistance Project*. Retrieved August 29, 2002 (www.vaw.umn.edu/Promise/PP3).

MacKinnon, C. E., Z. Stoneman, and G. H. Brody. 1984. "The Impact of Maternal Employment and Family Form on Children's Sex Role Stereotypes and Mothers' Traditional Attitudes." *Journal of Divorce* 8:51–60.

Manning, Wendy and Pamela Smock. 2002. "First Comes Cohabitation and then Comes Marriage? A Research Note." *Journal of Family Issues* 23(8):1065–87.

Maynard, Rebecca, ed. 1997. *Kids Having Kids: Economic Costs and Social Consequences of Teen Pregnancy*. Washington, DC: Urban Institute Press.

McKay, M. 1994. "The Link between Domestic Violence and Child Abuse: Assessment and Treatment Considerations." *Child Welfare* 73(1):29–39.

National Center on Elder Abuse. 2002a. "The Basics: What Is Elder Abuse?" Retrieved August 29, 2002 (www.elderabusecenter.org/basic/).

———. 2002b. *Sentinels: Reaching Hidden Victims-Final Report, May 2002*. Washington, DC: Author.

———. 2006. *The 2004 Survey of State Protective Services: Abuse of Adults 60 Years of Age and Older*. Washington, DC: Author.

National Clearing House on Child Abuse and Neglect Information. 2002. "What Is Child Maltreatment?" Retrieved August 27, 2002 (www.calib.com/nccanch/pubs/factsheets/childmal.cfm).

National Partnership for Women and Families. 2002. "What It Took to Pass the Family and Medical Leave Act: A Nine-Year Campaign Pays Off." Retrieved August 28, 2002 (http://www.nationalpartnership .org/content.cfm?L1=202&DBT=Documents&NewsItemID=275).

Navaie-Waliser, Maryam, P. Feldman, D. Gould, C. Levine, A. Kuebris, and Karen Donelan. 2002. "When the Caregiver Needs Care: The Plight of Vulnerable Caregivers." *American Journal of Public Health* 92(3):409–13.

New Jersey Safe Haven Protection Act. 2007. "FAQ's." Retrieved January 14, 2007 (http://www.njsafehaven .org/faq.html).

Newman, David. 2006. *Sociology: Exploring the Architecture of Everyday Life*. Thousand Oaks, CA: Pine Forge.

Popenoe, David. 1993. "American Family Decline, 1960–1990: A Review and Appraisal." *Journal of Marriage and the Family* 55:527–55.

———. 2006. "Marriage and Family in the Scandinavian Experience." *Society* 43(4):68–72.

Povoledo, Elisabetta. 2007. "Updating an Old Way to Leave the Baby on the Doorstep." *New York Times*, February 28, p. A4.

Radcliffe Public Policy Center. 2000. *Life's Work: Generational Attitudes Toward Work and Life Integration*. Cambridge, MA: Author.

Rennison, Callie Marie. 2003. *Intimate Partner Violence, 1993–2001* (NCJ 197838). Washington, DC: U.S. Department of Justice, Office of Justice Programs.

Roberts, Sam. 2007. "51% of Women Are Now Living Without a Spouse." *New York Times*, January 16, pp. A1, A19.

Rubin, Lillian. 1995. *Families on the Fault Line: America's Working Class Speaks About the Family, the Economy and Ethnicity*. New York: Harper Perennial.

Rutter, Michael and Marta Tienda. 2005. "The Multiple Facets of Ethnicity." Pp. 50–59 in *Ethnicity and Causal Mechanisms,* edited by Michael Rutter and Marta Tienda. New York: Cambridge University Press.

Selman, Peter and Caroline Glendinning. 1994. "Teenage Parenthood and Social Policy." *Youth and Policy* 47:39–58.

Simmons, T. and J. L. Dye. 2003. *Grandparents Living with Grandchildren: 2000* (Current Population Reports, C2KBR-31). Washington, DC: U.S. Census Bureau.

Simons, R. L. 1996. *Understanding Differences between Divorced and Intact Families: Stress, Interaction, and Child Outcomes*. Thousand Oaks, CA: Sage Publications.

SmithBattle, Lee. 2007. "'I Wanna Have a Good Future': Teen Mothers' Rise in Educational Aspirations, Competing Demands, and Limited School Support." *Youth & Society* 38(3):348–71.

Smock, Pamela J. 1994. "Gender and Short Run Economic Consequences of Marital Disruption." *Social Forces* 73:243–62.

Somers, Cheryl and Mariane Fahlman. 2001. "Effectiveness of the 'Baby Think It Over' Teen Pregnancy Prevention Program." *Journal of School Health* 71(5):188–207.

Stacey, Judith. 1996. *In the Name of the Family: Rethinking Family Values in the Post-Modern Age*. Boston, MA: Beacon Press.

Tjaden, Patricia and Nancy Thoennes. 2000. *Extent, Nature and Consequences of Intimate Partner Violence, NCJ 18186*. Washington, DC: National Institute of Justice, U.S. Department of Justice.

U.S. Bureau of Justice. 2006. "Intimate Partner Violence in the United States: Victim Characteristics." Retrieved March 16, 2007 (http://www.ojp.usdoj.gov/bjs/intimate/victims.htm).

U.S. Census Bureau. 1999. *Statistical Abstract of the United States: 1999* (Table No. 91). Washington, DC: Author.

U.S. Department of Health and Human Services. 1998. *A National Strategy to Prevent Teen Pregnancy: Annual Report 1997–1998*. Washington, DC: Author.

U.S. Department of Labor. 2007. *Federal vs. California Family and Medical Leave Laws.* Retrieved January 13, 2007 (http://www.dol.gov/esa/programs/whd/state/fmla/ca.htm).

Uunk, Wilfred. 2004. "The Economic Consequences of Divorce for Women in the European Union: The Impact of Welfare State Arrangements." *European Journal of Population* 20(3):251–85.

Ventura, Stephanie, Joyce Abma, William Mosher, and Stanley Henshaw. 2006. *Recent Trends in Teenage Pregnancy in the United States: State Trends, 1990–2002* (National Vital Statistics Reports). Hyattsville, MD: National Center for Health Statistics.

Videon, Tami. 2002. "The Effects of Parent-Adolescent Relationships and Parental Separation on Adolescent Well- Being." *Journal of Marriage and Family* 64(2):489–504.

Chapter 8

Aber, J. L., J. Brown, and C. C. Henrich. 1999. *Teaching Conflict Resolution: An Effective School-Based Approach to Violence Prevention.* Washington, DC: National Center for Children in Poverty.

Adams, M. S. and T. D. Evans. 1996. "Teacher Disapproval, Delinquent Peers, and Self-Reported Delinquency: A Longitudinal Test of Labeling Theory." *Urban Review* 28(3):199–211.

American Association of University Women (AAUW). 1992. *How Schools Shortchange Girls: The AAUW Report.* New York: Author.

———. 2001. *Hostile Hallways: Bullying, Teasing, and Sexual Harassment in School.* Washington, DC: Author.

Anderson, M., J. Kaufman, T. Simon, L. Barrios, L. Paulozzi, G. Ryan, R. Hammond, W. Modzeleski, T. Feucht, L. Potter, and the School-Associated Violent Deaths Study Group. 2001. "School Associated Violent Deaths in the United States, 1994–1999." *Journal of the American Medical Association* 286(21):2695–702.

Ansalone, George. 2001. "Schooling, Tracking and Inequality." *Journal of Children and Poverty* 7(1):33–49.

———. 2004. "Educational Opportunity and Access to Knowledge: Tracking in the U.S. and Japan." *Race, Gender & Class* 11(3):140–52.

Bauman, Kurt J. 2001. "Home Schooling in the United States: Trends and Characteristics" (Working Paper No. 530). Washington, DC: U.S. Census Bureau.

Berliner, David and Bruce Biddle. 1995. *The Manufactured Crisis: Myths, Frauds, and the Attack on America's Public Schools.* Reading, MA: Addison-Wesley.

Bjork, Christopher and Ryoko Tsuneyoshi. 2005. "Education Reform in Japan: Competing Visions for the Future." *Phi Delta Kappan* April: 619–26.

Bohan, J. and G. M. Russell. 1999. "Support Networks for Lesbian, Gay and Bisexual Students." Pp. 279–94 in *Coming into Her Own: Educational Successes in Girls and Women,* edited by S. N. Davis, M. Crawford, and J. Sebrechts. San Francisco, CA: Jossey-Bass.

Bowman, D. H. 2001. "Student Survey Sees 1 in 10 Peers as Potentially Violent." *Education Week* 21(1):9–10.

Bumiller, Elisabeth. 2002. "Bush Calls Ruling About Vouchers a 'Historic' Move." *New York Times,* July 2, pp. A1, A15.

Callan, Patrick. 2006. "Introduction: International Comparisons Highlight Educational Gaps Between Young and Older Americans." Washington, DC: National Center for Public Policy and Higher Education.

College Entrance Examination Board. 1999. *Reaching the Top: A Report of the National Task Force on Minority High Achievement.* New York: Author.

Denton, K. and J. West. 2002. "Children's Reading and Mathematics Achievement in Kindergarten and First Grade." *Educational Statistics Quarterly.* Retrieved January 18, 2004 (http://nces.ed.gov/pubs2002/2002125.pdf).

Dillon, Sam. 2004. "Some School Districts Challenge Bush's Signature Education Law." *New York Times,* January 2, pp. A1, A13.

Dillon, Sam. 2007. "Math Scores Rise, but Reading Is Mixed." *New York Times,* September 26, p. A20.

Dinkes, R., E. Cataldi, G. Kena, and K. Baum. 2006. *Indicators of School Crime and Safety: 2006* (NCES 2007-003/NCJ 214262). U.S. Departments of Education and Justice. Washington, DC: U.S. Government Printing Office.

Equal Employment Opportunity Commission. 2001. "Facts about Sexual Harassment." Retrieved October 10, 2002 (www.eeoc.gov/facts/fs-sex.html).

Fineran, Susan. 2002. "Sexual Harassment between Same-Sex Peers: Intersection of Mental Health, Homophobia, and Sexual Violence in Schools." *Social Work* 47(1):65–74.

Finn, Chester. 1997. "The Politics of Change." Pp. 226–50 in *New Schools for a New Century: The Redesign of Urban Education,* edited by D. Ravitch and J. Viteritti. New Haven, CT: Yale University Press.

Flannery, Daniel. 1997. *School Violence: Risk, Preventive Intervention, and Policy.* (ERIC Clearing House on Urban Education, Publication Number RR93002016).

Freeman, Catherine. 2004. *Trends in Educational Equity of Girls & Women: 2004.* (NCES 2005-016). U.S. Department of Education, National Center for Education Statistics. Washington, DC: U.S. Government Printing Office

"Goals 2000: Reforming Education to Improve Student Achievement." 1998. Retrieved October 1, 2002 (www.ed.gov/pubs/G2KReforming/index.html).

Good, T. L. and J. S. Braden. 2000. *The Great School Debate: Choice, Vouchers, and Charters.* Mahwah, NJ: Lawrence Erlbaum.

Grunbaum, J. A., L. Kann, S. Kinchen, J. Ross, J. Hawkins, R. Lowry, W. Harris, T. McManus, D. Chyen, and J. Collins. 2004. *Youth Risk Behavior Surveillance—United States, 2003* (Report 53[SS02]). Atlanta, GA: Centers for Disease Control.

Hallinan, Maureen. 1994. "Tracking: From Theory to P." *Sociology of Education* 67(2):79–91.

Holland, D. C. and M. A. Eisenhart. 1990. *Educated in Romance: Women, Involvement, and College Culture.* Chicago, IL: University of Chicago Press.

Human Rights Watch. 2001. *Hatred in the Hallways: Violence and Discrimination Against Lesbian, Gay, Bisexual, and Transgender Students in U.S. Schools.* New York: Author.

Jacobs, Jerry A. 1996. "Gender Inequality and Higher Education." *Annual Review of Sociology* 22:153–85.

Johnson, R. C. and D. Viadero. 2000. "Unmet Promise: Raising Minority Achievement." *Education Week* 17(43):1.

Kahlenberg, Richard. 2002. "Socioeconomic Integration." Presented at 20th Annual Magnet Schools of America Conference. Retrieved March 16, 2008 (http://www.tcf.org/list.asp?type=NC&pubid=906).

Kanter, Rosabeth Moss. 1972. "The Organization Child: Experience Management in a Nursery School." *Sociology of Education* 45:186–211.

Kennedy, S. 2001. "Privatizing Education." *Phi Delta Kappan* 82(6):450–57.

Kingery, P. M., B. E. Pruitt, G. Heuberger, and J. A. Brizzolara. 1993. "School Violence Reported by Adolescents in Rural Central Texas." Texas A&M University, College Station, TX. Unpublished manuscript.

Kirsch, Irwin, Ann Jungeblut, Lynn Jenkins, and Andrew Kolstad. 2002. *Adult Literacy in America: A First Look at the Results of the National Adult Literacy Survey.* Retrieved March 16, 2008 (http://nces.ed.gov/pubs93/93275.pdf).

Kopp, Wendy. 2001. *One Day, All Children: The Unlikely Triumph of Teach for America and What I Learned Along the Way.* New York: PublicAffairs.

Kosciw, J. G. and E. M. Diaz. 2006. *The 2005 National School Climate Survey: The Experiences of Lesbian, Gay, Bisexual and Transgender Youth in Our Nation's Schools.* New York: GLSEN.

Kozol, Jonathan. 1991. *Savage Inequalities: Children in America's Schools.* New York: Harper Perennial.
———. 2005. *The Shame of the Nation: The Restoration of Apartheid School in America.* New York: Crown.

Laird, J., S. Lew, M. DeBell, and C. Chapman. 2006. *Dropout Rates in the United States: 2002 and 2003 (NCES 2006-062).* U.S. Department of Education. Washington, DC: National Center for Education Statistics.

Linn, M. and C. Kessel. 1996. "Success in Mathematics: Increasing Talent and Gender Diversity Among College Majors." Pp. 83–100 in *Issues in Mathematics Education, Conference of the Mathematical Sciences.* Vol. 6, *Research in Collegiate Mathematics Education II,* edited by J. Kaput, A. H. Schoenfeld, and E. Dubinsky. Providence, RI: American Mathematical Society.

Literacy Volunteers of America. 2002. "Facts on Literacy in America." Retrieved September 5, 2003 (www .literacyvolunteers.org/about/faqs/facts.html).

Marger, Martin. 2008. *Social Inequality: Patterns and Processes.* Boston, MA: McGraw-Hill.

Marshak, David. 2003. "No Child Left Behind: A Foolish Race into the Past." *Phi Delta Kappan* 85(3):229–31.

Maruyama, G. 2003. "Disparities in Educational Opportunities and Outcomes: What Do You Know and What Can We Do?" *Journal of Social Issues* 59(3):653–76.

Meier, D. 1995. *The Power of Their Ideas.* Boston, MA: Beacon Press.

Mori, Rie. 2002. "Entrance Examinations and Remedial Education in Japanese Higher Education." *Higher Education* 43:27–42.

National Advisory Council on Violence Against Women. 2001. *Toolkit to End Violence Against Women.* Retrieved January 20, 2007 (http://toolkit.ncjrs.org/).

National Center for Educational Statistics. 2004. *1.1 Million Home Schooled in the United States in 2003.* NCES 2004-115. Washington, DC: U.S. Department of Education.

National Commission on Excellence in Education. 1983. *A Nation at Risk: The Imperatives for Educational Reform.* Washington, DC: U.S. Department of Education.

National Girls Collaborative Project. 2006. "Overview." Retrieved March 14, 2007 (http://www.puget soundcenter.org/ngcp/about/index.html).

Ono, Hiroshi. 2001. "Who Goes to College? Features of Institutional Tracking in Japanese Higher Education." *American Journal of Education* 109(2):161–95.

Perkins-Gough, Deborah. 2006. "Do We Really Have a 'Boy Crisis'?" *Educational Leadership* 64(1):93–94.

Popham, James. 2003. Interview with *Frontline.* Retrieved January 23, 2004 (http://www.pbs.org/wgbh/ pages/frontline/shows/schools/nochild/bush.html).

ProLiteracy Worldwide. 2006. *The State of Adult Literacy 2006.* New York: Author.

Ravitch, Diane. 1997. "Somebody's Children: Educational Opportunity for All American Children." Pp. 251–273 in *New Schools for a New Century: The Redesign of Urban Education,* edited by Diane Ravitch and Joseph P. Viteritti. New Haven, CT: Yale University Press.

Ravitch, Diane and Joseph P. Viteritti. 1997. "Introduction." Pp. 1–16 in *New Schools for a New Century: The Redesign of Urban Education,* edited by Diane Ravitch and Joseph P. Viteritti. New Haven, CT: Yale University Press.

Rohlen, Thomas. 1983. *Japan's High Schools.* Berkeley: University of California Press.

Sadker, Myra and David Sadker. 1994. *Failing at Fairness: How Our Schools Cheat Girls.* New York: Simon & Schuster.

Saporito, Salvatore. and Annette Lareau. 1999. "School Selection as a Process: The Multiple Dimensions of Race in Framing Educational Choice." *Social Problems* 46:418–39.

Schemo, Diana Jean. 2002. "Few Exercise New Right to Leave Failing Schools." *New York Times,* August 28, pp. A1, A14.

———. 2006. "Study of Test Scores Finds Charter Schools Lagging." *New York Times,* August 23, p. A14.

———. 2007. "Bush Proposes Broadening the No Child Left Behind Act." *New York Times,* January 25, p. A18.

Shaw, Margaret. 2001. *Promoting Safety in Schools: International Experience and Action* (Bureau of Justice Assistance Monograph, NCJ186937). Washington, DC: U.S. Department of Justice.

Silverman, L. K. 1986. "What Happens to the Gifted Girl?" Pp. 43–89 in *Critical Issues in Gifted Education.* Vol. 1, *Defensible Programs for the Gifted,* edited by C. J. Maker. Rockville, MD: Aspen.

Slaughter-Defoe, Diana and Henry Rubin. 2001. "A Longitudinal Case Study of Head Start Eligible Children: Implications for Urban Education." *Educational Psychologist* 36(1):31–44.

Smith, D. 2000. "Schooling for Inequality." *Signs* 25(4):1147–51.

Smith, R. S. 1995. "Giving Credit Where Credit Is Due: Dorothy Swaine Thomas and the 'Thomas Theorem.'" *American Sociologist* 26(4):9–29.

Spencer, S., Claude Steele, and D. Quinn. 1999. "Stereotype Threat and Women's Math Performance." *Journal of Experimental Social Psychology* 35:4–28.

Steele, Claude. 1997. "A Threat in the Air: How Stereotypes Shape Intellectual Identity and Performance." *American Psychologist* 52:613–29.

Steele, Claude and Joshua Aronson. 1995. "Stereotype Threat and the Intellectual Test Performance of African Americans." *Journal of Personality and Social Psychology* 69(5):797–811.

Stoops, Nicole. 2004. *Educational Attainment in the United States: 2003* (Current Population Report, P20–550). Washington, DC: U.S. Census Bureau.

Strand, K. and M. E. Mayfield. 2002. "Pedagogical Reform and College Women's Persistence in Mathematics." *Journal of Women and Minorities in Science and Engineering* 8:67–83.

Stombler, Mindy and Patricia Yancey Martin. 1994. "Bringing Women in, Keeping Women Down." *Journal of Contemporary Ethnography* 23(2):150–84.

Sum, A., T. Kirsch, and R. Taggart. 2002. *The Twin Challenges of Mediocrity and Inequality: Literacy in the U.S. from an International Perspective.* Princeton, NJ: Educational Testing Service.

Thomas, W. I. and Dorothy Swaine Thomas. 1928. *The Child in America: Behavior Problems and Programs.* New York: Knopf.

Traub, J. 2000. "What No School Can Do." *New York Times Magazine,* January 16, pp. 52–57, 68, 81, 90–91.

United Nations Educational, Scientific and Cultural Organization (UNESCO). 2006. "Literacy: The Core of Education for All." *Education for All Global Monitoring Report.* New York: Author.

U.S. Census Bureau. 2001. *School Enrollment in the United States—Social and Economic Characteristics of Students: October 1999* (Current Population Reports, P20–533, March 2001). Washington, DC: Author.

———. 2006. Current Population Survey, *2005 Annual Social and Economic Supplement.* Washington, DC: Author.

U.S. Department of Health and Human Services. 2005. *Head Start Impact Study: First Year Findings.* Washington, DC: Author.

Wagner, Alan. 2006. *Measuring up Internationally: Developing Skills and Knowledge for a Global Knowledge Economy.* National Center Report #06-07. Washington, DC: National Center for Public Policy and Higher Education.

Wall, S., E. Timberlake, M. Farber, C. Sabatino, H. Liebow, N. Smith, and N. Taylor. 2000. "Needs and Aspirations of the Working Poor: Early Head Start Program Applicants." *Families in Society* 81(4):412–21.

Washington, V. and U. J. Oyemade Bailey. 1995. *Project Head Start: Models and Strategies for the Twenty-first Century.* New York: Garland.

White House. 2007. *Fact Sheet: The No Child Left Behind Act: Five Years of Results for America's Children.* Retrieved March 16, 2008 (http://www.whitehouse.gov/news/releases/2007/01/20070108-6.html).

Wiland, Harry and Dale Bell. 2006. *Edens Lost & Found: How Ordinary Citizens Are Restoring Our Great American Cities.* White River Junction, VT: Chelsea Green.

Wilgoren, J. 2001. "Schools Are Now Marketers Where Choice Is Taking Hold." *New York Times,* April 20, pp. A1, A12.

Zernike, K. 2000. "American Education Gets an A for Effort." *New York Times,* June 2, p. A16.

Zigler, Edward and Susan Muenchow. 1992. *Head Start: The Inside Story of America's Most Successful Educational Experiment.* New York: Basic Books.

Chapter 9

Abell, Hilary. 1999. "Endangering Women's Health for Profit: Health and Safety in Mexico's Maquiladoras." *Development in Practice* 9(5):595–601.

AFL-CIO. 2003. "Who Are Low Wage Workers?" Retrieved April 13, 2003 (www.aflcio.org/issuespolitics/minimumwage/whoarelowwage.cfm).

Armas, Genaro. 2004. "Outearning Men Women's Toughest Job." *News Tribune,* June 4, p. 7.

Bahnisch, Mark. 2000. "Embodied Work, Divided Labour: Subjectivity and the Scientific Management of the Body in Frederick W. Taylor's 1907 'Lecture on Management.'" *Body & Society* 6(1):51–68.

Bernstein, Jared. 1997. "Low-Wage Labor Market Indicators by City and State: The Constraints Facing Welfare Reform" (EPI Working Paper No. 118). Washington, DC: Economic Policy Institute.

Bernstein, Jared, Heidi Hartmann, and John Schmitt. 1999. *The Minimum Wage Increase: A Working Woman's Issue* (EPI Issue Brief No. 133). Washington, DC: Economic Policy Institute.

Bertrand, M. and S. Mullainathan. 2003. "Are Emily and Greg More Employable Than Lakisha and Jamal? A Field Experiment in Market Discrimination" (NBER Working Paper No. 9873). Cambridge, MA: National Bureau of Economic Research.

Blanchard, Oliver. 2004. "Explaining European Unemployment." Retrieved July 31, 2007 (http://www.nber.org/reporter/summer04/blanchard.html).

Bluestone, Barry and Bennett Harrison. 1982. *The Deindustrialization of America*. New York: Basic Books.

Brady, D. and M. Wallace. 2001. "Deindustrialization and Poverty: Manufacturing Decline and AFDC Recipiency in Lake County, Indiana 1964–93." *Sociological Forum* 16(2):321–58.

Braverman, Harry. 1974. *Labor and Monopoly Capital: The Degradation of Work in the Twentieth Century*. New York: Monthly Press Review.

Brenner, Joanna. 1998. "On Gender and Class in U.S. Labor History." *Monthly Review: An Independent Socialist Magazine* 50(6):1–15.

Budig, Michelle. 2002. "Male Advantage and the Gender Composition of Jobs: Who Rides the Glass Escalator?" *Social Problems* 49(2):258–77.

Byrne, Geraldine and Robert Heyman. 1997. "Understanding Nurses' Communication with Patients in Accident and Emergency Departments Using a Symbolic Interactionist Perspective." *Journal of Advanced Nursing* 26:93–100.

Campbell, Jim. 2006. "Proportion of Workers in Selected Pay Ranges by Region and State, 2005." *Monthly Labor Review* December:66–69.

Carley, Michael. 2000. "Urban Partnerships, Governance and Regeneration of Britain's Cities." *International Planning Studies* 5(3):273–97.

Clark, Kim. 2002. "No Pink Slips at This Plant." *U.S. News and World Report* 132(4):40.

Co-Op America. 2003. "What Is a Sweatshop." Retrieved April 26, 2003 (www.sweatshops.org/educated/issue.html).

Creswell, Julie. 2006. "How Suite It Isn't: A Dearth of Female Bosses." *New York Times,* December 17, pp. Business 3–1, 9–10.

Darity, William A. 2003. "Employment Discrimination, Segregation and Health." *American Journal of Public Health* 93(2):226–32.

Davidson, Linda. 1999. "Temp Workers Want a Better Deal." *Workforce* 78(10):44–49.

DeJong, G. F. and A. Madamba. 2001. "A Double Disadvantage? Minority Group, Immigrant Status, and Underemployment in the United States." *Social Science Quarterly* 82(1):117–30.

DeNavas-Walt, Carmen, Bernadette Proctor, and Cheryl H. Lee. 2006. *Income, Poverty and Health Insurance Coverage in the United States: 2005* (Current Population Reports, P60–231). Washington, DC: U.S. Government Printing Office.

Dobbs, Lou. 2004. "Coming Up Empty." *U.S. News & World Report* January 26:46.

Ehrenreich, Barbara. 2001. *Nickled and Dimed: On (Not) Getting By in America*. New York: Metropolitan Book/Henry Holt.

Ehrenreich, Barbara and Thomas Geoghegan. 2002. "Lighting Labor's Fire." *The Nation* December 23. Retrieved February 4, 2007 (http://www.thenation.com/doc/20021223/ehrenreich).

Eisenberg, Daniel. 1999. "Rise of Permatemp." *Time* 153(2):48.

Eitzen, D. Stanley and Maxine Baca Zinn. 2006. "Globalization: An Introduction." Pp. 1–11 in *Globalization: The Transformation of Social Worlds,* edited by D. Stanley Eitzen and Maxine Baca Zinn. Belmont, CA: Thomson Wadsworth.

England, Paula and I. Browne. 1992. "Trends in Women's Economic Status." *Sociological Perspectives* 35(1):17–51.

Enron Corporation. 2000. "Enron Milestones." Retrieved May 1, 2003 (www.enron.com/corp/pressroom/milestones).

———. 2001. "Enron Named Most Innovative for Sixth Year." Retrieved September 3, 2004 (www.enron.com/corp/pressroom/releases/2001/ene/15-MostInnovative-02-06-01-LTR.html).

Equal Employment Opportunity Commission (EEOC). 2003. "Charge Statistics FY 1992 Through FY 2002." Retrieved March 22, 2003 (www.eeoc.gov/stats/charges).

Fantasia, Rick and Kim Voss. 2004. *Hard Work: Remaking the American Labor Movement*. Berkeley: University of California Press.

Foo, Lora Jo. 1994. "The Vulnerable and Exploitable Immigrant Workforce and the Need for Strengthening Worker Protection Legislation." *Yale Law Review* 103(8):2179–2212.

Fowler, Tom. 2001. "Formal Upgrade of Enron Investigation Gives Subpoena Power to SEC." Retrieved September 3, 2004 (www.chron.com/cs/CDA/ssistory.mpl/special/enron/oct01/1114532).

Fox, M. F. and S. Hesse-Biber 1984. *Women at Work*. Palo Alto, CA: Mayfield.

Frontline. 2002."Accounting Lessons." Retrieved May 1, 2003 (www.pbs.org/wgph/pages/frontline/shows/regulations/ lessons).

Fussell, Elizabeth. 2000. "Making Labor Flexible: The Recomposition of Tijuana's Maquiladora Female Labor Force." *Feminist Economics* 6(3):59–79.

General Accounting Office (GAO). 1994. "Garment Industry: Efforts to Address the Prevalence and Conditions of Sweatshops" (GAO/HEHS-95–29, November). Retrieved March 26, 2008 (http://www.gao.gov/archive/1995/he95029.pdf).

Gibson, C. and E. Lennon. 1999. "Historical Census Statistics on the Foreign-Born Population of the United States: 1850–1999" (Population Division Working Paper No. 29). Washington, DC: U.S. Census Bureau.

Gluck, Sherna Berger. 1987. *Rosie the Riveter Revisited: Women, the War, and Social Change*. Boston, MA: Twayne.

Gorman, Maryann. 2001. "The White Dog's Tale" *YES! Magazine,* Spring 2001. Retrieved March 27, 2008 (http://www.yesmagazine.org/article.asp?ID=425).

Greenhouse, Steven. 2007a. "Maryland Is First State to Require a Living Wage." *New York Times* May 9, p. A18.
————. 2007b. "Sharp Decline in Union Membership in '06." *New York Times,* January 26, p. A11.

Gruben, William C. 2001. "Was NAFTA Behind Mexico's High Maquiladora Growth?" *CATO Journal* 18(2).263–75.

Gunther, Marc. 2006. "Corporate America Backs Gay Rights." Retrieved February 3, 2007 (http://money.cnn.com/2006/04/25/magazines/fortune/pluggedin_fortune/).

Hall, Richard. 1994. *Sociology of Work: Perspectives, Analyses, and Issues*. Thousand Oaks, CA: Pine Forge.

Hard Hatted Women. 2006. "Adult Education." Retrieved February 4, 2007 (http://www.hardhattedwomen.org/adulted.asp).

Helft, Miguel. 2007. "Google, Master of Online Traffic, Helps Its Workers Beat the Rush." *New York Times,* March 10, A1, B9.

Hirschhorn, Larry. 1984. *Beyond Mechanization: Work and Technology in a Post-Industrial Age*. Cambridge, MA: MIT Press.

Human Rights Campaign. 2003. "ENDA Quickfacts: The Right Solution for a Real Need." Retrieved March 22, 2003 (www.hrc.org/issues/federal_lcg/cnda).

Information for Decision Making. 2000. "Minimum Wage Legislation and Living Wage Campaigns." Retrieved April 20, 2003 (www.financeprojectinfo.org/MWW/minimum.asp#effects).

King, Rachel. 2006. "Outsourcing: Beyond Bangalore." *Business Week Online*. December 11:6-6.

Kovach, Kenneth and Peter Millspaugh. 1996. "Employment Nondiscrimination Act: On the Cutting Edge of Public Policy." *Business Horizons* 39(4):65–74.

Lazes, P. and J. Savage. 2000. "Embracing the Future: Union Strategies for the 21st Century." *Journal for Quality and Participation* 23(4):18–24.

Levering, R. and M. Moskowitz. 2007. "The 100 Best Companies to Work For." *Fortune* 155(1):94.

Lindquist, Diane. 2001. "Rules Change for Maquiladoras." *Industry Week* 250(1):23–26.

Lollock, Lisa. 2001. *The Foreign Born Population in the United States: March 2000*. (Current Population Reports, P20–534). Washington, DC: U.S. Census Bureau.

McDonough, Sioban. 2003. "EEOC: Job Discrimination Up Since 9/11." Retrieved March 22, 2003 (www.softcom.net/webnews/wed/ch/Adiscrimination.RNTX_CD3).

Mills, C. Wright. 2000. *The Sociological Imagination*. New York: Oxford University Press. (Original work published 1959)

Minchin, Timothy. 2006. "'Just Like a Death': The Closing of the International Paper Company Mill in Mobile, Alabama, and the Deindustrialization of the South, 2000–2005." *Alabama Review* January:44–77.

Moffatt, Allison. 2005. "Murder, Mystery and Mistreatment in Mexican Maquiladoras." *Women and Environments International Magazine* 66/67:19–21.

Mosisa, Abraham. 2002. "The Role of Foreign-Born Workers in the U.S. Economy." *Monthly Labor Review* 125(5):3–15.

National Institute for Occupational Safety and Health (NIOSH). 2003. "Stress at Work." Retrieved May 3, 2003 (www.cdc.gov/niosh/stresswk.html).

National Public Radio. 2003. "The Fall of Enron." Retrieved May 2, 2003 (www.npr.org/news/specials/enron/).

Novak, Laura. 2007. "For Women, a Recipe to Create a Successful Business." *New York Times,* June 23, p. B4.

Parenti, Michael. 1988. *Democracy for the Few.* 5th ed. New York: St. Martin's Press.

Parker, Andrew. 2004. Two-Speed Europe: Why 1 Million Jobs Will Move Offshore. Cambridge, MA: Forester Research.

Pollina, Ronald. 2003. "Can We Maintain the American Dream?" *Economic Development Journal* 2(3):54–58.

Reskin, Barbara and Irene Padavic. 1994. *Women and Men at Work.* Thousand Oaks, CA: Pine Forge.

Ritzer, George. 1989. "Sociology of Work: A Metatheoretical Analysis." *Social Forces* 67(3):593–604.

———. 2000. *Sociological Theory.* Boston, MA: McGraw-Hill.

Rix, Sara. 2006. *Update on the Aged 55+ Worker: 2005.* AARP Public Policy Institute. Retrieved January 31, 2007 (http://assets.aarp.org/rgcenter/econ/dd136_worker.pdf).

Rodriguez, Eunice. 2001."Keeping the Unemployed Healthy." *American Journal of Public Health* 91(9):1403–12.

Rosser, Sue. 2005. "Through the Lenses of Feminist Theory." *Frontiers: A Journal of Women's Studies.* 26(1):1–23.

Saftner, T. J. 1998. "Temps for Hire." *Career World* 27(3):18–22.

Schmidt, D. and G. Duenas. 2002. "Incentives to Encourage Worker-Friendly Organizations." *Public Personnel Management* 31(3):293–309.

Smith, Gerri. 2004. "Made in the Maquilas—Again." *Business Week* 3896:45.

Sowinski, Lara. 2000. "Maquiladoras." *World Trade* 13(9):88–92.

Student Labor Action Project. 2007. "About SLAP." Retrieved March 16, 2008 (http://www.jwj.org/projects/slap.html).

Sullivan, Teresa. 2004. "Work-Related Social Problems." Pp. 193–208 in *Handbook of Social Problems: A Comparative International Perspective,* edited by George Ritzer. Thousand Oaks, CA: Sage Publications.

Sweatshop Watch. 2000. "Student Organizing." Retrieved September 5, 2004 (http://swatch.igc.org/swatch/codes/).

———.2003. "Frequently Asked Questions." Retrieved April 15, 2003 (www.sweatshopwatch.org/swatch/questions).

Taylor, Frederick W. 1911. *The Principles of Scientific Management.* New York: Harper.

Thottam, Jyoti. 2004. "Is Your Job Going Abroad?" *Time* 163(9):26–36.

Toossi, Mitra. 2005. "Labor Force Projections to 2014: Retiring Boomers." *Monthly Labor Review November 2005.* Washington, DC: Bureau of Labor Statistics.

Uchitelle, Louis. 2006. "Raising the Floor on Pay: States Leap Ahead of Congress in Acting on Minimum Wage." *New York Times,* December 20, pp. C1, C15.

UNICEF. 2007. *The State of the World's Children: 2007.* New York: United Nations.

UNITE. 2003. 2004. "Labor Unions UNITE and HERE to Merge." Retrieved February 29, 2004 (we3ww.uniteunion.org/pressbox/merger.cfm).

United Nations Office on Drugs and Crime—Anti-Human Trafficking Unit. 2006. *Trafficking in Persons: Global Patterns.* New York: United Nations Office on Drugs and Crime.

United Professionals. 2006. "About United Professionals." Retrieved February 4, 2007 (http://www.unitedprofessionals.org/about/).

U.S. Bureau of Labor Statistics. 2003a. *News: The Employment Situation: February 2002* (Report USDL 03–99). Washington, DC: U.S. Department of Labor.

———. 2003b. Table 3. Retrieved February 1, 2004 (www.bls.gov/emp/emplab2000–03.pdf).

———. 2004. "The Employment Situation Summary." Retrieved September 3, 2004 (www.bls.gov/news.release/empsit.nr0.htm).

———. 2005. "Contingent Workers." *Monthly Labor Review* 128(8):2.

U.S. Bureau of Labor Statistics. 2006a. *Highlights of Women's Earnings in 2005* (Report 995). Washington, DC: U.S. Department of Labor.

————. 2006b. "Census of Fatal Occupational Injuries Summary, 2005." Retrieved February 3, 2007 (http://www.bls.gov/news.release/cfoi.nr0.htm).

————. 2006c. "Workplace Injuries and Illnesses in 2005." Retrieved February 3, 2007 (http://www.bls.gov/news.release/pdf/osh.pdf).

————. 2007a. "The Employment Situation December 2006." Retrieved January 31, 2007 (http://www.bls.gov/news.release/empsit.nr0.htm).

————. 2007b. "Textile, Apparel and Furnishing Operations." Retrieved March 16, 2008 (http://www.bls.gov/oco/ocos233.htm).

U.S. Census Bureau. 1951. *Statistical Abstract of the United States* (Table 203). Washington, DC: Author.

————. 1960. *Statistical Abstract of the United States* (Table 274). Washington, DC: Author.

————. 1966. *Statistical Abstract of the United States* (Table 319). Washington, DC: Author.

U.S. Department of Labor. 1996. "Dynamic Change in the Garment Industry: How Firms and Workers Can Survive and Thrive." Retrieved April 26, 2003 (www.dol.gov/esa/forum/report.htm).

————. 2000. "Garment Enforcement Report: October 2000–December 2000." Retrieved April 27, 2003 (www.dol.gov/esa/garment/garment21.htm).

————. 2002. "Minimum Wage Laws in the States." Retrieved March 22, 2003 (www.dol.gov/esa/minwage/america.htm).

————. 2003. "WANTO Grant Award 9-30-03." Retrieved March 16, 2008 (http://www.dol.gov/wb/03awards.htm).

————. 2006. "Wage and Hour Collects $172 Million in Back Wages for over 246,000 Employees in Fiscal Year 2006." Retrieved February 3, 2007 (http://www.dol.gov/esa/whd/statistics/200631.htm).

U.S. Department of Labor, Women's Bureau. 2000. "Facts of Working Women: Hot Jobs for the 21st Century." Retrieved March 22, 2003 (www.dol.gov/wb/wb_pubs/hotjobs02).

————. 2005. "Employment Status of Men and Women in 2005." Retrieved January 31, 2007 (www.dol.gov/wb/factsheets/Qf-ESWM05.htm).

Waldman, Amy. 2003. "More 'Can I Help You?' Jobs Migrate from U.S. to India." *New York Times,* May 11, p. 4.

Weidenbaum, Murray. 2006. "Globalization: Wonder Land or Waste Land?" Pp. 53–60 in *Globalization: The Transformation of Social Worlds,* edited by D. Stanley Eitzen and Maxine Baca Zinn. Belmont, CA: Thomson Wadsworth.

Weller, Christian and Adam Hersh. 2006. "Free Markets and Poverty." Pp. 69–73 in *Globalization: The Transformation of Social Worlds,* edited by D. Stanley Eitzen and Maxine Baca Zinn. Belmont, CA: Thomson Wadsworth.

Yergin, Daniel. 2006. "Globalization Opens Door to New Dangers." Pp. 30–31 in *Globalization: The Transformation of Social Worlds,* edited by D. Stanley Eitzen and Maxine Baca Zinn. Belmont, CA: Thomson Wadsworth.

Zeitlin, Irving. 1997. *Ideology and the Development of Sociological Theory.* Englewood Cliffs, NJ: Prentice Hall.

Zhou, Min. 1993. "Underemployment and Economic Disparities Among Minority Groups." *Population Research and Policy Review* 12:139–57.

Chapter 10

Altman, Lawrence. 2006. "Children Slip Through Cracks of AIDS Efforts, W.H.O. Says." *New York Times,* August 17, p. A8.

Altman, Stuart H. and Uwe E. Reinhardt. 1996. "Where Does Health Care Reform Go From Here? An Unchartered Odyssey." Pp. xxi–xxxii in *Strategic Choices for a Changing Health Care System,* edited by Stuart H. Altman and Uwe E. Reinhardt. Chicago: Health Administration Press.

American Diabetes Association (ADA). 2003. "Type 2 Diabetes." Retrieved March 27, 2008 (http://www.diabetes.org/type-2-diabetes.jsp).

American Medical Association (AMA). 2003. "AMA's 2003 Federal Legislative Agenda." Retrieved January 5, 2003 (http://www.ama-assn.org/ama/pub/category/7334.html).

Beatrice, D. 1996. "States and Health Care Reform: The Importance of Program Implementation." Pp. 183–206 in *Strategic Choices for a Changing Health Care System,* edited by S. Altman and U. Reinhardt. Chicago: Health Administration Press.

Becker, G. S. and C. B. Mulligan. 1994. *On the Endogenous Determination of Time Preference.* Discussion Paper no. 94-2, Economics Research Center/National Opinion Research Center, July.

Begley, C. E., L. A. Aday, D. R. Lairson, and C. H. Slater. 2002. "Expanding the Scope of Health Reform: Application in the United States." *Social Science & Medicine* 55(7):1213–29.

Binder, C., L. Purvis, D. Gross, S. Schondelmeyer, and S. Raetzman. 2007. "Trends in Manufacturer Prices of Prescription Drugs Used by Older Americans." Retrieved February 9, 2007 (http://www.aarp.org/research/health/drugs/aresearch-import-869-2004-06--IB69.html).

Braveman, P. and E. Tarimo. 2002. "Social Inequalities in Health within Countries: Not Only an Issue for Affluent Countries." *Social Science and Medicine* 54:1621–35.

Broder, J., R. Pear, and M. Freudenheim. 2002. "Problem of Lost Health Benefits Is Reaching into the Middle Class." *New York Times,* November 25, pp. A1, A17.

Brumberg, J. J. 1988. *Fasting Girls.* Cambridge, MA: Harvard University Press.

Buse, Uwe. 2006. "A Woman's Fight to Save the Poor from Black Fever." *Spiegel Online,* December 21.

Calnan, M. 1987. *Health & Illness: The Lay Perspective.* London, England: Tavistock.

Centers for Disease Control. 2006. "Diabetes Fact Sheet." Retrieved February 9, 2007 (www.cdc.gov/diabetes/pubs/pdf/ndfs_2005.pdf).

———. 2007. "Key facts about Avian Influenza." Retrieved May 24, 2007 (http://www.cdc.gov/flu/avian/gen-info/facts.htm).

Centers for Medicare and Medicaid Services. 2006. "National Health Expenditure Data." Retrieved February 9, 2007 (www.cms.hhs.gov/NationalHealthExpendData/downloads/highlights.pdf).

Citizens Council on Health Care. 2003. "Summary of Minnesota's 1992 Health Care Reform Law." Retrieved March 11, 2003 (www.cchconline.org/privacy/mncaresumm.php3).

Cockerman, W. 2004. "Health as a Social Problem." Pp. 281–97 in *Handbook of Social Problems: A Comparative International Perspective,* edited by George Ritzer. Thousand Oaks, CA: Sage Publications.

Cockerman, W. and M. Glasser. 2001. "Epidemiology." Pp. 1–2 in *Readings in Medical Sociology,* edited by W. Cockerman and M. Glasser. Upper Saddle River, NJ: Prentice Hall.

Collins, K. S., D. L. Hughes, M. M. Doty, B. L. Ives, J. N. Edwards, and K. Tenney. 2002. *Diverse Communities, Common Concerns: Assessing Health Care Quality for Minority Americans: Findings from the Commonwealth Fund 2001 Health Care Quality Survey.* New York: Commonwealth Fund.

Commonwealth Fund. 2003a. "Hispanics Face High Rates of Unstable Health Care Coverage, Low Rates of Preventative Care." Retrieved March 27, 2008 (http://www.commonwealthfund.org/newsroom/newsroom_show.htm?doc_id=223566).

———. 2003b. "Minority Americans Lag Behind Whites on Nearly Every Measure of Health Care Quality." Retrieved March 27, 2008 (http://www.commonwealthfund.org/newsroom/newsroom_show.htm?doc_id=223608).

Conrad, Peter. 2001a. "General Introduction." Pp. 1–6 in *The Sociology of Health and Illness: Critical Perspectives,* edited by P. Conrad. New York: Worth.

———. 2001b. "The Social and Cultural Meanings of Illness." Pp. 91–93 in *The Sociology of Health and Illness: Critical Perspectives,* edited by P. Conrad. New York: Worth.

Conrad, Peter and Valerie Leiter. 2003. "Introduction." Pp. 1–6 in *Health and Health Care as Social Problems,* edited by Peter Conrad and Valerie Leiter. Landham, MD: Rowman & Littlefield.

Davis, Karen, Cathy Schoen, Stephen C. Schoenbaum, Michelle M. Doty, Alyssa L. Holmgren, Jennifer L. Kriss, and Katherine K. Shea. 2007. *Mirror, Mirror on the Wall: An International Update on the Comparative Performance of American Health Care.* Washington, DC: The Commonwealth Fund.

DeNavas-Walt, Carmen, Bernadette Proctor, and Cheryl H. Lee. 2006. *Income, Poverty and Health Insurance Coverage in the United States: 2005* (Current Population Reports, P60–231). Washington, DC: U.S. Government Printing Office.

Esmail, Nadeem and Michael Walker. 2006. *Waiting Your Turn 16th Edition: Hospital Waiting Lists in Canada.* Vancouver, British Columbia: Fraser Institute.

Families USA. 2003. *Going without Health Insurance: Nearly One in Three Non-Elderly Americans*. Washington, DC: Robert Wood Johnson Foundation.

Federal Interagency Forum on Child and Family Statistics. 2007. *America's Children 2007*. Washington, DC: U.S. Government Printing Office.

Freudenheim, M. 2003. "States Organizing a Nonprofit Group to Cut Drug Costs." *New York Times,* January 14, pp. A1, A20.

———. 2007. "Company Clinics Cut Health Costs." *New York Times,* January 14, pp. 1, 19.

Goffman, Erving. 1961. *Asylums: Essays on the Social Situation of Mental Health*. New York: Pantheon.

Grossman, Michael and Robert Kaestner. 1997. "Effects of Education on Health." Pp. 69–123 in *The Social Benefits of Education,* edited by J. Behrman and N. Stacey. Ann Arbor: University of Michigan Press.

Hamilton, Brady, Joyce Martin, and Stephanie Ventura. 2006. "Births: Preliminary Data for 2005." Retrieved February 9, 2007 (www.cdc.gov/nchs/products/pubs/pubd/hestats/prelimbirths05/prelimbirths05.htm).

Hamilton, J. A. 1994. "Feminist Theory and Health Psychology: Tools for an Egalitarian, Women-Centered Approach to Women's Health." Pp. 56–66 in *Reframing Women's Health: Multidisciplinary Research and Practice,* edited by A. Dan. Thousand Oaks, CA: Sage Publications.

Haney, D. 2003. "Big Study Dispels More Myths about Estrogen." *News Tribune,* March 18, p. A3.

Hargreaves, Margaret, Carolyne Arnold, and William Blot. 2006. "Community Health Centers." Pp. 485–94 in *Multicultural Medicine and Health Disparities,* edited by D. Satcher and R. Pamies. New York: McGraw Hill.

Health Canada. 2003. "About Health Canada." Retrieved March 23, 2004 (www.hc-sc.gc.ca/english/about/about.html).

Heffernan, Tim. 2005. "Victoria Hale, Exec of the Year." Retrieved March 11, 2007 (http://www.keepmedia.com/pubs/Esquire/2005/12/01/1078555).

House Energy and Commerce Subcommittee on Health and Environment. 1993. *Medicaid Sourcebook: Background Data and Analysis*. Washington, DC: U.S. Government Printing Office.

Kaiser Family Foundation. 2006. "Prescription Drug Trends." Retrieved February 9, 2007 (www.kff.org/rxdrugs/upload/3057-05.pdf).

———. 2007a. "Children's Coverage and SCHIP Reauthorization." Retrieved August 1, 2007 (http://www.kaiseredu.org/topics_im.asp?id=704&imID=1&parentID=65).

———. 2007b. "Insurance Premium Cost Sharing and Coverage Take Up." Retrieved February 9, 2007 (http://www.kff.org/insurance/snapshot/chcm020707oth.cfm).

Kleinfield, N. R. 2006. "Modern Ways Open India's Doors to Diabetes." *New York Times,* September 13, pp. A1, A12.

Kolata, Gina. 2007. "A Surprising Secret to Long Life: Stay in School." *New York Times,* January 3, p. A1.

Krauss, Clifford. 2003. "Long Lines Mar Canada's Low-Cost Health Care." *New York Times,* February 13, p. A3.

Krieger, N., D. R. Williams, and N. Moss. 1997. "Measuring Social Class in U.S. Public Health Research: Concepts, Methodologies, and Guidelines." *Annual Review of Public Health* 18:341–78.

Lambert, B. L., R. L. Street, D. J. Cegala, D. H. Smith, S. Kurtz, and T. Schofeld. 1997. "Provider-Patient Communication, Patient-Centered Care, and the Mangle of Practice." *Health Communication* 9(1):27–43.

Link, B. and J. Phelan. 2001. "Social Conditions as Fundamental Causes of Disease." Pp. 3–17 in *Readings in Medical Sociology,* 2nd ed., edited by W. Cockerman and M. Glasser. Upper Saddle River, NJ: Prentice Hall.

Livingston, M. 1998. "Update on Health Care in Canada: What's Right, What's Wrong, What's Left." *Journal of Public Health Policy* 19(3):267–88.

Lock, M. 1993. *Encounters with Aging: Mythologies of Menopause in Japan and North America*. Berkeley: University of California Press.

Lurie, Nicole and Tamara Dubowitz. 2007. "Health Disparities and Access to Health." *Journal of the American Medical Association* 297(10):1118–21.

Markens, S. 1996. "The Problem of 'Experience': A Political and Cultural Critique of PMS." *Gender and Society* 10:42–58.

Marmor, T. and J. Mashaw. 2001. "Canada's Health Insurance and Ours: The Real Lessons, the Big Choices." Pp. 470–80 in *The Sociology of Health and Illness,* edited by P. Conrad. New York: Worth.

Marmot, Michael. 2005. "Social Determinants of Health Inequalities." *Lancet* 365:1099–1104.

Men's Health Network. 2002. "The Men's Disease Awareness and Prevention Project." Retrieved December 8, 2002 (www.menshealthnetwork.org/ProgramAreas/Prevention).

Miniño, Arialdi, Melonie Heron, Sherry L. Murphy, and Kenneth D. Kochanek. 2006. "Deaths: Final Data for 2004." Retrieved February 9, 2007 (www.cdc.gov/nchs/products/pubs/pubd/hestats/finaldeaths04/finaldeaths04.htm).

Murphy, Michael, Martin Bobak, Amanda Nicholson, Michael Marmot, and Richard Rose. 2006. "The Widening Gap in Mortality by Educational Level in the Russian Federation, 1980–2001." *American Journal of Public Health* 96(7):1293–99.

National Center for Health Statistics. 2006. *Health, United States, 2006 with Chartbook on Trends in the Health of Americans*. Hyattsville, MD: Author.

National Women's Health Information Center (NWHIC). 2002. "Latina Women's Health." Retrieved December 8, 2002 (www.4woman.gov/faq).

Oberlander, J. 2002. "The U.S. Health Care System: On a Road to Nowhere." *Canadian Medical Association Journal* 167(2):163–68.

Organisation for Economic Co-operation and Development. 2007. *OECD Health Data 2007*. Paris, France: OECD.

Parsons, Talcott. 1951. *The Social System*. New York: Free Press.

Pear, Robert. 2002. "Pennsylvania Struggles to Repair Model Prescription Aid Program." *New York Times,* July 13, pp. A1, A8.

———. 2007a. "Lacking Health Coverage, Good Life Turns Fragile." *New York Times,* March 5, pp. A1, A17.

———. 2007b. "Governors Worry Over Money for Child Health Program." *New York Times,* February 25, p. A15.

———. 2007c. "Child Care Splits White House and States." *New York Times,* February 27, pp. A1, A14.

Peterson, S., M. Heesacker, and R. Schwartz. 2001. "Physical Illness: Social Construction or Biological Imperative?" *Journal of Community Health Nursing* 18(4):213–22.

Phalen, K. 2000. "A Community of Men: Project Brotherhood, a Black Men's Clinic." Retrieved March 26, 2003 (www.ama-assn.org/amednews/2000/10/02/hlsa1002.htm).

Project Brotherhood. 2006. "History." Retrieved February 8, 2007 (http://www.projectbrotherhood.net/beta/aboutus.htm).

Richman, J. and L. Jason. 2001. "Gender Biases Underlying the Social Construction of Illness States: The Case of Chronic Fatigue Syndrome." *Current Sociology* 49(3):15–40.

Roth, Kimberlee. 2006. "A Love of Science and a Vision to Save Millions of Lives Make Her Day." Retrieved March 11, 2007 (http://www.oneworldhealth.org/pdf/Chron_of_Phelan.pdf).

Sacks, H., T. Kutyla, and S. Silow-Carroll. 2002. *Toward Comprehensive Health Coverage for All: Summaries of 20 State Planning Grants* (U.S. Health Resources and Services Administration, Report 577). Washington, DC: Commonwealth Fund.

Save the Children. 2007. *State of the World's Mothers 2006*. New York: Save the Children.

Seccombe, K. and C. Amey. 2001. "Playing by the Rules and Losing: Health Insurance and the Working Poor." Pp. 323–39 in *Readings in Medical Sociology,* edited by W. Cockerman and M. Glasser. Upper Saddle River, NJ: Prentice Hall.

Silverstein, Ken. 1999. "Millions for Viagra, Pennies for Diseases of the Poor." *The Nation,* July 19, pp. 13–19.

Singer, Natasha. 2007. "Is the 'Mom Job' Really Necessary?" *New York Times,* October 4, pp. E1, E3.

Smith, B. 1987. *Digging Our Own Grave: Coal Miners and the Struggle over Black Lung Disease*. Chicago, IL: University of Chicago Press.

Starr, Paul. 1982. *The Social Transformation of American Medicine*. New York: Basic Books.

Street, R. 1991. "Information-Giving in Medical Consultations: The Influence of Patients' Communicative Styles and Personal Characteristics." *Social Science and Medicine* 32:541–48.

Szasz, Thomas. 1960. "The Myth of Mental Illness." *American Psychologist* 15:113–18.

Tanner, Lindsey. 2003. "Barbershop/Clinic Serves Black Men in Familiar Setting." *News Tribune,* November 27, p. A12.

Taylor, M. 1990. *Insuring National Health Care: The Canadian Experience*. Chapel Hill: University of North Carolina Press.

Testa, K. 2003. "More Drug Buyers Turn to Canada." *News Tribune,* December 10, p. A9.

U.S. Census Bureau. 2002."Motherhood: The Fertility of American Women, 2000." *Population Profile of the United States: 2000* (Internet Release). Retrieved February 22, 2003 (http://www.census.gov/population/popprofile/2000/chap04.pdf).

Waldron, I. 2001. "What Do We Know about Causes of Sex Differences in Mortality? A Review of the Literature." Pp. 37–49 in *The Sociology of Health and Illness: Critical Perspectives,* edited by A. Dan. New York: Worth.

Weitz, Rose. 2001. *The Sociology of Health, Illness, and Health Care: A Critical Approach.* Belmont, CA: Wadsworth/Thompson Learning.

Wells, Susan. 2006. "The Doctor Is In-House." *HR Magazine* April:48–54.

Wilkinson, R. G. 1996. *Unhealthy Societies: The Afflictions of Inequality.* London, England: Routledge.

World Health Organization. 2007. "Avian Influenza (Bird Flu) Fact Sheet." Retrieved May 24, 2007 (http://www.who.int/mediacentre/factsheets/avian_influenza/en/index.html).

Chapter 11

Action Coalition for Media Education. 2004. *Questioning the Media: Ten Basic Principles of Media Literacy Education.* Albuquerque, NM: Author.

Alexander, A. and J. Hanson. 1995. *Taking Sides: Clashing Views on Controversial Issues in Mass Media and Society.* Guilford, CT: Dushkin.

Altheide, David L. 1997. "The News Media, the Problem Frame, and the Production Fear." *Sociological Quarterly* 38(4):647–68.

American Academy of Pediatrics, Committee on Public Education. 2001a. "Children, Adolescents, and Television." *Pediatrics* 107(2):423–26.

———. 2001b. "Media Violence." *Pediatrics* 108(5):1222–26.

Anderson, C., L. Berkowitz, E. Donnerstein, L. R. Huesmann, J. Johnson, D. Linz, N. M. Malamuth, and E. Wartella. 2003. "The Influence of Media Violence on Youth." *Psychological Science in the Public Interest* 4(3):81–110.

Anderson, Craig and Brad Bushman. 2001. "Effects of Violent Video Games on Aggressive Behavior, Aggressive Cognition, Aggressive Affect, Physiological Arousal, and Prosocial Behavior." *Psychological Science* 12(5):353–59.

Bagdikian, Ben. 1997. *The Media Monopoly.* Boston, MA: Beacon Press.

Ball-Rokeach, Sandra and Muriel Cantor, eds. 1986. *Media, Audience, and Social Structure.* Beverly Hills, CA: Sage Publications.

Barner, Mark. 1999. "Sex-Role Stereotyping in FCC-Mandated Children's Educational Television." *Journal of Broadcasting and Electronic Media* 43:551–64.

Bauder, David. 2006. "Homes Have More TVs Than People." *News Tribune,* September 23, p. A11.

Case, Carl and Kimberly Young. 2002. "Employee Internet Management: Current Business Practices and Outcomes." *CyberPsychology and Behavior* 5(4):355–61.

Central Intelligence Agency. 2007. "Rank Order: Internet Users." Retrieved July 25, 2007 (https://www.cia.gov/library/publications/the-world-factbook/rankorder/2153rank.html).

Chen, W. 1999. "Web 547 Is Launched to Combat Pornography in Cyberspace." *China Times,* July 22, p. 7.

Chomsky, Noam. 1989. *Necessary Illusions: Thought Control in Democratic Societies.* Boston, MA: South End Press.

Cohen, Stanley. 2002. *Folk Devils and Moral Panics.* New York: Routledge.

Committee to Protect Journalists. 2008. "Journalists Killed in 2007." Retrieved March 21, 2008 (http://www.cpj.org/deadly/killed07.html).

Corliss, R. and J. McDowell. 2001. "Go Ahead, Make Her Day." *Time* 157:64–66.

Croteau, David and William Hoynes. 2000. *Media Society: Industries, Images, and Audiences.* Thousand Oaks, CA: Pine Forge.

———. 2001. *The Business of Media: Corporate Media and the Public Interest.* Thousand Oaks, CA: Pine Forge.

Davis, Geena. 2006. "Where Are the Girls?" Retrieved July 25, 2007 (http://www.commonsenseblog.org/archives/2006/02/where_are_the_g.php).

DeBell, Matthew and Chris Chapman. 2006. *Computer and Internet Use by Students in 2003 (NCES 2006-065)*. Washington, DC: National Center for Education Statistics.

Dietz, William and Steven Gortmaker. 1985. "Do We Fatten Our Children at the Television Set? Obesity and Television Viewing in Children and Adolescents." *Pediatrics* 75(5):807–12.

"Doctors Diagnose Extreme Internet Use." 2003. *Biotech Week,* August 27, pp. 512–15.

Drori, Gili. 2004. "The Internet as a Global Social Problem." Pp. 433–50 in *Handbook of Social Problems: A Comparative International Perspective,* edited by George Ritzer. Thousand Oaks, CA: Sage Publications.

Durham, Meenakshi Gigi 2003. "The Girling of America: Critical Reflections on Gender and Popular Communication." *Popular Communication* 1(1):23–31.

Epstein, Edward Jay. 1981. "The Selection of Reality." Pp. 119–32 in *What's News,* edited by E. Abel. San Francisco, CA: Institute for Contemporary Studies.

Eschholz, Sarah, Jana Bufkin, and Jenny Long. 2002. "Symbolic Reality Bites: Women and Racial/Ethnic Minorities in Modern Film." *Sociological Spectrum* 22:299–334.

Fairness and Accuracy in Reporting (FAIR). 2004. "What's FAIR?" Retrieved March 23, 2004 (http://www.fair.org/index.php?page=100).

Faith, M. S., N. Berman, M. Heo, A. Pietrobelli, D. Gallagher, L. H. Epstein, M. T. Eiden, and D. B. Allison. 2001. "Effects of Contingent Television on Physical Activity and Television Viewing on Obese Children." *Pediatrics* 107(5):1043–48.

Federal Communications Commission (FCC). 2004a. "About the FCC." Retrieved March 21, 2004 (www.fcc.gov/aboutus.html).

———. 2004b. "Obscene, Indecent, and Profane Broadcasts." Retrieved March 21, 2008 (http://www.fcc.gov/cgb/consumerfacts/obscene.html).

———. 2004c. "V-Chip: Viewing Television Responsibly." Retrieved March 21, 2004 (www.fcc.gov/vchip/# guidelines).

Finkelhor, David, Kimberly J. Mitchell, and Janis Wolak. 2000. *Online Victimization: A Report on the Nation's Youth*. Washington, DC: National Center for Missing and Exploited Children.

Galtung, J. and M. Ruge. 1973. "Structuring and Selecting News." Pp. 67–72 in *The Manufacture of News,* edited by S. Cohen and J. Young. London, England: Constable.

Gans, Herbert. 1979. *Deciding What's News: A Study of CBS Evening News, NBC Nightly News, Newsweek, and* Time. New York: Pantheon Books.

Gantz, Walter, Nancy Schwartz, James Angelini, and Victoria Rideout. 2007. *Food for Thought: Television Food Advertising for Children in the United States.* Menlo Park, CA: Henry J. Kaiser Family Foundation.

Glassner, Barry. 1997. *The Culture of Fear: Why Americans Are Afraid of the Wrong Things.* New York: Basic Books.

Griffiths, Mark. 2002. "Occupational Health Issues Concerning Internet Use in the Workplace." *Work and Stress* 16(4):283–86.

———. 2003. "Internet Abuse in the Workplace: Issues and Concerns for Employers and Employment Counselors." *Journal of Employment Counseling* 40:87–96.

Greeson, Larry and Rose Williams. 1986. "Social Implications of Music Videos for Youth: An Analysis of the Content and Effects of MTV." *Youth and Society* 18(2):177–89.

Guillén, Mauro and Sandra Suárez. 2005. "Explaining the Global Digital Divide: Economic, Political and Sociological Drivers of Cross-National Internet Use." *Social Forces* 84(2):681–708.

Gunkel, David J. 2003. "Second Thoughts: Towards a Critique of the Digital Divide." *New Media and Society* 5(4):499–522.

Gurevitch, Michael and Mark R. Levy. 1985. *Mass Communication Review Yearbook* (Vol. 5). Beverly Hills, CA: Sage Publications.

Hapgood, F. 1996. "Sex Sells." *Inc. Technology* 4:45–51.

The Henry J. Kaiser Family Foundation. 2001. *Generation Rx.com, Young People Use the Internet for Health Information*. Menlo Park, CA: Author.

Howard, I. 2002. "Power Sources: On Party, Gender, Race and Class, TV News Looks to the Most Powerful Groups." Retrieved March 20, 2003 (www.fair.org/extra/0205/power_sources.html).

Huesmann, L. R., J. Moise-Titus, C. L. Podolski, and L. D. Eron. 2003. "Longitudinal Relations between Children's Exposure to TV Violence and Their Aggressive and Violent Behavior in Young Adulthood: 1977–1992." *Developmental Psychology* 39(2):201–22.

Huesmann, L., L. E. Rowell, R. Klein, P. Brice, and P. Fischer. 1983. "Mitigating the Imitation of Aggressive Behaviors by Children's Attitudes about Media Violence." *Journal of Personality and Social Psychology* 44:899–910.

Huxley, Aldous. 1932. *Brave New World.* London: Chatto and Windus.

Kellner, Douglas M. 1995. *Media Culture: Cultural Studies, Identity, and Politics between the Modern and the Postmodern.* New York: Routledge.

Kelly, Joe and Stacy Smith. 2006. *Where the Girls Aren't: Gender Disparity Saturates G-Rated Films.* Duluth, MN: Dads and Daughters Foundation.

Kelly, K., K. Clark, and L. Kulman. 2004. "Trash TV." *U.S. News and World Report* 136(6):48–51.

Klite, P., R. A. Bardwell, and J. Salzman. 1997. "Local TV News: Getting Away with Murder." *Harvard International Journal of Press/Politics* 2:102–12.

Kuipers, Giselinde. 2006. "The Social Construction of Digital Danger: Debating, Defusing and Inflating the Moral Dangers on Online Humor and Pornography in the Netherlands and the United States." *New Media Society* 8:379–400.

Kunkel, D., E. Biely, K. Eyal, K. Cope-Farrar, E. Donnerstein, and R. Fandrich. 2003. *Sex on TV 3.* Washington, DC: Henry J. Kaiser Family Foundation.

Lazarus, Wendy and Francisco Mora. 2000. *Online Content for Low-Income and Underserved Americans: The Digital Divide's New Frontier.* Santa Monica, CA: Children's Partnership.

Li, J. 2000. "Cyberporn: The Controversy." Retrieved July 26, 2007 (http://www.firstmonday.org/issues/issue5_8/li/).

Lo, V. and R. Wei. 2002. "Third-Person Effect, Gender, and Pornography on the Internet." *Journal of Broadcasting & Electronic Media* 46(1):13–33.

MAGIC. 2004. "About Magic." Retrieved March 19, 2004 (www.unicef.org/magic/briefing/about.html).

Martin, Andrew. 2007. "Kellogg to Phase Out Some Food Ads to Children." *New York Times,* June 14, pp. C1-C2.

McNair, B. 1998. *The Sociology of Journalism.* London, England: Oxford University Press.

Media Awareness Network. 2004. "What Is Media Literacy?" Retrieved March 20, 2004 (www.media-awareness.ca/english/teachers/media_literacy/what_is_media_literacy.cfm).

Miller, Mark Crispin. 2002. "What's Wrong with This Picture?" *The Nation* 274(1):18–22.

Nathanson, Amy I. 1999. "Identifying and Explaining the Relationship between Parental Mediation and Children's Aggression." *Communication Research* 26(2):124–43.

Parekh, Angana. 2001. "Bringing Women's Stories to a Reluctant Mainstream Press." *Nieman Reports* 55(4):90–92.

Parents Television Council. 2003. "The Blue Tube: Foul Language on Prime Time Network TV." Retrieved March 14, 2004 (www.parentstv.org/ptc/publications/reports/stateindustrylanguage/main.asp).

Peterson, T. 1981. "Mass Media and Their Environments: A Journey into the Past." Pp. 13–32 in *What's News,* edited by E. Abel. San Francisco: Institute for Contemporary Studies.

Pew Research Center for the People and the Press. 2002. "News Media's Improved Image Proves Short-Lived." Retrieved March 7, 2004 (http://people-press.org/reports/display.php3?ReportID=159).

———. 2005. "People More Critical of Press, but Goodwill Persists." Retrieved August 1, 2007 (http://people-press.org/reports/display.php3?ReportID=248).

———. 2007. *What Americans Know: 1989–2007.* Retrieved March 21, 2008 (http://people-press.org/reports/pdf/319.pdf).

Podlas, K. 2000. "Mistresses of Their Domain: How Female Entrepreneurs in Cyberporn Are Initiating a Gender Power Shift." *CyberPsychology & Behavior* 3(5):847–54.

Porter, William E. 1981. "The Media Baronies: Bigger, Fewer, More Powerful." Pp. 97–118 in *What's News,* edited by E. Abel. San Francisco: Institute for Contemporary Studies.

Postman, N. 1989. *Amusing Ourselves to Death.* London, England: Methuen.

Potter, R. H. and L. A. Potter. 2001. "The Internet, Cyberporn, and Sexual Exploitation of Children: Media Moral Panics and Urban Myths for Middle-Class Parents?" *Sexuality and Culture* 5(3):31–51.

Powell, Kimberly and Lori Abels. 2002. "Sex Role Stereotypes in TV Programs Aimed at the Preschool Audience: An Analysis of Teletubbies and Barney and Friends." *Women and Language* 25(1):14–22.

Project for Excellence in Journalism. 2004. *The State of the News Media 2004*. Retrieved March 21, 2004 (www.stateofthenewsmedia.org/narrative_overview_publicattitudes.asp?media=1).

Schorr, Daniel. 1998. "Mother Teresa and Diana." *Christian Science Monitor* 90(193):15.

Schudson, M. 1986. "The Menu of Media Research." Pp. 43–48 in *Media, Audience, and Social Structure,* edited by S. Ball-Rokeach and Muriel Cantor. Beverly Hills, CA: Sage Publications.

Servon, L. 2002. *Bridging the Digital Divide: Technology, Community, and Public Policy*. Boston, MA: Blackwell.

Seymour-Ure, C. 1974. *The Political Impact of Mass Media*. Beverly Hills, CA: Sage Publications.

Shaw, D. L. and M. E. McCombs. 1997. *The Emergence of American Political Issues: The Agenda Setting Function of the Press*. St. Paul, MN: West.

Signiorelli, Nancy. 1990. *Sourcebook on Children and Television*. New York: Greenwood Press.

Snider, M. 2003. "Wired to Another World." *Maclean's* 116(9):23–24.

TV Turnoff Network. 2004. "TV Facts and Figures." Retrieved March 1, 2004 (www.tvturnoff.org/facts.htm).

United Nations Development Programme. 2001. *Human Development Report 2001: Making New Technologies Work for Human Development*. New York: Author.

U.S. Census Bureau. 2003. *Statistical Abstract of the United States*. Washington, DC: Author.

———. 2007. *Statistical Abstract of the United States*. Washington, DC: Author.

VandeBerg, L. R. 1991. "Using Television to Teach Courses in Gender and Communication." *Communication Education* 40(1):105–11.

Van den Bulck, J. 2000. "Is Television Bad for Your Health? Behavior and Body Image of the Adolescent 'Couch Potato.'" *Journal of Youth and Adolescence* 29(3):273–88.

Vartanova, Elena. 2002. "Digital Divide and the Changing Political/Media Environment of Post-Socialist Europe." *International Communication Gazette* 64:449–65.

Vessey, J. and J. Lee. 2000. "Violent Video Games Affecting Our Children." *Pediatric Nursing* 26(6):607–10.

Warschauer, M. 2003. *Technology and Social Inclusion*. Cambridge, MA: MIT Press.

Chapter 12

Abatemarco, Diane, Bernadette West, Vesna Zec, Andrea Russo, Persis Sosiak, and Vedran Mardesic. 2004. "Project Northland in Croatia: A Community Based Adolescent Alcohol Prevention Intervention." *Journal of Drug Education* 34(2):167–78.

Abramsky, Sasha. 2003. "The Drug War Goes Up in Smoke." *The Nation* 277(5):25–28.

Alaniz, Maria Luisa. 1998. "Alcohol Availability and Targeted Advertising in Racial/Ethnic Minorities Communities." *Alcohol Health and Research World* 22(4):286–89.

Alcohol Information Scotland. 2007. *Alcohol Information Scotland, 2007*. Edinburgh, Scotland: IDS Publications.

Alcohol Policies Project. 2001. "National Poll Shows 'Alcopop' Drinks Lure Teens." Retrieved April 27, 2007 (http://www.cspinet.org/booze/alcopops_press.htm).

American Lung Association. 2006. "Smoking Fact Sheet." Retrieved February 21, 2007 (www.lungusa.org/site/pp.asp?c=dvLUK9O0E&b=39853).

American Medical Association. 2004. "Girlie Drinks . . . Women's Diseases." Retrieved April 27, 2007 (http://www.cslep.org/CSLEP/publications/girlie_drinks_survey.pdf).

Becker, Howard. 1963. *Outsiders: Studies in the Sociology of Deviance*. New York: Free Press.

Beckett, Katherine. 1995. "Fetal Rights and 'Crack Moms': Pregnant Women in the War on Drugs." *Contemporary Drug Problems* 22:587–612.

Boyd, Carol, Sean Esteban McCabe, and Hannah d'Arcy. 2003. "Ecstasy Use among College Undergraduates: Gender, Race, and Sexual Identity." *Journal of Substance Abuse Treatment* 24:209–15.

Brower, Aaron. 2002. "Are College Students Alcoholics?" *Journal of American College Health* 50:253–55.

Bureau of Justice Statistics. 2006. "Drug and Crime Facts: Enforcement." Retrieved February 21, 2007 (http://www.ojp.usdoj.gov/bjs/dcf/tables/arrtot.htm).

Butler, E. R. 1993. "Alcohol Use by College Students: A Rite of Passage Ritual." *NASPA Journal* 31(1):48–55.

Caetano, Raul, Catherine Clark, and Tammy Tam. 1998. "Alcohol Consumption among Racial/Ethnic Minorities: Theory and Research." *Alcohol Health and Research World* 22(4):233–38.

Caetano Raul and L. A. Kaskutas. 1995. "Changes in Drinking Patterns among Whites, Blacks, and Hispanics, 1984–1992." *Journal of Studies on Alcohol* 50:15–23.

Caetano, Raul, John Schafer, and Carol Cunradi. 2001. "Alcohol Related Intimate Partner Violence among White, Black, and Hispanic Couples in the United States." *Alcohol Research and Health* 25(1):58–65.

Campbell-Grossman, Christie, Diane Hudson, and Margaret Fleck. 2003. "Chewing Tobacco Use: Perceptions and Knowledge in Rural Adolescent Youths." *Issues in Comprehensive Pediatric Nursing* 26:13–21.

Chen, Kevin and Denise B. Kandel. 1995. "The Natural History of Drug Use from Adolescence to the Mid-Thirties in a General Population Sample." *American Journal of Public Health* 85:41–47.

Clapp, John, Audrey Shillington, and Lance Segars. 2000. "Deconstructing Contexts of Binge Drinking among College Students." *American Journal of Drug and Alcohol Abuse* 26(1):139–54.

Collins, R. Lorraine and Lily McNair. 2003. "Minority Women and Alcohol Use." Retrieved October 5, 2003 (http://pubs.niaaa.nih.gov/publications/arh26-4/251-256.htm).

Community Anti-Drug Coalitions of America. 2003. "Beloit's Safe and Drug Free Schools Program." Retrieved August 10, 2003 (www.cadca.org/images/PromPromise.gif).

Crispo, A., P. Brennan, K. H. Jockel, A. Schaffrath-Rosario, H.E. Wichmann, F. Nyberg, L. Simonato, F. Mereletti, F. Forastiere, P. Boffetta, and S. Darby. 2004. "The Cumulative Risk of Lung Cancer among Current, Ex- and Never-smokers in European Men." *British Journal of Cancer* 91:1280–86.

Cussen, Meaghan and Walter Block. 2000. "Legalize Drugs Now! An Analysis of the Benefits of Legalized Drugs." *American Journal of Economics and Sociology* 59(3):525–36.

Drug Policy Alliance. 2003a. "About the Alliance." Retrieved August 2, 2003 (www.drugpolicy.org/about/).

———. 2003b. "Women and the War on Drugs." Retrieved August 2, 2003 (www.drugpolicy.org/communities/women).

European Lung Association. 2006. "First Ever EU Figures on Passive Smoking Deaths." Retrieved April 28, 2007 (http://www.european-lung-foundation.org/uploads/Document/WEB_CHEMIN_285_1142589119.pdf).

Fellner, Jamie. 2000. *Punishment and Prejudice: Racial Disparities in the War on Drugs* (Vol. 2, No. 2). New York: Human Rights Watch.

Filmore, K. M., J. M. Golding, E. V. Leino, M. Motoyoshi, C. Shoemaker, H. Terry, C. Ager, and H. Ferrer. 1997. "Patterns and Trends in Women's and Men's Drinking." Pp. 21–48 in *Gender and Alcohol,* edited by R. W. Wilsnack and S. C. Wilsnack. New Brunswick, NJ: Rutgers Center of Alcohol Studies.

Frone, M. R. 2004. "Alcohol, Drugs and Workplace Safety Outcomes: A View from a General Model of Employee Substance Use and Productivity." Pp. 127–56 in *The Psychology of Workplace Safety,* edited by J. Barling and M. R. Frone. Washington, DC: American Psychological Association.

Glenn, B. 2000. "A Crusader against Underage Drinking." Retrieved June 9, 2004 (www.madd.org/activism/0,1056,4028_print,00.html).

Goode, Erich. 2004. "Drug Use as a Global Social Problem." Pp. 494–520 in *Handbook of Social Problems: A Comparative international Perspective,* edited by George Ritzer. Thousand Oaks, CA: Sage Publications.

Greenfeld, Lawrence and Maureen Henneberg. 2001. "Victim and Offender Self Reports of Alcohol Involvement in Crime." *Alcohol Research and Health* 25(1):20–31.

Hawken, Paul. 1993. *The Ecology of Commerce: A Declaration of Sustainability.* New York: HarperCollins.

Herd, Denise and Joel Grube. 1996. "Black Identity and Drinking in the U.S.: A National Study." *Addiction* 91(6):845–57.

Hesselbrock, M. N. and V. M. Hesselbrock. 1997. "Gender, Alcoholism, and Psychiatric Comorbidity." Pp. 49–71 in *Gender and Alcohol,* edited by R. W. Wilsnack and S. C. Wilsnack. New Brunswick, NJ: Rutgers Center of Alcohol Studies.

Hoffman, John and Cindy Larison. 1999. "Worker Drug Use and Workplace Drug Testing Programs: Results from the 1994 National Household Survey on Drug Use." *Contemporary Drug Problems* 26:331–54.

Holmes, Malcolm and Judith Antell. 2001. "The Social Construction of American Indian Drinking: Perceptions of American Indian and White Officials." *Sociological Quarterly* 42(2):151–73.

Inciardi, James. 1999. "American Drug Policy: The Continuing Debate." Pp. 1–8 in *The Drug Legalization Debate,* edited by James Inciardi. Thousand Oaks, CA: Sage Publications.

Johnston, L. D., P. M. O'Malley, and J. G. Bachman. 2001. *Monitoring the future: National Survey Results on Drug Use, 1975–2000* (Vol. II, NIH Publication No. 01–4925). Bethesda, MD: National Institute on Drug Abuse.

Klingner, Donald, Gary Roberts, and Valerie Patterson. 1998. "The Miami Coalition Surveys of Employee Drug Use and Attitudes: A Five-Year Retrospective." *Public Personnel Management* 27(2):201–22.

Larson, S. L., J. Eyerman, M. S. Foster, and J. C. Gfroerer. 2007. *Worker Substance Use and Workplace Policies and Programs (DHHS Publication SMA 07-4273).* Rockville, MD: Substance Abuse and Mental Health Services Administration, Office of Applied Studies.

Lederman, Linda, Lea Stewart, Fern Goodhart, and Lisa Laitman. 2003. "A Case against 'Binge' as a Term of Choice: Convincing College Students to Personalize Messages about Dangerous Drinking." *Journal of Health Communications* 8:79–91.

Leon, Dean and Jim McCambridge. 2006. "Liver Cirrhosis Mortality Rates in Britain from 1950 to 2002: An Analysis of Routine Data." *Lancet* 367:52–56.

Leonard, K. E. and J. C. Rothbard. 1999. "Alcohol and the Marriage Effect." *Journal of Studies on Alcohol* (Suppl. 13):139–46.

Lewinsohn, Peter, Richard Brown, John Seeley, and Susan Ramsey. 2000. "Psychosocial Correlates of Cigarette Smoking Abstinence, Experimentation, Persistence and Frequency during Adolescence." *Nicotine and Tobacco Research* 2:121–31.

MacDonald, Scott, Samantha Wells, and T. Cameron Wild. 1999. "Occupational Risk Factors Associated with Alcohol and Drug Problems." *American Journal of Drug and Alcohol Abuse* 25(2):351–69.

MacDonald, Sharon. 1994. "Whisky, Women, and the Scottish Drink Problem: A View from the Highlands." Pp. 125–44 in *Gender, Drink, and Drugs,* edited by M. McDonald. Oxford, England: Berg.

Marshall, Carolyn. 2007. "Drinks with Youth Appeal Draw Growing Opposition." *New York Times,* April 13, p. A12.

McBride, Duane, Yvonne Terry, and James Inciardi. 1999. "Alternative Perspectives on the Drug Policy Debate." Pp. 9–54 in *The Drug Legalization Debate,* edited by James Inciardi. Thousand Oaks, CA: Sage Publications.

Muthen, L. K. and B. Muthen. 2000. "The Development of Heavy Drinking and Alcohol-Related Problems from Ages 18 to 37 in a U.S. National Drinking Sample." *Journal of Studies on Alcohol* 61:290–300.

National Cancer Institute. 2002. "Smokeless Tobacco: Health and Other Effects." Retrieved October 5, 2003 (http://dccps.nci.nih.gov/tcrb/less_effects.html).

National Clearinghouse for Alcohol and Drug Information (NCADI). 2003a. "Intensive Outpatient Treatment for Alcohol and Other Drug Abuse: The Treatment Needs of Special Groups." Retrieved August 2, 2003 (www.health.org/govpubs/bkd139/8g.aspx).

———. 2003b. "Cigarettes and Other Nicotine Products." Retrieved August 2, 2003 (http://store.health .org/catalog/facts.aspx?topic=9).

National Institute on Alcohol Abuse and Alcoholism (NIAAA). 2003a. "What Is Alcoholism?" Retrieved July 22, 2003 (www.niaaa.nih.gov/faq/q-a.htm#question1).

———. 2003b. "The Creation of the National Institute on Alcohol Abuse and Alcoholism." Retrieved March 21, 2008 (http://www.niaaa.nih.gov/AboutNIAAA/OrganizationalInformation/History.htm).

———. 2003c. "NIAAA's Purpose." Retrieved July 31, 2003 (www.niaaa.nih.gov/about/purpose.htm).

———. 2004. "Alcoholism: Getting the Facts." Retrieved February 21, 2007 (pubs.niaaa.nih.gov/publications/ GettheFacts_HTML/facts.htm).

———. 2005. "Alcoholism: Natural History and Background." Retrieved February 21, 2007 (http://pubs .niaaa.nih.gov/publications/HealthDisparities/Alcoholism1.htm).

National Institute of Drug Abuse (NIDA). 1998. "Nicotine Addiction." Retrieved August 2, 2003 (www .drugabuse.gov/researchreorts/nicotine/nicotine.html).

———. 2002a. *Marijuana Abuse* (NIH Publication No. 02–3859, October 2002). Rockville, MD: Author.

National Institute of Drug Abuse (NIDA). 2002b. "What Are the Medical Complications of Cocaine Abuse?" Retrieved July 27, 2003 (www.drugabuse.gov/ResearchReports/Cocaine/cocaine3.html#medical).

———. 2002c. "What Is the Effect of Maternal Cocaine Use?" Retrieved July 27, 2003 (www.drugabuse.gov/ResearchReports/Cocaine/cocaine4.html#maternal).

———. 2003a. "NIDA Info Facts: MDMA (Ecstasy)." Retrieved July 22, 2003 (www.drugabuse.gov/Infofax/ecstasy.html).

———. 2003b. "About NIDA." Retrieved July 31, 2003 (www.drugabuse.gov/about/AboutNIDA.html).

———. 2003c. "Drug Addiction Treatment Methods." Retrieved August 2, 2003 (www.nida.nih.gov/infofax/treatmeth.html).

———. 2006. "Cigarettes and other tobacco products." Retrieved February 21, 2007 (http://www.nida.nih.gov/infofacts/tobacco.html).

———. 2007. "Crack and Cocaine." Retrieved June 27, 2007 (http://www.nida.nih.gov/Infofacts/cocaine.html).

Newman, I. 1999. *Adolescent Tobacco Use in Nebraska from Nebraska.* Lincoln: Nebraska Prevention Center for Alcohol and Drug Abuse.

Office of National Drug Control Policy (ONDCP). 2003a. "Marijuana." Retrieved July 23, 2003 (www.whitehousedrugpolicy.gov/drugfact/marijuana/index.html).

———. 2003b. "Methamphetamine." Retrieved July 23, 2003 (www.whitehousedrugpolicy.gov/drugfact/methamphetamine/index.html).

———. 2003c. "Enabling Legislation." Retrieved July 31, 2003 (www.whitehousedrugpolicy.gov/about/legislation.html).

———. 2006a. "Marijuana Overview." Retrieved February 21, 2007 (http://www.whitehousedrugpolicy.gov/drugfact/marijuana/index.html#extentofuse).

———. 2006b. "Methamphetamine." Retrieved February 21, 2007 (http://www.whitehousedrugpolicy.gov/drugfact/methamphetamine/index.html).

———. 2006c. "Club Drugs." Retrieved February 21, 2007 (http://www.whitehousedrugpolicy.gov/drugfact/club/index.html).

———. 2006d. "MDMA Abuse." Retrieved February 21, 2007 (http://www.drugabuse.gov/ResearchReports/MDMA/default.html).

O'Malley, Patrick, Lloyd Johnston, and Jerald Bachman. 1998. "Alcohol Use among Adolescents." *Alcohol Health and Research World* 22(2):85–94.

PBS. 2000. "Frontline: Drug Wars." Retrieved August 2, 2003 (http://www.pbs.org/wgbh/pages/frontline/shows/drugs/).

Pennell, Susan, Joe Ellett, Cynthia Rienick, and Jackie Grimes. 1999. *Meth Matters: Report on Methamphetamine Users in Five Western Cities.* Washington, DC: U.S. Department of Justice, Office of Justice Programs, National Institute of Justice.

Perry, C. L., C. L. Williams, K. A. Komro. 2002. "Project Northland: Long-Term Outcomes of Community Action to Reduce Adolescent Alcohol Use." *Health Education Research* 17:117–32.

Pew Research Center. 2001. "Interdiction and Incarceration Still Top Remedies." Retrieved March 21, 2008 (http://people-press.org/reports/display.php3?ReportID=16).

Plant, M. A., P. Miller, and M. L. Plant. 2005. "Trends in Drinking, Smoking and Illicit Drug Use among 15- and 16-Year Olds in the UK (1995–2003)." *Journal of Substance Use* 10(6):331–39.

Reinarman, Craig and Harry G. Levine. 1997. *Crack in America: Demon Drugs and Social Justice.* Berkeley: University of California Press.

Ritson, Bruce. 2002. "Editorial: Scotland's National Plan on Alcohol Problems." *Drugs: Education, Prevention, and Policy* 9(3):217–20.

Roberts, Dorothy. 1991. "Punishing Drug Addicts Who Have Babies: Women of Color, Equality, and the Right of Privacy." *Harvard Law Review* 104(7):1419–82.

Roman, Paul and Terry Blum. 2002. "The Workplace and Alcohol Problem Prevention." *Alcohol Research and Health* 26(1):49–57.

Scalia, John. 2001. *Federal Drug Offenders, 1999 with Trends 1984–1999* (August 2001, NCJ 187285). Washington, DC: U.S. Department of Justice, Office of Justice Programs.

Silbering, Robert. 2001. "The 'War on Drugs': A View from the Trenches." *Social Research* 68(3):890–96.

Snell, Marilyn Berlin. 2001. "Welcome to Meth Country." Retrieved July 27, 2003 (www.sierraclub.org/sierra/200101/Meth.asp).

Social Issues Research Center. 1998. *Social and Cultural Aspects of Drinking*. Retrieved September 2, 2007 (http://www.sirc.org/publik/drinking3.html#_VPID_5).

Substance Abuse and Mental Health Services Administration (SAMHSA). 2003a. *Overview of Findings from the 2002 National Survey on Drug Use and Health* (Office of Applied Statistics, NHSDA Series H-21, DHHS Publication No. SMA 03–3774). Rockville, MD: U.S. Department of Health and Human Services.

———. 2003b. "Drug Free Workplace Programs." Retrieved July 31, 2003 (www.gov/DrugFreeWP/Legal.html).

———. 2003c. "Benefits and Costs." Retrieved July 31, 2003 (www.samhsa.gov/DrugFreeWP/Benefits.html).

———. 2004. "Table B.1 Any Illicit Drug Use in the Past Month, by Age Group and State." Retrieved February 21, 2007 (http://www.oas.samhsa.gov/2k4State/appB.htm#TabB.1).

———. 2005. *National Survey on Drug Use and Health: National Findings*. Rockville, MD: U.S. Department of Health and Human Services.

Sutherland, Edwin. 1939. *Principles of Criminology*. 3rd ed. Philadelphia, PA: J. B. Lippincott.

Tanner, Lindsey. 2003. "Underage Drinkers Consume 20 Percent of Booze, Study Says." *News Tribune,* February 26, p. A7.

Task Force of the National Advisory Council on Alcohol Abuse and Alcoholism. 2002. *A Call to Action: Changing the Culture of Drinking at U.S. Colleges*. Washington, DC: National Institutes of Health, U.S. Department of Health and Human Services.

Tonry, Michael. 2004. "Crime." Pp. 465–79 in *Handbook of Social Problems: A Comparative International Perspective,* edited by George Ritzer. Thousand Oaks, CA: Sage Publications.

Trevino, Roberto and Alan Richard. 2002. "Attitudes toward Drug Legalization among Drug Users." *American Journal of Drug and Alcohol Abuse* 28(1):91–108.

United Nations Office on Drugs and Crime (UNODC). 2006. "Who Is Doing Drugs?" Retrieved June 27, 2007 (http://www.unodc.org/unodc/drug_demand_who.html).

Unterberger, Gail. 1989. "Twelve Steps for Women Alcoholics." *Christian Century* 106(37):1150–52.

U.S. Department of Labor. 2003. "Working Partners: Small Business Workplace Kit: Facts and Figures." Retrieved July 31, 2003 (www.dol.gov/asp/programs/drugs/workingpartners/Screen15.htm).

Vaughn, Christy. 2002. "Ecstasy: More Deadly Than Many Young People Know." Retrieved August 2, 2003 (www.health.org/newsroom/rep/182.aspx).

Wechsler, Henry. 1996. "Alcohol and the American College Campus." *Change* 28(4):20–25.

Wechsler, Henry, G. W. Dowdall, A. Davenport, and S. Castillo. 1995. "Correlates of College Student Binge Drinking." *American Journal of Public Health* 85:921–26.

Weisheit, Ralph and Kathrine Johnson. 1992. "Exploring the Dimensions of Support for Decriminalizing Drugs." *Journal of Drug Issues* 92(22):53–75.

Weitzman, Elissa, Alison Folkman, Kerry Folkman, and Henry Wechsler. 2003. "The Relationship of Alcohol Outlet Density to Heavy and Frequent Drinking and Drinking-Related Problems among College Students at Eight Universities." *Health and Place* 9:1–6.

The White House. 2007. "Drug Control Strategy FY 2008 Budget Summary." Retrieved February 21, 2007 (http://www.whitehousedrugpolicy.gov/publications/policy/08budget/08budget.pdf).

Widom, Cathy Spatz and Susanne Hiller-Sturmhofel. 2001. "Alcohol Abuse as a Risk Factor for and Consequence of Child Abuse." *Alcohol Research and Health* 25(1):52–57.

Williams, Carolyn and Cheryl Perry. 1998. "Lesson from Project Northland: Preventing Alcohol Problems during Adolescence." *Alcohol Health and Research World* 22(2):107–16.

Wilsnack, R. W. and S. C. Wilsnack. 1992. "Women, Work, and Alcohol: Failures of Simple Theories." *Alcoholism: Clinical and Experimental Research* 16:172–79.

Wilsnack, S. C., N. D. Vogeltanz, A. D. Klassen, and T. R. Harris. 1997. "Childhood Sexual Abuse and Women's Substance Abuse: National Survey Findings." *Journal of Studies on Alcohol* 58:264–71.

Winter, Greg. 2003. "Study Finds No Sign That Testing Deters Student Drug Use." *New York Times,* May 17, pp. A1, A12.

World Health Organization (WHO). 2004. *Global Status Report on Alcohol 2004*. Singapore: World Health Organization.

———. 2006. "Management of Substance Abuse: Facts and Figures." Retrieved February 21, 2007 (www.who.int/substance_abuse/facts/en/).

World Health Organization (WHO). 2007a. "Why Is Tobacco a Public Health Priority?" Retrieved August 8, 2007 (http://www.who.int/tobacco/health_priority/en/index.html).

———. 2007b. "Cocaine." Retrieved August 8, 2007 (http://www.who.int/substance_abuse/facts/cocaine/en/).

Yamaguchi, Ryoko, Lloyd Johnston, and Patrick O'Malley. 2003. "Relationship between Student Illicit Drug Use and School Drug-Testing Policies." *Journal of School Health* 73(4):159–64.

Ziemelis, Andris, Ronald Buckman, and Abdulaziz Elfessi. 2002. "Prevention Efforts Underlying Decreases in Binge Drinking at Institutions of Higher Education." *Journal of American College Health* 50(5):238–52.

Chapter 13

Ackerman, W. 1998. "Socioeconomic Correlates of Increasing Crime Rates in Smaller Communities." *Professional Geographer* 50(3):372–87.

Adler, Freda. 1975. *Sisters in Crime*. New York: McGraw-Hill.

Adler, Freda, Gerhard Mueller, and William Laufer. 1991. *Criminology*. New York: McGraw-Hill.

Akers, Ronald. 1997. *Criminological Theories: Introduction and Evaluation*. Los Angeles: Roxbury.

Albrecht, Hans-Jorg. 2001. "Post-Adjudication Dispositions in Comparative Perspective." Pp. 293–30 in *Sentencing and Sanctions in Western Countries,* edited by Michael Tonry and Richard Frase. New York: Oxford University Press.

American Civil Liberties Union (ACLU). 1999. "Prisoners' Rights" (ACLU Position Paper). New York: Author.

Amnesty International. 2000. "Pregnant and Imprisoned in the United States." *Birth* 27(4):266–71.

———. 2007. "Abolitionist and Retentionist Countries." Retrieved July 15, 2007 (http://web.amnesty.org/pages/deathpenalty-countries-eng).

Anderson, C. 2003. "Prison Populations Challenge Already Cash-Strapped States." *News Tribune*, July 28, p. A7.

Bales, William, Laura Bedard, Susan Quinn, David Ensley, and Glen Holley. 2005. "Recidivism of Public and Private State Inmates in Florida." *Criminology and Public Policy* 4(1):57–82.

Barnett, C. n.d. *The Measurement of White Collar Crime Using Uniform Crime Reporting (UCR) Data*. Washington, DC: U.S. Government Printing Office.

Beckett, Katherine and Theodore Sasson. 2000. *The Politics of Injustice*. Thousand Oaks, CA: Pine Forge.

Bigda, C. 2001. "College Students Oppose Private Prisons." Retrieved March 20, 2004 (www.dollarsandsense.org/archives/2001/0901bigda.html).

Blumstein, A. and R. Rosenfeld. 1998. "Explaining Recent Trends in U.S. Homicide Rates." *The Journal of Criminal Law and Criminology* 88(4):1175–1216.

Braithwaite, John. 1989. *Crime, Shame and Reintegration*. Cambridge, England: Cambridge University Press.

Brownstein, Henry. 2001. *The Social Reality of Violence and Violent Crime*. Boston, MA: Allyn & Bacon.

Bureau of Justice Statistics. 2003a. "Crime Characteristics." Retrieved July 8, 2003 (www.ojp.usdoj.gov/bjs/cvict_c.htm).

———. 2003b. "Criminal Offenders Statistics." Retrieved July 8, 2003 (www.ojp.usdoj.gov/bjs/crimoff.htm).

———. 2003c. "State and Local Law Enforcement Statistics." Retrieved July 14, 2003 (wwww.ojp.usdoj.gov/bjs/sandlle.htm).

———. 2006. "Federal Law Enforcement Statistics." Retrieved May 16, 2007 (www.ojd.usdoj.gov/bjs/fedle.htm).

———. 2007. "State and Local Law Enforcement Statistics." Retrieved May 16, 2007 (http://www.ojp.usdoj.gov/bjs/sandlle.htm#personnel).

Caldeira, T. and J. Holston. 1999. "Democracy and Violence in Brazil." *Society for Comparative Study of Society and History* 41(4):691–729.

Camp, S. and Gerald Gaes. 2002. "Growth and Quality of U.S. Private Prisons: Evidence from a National Survey." *Criminology and Public Policy* 1(3):427–50.

Cao, Liqun, Anthony Troy Adams, and Vickie Jensen. 2000. "The Empirical Status of the Black-Subculture-of-Violence Thesis." Pp. 47–62 in *The System in Black and White,* edited by M. Markowitz and D. Jones-Brown. Westport, CT: Praeger.

Catalano, Shannan. 2006a. *Criminal Victimization, 2005* (NCJ 214644). Washington, DC: U.S. Department of Justice, Office of Justice Programs.

———. 2006b. "Intimate Partner Violence in the United States." Retrieved February 25, 2007 (http://www.ojp.usdoj.gov/bjs/intimate/ipv.htm).

Chambliss, William. 1988. *Exploring Criminology.* New York: Macmillan.

———. 1995. "The Saints and the Roughnecks." Pp. 254–67 in *Down to Earth Sociology,* edited by J. Henslin. New York: Free Press. (Original work published in 1973)

Chesney-Lind, M. and L. Pasko. 2004. *The Female Offender: Girls, Women, and Crime.* Thousand Oaks, CA: Sage Publications.

Citizens United for Alternatives to the Death Penalty. 2003a. "About CUADP." Retrieved August 2, 2003 (www.cuadp.org/about.html).

———. 2003b. "Potential Cases for Wrongful Conviction." Retrieved August 2, 2003 (www.cuadp.org/pris/pot.html).

Community Oriented Policing Services (COPS). 2003. "Who We Are." Retrieved July 22, 2003 (http://www.cops.usdoj.gov/Default.asp?Item=35).

Cureton, Steven. 2001. "Determinants of Black-to-White Arrest Differentials: A Review of the Literature." Pp. 65–71 in *The System in Black and White,* edited by M. Markowitz and D. Jones-Brown. Westport, CT: Praeger.

Davis, R., B. Taylor, and R. Titus. 1997. "Implications for Victim Services and Crime Prevention." Pp. 167–82 in *Victims of Crime,* edited by R. Davis, A. Lurigio, and W. Skogan. Thousand Oaks, CA: Sage Publications.

Death Penalty Information Center. 2007. "Facts about the Death Penalty, May 15, 2007." Retrieved May 16, 2007 (http://www.deathpenaltyinfo.org/FactSheet.pdf).

Durose, Matthew, Erica Smith, and Patrick Langan. 2007. *Contacts between Police and the Public, 2005 (NCJ215243).* Washington, DC: U.S. Department of Justice, Office of Justice Programs.

Elliot, D. S., S. Ageton, and R. Canter. 1979. "An Integrated Perspective on Delinquent Behavior." *Journal of Research in Crime and Delinquency* 16:3–27.

Erickson, Kai. 1964. "Notes on the Sociology of Deviance." Pp. 9–21 in *The Other Side: Perspectives of Deviance,* edited by Howard Becker. New York: Free Press.

Federal Bureau of Investigation. 1989. *White Collar Crime: A Report to the Public.* Washington, DC: Government Printing Office.

———. 2006. *Law Enforcement Officers Killed and Assaulted.* Retrieved May 16, 2007 (http://www.fbi.gov/ucr/killed/2005/feloniouslykilled.htm).

Flavin, Jeanne. 2001. "Feminism for the Mainstream Criminologist: An Invitation." *Journal of Criminal Justice* 29:271–85.

Fox, James Alan and Marianne Zawitz. 2006. "Homicide trends in the United States." Retrieved February 25, 2007 (http://www.ojp.usdoj.gov/bjs/pub/pdf/htius.pdf).

Gaes, Gerald, Timothy Flannagan, Laurence Motiuk, and Lynn Stewart. 1999. "Adult Correctional Treatment." Pp. 361–426 in *Crime and Criminal Justice: A Review of Research,* Vol. 26, *Prisons,* edited by Michael Tonry and Joan Petersilia. Chicago, IL: University of Chicago Press.

Goldstein, H. 1990. *Problem-Oriented Policing.* Boston, MA: McGraw-Hill.

Greene, J. R. and W. V. Pelfrey. 1997. "Shifting the Balance of Power between Police and Community: Responsibility for Crime Control." Pp. 393–423 in *Critical Issues in Policing: Contemporary Readings,* edited by R. Dunham and G. Alpert. Prospect Heights, IL: Waveland Press.

Hannon, L. and J. Defronzo. 1998. "The Truly Disadvantaged, Public Assistance, and Crime." *Social Problems* 45(3):383–92.

Harlem Live. 1999. "The Blue Nile Rites of Passage." Retrieved August 7, 2003 (www.harlemlive.org/community/pop/10-18-99).

Harrison, Paige and Allen Beck. 2006. *Prisoners in 2005* (NCJ 215092). Washington, DC: U.S. Department of Justice, Office of Justice Programs.

Hautzinger, S. 1997. "'Calling a State a State': Feminist Politics and the Policing of Violence against Women in Brazil." *Feminist Issues* 15(1):3–30.

Hautzinger, S. 2002. "Criminalizing Male Violence in Brazil's Women's Police Stations: From Flawed Essentialism to Imagined Communities." *Journal of Gender Studies* 11(3):243–51.

Hemenway, David. 2004. "A Public Health Approach to Firearms Policy." Pp. 85–98 in *Policy Challenges in Modern Health Care,* edited by David Mechanic, Lynn B. Rogut, David Colby, and James Knickman. New Brunswick, NJ: Rutgers University Press.

Hirschi, Travis. 1969. *Causes of Delinquency.* Berkeley: University of California Press.

Huggins, M. 1997. "From Bureaucratic Consolidation to Structural Devolution: Police Death Squads in Brazil." *Policing and Society* 7:207–34.

Human Rights Watch. 1995. "Global Report on Women's Human Rights 1990 to 1995." Retrieved July 30, 2003 (www.hrw.org/about/projects/womrep/General-187.htm#P2966_897325).

———. 1997. *Police Brutality in Urban Brazil.* Retrieved July 30, 2003 (http://www.hrw.org/reports/1997/brazil/).

———. 2007. "Brazil: Investigate Deaths in Rio Police Operation." Retrieved July 15, 2007 (http://hrw.org/english/docs/2007/06/29/brazil16298.htm).

Innocence Project. 2007a. "What Is the Innocence Project?" Retrieved July 15, 2007 (http://www.innocenceproject.org/Content/9.php).

———. 2007b. "Byron Halsey Is Fully Exonerated in New Jersey after DNA Proves His Innocence in 1985 Child Rapes and Murders." Retrieved July 15, 2007 (http://www.innocenceproject.org/Content/689.php).

Irwin, Darrell. 2002. "Alternatives to Delinquency in Harlem: A Study of Faith-Based Community Mentoring." *Justice Professional* 15(2):29–36.

Jensen, Vickie. 2001. *Why Women Kill: Homicide and Gender Equality.* Boulder, CO: Lynne Rienner.

Joseph, J. 2000. "Overrepresentation of Minority Youth in the Juvenile Justice System: Discrimination or Disproportionality of Delinquent Acts? Status of the Black-Subculture-of-Violence Thesis." Pp. 227–40 in *The System in Black and White,* edited by M. Markowitz and D. Jones-Brown. Westport, CT: Praeger.

JUMP. 1998. *Juvenile Mentoring Program: 1998 Report to Congress.* Washington, DC: U.S. Department of Justice, Office Justice Programs, Office of Juvenile Justice and Delinquency Programs.

———. 2003. "Overview and Purpose." Retrieved July 27, 2003 (http://ojjdp.ncjrs.org/jump/oview.html).

Kappeler, Victor, Mark Blumberg, and Gary Potter. 2000. *The Mythologies of Crime and Criminal Justice.* Prospect Heights, IL: Waveland Press.

Klaus, Patsy A. 2004. *Crime and the Nation's Households, 2002* (Bureau of Justice Statistics Bulletin, NCJ 201797). Washington, DC: U.S. Department of Justice, Office of Justice Programs.

LaFree, Gary and Gwen Hunnicutt. 2006. "Female and Male Homicide Victimization Trends." Pp. 195–239 in *Gender and Crime: Patterns of Victimization and Offending,* edited by K. Heimer and C. Krusttschnitt. New York: New York University Press.

Lane, Barbara Parsons. 2003. "Puzzle Pieces." Pp. 211–44 in *Couldn't Keep It to Myself,* edited by W. Lamb. New York: Regan Books.

Langan, Patrick and David Levin. 2002. *Recidivism of Prisoners Released in 1994* (NCJ 193427). Washington, DC: U.S. Department of Justice, Office of Justice Programs.

Larason Schneider, A. 1999. "Public-Private Partnerships in the U.S. Prison System." *American Behavioral Scientist* 43(1):192–208.

Laub, John. 1983. "Urbanism, Race, and Crime." *Journal of Research of Crime and Delinquency* 20:183–98.

———. 1997. "Patterns of Criminal Victimization in the United States." Pp. 9–26 in *Victims of Crime,* edited by R. Davis, A. Lurigio, and W. Skogan. Thousand Oaks, CA: Sage Publications.

Lemert, Edwin M. 1967. *Human Deviance, Social Problems, and Social Control.* Englewood Cliffs, NJ: Prentice Hall.

MacDonald, J. 2002. "The Effectiveness of Community Policing in Reducing Urban Violence." *Crime and Delinquency* 48(4):592–618.

Malik, Nazneen. 2005. "The Bard College Prison Initiative." *Education Update* 10(9):19.

Merton, Robert. 1957. *Social Theory and Social Structure.* Glencoe: IL: Free Press.

Mitchell, M. and C. Wood. 1999. "Ironies of Citizenship: Skin Color, Police Brutality, and the Challenge of Democracy in Brazil." *Social Forces* 77(3):1001–20.

Moore, M. 1999. "Security and Community Development." Pp. 293–337 in *Urban Problems and Community Development,* edited by R. F. Ferguson and W. T. Dickens. Washington, DC: Brookings Institution.

Moore, Solomon. 2007. "States Export Their Inmates at Prisons Fill." *New York Times,* July 31, pp. A1, A14.

Morenoff, Jeffrey. 2005. "Racial and Ethnic Disparities in Crime in the United States." Pp. 139–73 in *Ethnicity and Causal Mechanisms,* edited by M. Rutter and M. Tienda. New York: Cambridge University Press.

Naffine, Ngaire. 1996. *Feminism and Criminology.* Philadelphia: Temple University Press.

National Night Out. 2003. "The History of NATW and National Night Out." Retrieved August 6, 2003 (www.nationaltownwatch.org/nno/history.html).

National White Collar Crime Center. 2002a. *WCC Issue: Check Fraud.* Morgantown, WV: Author.

———. 2002b. *WCC Issue: Embezzlement/Employee Theft.* Morgantown, WV: Author.

National White Collar Crime Center and the Federal Bureau of Investigation. 2006. *Internet Crime Report January 1, 2006–December 31, 2006.* Washington, DC: National White Collar Crime Center.

Norris, Clive, Nigel Fielding, Clark Kemp, and Jane Fielding. 1992. "Black and Blue: An Analysis of the Influence on Being Stopped by the Police." *British Journal of Sociology* 43(2):207–24.

Novak, Kenneth, Leanne Alarid, and Wayne Lucas. 2003. "Exploring Officers' Acceptance of Community Policing: Implications for Policy Implementation." *Journal of Criminal Justice* 31:57–71.

Office on Violence Against Women. 2003. "About the Office on Violence Against Women." Retrieved July 27, 2003 (www.ojp.usdoj.gov/vawo/about.htm).

Pratt, T. and J. Maahs. 1999. "Are Private Prisons More Cost-Effective Than Public Prisons? A Meta-Analysis of Evaluation Research Studies." *Crime and Delinquency* 45(3):358–71.

Pressman, D., R. Chapman, and L. Rosen. 2002. *Creative Partnerships: Supporting Youth, Building Communities.* Washington, DC: U.S. Department of Justice, Office of Community Oriented Policing Services.

Radalet, Michael and Glenn Pierce. 2005. "The Impact of Legally Inappropriate Factors on Death Sentencing for California Homicides 1990–1999." *Santa Clara Law Review* 49:1–47.

Reiman, Jeffrey. 1998. *The Rich Get Richer and the Poor Get Prison.* Boston, MA: Allyn & Bacon.

Rich, John and Marguerite Ro. 2002. *A Poor Man's Plight: Uncovering the Disparity in Men's Health.* Battle Creek, MI: W. K. Kellogg Foundation.

Sampson, Robert J. and William Julius Wilson. 1995. "Towards a Theory of Race, Crime, and Urban Inequality." Pp. 37–54 in *Crime and Inequality,* edited by John Hagen and Ruth Peterson. Stanford, CA: Stanford University Press.

Sanders, W. 1981. *Juvenile Delinquency: Causes, Patterns, and Reactions.* New York: Holt, Rinehart and Winston.

Simon, Thomas, James Mercy, and Craig Perkins. 2001. *Injuries from Violent Crime, 1992–1998* (NCJ 168633). Washington, DC: U.S. Department of Justice, Office of Justice Programs.

Snyder, Howard and Melissa Sickmund. 2006. *Juvenile Offenders and Victims: 2006 National Report.* Washington, DC: U.S. Department of Justice, Office of Justice Programs, Office of Juvenile Justice and Delinquency Prevention.

Stafford, Mark C. 2004. "Juvenile Deliquency." Pp. 480–93 in *Handbook of Social Problems: A Comparative International Perspective,* edited by George Ritzer. Thousand Oaks, CA: Sage Publications.

Steffensmeier, Darrell and Emilie Allan. 1996. "Gender and Crime: Toward a Gendered Theory of Female Offending." *Annual Review of Sociology* 22:459–87.

Stephan, James. 2004. *State Prison Expenditures, 2001* (NCJ 202949). Washington, DC: U.S. Department of Justice, Bureau of Justice Statistics.

Stoutland, Sara. 2001. "The Multiple Dimensions of Trust in Resident/Police Relations in Boston." *Journal of Research in Crime and Delinquency* 38(3):226–56.

Subcommittee on Civil and Constitutional Rights, Committee on the Judiciary. 1994. "Racial Disparities in Federal Death Penalty Prosecutions 1988–1994." Retrieved July 26, 2003 (www.deathpenaltyinfo.org/article.php?scid=45&did=528).

Sutherland, Edwin H. 1949. *White Collar Crime.* New York: Dryden Press.

Tonry, Michael. 2004. "Crime." Pp. 465–79 in *Handbook of Social Problems: A Comparative International Perspective,* edited by George Ritzer. Thousand Oaks, CA: Sage Publications.

Tonry, Michael and David P. Farrington. 1995. "Strategic Approaches to Crime Prevention." Pp. 1–20 in *Crime and Justice: A Review of Research,* Vol. 19, *Building a Safer Society—Strategic Approaches to Crime Prevention,* edited by Michael Tonry and David P. Farrington. Chicago, IL: University of Chicago Press.

Turk, Austin. 1969. *Criminality and Legal Order.* Chicago: Rand McNally.

———. 1976. "Law as a Weapon in Social Conflict." *Social Problems* 23:276–91.

U.S. Congress. 2002. Public Law 107–273, Title II Juvenile Justice. Retrieved August 29, 2003 (ojjdp.ncjrs .org/about/PL_107_273.html).

U.S. Department of Justice. 2002. "DOJ Seal—History and Motto." Retrieved July 22, 2003 (http://www .usdoj.gov/jmd/ls/dojseal.htm).

Walmsley, Roy. 2006. "World Female Imprisonment List." Retrieved February 25, 2007 (http://www.kcl .ac.uk/depsta/rel/icps/women-prison-list-2006.pdf).

Walsh, Anthony and Lee Ellis. 2007. *Criminology: An Interdisciplinary Approach.* Thousand Oaks, CA: Sage Publications.

Wilson, William Julius. 1996. *When Work Disappears: The World of the New Urban Poor.* New York: Knopf.

Wolf, J. 2000. "War Games Meets the Internet: Chasing 21st Century Cybercriminals with Old Laws and Little Money." *American Journal of Criminal Law* 28:95–117.

Wolfgang, Marvin and Franco Ferracuti. 1967. *The Subculture of Violence: Towards an Integrated Theory in Criminology.* New York: Tavistock.

World Health Organization. 2005. *WHO Multi-Country Study on Women's Health and Domestic Partner Violence against Women: Summary Report of Initial Results on Prevalence, Health Outcomes and Women's Responses.* Geneva, Switzerland: World Health Organization.

Zhao, Jihong, M. Scheider, and Quint Thurman. 2002. "Funding Community Policing to Reduce Crime: Have Cops Grants Made a Difference?" *Criminology and Public Policy* 2(1):7–32.

Chapter 14

Appleton, Lynn M. 1995. "The Gender Regimes in American Cities." Pp. 44–59 in *Gender in Urban Research,* edited by Judith A. Garber and Robyne S. Turner. Thousand Oaks, CA: Sage Publications.

Applied Real Estate Analysis. 1998. *Making Home Ownership a Reality: Survey of Habitat for Humanity International (HFHI), Inc. Homeowners and Affiliates, April 1998* (Report prepared for Office of Policy Development and Research). Washington, DC: U.S. Department of Housing and Urban Development.

Armas, Genaro. 2003. "Public Housing Revitalization Program May Be Scrapped." *News Tribune,* February 7, p. A9.

Banerjee, Tridib and William C. Baer. 1984. *Beyond the Neighborhood Unit: Residential Environments and Public Policy.* New York: Plenum Press.

Boddy, Sharon. 2000. "Car-Free and Carefree." *E Magazine* 11(2):14–18.

Bunce, H., S. G. Neal, W. Reeder, and R. J. Sepanik. 1996. *New Trends in American Homeownership.* Washington, DC: U.S. Department of Housing and Urban Development.

Butler, S. M. 1991. "The Conceptualization of Enterprise Zones." Pp. 27–40 in *Enterprise Zones: New Directions in Economic Development,* edited by R. E. Green. Newbury Park, CA: Sage Publications.

Center for Community Change. 2003. *A HOPE Unseen: Voices from the Other Side of HOPE VI.* Washington, DC: Author.

Choi, N. and L. Snyder. 1999. *Homeless Families with Children: A Subjective Experience of Homelessness.* New York: Springer.

Clark, W., M. Deurloo, and F. Dieleman. 2000. "Housing Consumption and Residential Crowding in U.S. Housing Markets." *Journal of Urban Affairs* 22(1):49–64.

Cohen, H. and M. Northridge. 2000. "Getting Political: Racism and Urban Health." *American Journal of Public Health* 90(6):841–43.

Corvin, A. 2001. "Urban Sprawl Creates Belly Sprawl, CDC Suggests." *News Tribune,* November 2, p. A11.

Crenson, Matt. 2003. "He Wants to Reclaim Towns for Pedestrians." *Christian Science Monitor,* October 15. Retrieved March 19, 2004 (www.csmonitor.com/2003/1015/p13s02-lihc.htm).

De Pommereau, Isabelle. 2006. "New German Community Models for Car-Free Living." *Christian Science Monitor* 99(18):1, 11.

Desai, Sonalde. 2004. "Population Change." Pp. 69–86 in *Handbook of Social Problems: A Comparative International Perspective,* edited by George Ritzer. Thousand Oaks, CA: Sage Publications.

Dreier, Peter. 1996. "America's Urban Crisis: Symptoms, Causes, and Solutions." Pp. 79–141 in *Race, Poverty, and American Cities,* edited by J. C. Boger and J. W. Wegner. Chapel Hill: University of North Carolina Press.

Dretzka, G. 1995. "Filling Void, Real and Symbolic Ex-Laker 'Magic,' Sony Bring First-Run Movies, Hope to South-Central L.A." *Chicago Tribune,* July 21, Business section, p. 1.

Drew, Bettina. 1998. *Crossing the Expendable Landscape.* St. Paul, MN: Graywolf Press.

Eisenhower, Dwight. 1963. *Mandate for Change 1953–1956.* Garden City, NY: Doubleday.

Evans, G., S. Saegert, and R. Harris, R. 2001. "Residential Density and Psychological Health among Children in Low-Income Families." *Environment and Behavior* 33(2):165–80.

Ewing, Reed, Rolf Pendall, and Don Chen. 2002. *Measuring Urban Sprawl and Its Impact.* Washington, DC: Smart Growth America.

Fair Housing Law. 2004. "About Fair Housing Campaign." Retrieved February 28, 2004 (http://www.fairhousinglaw.org/about_us/).

Feagin, Joe R. 1998a. "Introduction." Pp. 1–24 in *The New Urban Paradigm,* edited by Joe R. Feagin. Lanham, MD: Rowman & Littlefield.

———. 1998b. "Urban Real Estate Speculation." Pp. 133–58 in *The New Urban Paradigm,* edited by Joe R. Feagin. Lanham, MD: Rowman & Littlefield.

Freeman, Lance. 2002. "America's Affordable Housing Crisis: A Contract Unfulfilled." *American Journal of Public Health* 92(5):709–13.

French, Howard. 2007. "Big, Gritty Chongqing, City of 12 million Is China's Model for Future." *New York Times,* June 1, p. A13.

Garber, Judith A. and Robyne S. Turner. 1995. "Introduction." Pp. x–xxvi in *Gender in Urban Research,* edited by Judith A. Garber and Robyne S. Turner. Thousand Oaks, CA: Sage Publications.

Gordon, J. S. 2001. "The Business of America." *American Heritage* 52(4):6–59.

Gottdiener, Mark. 1977. *Planned Sprawl: Private and Public Interests in Suburbia.* Beverly Hills, CA: Sage Publications.

Grass Roots.org. 2003a. "Groups That Change Communities: Light Street Housing Corp." Retrieved March 24, 2003 (http://www.grass-roots.org/usa/lightst.shtml).

———. 2003b. "Groups That Change Communities: Trinity Housing Corp." Retrieved March 24, 2003 (http://www.grass-roots.org/usa/trinhous.shtml).

Institute for Government Innovation. 2000. "Awards Recipients: Hope VI Mixed Finance Public Housing." Retrieved March 11, 2003 (www.innovations.harvard.edu/).

Johnson, Roy S. 2003. "It Must Be Magic." *Savoy,* February:70–71, 72–78, 80.

Johnson, William A. 2007. "Sprawl and Civil Rights: A Mayor's Reflections." Pp. 103–23 in *Growing Smarter: Achieving Livable Communities, Environmental Justice and Regional Equity,* edited by R. Bullard. Cambridge, MA: MIT Press.

Karp, David, Greg Stone, and William Yoels. 1991. *Being Urban: A Sociology of City Life.* London, England: Greenwood.

Kennedy, Maureen and Paul Leonard. 2001. "Dealing with Neighborhood Change: A Primer on Gentrification and Policy Choices" (Discussion paper). Washington, DC: Brookings Institution Center on Urban and Metropolitan Policy and Policy Link.

Kim, Chigon and Mark Gottdiener. 2004. "Urban Problems in Global Problems." Pp. 172–92 in *Handbook of Social Problems: A Comparative International Perspective,* edited by George Ritzer. Thousand Oaks, CA: Sage Publications.

Krieger, J. and D. Higgins. 2002. "Housing and Health: Time Again for Public Health Action." *American Journal of Public Health* 92(5):758–68.

Loven, J. 2001. "Affordable Housing Shortage Getting Worse." *News Tribune,* December 24, p. A4.

Magin, Janis. 2006. "For 1,000 or More Homeless in Hawaii, Beaches are the Best Option." *New York Times,* December 5, p. A16.

Martinez, S. 2000. "The Housing Act of 1949: Its Place in the Realization of the American Dream of Homeownership." *Housing Policy Debate* 11(2):467–87.

Massey, Douglas. 2001. "Residential Segregation and Neighborhood Conditions on U.S. Metropolitan Areas." Pp. 391–434 in *America Becoming: Racial Trends and Their Consequences,* Vol. 1, edited by N. Smelser, W. J. Wilson, and F. Mitchell. Washington, DC: National Academy Press.

Massey, Douglas and Mitchell Eggers. 1993. "The Spatial Concentration of Affluence and Poverty during the 1970s." *Urban Affairs Quarterly* 29(2):299–315.

Masson, D. 1984. "Les Femmes dans les Structures Urbanies: Apercu d'un Nouveau Champ de Recherché." *Canadian Journal of Political Science* 17:753–82.

Molotch, H. 1976. "The City as a Growth Machine." *American Journal of Sociology* 82:309–32.

Murphy, D. E. 2003."New Californian Identity Predicted by Researchers." *New York Times,* February 17, p. A13.

Myers, Dowell and Seong Woo Lee. 1996. "Immigration Cohorts and Residential Overcrowding in Southern California." *Demography* 33:51–65.

Naparstek, A., S. Freis, G. T. Kingsley, D. Dooley, and H. Lewis. 2000. *Hope VI: Community Building Makes a Difference.* Washington, DC: U.S. Department of Housing and Urban Development.

National Coalition for the Homeless. 2002. "How Many People Experience Homelessness?" Retrieved June 10, 2003 (http://www.nationalhomeless.org/publications/facts/How_Many.pdf).

National Law Center on Homelessness and Poverty. 1999. *Out of Sight Out of Mind? A Report on Anti Homeless Laws, Litigation, and Alternatives in 50 United States Cities.* Washington, DC: Author.

National Low Income Housing Coalition. 2007. *Why the National Housing Trust Fund Primarily Targets the Lowest Income Renters.* Washington, DC: Author.

Norris, Tyler. 2001. "Civic Gemstones: The Emergent Communities Movement." *National Civic Review* 90(4):307–18.

North Carolina State University. 2007. "Mayday 23: World Population Becomes More Urban than Rural." Retrieved August 8, 2007 (http://news.ncsu.edu/releases/2007/may/104.html).

Ohlemacher, Stephen. 2007. "In Nation of Sprawl, We're All Just Single Dots in Our Cars." *News Tribune,* June 14, p. A3.

Oliver, Melvin. 2003. "American Dream? How Government Initiatives Made Blacks House Poor." *The New Crisis* 110(5):17–19.

Pastor, M. 2001. "Geography and Opportunity." Pp. 435–68 in *American Becoming: Racial Trends and Their Consequences,* Vol. 1, edited by N. J. Smelser, W. J. Wilson, and F. Mitchell. Washington, DC: National Academy Press.

Perry, Marc. 2006. *Domestic Net Migration in the United States: 2000 to 2004* (Current Population Reports, P25-1135). Washington, DC: U.S. Census Bureau.

Powell, John. 2007. "Race, Poverty and Urban Sprawl: Access to Opportunities through Regional Strategies." Pp. 51–71 in *Growing Smarter: Achieving Livable Communities, Environmental Justice and Regional Equity,* edited by R. Bullard. Cambridge, MA: MIT Press.

Pugh, Tony. 2007. "Housing Bleaker for Nation's Poor, Study Finds." *News Tribune,* July 15, p. A13.

Ramirez, Roberto and G. Patricia de la Cruz. 2003. *The Hispanic Population in the United States: March 2002* (Current Population Reports, P20–545). Washington, DC: U.S. Census Bureau.

Rast, J. 2001. "Manufacturing Industrial Decline: The Politics of Economic Change in Chicago, 1955–1998." *Journal of Urban Affairs* 23(2):175–90.

Roberts, Sam. 2007. "New Demographic Racial Gap Emerges." *New York Times,* May 17, p. A19.

Rossi, Peter. 1989. *Down and Out in America: The Origins of Homelessness.* Chicago, IL: University of Chicago Press.

Savage, H. A. 1999. *Who Could Afford to Buy a House in 1995?* (U.S. Census Bureau Current Housing Reports, H121/99–1). Washington, DC: U.S. Census Bureau.

Schachter, J. 2004. *Geographical Mobility: 2002 to 2003* (Current Population Reports, P20–549). Washington, DC: U.S. Census Bureau.

Simmel, Georg. 1997. "The Metropolis and Mental Life." Pp. 174–86 in *Simmel on Culture: Selected Writings,* edited by D. Frisby and M. Featherstone. London, England: Sage. (Original work dated 1903)

Smith, Eric L. 1999. "The Magic Touch." *Black Enterprise,* May:74–82.

Smith, Neil. 1986. "Gentrification, the Frontier, and the Restructuring of Urban Space." Pp. 15–34 in *Gentrification of the City,* edited by Neil Smith and Peter Williams. Boston, MA: Allen and Unwin.

———. 2002. "New Globalism, New Urbanism: Gentrification as Global Urban Strategy. *Antipode* 34(3):427–50.

Stanback, T. M., Jr. 1991. *The New Suburbanization: Challenge to the Central City.* Boulder, CO: Westview Press.

Stein, R. 2003. "Waistlines Sprawl with Suburbs, Study Finds." *News Tribune,* August 29, p. A3.

Stock, Mathis. 2006. "European Cities: Toward a Recreational Turn?" *Studies in Culture, Polity and Identities* 7(1):1–19.

Suro, Roberto and Audrey Singer. 2002. *Latino Growth in Metropolitan America: Changing Patterns, New Locations* (Brookings Institution Center on Urban and Metropolitan Policy and Pew Hispanic Center, Survey Series). Washington, DC: Brookings Institution.

Sustainable Communities. 2002. *Urban Resources Initiative: Detroit, MI.* Retrieved February 27, 2003 (www.sustainable.org/casestudies/SIA_PDFs/SIA_michican.pdf).

Toro, Paul. 2007. "Toward an International Understanding of Homelessness." *Journal of Social Issues* 63(3):461–81.

Turner, M. A., S. L. Ross, G. Galster, and J. Yinger. 2002. *Discrimination in Metropolitan Housing Markets: National Results from Phase 1 HDS 2000.* Washington, DC: Urban Institute and U.S. Department of Housing and Urban Development.

Turner, Robyne S. 1995. "Concern for Gender in Central-City Development." Pp. 271–88 in *Gender in Urban Research,* edited by Judith A. Garber and Robyne S. Turner. Thousand Oaks, CA: Sage Publications.

United Nations Population Division. 2006. "Fact Sheet 7: Mega-Cities." Retrieved August 8, 2007 (http://www.un.org/esa/population/publications/WUP2005/2005WUP_FS7.pdf).

United Nations Population Fund. 1996. *The State of the World's Population: Changing Places: Population, Development and the Urban Future.* New York: Author.

———. 2007. *The State of the World's Population: Unleashing the Potential of Urban Growth.* New York: Author.

U.S. Census Bureau. 1995. "Urban and Rural Populations 1900 to 1990." Retrieved March 23, 2008 (http://www.census.gov/population/censusdata/urpop0090.txt).

———. 2002. "American Housing Survey, 2002. Detailed Tables for Total Occupied Housing Units, Black Occupied Housing Units, and Households of Hispanic Origin." Retrieved March 23, 2003 (www.ccnsus.gov/hhcs/www/housing/ahs/01ddtchrt).

———. 2005. "Americans Spend More that 100 Hours Commuting to Work Each Year." Retrieved October 12, 2007 (http://www.census.gov/Press-Release/www/releases/archives/american_community_survey_acs/004489.html).

———. 2006. "Housing Vacancies and Home Ownership: Annual Statistics 2006." Retrieved October 2, 2007 (http://www.census.gov/hhes/www/housing/hvs/annual06/ann06t20.html).

———. 2007. *Statistical Abstract of the United States.* Washington, DC: Author.

U.S. Conference of Mayors. 2006. *A Status Report on Hunger and Homelessness in America's Cities.* Washington, DC: Author.

U.S. Department of Housing and Urban Development (HUD). 1999. "The State of Cities, 1999." Retrieved March 25, 2004 (http://www.huduser.org/publications/polleg/tsoc99/summ-03.html).

———. 2003a. "About HOPE VI." Retrieved March 11, 2003 (www.hud.gov:80/offices/pih/programs/ph/hope6/about/index.cfm).

———. 2003b. "HUD's History." Retrieved March 11, 2003 (www.hud.gov/library/bookshelf18/hudhistory.cfm).

———. 2003c. "Welcome to the Community Renewal Initiative." Retrieved June 24, 2004 (http://www.hud.gov/offices/cpd/economicdevelopment/programs/rc/index.cfm).

———. 2004. "Affordable Housing." Retrieved February 28, 2004 (www.hud/gov/offices/cpd/affordable housing.index.cfm).

U.S. Department of Transportation. 2001. "Household, Individual and Vehicle Characteristics." Retrieved October 6, 2007 (http://www.bts.gov/publications/highlights_of_the_2001_national_household_travel_survey/html/section_01.html).

Wilborn, P. 2002. "Magic Johnson Now Winning at Business; Politics Could Be Next." *Cincinnati Enquirer On Line Edition,* May 26. Retrieved October 8, 2002 (http://enquirer.com/editions/2002/05/26/spt_Magic_johnson_now.html).

Williams, D. C. 2000. *Urban Sprawl: A Reference Handbook.* Santa Barbara, CA: ABC-CLIO.

Women's International Network News. 1999. "Women and the Urban Environment." *Women's International Network News* 25(1):60–61.

Wood, Daniel. 2007. "On the Rise in American Cities: The Car-Free Zone." *Christian Science Monitor* 99(109):1, 12.

Chapter 15

American Rivers. 2002. "63 Dams in 16 States to Be Removed in 2002." Retrieved August 16, 2003 (www.amrivers.org/pressrelease/damremova1071802.htm).

Arkansas Wildlife Federation. 2003a. "About the Arkansas Wildlife Federation." Retrieved June 20, 2003 (http://www.arkansaswildlifefederation.org/about.html).

———. 2003b. "Educational Projects." Retrieved June 20, 2003 (http://www.arkansaswildlifefederation.org/education.html).

Babbitt, Bruce. 1998. Speech to the Ecological Society of America, August 4, 1998. Retrieved August 16, 2003 (http://www.hetchhetchy.org/babbitt_on_dams_9_4-98.html).

Barcott, Bruce. 2003. "What's a River For?" *Mother Jones* 28(3):44–51.

Blacksmith Institute. 2007. *The World's Most Polluted Places.* New York: Author.

Blumenthal, Ralph. 2006. "Texas Lawsuit Includes a Mix of Race and Water." *New York Times,* July 9, p. 11.

Breton, M. J. 1998. *Women Pioneers for the Environment.* Boston, MA: Northeastern University Press.

Broder, John. 2007. "Governors Join in Creating Regional Pacts on Climate Change." *New York Times,* November 15, p. A16.

Broder, John and Marjorie Connelly. 2007. "Public Says Warming Is a Problem, but Remains Split on Response." *New York Times,* April 27, p. A23.

Bullard, Robert. 1994. *Dumping in Dixie: Race, Class, and Environmental Quality.* Boulder, CO: Westview Press.

Cable, Sherry and Charles Cable. 1995. *Environmental Problems, Grassroots Solutions: The Politics of Grassroots Environmental Conflict.* New York: St. Martin's Press.

Caldwell, L. 1970. *Environment: Challenge to Modern Society.* Garden City, NY: Natural History Press.

———. 1997. "Environment as a Problem for Policy." P. 118 in *Environmental Policy: Transnational Issues and National Trends,* edited by L. Caldwell and R. Bartlett. Westport, CT: Quorum Books.

Casey, Michael. 2007. "World's Reliance on Coal Leaves Environmental Mess." *News Tribune,* November 5, p. A5.

Caulfield, H. 1989. "The Conservation and Environmental Movements: A Historical Analysis." Pp. 13–56 in *Environmental Politics and Policy,* edited by J. Lester. Durham, NC: Duke University Press.

Center for Health, Environment, and Justice. 2001. *Poisoned Schools: Invisible Threats, Visible Actions.* Falls Church, VA: Child Proofing Our Communities Campaign.

Clean Air Network. 2003. *Danger in the Air: Unhealthy Levels of Smog in 2002.* Washington, DC: U.S. Public Interest Research Group Education Fund.

Davies, J. Clarence. 1970. *The Politics of Pollution.* New York: Pegasus.

Deutsch, Claudia. 2007. "College Leaders Push for Carbon Neutrality." *New York Times,* June 13, p. A21.

Dunlap, R. 1997. "The Evolution of Environmental Sociology: A Brief History and Assessment of the American Experience." Pp. 21–39 in *The International Handbook of Environmental Sociology,* edited by M. R. Redclift and G. Woodgate. Cheltenham, England: Edward Elgar.

Dunlap, Riley and William Catton. 1994. "Struggling with Human Exemptionalism: The Rise, Decline, and Revitalization of Environmental Sociology." *American Sociologist* 25(1):5–30.

Dunlap, Riley and Angela Mertig. 1992. *American Environmentalism: The U.S. Environmental Movement, 1970–1990.* Washington, DC: Taylor & Francis.

Earth Island Institute. 2003. "About EII: Origins and Purpose." Retrieved July 27, 2003 (http://www.earthis land.org/abouteii/).

———. 2007. "2007 Brower Youth Awards." Retrieved October 30, 2007 (http://www.broweryoutha wards.org/userdata_display.php?modin=50&sortby=&custom2=2007).

Ehrlich, Paul and Anne Ehrlich. 1990. *The Population Explosion.* New York: Touchstone/Simon & Schuster.

———. 1996. *Betrayal of Science and Reason: How Anti-Environmental Rhetoric Threatens Our Future.* Washington, DC: Island Press.

Ehrlich, Paul, Anne Ehrlich, and John Holdren. 1973. *Human Ecology: Problems and Solutions.* San Francisco, CA: W. H. Freeman.

Eisele, K. 2003. "With Every Breath You Take." Retrieved March 2, 3003 (http://www.nrdc.org/onearth/03win/asthma1.asp).

Environmental Protection Agency (EPA). 1999a. *Smog—Who Does It Hurt? What You Need to Know about Ozone and Your Health* (EPA-452/K-99–001). Washington, DC: Author.

———. 1999b. "Water on Tap: A Consumer's Guide to the Nation's Drinking Water. Retrieved June 27, 2003 (www.epa.gov/safewater/wot/introtap.html).

———. 2003a. "EPA Announces $73.1 Million in National Brownfields Grants in 37 States and Seven Tribal Communities." Retrieved August 26, 2003 (www.epa.gov/brownfields/news/pr062003.htm).

———. 2003b. *Water on Tap: What You Need to Know.* Washington, DC: Author.

———. 2004. "About EPA." Retrieved June 28, 2004 (www.epa.gov/epahome/aboutepa.htm).

———. 2007a. "Particulate Matter." Retrieved October 23, 2007 (http://www.epa.gov/air/particles/index.html).

———. 2007b. "Cleaning up the Nation's Hazardous Wastes Sites." Retrieved October 20, 2007 (http://www.epa.gov/superfund/).

———. 2007c. *Complete Assessment Needed to Ensure Rural Texas Community Has Safe Drinking Water, Report No. 2007-P-00034.* Washington, DC: Author.

———. 2007d. State Actions. Retrieved October 24, 2007 (http://www.epa.gov/climatechange/wycd/stateandlocalgov/state.html).

———. 2007e. State Action Plan Recommendations Matrix. Retrieved October 24, 2007 (http://yosemite.epa.gov/gw/StatePolicyActions.nsf/matrices/0?opendocument).

Epstein, B. 1995. "Grassroots Environmentalism and Strategies for Social Change." *New Political Science* 32(Summer):1–24.

Freudenberg, Nicolas and Carol Steinsapir. 1992. "Not in Our Backyard: The Grassroots Environmental Movement." Pp. 27–35 in *American Environmentalism: The U.S. Environmental Movement, 1970–1990,* edited by Riley Dunlap and Angela Mertig. Washington, DC: Taylor & Francis.

Hamilton, Cynthia. 1994. "Concerned Citizens of South Central L.A." Pp. 207–19 in *Unequal Protection,* edited by Robert Bullard. San Francisco, CA: Sierra Club Books.

Hannigan, John. 1995. *Environmental Sociology: A Social Constructionist Perspective.* London, England: Routledge.

Hawken, Paul. 1993. *The Ecology of Commerce: A Declaration of Sustainability.* New York: HarperCollins.

———. 1997. "Natural Capitalism." *Mother Jones* 22(2):40–58.

Ingram, H. and D. Mann. 1989. "Interest Groups and Environmental Policy." Pp. 135–57 in *Environmental Politics and Policy,* edited by J. Lester. Durham, NC: Duke University Press.

Irwin, Alan. 2001. *Sociology and the Environment.* Cambridge, England: Polity Press.

Kahn, Joseph and Jim Yardley. 2007. "As China Roars, Pollution Reaches Deadly Extremes." *New York Times,* August 26, pp. 1, 6–7.

Kanter, James. 2007. "U.N. Warns of Rapid Decay of Environment." *New York Times,* October 26, p. A8.

Kanter, James and Andrew Revkin. 2007. "Scientists Detail Climate Changes, Poles to Tropics." *New York Times,* April 7, pp. A1, A5.

Ketcham, Christopher. 2007. "The Hundred Mile Diet." Retrieved October 30, 2007 (http://www.thenation.com/doc/20070910/ketcham).

Korten, David. 1995. *When Corporations Rule the World.* West Hartford, CT: Kumarian Press.

Kraft, M. and N. Vig. 2000. "Environmental Policy from the 1970s to 2000: An Overview." Pp. 1–31 in *Environmental Policy,* edited by N. Vig and M. Kraft. Washington, DC: Congressional Quarterly Press.

Lee, J. 2003. "7 States to Sue EPA over Standards on Air Pollution." *New York Times,* February 21, p. A24.

Lipka, Sara. 2006. "Students Call for Action on Campuses." *Chronicle of Higher Education* 53(9):11.

Littig, B. 2001. *Feminist Perspectives on Environment and Society.* Harlow, England: Prentice Education.

Locavores. 2007. "Locavores." Retrieved October 30, 2007 (http://locavores.com/).

Lueck, Thomas J. 2007. "The Mayor Draws a Blueprint for a Greener City." *New York Times,* April 23, p. A18.

Mazmanian, D. and M. Kraft. 1999. "The Three Epochs of the Environmental Movement." Pp. 3–42 in *Toward Sustainable Communities: Transition and Transformations in Environmental Policy,* edited by D. Mazmanian and M. Kraft. Cambridge, MA: MIT Press.

McRandle, P. W. and Sara Smiley Smith. 2006. "The Top Ten Green Cities in the U.S.: 2006." Retrieved October 24, 2007 (http://www.thegreenguide.com/doc/113/top10cities).

Morland, Kimberly and Steve Wing. 2007. "Food Justice and Health in Communities of Color." Pp. 171–88 in *Growing Smarter: Achieving Livable Communities, Environmental Justice and Regional Equity,* edited by Robert Bullard. Cambridge, MA: MIT Press.

National Academies. 2007. "Press Release February 21,2007." Retrieved October 30, 2007 (http://www8 .nationalacademies.org/onpinews/newsitem.aspx?RecordID=11857).

National Parks Conservation Association. 2003. "Ten Most Endangered." Retrieved June 27, 2003 (http:// www.npca.org/media_center/press_releases/2003/page-27599890.html).

National Resources Defense Council (NRDC). 1996. "Particulate Pollution." Retrieved March 24, 2008 (www.nrdc.org/air/pollution/qbreath.asp).

National Wilderness Preservation System. 2004. "Fast Facts." Retrieved June 28, 2004 (www.wilderness .net/index.cfm?fuse=NWPS&sec=fastFacts).

Newman, R. 2001. "Making Environmental Politics: Women and Love Canal Activism." *Women's Studies Quarterly* 1–2:65–84.

NewsMax.com. 2001. "Bush Defends Rejection of Kyoto Treaty." Retrieved June 25, 2003 (http://archive .newsmax.com/archives/articles/2001/3/29/164418.shtml).

Palmer, C. 1997. *Environmental Ethics.* Santa Barbara, CA: ABC-CLIO.

Power, Matthew. 2006. "The Magic Mountain." *Harper's Magazine* December:57–68.

Pregracke, Chad and Jeff Barrow. 2006. *From the Bottom Up: One's Man Crusade to Clean America's Rivers.* Washington, DC: National Geographic.

Revkin, Andrew. 2003. "U.S. Is Urged to Overhaul Its Approach to Protecting Oceans." *New York Times,* June 5, p. A32.

Ringquist, E. J. 2000. "Environmental Justice: Normative Concerns and Empirical Evidence." Pp. 232–56 in *Environmental Policy,* edited by N. Vig and M. Kraft. Washington, DC: Congressional Quarterly Press.

Rockler-Gladen, Naomi. 2007. "Green Tips for College Students." Retrieved November 17, 2007 (http:// collegeuniversity.suite101.com/article.cfm/green_tips_for_college_students).

Roosevelt, M. 2003. "Fresh off the Farm." *Time,* November 3, pp. 60–61.

Rosenthal, Elisabeth. 2007. "Parents and Health Experts Try to Ease Italy's Pollution." *New York Times,* June 12, p. A3.

Scarce, Rik. 1990. *Eco-Warriors: Understanding the Radical Environmental Movement.* Chicago, IL: Noble Press.

Stolberg, Sheryl Gay. 2007. "Bush Proposes Goal to Reduce Greenhouse Gas." *New York Times,* June 1, pp. A1, A12.

Tobin, R. 2000. "Environment, Population, and the Developing World." Pp. 326–49 in *Environmental Policy,* edited by N. Vig and M. Kraft. Washington, DC: Congressional Quarterly Press.

U.S. Fish and Wildlife Service. 2007. "Threatened and Endangered Species System." Retrieved October 20, 2007 (http://ecos.fws.gov/tess_public/Boxscore.do).

Vrijheid, M. 2000. "Health Effects of Residence near Hazardous Waste Landfill Sites: A Review of Epidemiologic Literature." *Environmental Health Perspectives I* 108 (Suppl 1):101–12.

Weiss, K. 2003. "Life in U.S. Ocean Waters in Death Spiral, Study Says." *News Tribune,* June 5, p. A3.

Wenz, P. 2001. *Environmental Ethics Today.* New York: Oxford University Press.

World Health Organization. 2006. "Asthma." Retrieved March 24, 2008 (http://www.who.int/mediacentre/factsheets/fs307/en/index.html).

Yardley, William. 2007. "Climate Change Adds New Twist to Debate Over Dams." *New York Times,* April 23, p. A12.

Chapter 16

Ahern, B. 2003. "In Search of Peace: The Fate and Legacy of the Good Friday Agreement." *Harvard International Review* Winter:26–31.

Ali, M. and I. Shah. 2000. "Sanctions and Childhood Mortality in Iraq." *Lancet* 355(9218):1851–57.

Ayubi, Shaheen, R. E. Bissell, N. Korsah, and L. Lerner. 1982. *Economic Sanctions in U.S. Foreign Policy.* Philadelphia, PA: Foreign Policy Research Institute.

Betts, Richard K. 2001. "Intelligence Test: The Limits of Prevention." Pp. 145–62 in *How Did This Happen? Terrorism and the New War,* edited by James F. Hoge, Jr., and Gideon Rose. New York: PublicAffairs.

Block, Fred. 2007. "Why Is the U.S. Fighting in Iraq?" *Context* 6(3):33–37.

Booth, K. and T. Dunne. 2002. *Worlds in Collision: Terror and the Future of Global Order.* New York: Palgrave Macmillan.

Bush, G. W. 2001. Statement by the President in His Address to the Nation. September 11, 2001. Retrieved September 17, 2003 (http://www.whitehouse.gov/news/releases/2001/09/20010911-16.html).

Central Asia Institute. 2003. "When Is a Book More Powerful Than a Bomb?" (Press Release, July). Retrieved November 9, 2003 (www.ikat.org/pressrelease.html).

———. 2007. "Central Asia Institute Projects." Retrieved November 4, 2007 (http://www.ikat.org/projects.html).

Center for Defense Information. 2003. "The World's Nuclear Arsenals." Retrieved August 17, 2003 (www.cdi.org/issues/nukef&f/database/nukearsenals.cfm).

Centers for Disease Control. 2003a. *Veterans' Health Activities.* Retrieved August 15, 2003 (www.cdc.gov/nceh/veterans/vet_hlth_actvy.pdf).

———. 2003b. "Anthrax: What You Need to Know." Retrieved August 13, 2003 (www.bt.cdc.gov/agent/anthrax/needtoknow.asp).

Chatfield, C. 1992. *The American Peace Movement.* New York: Twayne.

Cole, D. and J. Dempsey. 2002. *Terrorism and the Constitution.* New York: New Press.

Crenshaw, Martha. 1983. "Introduction: Reflections on the Effects of Terrorism" Pp. 1–37 in *Terrorism, Legitimacy, and Power,* edited by Martha Crenshaw. Middletown, CT: Wesleyan University Press.

———. 1995. *Terrorism in Context.* University Park: Pennsylvania State University Press.

———. 1998. "The Logic of Terrorism: Terrorist Behavior as a Product of Strategic Choice." Pp. 7–24 in *Origins of Terrorism: Psychologies, Ideologies, States of Mind,* edited by W. Reich. Washington, DC: Woodrow Wilson International Center for Scholars and Cambridge University Press.

Cunningham, Kenneth. 2004. "Permanent War? The Domestic Hegemony of the New American Militarism." *New Political Science* 26(4):551–67.

Cuomo, C. 1996. "War Is Not Just an Event: Reflections on the Significance of Everyday Violence." *Hypatia* 11(4):30–45.

Donn, Jeff and Kimberly Hefling. 2007. "Wounded Vets Return to Empty Pockets." *News Tribune,* September 30, pp. A6–7.

Easterbrook, Gregg. 2001. "The All-Too-Friendly Skies: Security as an Afterthought." Pp. 163–82 in *How Did This Happen? Terrorism and the New War,* edited by James F. Hoge, Jr., and Gideon Rose. New York: PublicAffairs.

Egan, T. 2004. "Sensing the Eyes of Big Brother, and Pushing Back." *New York Times,* August 8, p. 16.

Eisenhower, Dwight. 1953. Speech to the American Society of Newspaper Editors, April 16, 1953, Washington, DC. Retrieved March 24, 2008 (http://www.eisenhowermemorial.org/speeches/19530416%20Chance%20for%20Peace.htm).

Enloe, Cynthia. 1990. *Bananas, Beaches, and Bases: Making Feminist Sense of International Politics.* Berkeley: University of California Press.

"The Events of 11 September (2001) and Beyond." 2002. *International Feminist Journal of Politics* 4(1):95–113.

FBI Counterterrorism Unit, Counterterrorism Threat Assessment and Warning Unit. 1999. *Terrorism in the United States: 1999.* Washington, DC: U.S. Department of Justice, Federal Bureau of Investigation.

Featherstone, Liza. 2003. "Students Wrestle with War." *The Nation* 273(20):18–20.

Fedarko, Kevin. 2003. "He Fights Terror with Books." *Parade,* April 6, pp. 4–6.

Flores, D. A. 1981. "Note: Export Controls and the US Effort to Combat International Terrorism." *Law and Policy in International Business* 13(2):521–90.

Flynn, Stephen E. 2001. "The Unguarded Homeland: A Study in Malign Neglect." Pp. 183–98 in *How Did This Happen? Terrorism and the New War,* edited by James F. Hoge, Jr., and Gideon Rose. New York: PublicAffairs.

Freeh, Louis J. 2001. "Threat of Terrorism to the United States." Testimony before the U.S. Senate Committees on Appropriations, Armed Services, and Select Committee on Intelligence, May 10, 2001. Retrieved August 14, 2003 (http://www.fbi.gov/congress/congress01/freeh051001.htm).

Garfield, R. 2002. "Economic Sanctions, Humanitarianism, and Conflict after the Cold War." *Social Justice* 29(3):94–107.

Gibbs, J. 1989. "Conceptualization of Terrorism." *American Sociological Review* 54:329–40.

Haas, Richard. 1997. "Sanctioning Madness." *Foreign Affairs* November/December:74–85.

Hamilton, K. 2003. "Activists for the New Millenium." *Black Issues in Higher Education* 20(5):16–21.

Harmon, C. 2000. *Terrorism Today.* London, England: Frank Cass.

Harvard Study Team. 1991. "The Effect of the Gulf Crisis on the Children of Iraq." *New England Journal of Medicine* 325(13):977–80.

Hoge, James F., Jr. and Gideon Rose. 2001. "Introduction." Pp. ix–xiv in *How Did This Happen? Terrorism and the New War,* edited by James F. Hoge, Jr., and Gideon Rose. New York: PublicAffairs.

Jansen, S. C. 2002. "Media in Crisis: Gender and Terror, September 2001." *Feminist Media Studies* 2(1):139–41.

Jarboe, James. 2002. "The Threat of Eco-Terrorism." Retrieved March 18, 2007 (http://www.fbi.gov/congress/congress02/jarboe021202.htm).

Jenkins, Brian Michael. 1980. *The Study of Terrorism: Definitional Problems* (P-6563). Santa Monica: CA: RAND Corporation.

———. 1988. "Future Trends in International Terrorism." Pp. 246–66 in *Current Perspectives on International Terrorism,* edited by R. Slater and M. Stohl. New York: St. Martin's Press.

Kaplan, Laura Duhan. 1994. "Women as Caretaker: An Archetype That Supports Patriarchal Militarism." *Hypatia* 9(2):123–33.

Kean, Thomas and Lee Hamilton. 2007. "Are We Safer Today?" *Washington Post,* September 9, p. B01.

Knickerbocker, B. 2002. "Return of the Military-Industrial Complex?" *Christian Science Monitor* 94(52):2.

Knott, Paul D. 1971. *Student Activism.* Dubuque, IA: William C. Brown.

Lemish, Dafna. 2005. "The Media Gendering of War and Conflict." *Feminist Media Studies* 5(3):367–70.

Lorber, Judith. 2002. "Heroes, Warriors, and Burquas: A Feminist Sociologist's Reflections on September 11." *Sociological Forum* 17(3):377–96.

Lyall, Sarah and Eamon Quinn. 2007. "At the Polls, Northern Ireland Tries to Resurrect Self-Rule." *New York Times,* March 8, p. A9.

Mandelbaum, Michael. 2003. "Diplomacy in War Time: New Priorities and Alignments." Pp. 255–68 in *How Did This Happen? Terrorism and the New War,* edited by James F. Hoge, Jr., and Gideon Rose. New York: PublicAffairs.

Marks, S. 1999. "Economic Sanctions as Human Rights Violations: Reconciling Political and Public Health Imperatives." *American Journal of Public Health* 89(10):1509–13.

Martin, Gus. 2006. *Understanding Terrorism: Challenges, Perspectives and Issues.* Thousand Oaks, CA: Sage Publications.

National Center for PTSD. 2003a. "What Is Posttraumatic Stress Disorder?" Retrieved August 15, 2003 (www.ncptsd.org/facts/general/fs_what_is_ptsd.html).

National Center for PTSD. 2003b. "Epidemiological Facts about PTSD." Retrieved August 15, 2003 (www
.ncptsd.org/facts/general/fs_epidemiological.html).

National Intelligence Council. 2007. *National Intelligence Estimated—The Terrorist Threat to the US Homeland.* Washington, DC: Author.

National Priorities Project. 2007. "Federal Budget Year in Review 2007." Retrieved March 25, 2008 (http://www.nationalpriorities.org/yearinreview2007).

Neier, A. 1993. "Watching Rights." *The Nation* 257(19):683.

O'Conner, Anahad. 2004. "The Reach of War: The Soldiers." *New York Times,* July 1.

Perkins, B., T. Popovic, and K. Yeskey. 2002. "Public Health at a Time of Bioterrorism." *Emerging Infectious Diseases* (serial online) 8(10). Retrieved August 13, 2003 (www.cdc.gov/ncidod/EID/vol8no10/pdf/02-0444.pdf).

Pillar, Paul R. 2001. *Terrorism and U.S. Foreign Policy.* Washington, DC: Brookings Institution Press.

Reich, W. 1998. "Understanding Terrorist Behavior: The Limits and Opportunities of Psychological Inquiry." Pp. 261–80 in *Origins of Terrorism: Psychologies, Ideologies, States of Mind,* edited by W. Reich. Cambridge, MA: Woodrow Wilson International Center for Scholars and Cambridge University Press.

Rhoads, H. 1991. "Activism Revives on Campus." *The Progressive* 55(3):15–17.

Sanger, D. 2003. "N. Korea Says It Has Material for Nukes." *News Tribune,* July 15, pp. A1–A9.

Smith, Brent. 1994. *Terrorism in America: Pipe Bombs and Pipe Dreams.* New York: State University of New York Press.

Southern Poverty Law Center. 2007. "Active Hate Groups in 2005." Retrieved March 17, 2007 (http://www.splcenter.org/intel/map/hate.jsp).

Stockholm International Peace Research Institute. 2007. "Recent Trends in Military Expenditure." Retrieved July 4, 2007 (http://www.sipri.org/contents/milap/milex/mex_trends.html).

Tétreault, Mary Ann. 2006. "The Sexual Politics of Abu Ghraib: Hegemony, Spectacle, and the Global War on Terror." *NWSA Journal* 18(3):33–50.

Tickner, J. A. 2002. "Feminist Perspectives on 9/11." *International Studies Perspectives* 3:333–50.

U.S. Commission on National Security in the 21st Century. 2001. "Road Map for National Security: Imperative for Change." Retrieved August 17, 2003 (http://govinfo.library.unt.edu/nssg/Reports/reports.htm).

U.S. Department of State. 2003. "Diplomacy: The State Department at Work." Retrieved October 31, 2003 (www.state.gov/r/pa/ei/rls/dos/4078.htm).

U.S. Department of Veterans Affairs. 2007. "Factsheet: America's Wars." Retrieved March 25, 2008 (http://www1.va.gov/opa/fact/amwars.asp).

Vedantam, Shankar. 2006. "Veterans Report Mental Illness." *Washington Post,* March 1, p. A01.

Vellela, T. 1988. *New Voices: Student Activism in the 80s and 90s.* Boston, MA: South End Press.

Walter, E. V. 1964. "Violence and the Process of Terror." *American Sociological Review* 29(2):248–57.

Wardlaw, Grant. 1988. "State Response to International Terrorism." Pp. 206–45 in *Current Perspectives on International Terrorism,* edited by R. Slater and M. Stohl. New York: St. Martin's Press.

Whittaker, David. 2002. *Terrorism: Understanding the Global Threat.* London, England: Longman.

Wilkinson, Paul. 1993. "The Orange and the Green: Extremism in Northern Ireland." Pp. 105–23 in *Terrorism, Legitimacy, and Power,* edited by Martha Crenshaw. Middleton, CT: Wesleyan University Press.

Young, Nigel. 1999. "Peace Movements in History." Pp. 228–36 in *Approaches to Peace,* edited by David P. Barash. New York: Oxford University Press.

Zoroya, Greg. 2005. "Key Iraq Wound: Brain Trauma." Retrieved October 7, 2007 (http://www.usatoday.com/news/nation/2005-03-03-brain-trauma-lcdc_x.htm).

Chapter 17

Berger, P. 1963. *Invitation to Sociology.* New York: Doubleday.

Close Up Foundation. 2004. *The 26th Amendment: Pathway to Participation.* Alexandria, VA: Author.

Davies, James. 1974. "The J-curve and Power Struggle Theories of Collective Violence." *American Sociological Review* 87:363–87.

Engle, Shaena. 2006. "More College Freshmen Committed to Social and Civic Responsibility, UCLA Survey Reveals." Retrieved October 14, 2007 (http://newsroom.ucla.edu/portal/ucla/More-College-Freshmen-Committed-6754.aspx?RelNum=6754).

Giugni, M. 1999. "Introduction." Pp. xiii–xxxiii in *How Social Movements Matter,* edited by M. Giugni, D. McAdam, and C. Tilly. Minnesota: University of Minnesota Press.

Hannigan, John. 1991. "Social Movement Theory and the Sociology of Religion: Toward a New Synthesis." *Sociological Analysis* 52(4):311–31.

Harper, Charles and Kevin Leicht. 2002. *Exploring Social Change: America and the World.* Upper Saddle River, NJ: Prentice Hall.

Hollender, J. and L. Catling. 1996. *How to Make the World a Better Place.* New York: Norton.

Lauer, R. H. 1976. "Introduction: Social Movements and Social Change: The Interrelationships." Pp. xi–xxviii in *Social Movements and Social Change,* edited by R. Lauer. Carbondale: Southern Illinois University

Lemert, Charles. 1997. *Social Things: An Introduction to the Sociological Life.* Lanham, MD: Rowman & Littlefield.

Levine, A. and J. Cureton. 1998. *When Hope and Fear Collide: A Portrait of Today's College Student.* San Francisco, CA: Jossey-Bass.

Levine, P. and M. Lopez. 2002. *Voter Turnout Has Declined, by Any Measure.* College Park, MD: Center for Information and Research on Civic Learning and Engagement.

Liazos, A. 1972. "Nuts, Sluts, and Perverts: The Sociology of Deviance." *Social Problems* 20:103–20.

Loeb, Paul Rogat. 1994. *Generation at the Crossroads.* New Brunswick, NJ: Rutgers University Press.

Loseke, Denise. 2003. *Thinking about Social Problems.* New York: Aldine de Gruyter.

Mannheim, K. 1940. *Man and Society in an Age of Reconstruction.* New York: Harcourt Brace.

McAdam, Doug. 1982. *Political Process and the Development of Black Insurgency.* Chicago, IL: University of Chicago Press.

McAdam, Doug, J. McCarthy, and M. Zald. 1988. "Social Movements." Pp. 695–737 in *Handbook of Sociology,* edited by N. Smelser. Newbury Park, CA: Sage Publications.

———. 1996. *Comparative Perspectives on Social Movement: Political Opportunities, Mobilizing Structures, and Cultural Framings.* Thousand Oaks, CA: Sage Publications.

McCarthy, J. and M. Zald. 1977. "Resource Mobilization and Social Movements: A Partial Theory." *American Journal of Sociology* 82:1212–41.

Meyer, D. 2000. "Social Movements: Creating Communities of Change." Pp. 33–55 in *Feminist Approaches to Social Movements, Community, and Power,* Vol. 1, edited by R. Teske and M. A. Tetreault. Columbia: University of South Carolina Press.

Meyer, David and Suzanne Staggenbord. 1996. "Movements, Countermovements, and the Structure of Political Opportunity." *American Journal of Sociology* 101(6):1628–60.

Nader, Ralph. 1972. *Action for a Change.* New York: Grossman.

Norris, Pippa. 2002. *Democratic Phoenix: Reinventing Political Activism.* New York: Cambridge University Press.

Piven, Francis and Richard Cloward. 1979. *Poor People's Movements: Why They Succeed, How They Fail.* New York: Vintage Books.

Plotke, David. 1995. "What's So New about New Social Movements?" Pp. 113–36 in *Social Movements: Critique, Concepts, Case-Studies,* edited by S. Lyman. New York: New York University Press.

Rayside, David. 1998. *On the Fringe: Gays and Lesbians in Politics.* Ithaca, NY: Cornell University.

Ritzer, George. 2000. *Sociological Theory.* New York: McGraw-Hill.

Rock the Vote. 2004. "About Us." Retrieved February 29, 2004 (www.rockthevote.org/rtv_about.php).

Skocpol, Theda. 2000. *The Missing Middle.* New York: Norton.

Smelser, Neil. 1963. *Theory of Collective Behavior.* New York: Free Press.

Sztompka, Piotr. 1994. *The Sociology of Social Change.* Cambridge, MA: Blackwell.

Taylor, V. and N. Whittier. 1995. "Analytical Approaches to Social Movement Culture: The Culture of the Women's Movement." Pp. 163–87 in *Social Movements and Culture,* edited by H. Johnson and B. Klandermans. Minneapolis: University of Minnesota Press.

Wilson, J. 1973. *Introduction to Social Movements.* New York: Basic Books.

Wilson, K. and A. Orum. 1976. "Mobilizing People for Collective Political Action." *Journal of Political and Military Sociology* 4:187–202.

Credits

Image, **In Focus features**. © iStockphoto.com/blackred.

Image, **Taking a World View features**. © iStockphoto.com/ Alex Slobodkin.

Part I

Part-opening photo, **pages xxii–1**. © Alan Schein Photograph/Corbis.

Visual Essay I

Photos, **pages 2 and 3**. © Ted Streshinsky/Corbis.

Chapter 1

Chapter-opening photo, **page 4**. © Don Mason/Corbis.

Photo 1.1a, **page 5**. © Stefan Zaklin/epa/Corbis.

Photo 1.1b, **page 5**. © Louise Gubb/Corbis SABA.

Photo 1.2, **page 8**. © Roger Ressmeyer/Corbis.

Taking a World View: A Social Constructionist Approach to AIDS in Africa, pages 10–11. Reprinted with permission from Lichtenstein 2004.

Photo 1.3, **page 12**. © AFP/Getty Images.

Photo 1.4, **page 20**. © Wally McNamee/Corbis.

Photo 1.5, **page 21**. © Coby Burns/ZUMA/Corbis.

Photo 1.6, **page 22**. © Getty Images.

Part II

Part-opening photo, **pages 26–27**. © Mona Reeder/Dallas Morning News/Dallas Morning News/Corbis.

Visual Essay II

Photo, **page 30**. © Steve Chenn/Corbis.

Photo, **page 31, top**. © Ariel Skelley/Corbis.

Photo, **page 31, bottom**. © Kristy-Anne Glubish/Design Pics/Corbis.

Photo, **pages 32–33**. © Joel Stettenheim/Corbis.

Photo, **page 34**. © Gabe Palmer/Corbis.

Chapter 2

Chapter-opening photo, **page 35**. © Paul Colangelo/Corbis.

Photo 2.1, **page 36**. © Getty Images.

Photo 2.2, **page 38**. © Warren Toda/epa/Corbis.

Photo 2.3a, **page 39**. © Peter Gridley/Getty Images.

Photo 2.3b, **page 39**. © Getty Images.

Photo 2.4, **page 40**. © Louie Psihoyos/Corbis.

Chapter 3

Chapter-opening photo, page 63. © Lucy Nicholson/Reuters/Corbis.

Photo 3.1, page 64. © Tannen Maury/epa/Corbis.

Photo 3.2, page 67. © Christophe Calais/Corbis.

Taking a World View: Caste Discrimination in India, page 68. Adapted from Human Rights Watch 2001.

Photo 3.3, page 69. © Bettmann/Corbis.

Photo 3.4, page 74. © Associated Press.

Photo 3.5, page 78. © Andrew Lichtenstein/Corbis.

Photo 3.6, page 79. © Christopher Morris/Corbis.

Photo 3.7, page 80. © Associated Press.

Chapter 4

Chapter-opening photo, page 87. © Daniel Mirer/Corbis.

Photo 4.1, page 88. © Congressional Quarterly/Getty Images.

Photo 4.2, page 91. © moodboard/Corbis.

Photo 4.3, page 92. © Elyse Lewin/Brand X/Corbis.

Photo 4.4, page 95. © John B. Boykin/Corbis.

Taking a World View: Violence Against Women, page 100. United Nations 2006a and 2006b.

Photo 4.5, page 102. © Bart Young/NewSport/Corbis.

Voices in the Community: Bernice R. Sandler, pages 103–104. Sandler 1997. Reprinted with permission from Bernice Resnick Sandler, Senior Scholar, Women's Research and Education Institute, Washington, DC. E-mail: sandler@bernicesandler.com; Web site: www.bernicesandler.com.

Chapter 5

Chapter-opening photo, page 107. © Rick Friedman/Corbis.

Photo 5.1, page 108. © Colin McPherson/Corbis.

Photo 5.2, page 114. © Liz Mangelsdorf/San Francisco Chronicle/Corbis.

Photo 5.3, page 115. © Ed Quinn/Corbis.

In Focus: Gay Friendly Campuses, page 116. Windmeyer 2006.

Photo 5.4, page 120. © Getty Images.

Chapter 6

Chapter-opening photo, page 123. © Viviane Moos/Corbis.

Photo 6.1, page 124. © Karen Kasmauski/Corbis.

Photo 6.2, page 125. © Getty Images.

Photo 6.3, page 129. © Getty Images.

Photo 6.4, page 131. © Guido Manuilo/epa/Corbis.

Photo 6.5, page 138. © Larry Downing/Reuters/Corbis.

In Focus: Senior Political Power, pages 139–140. Binstock 2005 and 2006–2007.

Photo 6.6, page 140. © Noah K. Murray/Star Ledger/Corbis.

Voices in the Community: Doris Haddock, pages 141–142. Landphair 2007/Voice of America.

Part III

Part-opening photo, pages 144–145. © Dan Lamont/Corbis.

Visual Essay III

Photo, page 141, top. © Image Source/Corbis.

Photo, page 147, bottom. © John Henley/Corbis.

Photo, page 148, top left. © Laura Dwight/Corbis.

Photo, page 148, right. © Peter M. Fisher/Corbis.

Photo, page 148, bottom left. © Ronnie Kaufman/Corbis.

Chapter 7

Chapter-opening photo, page 149. © Alison Wright/Corbis.

Photo 7.1, page 150. © Associated Press.

Taking a World View: The Family in Sweden, page 156. Popenoe 2006.

Photo 7.2, page 160. © Bruce Ayres/Getty Images.

Photo 7.3, page 161. © Getty Images.

In Focus: Teen Parenting and Education, pages 162–163. Reprinted with permission from SmithBattle 2007.

Photo 7.4, page 164. © Ashley Cooper/Corbis.

Photo 7.5, page 172. © Ryuichi Sato/Getty Images.

Chapter 8

Chapter-opening photo, page 175. © Aristide Economopoulos/Star Ledger/Corbis.

Photo 8.1, page 177. © Mark Scott/Getty Images.

Photo 8.2, page 179. © Sean Justice/Getty Images.

Photo 8.3, page 181. © Uppercut Images/Getty.

Photo 8.4, page 185. © Ron Chapple Stock/Corbis.

Photo 8.5, page 190. © Getty Images.

Voices in the Community: Wendy Kopp, pages 195–196. Kopp 2001: 5–6, 10. Reprinted by permission of PublicAffairs, a member of Perseus Books Group.

Photo 8.6, page 196. © Associated Press.

Chapter 9

Chapter-opening photo, page 203. © Helen King/Corbis.

Photo 9.1, page 205. © Getty Images.

In Focus: The Economic and Social Impacts of Deindustrialization, page 206. Minchin 2006.

Photo 9.2, page 207. © Image Source Pink/Getty Images.

Photo 9.3, page 216. © AFP/Getty Images.

Photo 9.4, page 221. © Deanne Fitzmaurice/San Francisco Chronicle/Corbis.

Voices in the Community: Judy Wicks, pages 225–226. Gorman 2001. Reprinted from *YES! Magazine,* P.O. Box 10818, Bainbridge Island, WA. www.yesmagazine.org.

Photo 9.5, page 227. © Getty Images.

Chapter 10

Chapter-opening photo, page 230. © Helen King/Corbis.

Photo 10.1, page 232. © AFP/Getty Images.

Photo 10.2, page 235. © Getty Images.

Photo 10.3, page 243. © Jose Luis Pelaez Inc./Getty Images.

Photo 10.4, page 248. UPI/drr.net.

In Focus: Bird Flu Pandemic, page 250. Centers for Disease Control 2007; World Health Organization 2007.

Photo 10.5, page 252. © Kim Kulish/Corbis.

Photo 10.6, page 253. © Getty Images.

Chapter 11

Chapter-opening photo, page 257. © Justin Lane/epa/Corbis.

Photo 11.1, page 258. © Antonio Mo/Getty Images.

Photo 11.2, page 262. © Getty Images.

Photo 11.3, page 263. © Associated PRESS.

Taking a World View: Women's Feature Service, New Delhi, India, page 264. Parekh 2001. Reprinted with permission of the *Nieman Reports.*

Photo 11.4, page 265. © Getty Images.

Photo 11.5, page 274. © Digital Vision/Getty Images.

Part IV

Part-opening photo, pages 282–283. © Blaine Harrington III/Corbis.

Visual Essay IV

Photo, page 286, top. © Brenda Ann Kenneally/Corbis.

Photo, page 286, middle. © Brenda Ann Kenneally/Corbis.

Photo, page 286, bottom. © Brenda Ann Kenneally/Corbis.

Photo, page 287, top. © Brenda Ann Kenneally/Corbis.

Photo, page 287, bottom. © Brenda Ann Kenneally/Corbis.

Chapter 12

Chapter-opening photo, page 288. © Scott Houston/Sygma/Corbis.

Photo 12.1, page 291. © Louis Fox/Getty Images.

Photo 12.2, page 299. © Getty Images.

Photo 12.3, page 303. © Getty Images.

Photo 12.4, page 304. © Peter Dench/Corbis.

Photo 12.5, page 307. © Jeffrey L. Rotman/Corbis.

Chapter 13

Chapter-opening photo, page 315. © Michael Ainsworth/Dallas Morning News/Corbis.

Photo 13.1, page 317. © Getty Images.

Photo 13.2, page 320. © Scott Houston/Corbis.

Photo 13.3, page 331. © Douglas Kirkland/Corbis.

Photo 13.4, page 334. © Gus Ruelas/Reuters/Corbis.

In Focus: A Public Health Approach to Gun Control, pages 338–339. Hemenway 2004. Mechanic, David, et al., eds., *Policy Challenges in Modern Health Care,* Copyright © 2005 by Rutgers, The State University. Reprinted with permission of Rutgers University Press.

Voices in the Community: Max Kenner, pages 339–340. Malik 2005. Reprinted with permission of Dr. Pola Rosen, www.educationupdate.com.

Photo 13.5, page 342. © Jerry McCrea/Star Ledger/Corbis.

Chapter 14

Chapter-opening photo, page 345. © Daniella Nowtiz/Corbis.

Photo 14.1, page 348. © Lindsay Hebberd/Corbis.

Photo 14.2, page 353. © Tokyo Space Club/Corbis.

Photo 14.3, page 361. © Ted Soqui/Corbis.

Photo 14.4, page 362. © Antoine Gyori/AGP/Corbis.

Photo 14.5, page 365. © Brown, Amanda/Star Ledger/Corbis.

Chapter 15

Chapter-opening photo, page 372. © Andrew Lichtenstein/Corbis.

Photo 15.1, page 374. © Getty Images.

Photo 15.2, page 376. © Diego Azubel/epa/Corbis.

Photo 15.3, page 380. © Momatiuk - Eastcott/Corbis.

Photo 15.4, page 382. © David Gray/Reuters/Corbis.

Photo 15.5, page 386. © Mark E. Gibson/Corbis.

Chapter 16

Chapter-opening photo, page 399. © Reuters/Corbis.

Photo 16.1, page 401. © David Leeson/Dallas Morning News/Corbis Sygma.

Photo 16.2, page 404. © Jeff Mitchell/Reuters/Corbis.

Photo 16.3, page 409. © Sergio Barrenechea/epa/Corbis.

Voices in the Community: Greg Mortenson, pages 417–418. From Fedarko, K. "He Fights Terror with Books," in *Parade,* April 6, 2003. Reprinted with permission from *Parade,* copyright © 2003.

Photo 16.4, page 418. © Teru Kuwayama/Corbis.

Photo 16.5, page 420. © Shawn Thew/epa/Corbis.

Part V

Part-opening photo, pages 424–425. © STAFF/Reuters/Corbis.

Visual Essay V

Photo, page 426. © Bettmann/Corbis.

Photo, page 427, top. © Ed Kashi/Corbis.

Photo, page 427, bottom. © Bettmann/Corbis.

Photo, page 428, top. © Corbis.

Photo, page 428, bottom. © Bettmann/Corbis.

Photo, page 429. © David Butow/Corbis.

Chapter 17

Chapter-opening photo, page 430. © Bob Daemmrich/Corbis.

Photo 17.1, page 433. © AFP/Getty Images.

Photo 17.2, page 435. © Steve Schapiro/Corbis.

Photo 17.3, page 439. © Jack Moebes/Corbis.

Photo 17.4, page 440. © Bernd Obermann/Corbis.

Glossary/Index

About the Author

Anna Leon-Guerrero is Professor of Sociology at Pacific Lutheran University, Tacoma, Washington. A recipient of the university's Faculty Excellence Award, she teaches courses on statistics, sociological theory, and social problems. As a social service program evaluator and consultant, her research has focused on welfare reform, employment strategies for the working poor, and program assessment. She is the coauthor of *Social Statistics for a Diverse Society* (with Chava Frankfort-Nachmias).